Fundamentals of Signal Processing in Generalized Metric Spaces

Fundamentals of Signal Processing in Generalized Metric Spaces

Algorithms and Applications

Andrey Popoff

CRC Press
Taylor & Francis Group
Boca Raton London New York

CRC Press is an imprint of the
Taylor & Francis Group, an **informa** business

First edition published 2022
by CRC Press
6000 Broken Sound Parkway NW, Suite 300, Boca Raton, FL 33487-2742

and by CRC Press
4 Park Square, Milton Park, Abingdon, Oxon, OX14 4RN

CRC Press is an imprint of Taylor & Francis Group, LLC

© 2022 Andrey Popoff

Library of Congress Cataloging-in-Publication Data

Names: Popoff, Andrey, author.
Title: Fundamentals of signal processing in generalized metric spaces :
algorithms and applications / Andrey Popoff.
Description: First edition. | Boca Raton : CRC Press, 2022. | Includes
bibliographical references and index.
Identifiers: LCCN 2021054038 (print) | LCCN 2021054039 (ebook) | ISBN
9781032231259 (hardback) | ISBN 9781032231273 (paperback) | ISBN
9781003275855 (ebook)
Subjects: LCSH: Signal processing--Mathematics. | Metric spaces. | Lattice
ordered groups. | Order statistics. | Algorithms.
Classification: LCC TK5102.9 .P654 2022 (print) | LCC TK5102.9 (ebook) |
DDC 621.382/2--dc23/eng/20220105
LC record available at https://lccn.loc.gov/2021054038
LC ebook record available at https://lccn.loc.gov/2021054039

ISBN: 978-1-032-23125-9 (hbk)
ISBN: 978-1-032-23127-3 (pbk)
ISBN: 978-1-003-27585-5 (ebk)

DOI: 10.1201/9781003275855

Publisher's note: This book has been prepared from camera-ready copy provided by the authors.

Contents

Preface

Signal Processing Theory is evolved within close interaction with mathematics. Studying and solving the applied problems of signal processing require exploiting corresponding mathematical apparatus. On the other hand, daily practice of developed approaches, signal processing methods, and algorithms stimulate the development of mathematics. This can be seen by looking through the sections of Mathematics Subject Classification concerned with Signal Processing Theory.

We distinguish several interrelated directions, along which the Signal Processing Theory is developed:

1. Analysis of probabilistic-statistic characteristics of stochastic signals, interference (noise), their transformations and their influence on the operation of studied signal processing systems.

2. Optimization and analysis of useful signals sets used to solve the main signal processing problems in electronic systems of various functionality.

3. Synthesis and analysis of signal processing algorithms considered within solving the main signal processing problems: signal detection (including multiple-alternative detection), classification of signals, signal extraction and filtering, signal parameter estimation, resolution of signals and recognition of signals, etc.

4. Development of mathematical foundations of Signal Processing Theory.

In this book we consider the questions concerned with the 1st, 3rd, and 4th aforementioned directions.

The goal of the book is stating the methods and algorithms of signal processing in generalized metric spaces with lattice-ordered group (L-group) properties that provide a required signal processing efficiency under parametric and/or nonparametric prior uncertainty conditions and increase a processing rate by exploiting multiplication-free operations.

When stating the book material, we not only acquaint the reader with signal processing in L-groups but also prepare the reader to active utilizing the stated mathematical apparatus known in mathematics during rather long time, that until now is not directly used in theory and practice of signal processing.

Though the book does not have direct analogues, it differs from the closest monographs in the following points:

(a) We use generalized metric spaces with L-group properties as the signal spaces.

(b) L-group signal processing algorithms investigated in the book can be related to the following types: robust and nonparametric; nonlinear, quasilinear, and linear (as a degenerated case); invariant and adaptive. These algorithms are developed within metric, pseudometric, and semimetric spaces.

(c) Main useful properties of the suggested algorithms are the following: robustness or nonparametricity; multiplication-free; high processing rate; lesser requirements concerning computational resources of processing system; quasilinearity; invariance or adaptivity.

(d) The discussed algorithms are defined by some functions of order statistics and/or generalized metrics.

(e) The proposed algorithms are considered within comparative analysis with known optimal algorithms of signal processing.

(f) The suggested algebraic approach allows from the unified position considering signal processing algorithms based on order, nonparametric, robust, and other statistics.

This book is intended for professors, researchers, post-graduate and PhD students, for specialists in signal processing and information theories, electronics, radiophysics, telecommunications, various engineering disciplines including radioengineering, and information technology. It presents alternative approach to constructing Signal Processing Theory based upon Lattice Theory and L-groups. The book may be useful for mathematicians and physicists interested in applied problems in their studied areas. The material contained in the book differs from the traditional approaches and may interest pre-graduate students specializing in the directions: "radiophysics", "telecommunications", "electronic systems", "system analysis and control", "automation and control", "robotics", "electronics and communication engineering", "control systems engineering", "electronics technology", and others.

R&D specialists as well as engineers of electronic systems with various functionality must attentively trace the results of fundamental research and take them into account when solving the applied problems. When developing such systems one must take into consideration an operation experience and evolution perspectives of both known existing systems and competitive systems that are being developed, so that both the developer and engineer must have a clear understanding concerning the developed systems. The same thesis could be addressed to university professors that ought from time to time to correct curricula taking into account long-term outlook for signal processing, and also to graduate and postgraduate students which must be interested in new approaches and methods that have not yet been included in curricula, but being rather promising, they can be demanded in the nearest future.

Knowledge of standard university mathematics course is sufficient to understand the material involved in the book. When writing the book the author tried providing its practical orientation. Taking into account this circumstance, the author aspired to help the reader learn and exploit the existing mathematical apparatus used for analysis and synthesis of signal processing systems, and on the other hand, not to overload the stated material with mathematical inferences and proofs. To provide this, the author as far as possible used the simplified mathematical models within which the obtained results are enough comprehensible. The author, mindful of Alice's question on a book and pictures*, made every effort to accompany each example with helpful illustrations.

We place the emphasis on the essence of L-group algorithms leading to the ultimate aim—creating the system that meets necessary requirements. The material is constrained by the relationships that allow the reader to carry out the simulation of created signal processing systems, obtain necessary assessments, and make corresponding calculations. We discuss those applied problems that are simultaneously solved by using both known classic and proposed approaches, as well as their inferences concerning signal processing.

Since we mainly deal with zero mean stochastic process (signals), notions of correlation and covariance are coincide, so we use terms «correlation function», «correlation matrix» instead of analogous terms based on covariance.

The book contains eight chapters. The first chapter has an introductory character and must prepare the reader for perceiving the main material of the book. The material of Chapter 1 is sufficient for performing independent original research in the branch of methods and algorithms based on L-group operations within the applications corresponding to the reader's interest. We consider the main algebraic properties of L-group, so that we show interrelation between L-group and linear space, and on the other hand, we discuss interrelation between L-group and algebraic lattice. We establish the relationship between lattice operations and maximum/minimum functions. Physical essence of lattice operations is illustrated by examples of interactions of various signals. We demonstrate interrelation between L-group operations and known nonlinear functions, in particular, step functions, Huber function, and Hampel influence functions. We discuss the features of L-group operations when synthesizing the limiters with different amplitude characteristics, in particular, we consider analytical representation of amplitude characteristics of signal quantizers and companders on the basis of L-group operations. We discuss the examples of generalized metric spaces with L-group properties. We introduce the notion of sample space with L-group properties, within which some generalized metrics are considered, namely, pseudometric, l_1-metric, and semimetric. The corresponding normalized generalized metrics between samples are discussed. In sample space with L-group properties, we introduce measures of statistical interrelation (MSI) and normalized measures of statistical interrelation between samples. We establish relationships between the introduced MSIs and generalized metrics. On the basis of MSI, we introduce the notion of orthogonality for discrete signals in L-groups. We establish the estimators of sample MSIs of discrete signals in sample space with L-group properties. We consider the features of obtained metric characteristics of stochastic samples with arbitrary distributions in sample spaces with

---

*"What is the use of a book without pictures...?" L. Carrol. Alice in Wonderland

L-group properties. We investigate hyperspectral representations of discrete signals based on generalized metrics in sample spaces with L-group properties. We draw the parallel between discrete Fourier transform (DFT) and hyperspectral representation of discrete signals. We consider statistical demultiplexing in communication channel based on generalized metrics on L-group. We emphasize that statistical demultiplexing in sample space with L-group properties is a pithy analogue of known frequency, time, and code division demultiplexing in linear sample space with scalar product. We consider digital signal filtering based on MSI in sample space with L-group properties. We introduce the notion of digital hyperfilter of discrete signal in signal space with L-group properties. We show that digital hyperfilter, unlike classic digital filter realizing convolution operation, allows realizing signal filtering without using operation of multiplication.

Chapter 2 deals with signal parameter estimation in spaces with L-group properties. We consider methods and algorithms of sample ordering based on operations of lattice. It is shown that these algorithms of ordered sample forming can be realized with help of systolic processors. We depict block diagrams of systolic processors built on both binary and ternary elements, so that each element of the processor performs operations of join and meet of lattice. We consider algorithms of forming sample median based on L-group operations. We explore algorithms of forming M-, L-, and R-estimators based on L-group operations. We bring the results of relative efficiency of truncated mean estimator and Hodges-Lehmann estimator with respect to sample mean estimator for some most frequently used distributions of errors. We consider algorithms of signal parameter estimation in spaces with L-group properties pointing out the results of comparative analysis of efficiency of location parameter estimation by median filter, Huber filter, truncated mean filter, and Hodges-Lehmann filter with respect to homogeneous filter (moving average filter). When exploring location parameter estimation, estimation errors are described by four following symmetric distributions: Tukey model; logistic; Laplace (double exponential); and Cauchy. Besides, we discuss the results of comparative experimental analysis of aforementioned filters efficiency obtained in the presence of estimation errors with envelopes distributed on three wide classes described by Weibull, lognormal, and gamma distributions. We show that filters, built on the estimation algorithms in sample space with L-group properties, provide some gain in estimation accuracy as against homogeneous filter (moving average filter) on a wide class of distributions which describe interference (noise) of impulse type, and, vice versa, lose to the latter when estimating in the presence of interference (noise) of harmonic (sinusoidal) type.

Chapter 3 considers algorithms of discrete signal filtering in spaces with L-group properties. We investigate signal filtering (extracting) algorithms in spaces with L-group properties that are robust with respect to interference (noise) distribution, namely, algorithms realized by median filter, Huber filter, truncated mean filter, and Hodges-Lehmann filter. When exploring signal filtering algorithms, additive noise is described by four following symmetric distributions: Tukey model; logistic; Laplace (double exponential); and Cauchy. Besides, we discuss the results of comparative experimental analysis of aforementioned filters efficiency obtained in the presence of additive noises with envelopes distributed on three wide classes described by Weibull, lognormal, and gamma distributions. We show that filters, built on the signal filtering algorithms in sample space with L-group properties, provide some gain in filtering error as against homogeneous filter on a wide class of distributions which describe interference (noise) of impulse type, and, vice versa, lose to the latter when filtering in the presence of interference (noise) of harmonic type. We consider adaptive signal filtering algorithms built on L-group operations. Among these algorithms we investigate: (1) algorithms built on the method of mapping of linear space into space with lattice properties; (2) algorithms based on MSIs; (3) algorithms realized by Wiener filters based on L-group operations; (4) LMS-algorithm based on L-group operations; (5) RLS-algorithm based on L-group operations; (6) algorithm realized by Kalman filter built on the method of mapping of linear space into space with lattice properties. Besides, we investigate algorithm realized by composite filter built on the method of mapping of linear space into the signal space with lattice properties, as well as signal filtering algorithms based on MSIs that are robust with respect to parametric prior uncertainty conditions.

Chapter 4 explores the problems of signal detection, classification, and resolution in generalized metric spaces with L-group properties. To introduce the reader into the circle of the related problems, we consider known nonparametric algorithms of signal detection formulated in terms of L-group operations. Among these algorithms we examine the following: sign; rank; and sign-rank algorithms of signal detection. We describe methodology of obtaining their asymptotic relative efficiencies (ARE) of signal detection in the presence of interference (noise) with arbitrary distribution. We demonstrate the results of calculating ARE of these algorithms for the most frequently used distributions of interference (noise). We introduce the notion of generalized matched filter that allows us from the unified position to formulate algorithms of signal processing in space with arbitrary metric and algebraic properties. We compare numbers of algebraic operations required for performing generalized matched filtering algorithms in sample spaces with l_2-metric, pseudometric, l_1-metric, and semimetric; we find out that generalized matched filtering L-group algorithms do not require performing operation of multiplication, i.e. they are multiplication-free. We investigate algorithms of signal detection based on MSIs related with pseudometric, l_1-metric, and semimetric. We emphasize that these algorithms can be realized without operation of multiplication. It is shown that signal detection algorithms based on MSIs and related to pseudometric and l_1-metric belong to a class of nonparametric algorithms. We establish the relationships that allow evaluating quality indices (conditional probabilities D, F of correct detection and false alarm, correspondingly) of these signal detection algorithms in the presence of interference (noise) with arbitrary distribution. We describe methodology of evaluating their AREs with respect to classic algorithm of signal detection in linear space in the presence of interference (noise) with arbitrary distribution. We demonstrate the results of calculating ARE of these algorithms for the most frequently used distributions of interference (noise).

We explore the algorithms of orthogonal signal classification in generalized metric spaces with L-group properties, namely, algorithms based on MSIs related to pseudometric, l_1-metric, and semimetric. We underline that these algorithms can be realized without operation of multiplication. We establish the relationships that allow evaluating probability of bit error when classifying orthogonal signals in interference (noise) with arbitrary distribution. We obtain the relationships that permit evaluating AREs of suggested algorithms of signal classification with respect to classic algorithm of signal classification in linear space with scalar product in the presence of interference (noise) with arbitrary distribution.

We investigate algorithms of signal resolution in space with L-group properties. It is shown that the "signal" and "filter" approaches to signal resolution are based on the notions of ambiguity function and mismatching function, respectively. The aforementioned "signal" and "filter" approaches to signal resolution represent the points of view of Signal Theory and Signal Processing Theory, correspondingly. To characterize resolution of signals we introduce the notions of time-frequency mismatching function of deterministic signal for linear space, and for the spaces with pseudometric, l_1-metric, and semimetric as well. We obtain the relationships determining potential resolutions of considered generalized matched filters.

Chapter 5 contains the questions related to spectral estimation and spectral analysis based on L-group operations. We consider both correlogram and periodogram methods of spectral estimation based on L-group operations. Within the periodogram method of spectral estimation we use Bartlett (Bartlett M.S.) and Welch (Welch P.D.) approaches. We explore spectral estimation methods based on MSI matrix within both correlogram method and minimum variance method. We consider a group of spectral estimation algorithms based on eigenvectors of MSI matrix estimator, namely, eigenvector method, MUSIC method, minimum norm method, and ESPRIT method. All of them relate to spectral estimation L-groups algorithms.

Chapter 6 explores multichannel signal processing based on L-group operations in antenna arrays. We consider multichannel antenna systems with logical signal processing based on L-group operations. Starting the exploration from two-channel antenna system with logical both coherent and incoherent signal processing based on L-group operations, further we consider logical signal processing based on L-group operations in linear antenna array and Mills cross array. We discuss the advantages and disadvantages of multichannel antenna systems with logical signal processing based on L-group operations. When exploring spatial filtering based on L-group operations in linear antenna array we consider algorithms of direct inversion of correlation matrix, algorithms that are invariant with respect to receiving signals, and also suboptimal algorithms. It is shown that the efficiency of suggested algorithms of array signal processing based on L-group operations can be equivalent to known optimal signal processing algorithms that are synthesized for antenna arrays within linear signal space with scalar product. It is shown that when amplitude characteristics of antenna array channels are nonlinear, suboptimal algorithms of spatial filtering based on L-group operations can provide higher signal processing efficiency than optimal algorithm synthesized on the basis of minimum variance criterion. We investigate spatial filtering algorithms based on MSI matrix in both linear and circular antenna arrays. We consider direction of arrival (DOA) estimation algorithms based on L-group operations in antenna arrays exploiting the following methods: minimum variance; eigenvectors; MUSIC; minimum norm; ESPRIT. All of them relate to DOA estimation L-group algorithms. We explore the main features of wideband array signal processing based on L-group operations considering algorithms of spatial filtering in wideband linear array and wideband circular array; DOA estimation algorithms in wideband linear array and wideband circular array. We investigate adaptive spatial filtering algorithms based on L-group operations. We consider adaptive LMS-, and RLS-algorithms of spatial filtering based on L-group operations, and also adaptive spatial filtering algorithm based on the method of recursive forming MSI matrix estimator for spatial filtering.

Chapter 7 discusses signal processing algorithms based on L-group operations for multichannel communication systems and multi-station networks. We consider in brief the basics of multichannel communication systems and multi-station networks, and also the main types of orthogonal signal systems used in them. Depending on concrete multiplexing (multiple access) scheme, we develop L-group signal processing algorithms for demultiplexing and demodulating the signals in multichannel communication systems (multi-station networks) of the following sorts: TDM(A)-QPSK; DS-CDM(A)-QPSK; CP-MFSK-CDM(A)-QPSK; OFDM(A)-QPSK; OFDM-CDM(A)-QPSK. To compare the proposed L-group signal processing algorithms with their known analogues, we demonstrate I-Q diagrams of demodulation results.

In Chapter 8 we discuss the algorithms of discrete wavelet transform based on L-group operations. Starting from the consideration of classic continuous and discrete wavelet transforms, as well as their main properties, we define discrete wavelet transforms (DWT) in the sample space with L-group properties (L-group DWT). We consider L-group discrete wavelet transforms defined in the sample spaces with pseudometric, l_1-metric, and semimetric; L-group discrete wavelet transforms of harmonic signals in such sample spaces; L-group discrete wavelet transforms of BPSK, QPSK, V-LFM, FSK signals in sample space with l_1-metric. We explore the features of L-group DWTs of the signals with finite duration in sample spaces with pseudometric and semimetric. We compare numbers of algebraic operations required for performing DWTs in sample spaces with l_2-metric, pseudometric, l_1-metric, and semimetric; we find out that L-group DWTs do not require performing operation of multiplication, i.e. they are multiplication-free. We show that data processing rate when performing discrete wavelet transforms in sample spaces with L-group properties, unlike their analogue from linear space, does not depend on data width.

We introduce multiscale image decomposition on L-groups. We consider multiscale image decompositions based on Hadamard matrix defined in both linear space with scalar product and space with L-group properties. We consider fast 2D discrete wavelet transforms using both known linear algorithm and algorithm based on L-group operations.

Every chapter can be read by prepared reader independently of other chapters. For the readers unacquainted with discussed algebraic systems (L-groups and lattices) we recommend to look through the Chapter 1 before reading other chapters.

There is a triple numeration of formulas in the book: the first number corresponds to the chapter number, the second one denotes the section number, the third one corresponds to the relationship number within the current section, for example, (2.3.4) indicates the 4th formula in Section 2.3 of Chapter 2. The similar numeration system is used for theorems, lemmas, examples, definitions, figures, and tables. Endings of proofs of theorems and lemmas are denoted by $\square$ symbols. Endings of examples are indicated by the $\triangledown$ symbols. Outlines generalizing the obtained results are placed at the end of the corresponding sections if necessary. In general, the chapters are not summarized.

The author would like to extend his sincere appreciation, thanks, and gratitude to Victor Astapenya, C.Sc.; Alexander Geleseff, D.Sc.; Vladimir Horoshko, D.Sc.; Alexey Nalapko, Ph.D.; Vladimir Rudakov, D.Sc.; Victor Seletkov, D.Sc.; and Sergey Zibin, C.Sc. for attention, support, and versatile help that contributed greatly to expediting the writing of this book and improving its content.

The author would like to express his frank acknowledgment to Allerton Press, Inc. and also to its Senior Vice President Ruben de Semprun for granted permissions that allow using the material published by the author in *Radioelectronics and Communications System*.

The author would like to acknowledge understanding, patience, and support provided within Taylor & Francis Group by its staff and the assistance from Nora Konopka, Robin Lloyd-Starkes, and Prachi Mishra. The author would like also to thank Meeta Singh and her team from KnowledgeWorks Global Ltd. for providing help with the final text preparation.

The author would like to express his thanks to all LaTeX developers whose tremendous efforts greatly lighten STEM book author's burden.

Finally, seizing the opportunity, the author thanks all the readers that devoted their precious time to the first book of this series, and also express his deep appreciation to those of them, who formulated their critical remarks and gave valuable suggestions that will be taken into account when republishing the book.

ANDREY POPOFF

Introduction

The development of Signal Processing Theory resembles the evolution of astronomy using first telescopes. Initially, there appeared refracting telescopes, then, a little while later there appeared reflecting telescopes giving upside-down view of celestial objects, that was rather unusual at first glance. But as time goes by, reflecting telescopes little by little replaced refracting telescopes from observational astronomy, inasmuch as providing the required size of telescope objective on the basis of a mirror was found to be easier and cheaper, moreover there was no chromatic aberration.

Sooner or later the known signal processing algorithms will not be able to satisfy mankind needs in information transmitting, receiving, and processing in all spheres of human activity. By this moment one must create the groundwork in Signal Processing Theory which would allow expanding the framework of rather strict constraints imposed on consumer properties of known signal processing algorithms.

This work represents the logical continuation of a book series that started with "Fundamentals of Signal Processing in Metric Spaces with Lattice Properties" published by CRC Press in 2017. The first book contains the material that can be proposed by modern mathematics to provide proper solutions for the needs of both Information Theory and Signal Processing Theory. These solutions are obtained on the basis of algebraic systems with lattice properties, namely, Boolean algebra, lattice, and lattice-ordered groups. The first and the second books from this series are prepared within a recent general trend of researches (see, for instance, [1–10]) exploring the interrelation between algebraic, geometric, and informational approaches to Signal Processing Theory, though an idea basis of these two monographs is absolutely another.

Writing this book began with searching a response to the question: "Does there exist something other than just scalar product used almost everywhere in signal processing algorithms?" If the answer is appeared to be affirmative, one should formulate and answer the following additional questions: (1) what are these unusual signal spaces; (2) what are their metric properties and measures of distances between signals; (3) what are their algebraic properties; and (4) what are the measures of closeness between two signals from these spaces that correspond to the measures of distances and are acceptable for substituting scalar product in signal processing algorithms?

In this work we offer a new toolkit for signal processing which contains a set of approaches, methods, and algorithms that ideologically differ from those that rely on exploiting linear signal space with scalar product. Here the adjective "new" is used in the sense that for the first time we state the Fundamentals of Signal

Processing on the basis of generalized metric spaces with lattice properties, though, for instance, some of statistical methods and algorithms of data processing, that are known for a long time, suppose using the relation of order in sample spaces.

Considered signal processing algorithms are based on some functions of generalized metrics and/or order statistics. Order statistics are well presented in the corresponding literature. This circumstance predetermines the nonlinearity of discussed algorithms. From the standpoint of providing the required quality indices of signal processing, exploiting nonlinear signal processing is expedient in those cases when signal and/or interference (noise) distribution differs from normal one. This take place, for instance, when processing two dimensional signals (images) whose statistical properties essentially differ from normal ones. This distinction takes place, for instance, in normal noise that is exposed to nonlinear transformation in the receiving set with nonlinear amplitude characteristic; in jamming conditions, when a jamming signal possesses non-Gaussian distribution; in the presence of chaotic impulse interference; in the presence of clutter from sea and/or earth surface; in the presence of wave reflections from hydrometeors and radar chaff; in ionospheric propagation of decametric band signals, etc.

In addition to the fact that there exist noises whose distributions differ from Gaussian ones, sometimes one have to decline Gaussian model for the noises that traditionally considered to be normal. The point is that accuracy of approximation of an actual distribution by a normal one is rather high within a middle part of probability density function, whereas on the tails of distribution the accuracy quickly decreases when moving away from its middle part. Some systems of signal processing (for instance, radar and communication ones) are characterized by a probability of error (conditional probability of false alarm or probability of bit error, respectively) that is expressed by rather small quantity $10^{-4} \ldots 10^{-12}$, that is not inherent to those mathematical statistics usually deals with. These small probabilities correspond to the tails of interference (noise) distribution, where its normal approximation does not seem to be satisfactory. Exploiting nonlinear signal processing is considered to be expedient when dealing with normal noise of unknown or variable intensity, when it is hard to realize stability of receiving set operation or estimation of noise power with a required accuracy.

Most of algorithms and methods of discrete (digital) signal processing considered in textbooks and scientific literature are based on the notion of linear space with scalar product, and correspondingly, Euclidean metric between the signals. We usually refer to them well known algorithms of signal extraction, filtering, detection, classification, resolution, spectral analysis, recognition, and also signal parameter estimation in the presence of Gaussian interference and noise. There appear two essential problems here.

The first one relates to classic parametric statistics that determines optimal procedures and algorithms for exact parametric models, however, if there exist even an insignificant discrepancy in Gaussianity, optimality of classic approaches is lost. Thus, for instance, having analyzed asymptotic relative efficiency (see [11, Fig. 1.1.1]) of the estimators [11, (1.3),(1.4)] obtained on the basis of least squares method (LSM) and least moduli method (LMM) (or least absolute deviations method (LAD method)), for the Tukey model of ε-contaminated distribution (John W. Tukey): $F = (1-\varepsilon)N(m, D)+\varepsilon N(m, 9D)$, P. Huber (Peter J. Huber) fairly gives a preference to LMM-estimator, since it is found to be better for all ε from the interval $\varepsilon \in [0.002, 0.5]$. Here $N(m, D)$ denotes normal distribution with expectation m and variance D.

The second problem directly deals with the first one (namely, with a choice of metrics (generalized metrics) for the purpose of estimation), and also with computational aspects of signal processing algorithms based on operation of scalar product which requires performing N operations of multiplication (where N is a sample size). Thus, for instance, performing either operation of convolution or discrete Fourier transform requires N^2 and $2N^2$ operations of multiplication, respectively, whereas forming correlation matrix of a dimensionality $K \times K$ requires K^2N operations of multiplication. All aforementioned holds for real signals, but the case of complex signals just aggravates the situation.

Reflecting upon the first and the second problems leads to a seditious question: first, is it possible exploiting some another metric within a discrete signal space except Euclidean one (or, in other words, using non-Euclidean signal space); and, respectively, second, is it possible exploiting some other mathematical operations except scalar product and operation of multiplication related with latter? In addition to this, it is extremely desirable to maintain an efficiency of signal processing at a level that is above a given one in a wide class of interference (noise) distributions, so that signal processing would not use operation of multiplication at all. The answer on the formulated question is affirmative and is contained in the book. Exploring the approaches that do not require operation of multiplication for various applications of signal processing is included, for instance, in the following works [12–20].

The subject of consideration contained in this book are the methods and algorithms of signal processing based on lattice-ordered group (L-group) operations. In a great extent these methods and algorithms rely on the known l_1-metric that is determined by the relationship

$$\sum_i |a_i - b_i| = \sum_i (a_i \vee b_i - a_i \wedge b_i),$$

and on some other generalized metrics as well. Here the symbols $\vee$, $\wedge$ denote operations of join and meet of L-group, respectively, so that the following identities hold:

$$a \vee b = \max[a, b]; \quad a \wedge b = \min[a, b].$$

The results of exploring the algorithms based on l_1-metric are contained, for instance, in works [12, 16, 19, 21–23]. On the other hand, the discussed methods and algorithms of signal processing are based on further developing the notion of measure of statistical interrelation (MSI) introduced in [24, Section 3.2]. Here it is represented in the form of some function of order statistic that is defined by generalized metric of signal space. Indirectly this research direction relates to J. Tukey's approach titled "Exploratory Data Analysis" which uses five statistics:

median, two extreme values (maximum and minimum), and also lower and upper quartiles of distribution [25]. Related works published recently and discussing signal processing on lattices are the conference papers [26, 27] and monograph [28].

The emphasis in the book is focused rather on its idea content than on mathematical completeness. Unlike the previous work [24], where the statement is performed on the basis of the principle

lattice or lattice-ordered group vs linear space,

in this book, the material is stated in compliance with formula

lattice-ordered group vs linear space,

with the difference that in [24, Chapter 7] interaction between signals is described by *join* and/ or *meet operations* of lattice, whereas in this book signal interaction is defined by *operation of addition* of lattice-ordered group (L-group). The last circumstance provides equivalence of signal interaction representation in both signal space with L-group properties and linear signal space.

L-group algorithms relate to a wide class of nonlinear signal processing algorithms. So here we point out, first, what algebraic systems they relate to, second, what generalized metric spaces they are realized in, and third, what statistics they relate with. This circumstance determines the corresponding place of the suggested algorithms (are marked by a wavy line) discussed in the book within a general classification of signal processing algorithms shown in the table I.1.

TABLE I.1   General classification of signal processing algorithms

Element of classification	Types of algorithms
Algebraic system of signal space	group, lattice, L-group, others
Generalized metric of signal space	metric, pseudometric, semimetric, others
Used statistics	Bayesian, nonparametric, robust, order
Signal processing problem	detection, classification, resolution, recognition, filtering, parameter estimation, others
Superposition principle	linear, nonlinear, quasi-linear
Distributions of processed signals	Gaussian, non-Gaussian, model-based, model-free, others
Domains of processed signals	continuous, discrete; time, frequency, spatial, others
Prior uncertainty overcoming	adaptive, invariant, others
Phase difference between signals	coherent, incoherent
Kind of computational process	serial (linear), parallel, cyclic, recursive, branch and bound, others
Applications	array signal processing, image processing, pattern recognition, spectral analysis, wavelet analysis, data compression, radars, networks, others
Physical origin of signals	acoustic, optic, radiofrequency, others

Here we notice that depending on signal processing problems, we differ two groups of algorithms. Thus, signal detection, classification, resolution, and recognition relate to the group of decision-making algorithms, while signal extraction, filtering (including interpolation and extrapolation), and parameter estimation form the group of estimation algorithms.

Methods and algorithms based on L-group operations could be conditionally separated into four groups. *The first group* include the known algorithms based on obtaining the estimators that are robust with respect to influence of non-Gaussian interference (noise) and determined in terms of L-groups. *The second group* is represented by signal processing algorithms based on generalized metrics (including l_1-metric) and designed for signal extraction, filtering, detection, classification, resolution, spectral analysis, signal parameter estimation, demultiplexing and demodulation, wavelet analysis, etc., so that they do not require performing operation of multiplication. These algorithms are slightly inferior to known classic optimal signal processing algorithms synthesized within linear signal space in their efficiency in the presence of Gaussian interference (noise), but the aforementioned circumstance (i.e. multiplication-free property) provides a higher signal processing rate, and respectively, lesser requirements with respect to computational resources of processing system. Some of these algorithms relate to the class of robust ones, and some of them relate to the class of nonparametric algorithms. The third and forth groups of algorithms are formed depending on their relation to superposition principle. Thus, *the third group* is represented by quasi-linear algorithms, for which superposition principle holds with sufficient degree of accuracy. Most of algorithms from the second group can be related with the third group. At last, *the fourth group* is formed by linear signal processing algorithms which are formulated in terms of L-group operations. Their efficiency is appeared to be equivalent to known optimal algorithms–analogues that are synthesized within linear signal space.

Mainly, statement of the material is realized on the basis of comparative analysis of known optimal and suggested algorithms of signal processing. This comparative analysis is represented in the form that is convenient for visual perception. The reader can easily orient oneself with a help of offered graphic and tabular material, not going into details of mathematical elucidation, if it is not required.

All algorithms presented in the book are 100% tested with these or those software means. L-group algorithms proposed in the book are not oriented to some specific software developer, and described in such a way that can be realized by an arbitrary software product depending on reader's preferences.

List of Abbreviations

Notion	Abbreviation
Address signal coder	ASC
Address signal decoder	ASD
Address signal generator	ASG
Address signal selector	ASS
Asymptotic relative efficiency	ARE
Autocorrelation function	ACF
Autocorrelation matrix	ACM
Base station	BS
Binary phase shift keying	BPSK
Channel signal generator	CSG
Channel signal coder	CC
Channel signal decoder	CD
Channel signal selector	CS
Characteristic function	CF
Chip	ch
Code division multiplexing	CDM
Code division multiple access	CDMA
Continuous phase	CP
Cumulative distribution function	CDF
Demodulator	D
Direct sequence	DS
Discrete cosine transform	DCT
Discrete sine transform	DST
Discrete wavelet transform	DWT
Eigenvectors	EV
Estimation of signal parameters via rotational invariance techniques	ESPRIT
Frame	fr
Frequency division multiplexing	FDM
Frequency division multiple access	FDMA
Frequency shift keying	FSK
Group	g
Influence function	IF
Inphase	I
Inverse discrete wavelet transform	IDWT
Hodges-Lehmann filter	HLF
Homogeneous filter	hmF
Huber function	HF

Continued on next page

Notion	Abbreviation
Hyperspectral density	HSD
Hyperspectral transform	HST
Least mean squares	LMS
Least moduli method	LMM
Least squares method	LSM
Linear correlation (algorithm)	LC
Linear frequency modulation	LFM
Matched filter	MF
M-ary frequency shift keying	MFSK
Minimum norm	MN
Minimum variance	MV
Measure of statistical interrelation	MSI
Metric	m
Mobile station	MS
Modulator	M
MUltiple SIgnal Classification	MUSIC
Multiscale analysis	MSA
Normalized measure of statistical interrelation	NMSI
Orthogonal frequency division multiplexing	OFDM
Phase shift keying	PSK
Power spectral density	PSD
Probability density function	PDF
Pseudometric	pm
Quadrature	Q
Quadrature phase shift keying	QPSK
Radio frequency	RF
Rank (algorithm)	R
Rank Wilcoxon (algorithm)	RW
Reciever	Rx
Recursive least squares	RLS
Selector signal generator	SSG
Semimetric	sm
Sign correlation (algorithm)	SC
Sign-rank (algorithm)	SR
Signal-to-noise ratio	SNR
Slot	sl
Source of messages	Src
Time division multiplexing	TDM
Time division multiple access	TDMA
Time-frequency mismatching function	TFMF
Transmitter	Tx
Truncated mean filter	trMF
User of message	U
Wiener filter	WF

Notation System

Notion	Notation
ABSTRACT ALGEBRA	
Lattice	$\mathcal{L}(\vee, \wedge)$
L-group	$\mathcal{L}(+, \vee, \wedge)$
Operation of join	$a \vee b$
Operation of meet	$a \wedge b$
Partly ordered set	Γ
GENERALIZED METRIC SPACES AND NORMED SPACES	
Generalized metric	$\mathrm{M}(x, y)$
l_1-metric	$\mathrm{M}_{l_1}(x, y)$
l_2-metric	$\mathrm{M}_{l_2}(x, y)$
l_p-norm	$\|x\|_p$
Normalized l_1-metric	$\mu_{l_1}(x, y)$
Normalized l_2-metric	$\mu_{l_2}(x, y)$
Normalized pseudometric	$\mu_p(x, y)$
Normalized semimetric	$\mu_s(x, y)$
Pseudometric	$\mathrm{M}_p(x, y)$
Semimetric	$\mathrm{M}_s(x, y)$
PROBABILITY THEORY	
Bivariate cumulative distribution function	$F_{\xi\eta}(x, y)$
Bivariate probability density function	$p_{\xi\eta}(x, y)$
Cauchy distribution	$C(a, b)$
Characteristic function	$\Phi_\xi(u)$
Correlation coefficient between random variables ξ, η	$r_{\xi\eta}$
Cumulant of k-th order	κ_k
Double exponential (Laplace) distribution	$DE(a, b)$
Gamma distribution	$\Gamma(\beta, \alpha)$
First and second moments, respectively	m_1, m_2
Logistic distribution	$L(a, b)$
Log-normal distribution	$Ln(c, \sigma)$
Mathematical expectation of random variable ξ	$\mathrm{M}\{\xi\}$
Metric between the samples ξ_t, η_t	$\mu(\xi_t, \eta_t)$
Negative part of stochastic process $v(t)$	$v_-(t)$
Normal distribution	$N(a, b)$
Normalized autocorrelation function	$r_\xi(t_1, t_2)$, $r_\xi(\tau)$
Normalized measure of statistical interrelation	$\nu(\xi, \eta)$
Normalized variance function,	$\theta_\xi(t_1, t_2)$
Positive part of stochastic process $v(t)$	$v_+(t)$
Random variable	ξ, η

Continued on next page

Notion	Notation
Probability	$\mathbf{P}$
Stochastic process	$\xi(t)$, $\eta(t)$
Standard normal distribution	$N(0,1)$
Student distribution	$St(a,b,\nu)$
Tukey's «ε-contaminated» distribution	$T(a,\varepsilon,\tau)$
Univariate cumulative distribution function	$F_\xi(x)$
Univariate probability density function	$p_\xi(x)$
Weibull distribution	$W(c,\alpha)$
α-quantile of distribution	x_α
MATHEMATICAL STATISTICS	
Estimator of parameter λ	$\hat{\lambda}$
Linear sample space	$\mathcal{LS}(\mathcal{X},\mathcal{B}_\mathcal{X};+)$
Measure of statistical interrelation in l_1-metric space	$\mathrm{N}_{l_1}(x,y)$
Measure of statistical interrelation in pseudometric space	$\mathrm{N}_p(x,y)$
Measure of statistical interrelation in semimetric space	$\mathrm{N}_s(x,y)$
Null sample	$\mathbf{0}$
Random sample	$X=(X_1,\ldots,X_n)$, $Y=(Y_1,\ldots,Y_n)$
Realizations of random samples X, Y	$x=(x_1,\ldots,x_n)$, $y=(y_1,\ldots,y_n)$
Sample space with lattice properties	$\mathcal{L}(\mathcal{Y},\mathcal{B}_\mathcal{Y};\vee,\wedge)$
Sample space with L-group properties	$\mathcal{L}(\mathcal{X},\mathcal{B}_\mathcal{X};+,\vee,\wedge)$
Sample mean	$\bar{X}$
Sample median	$\widetilde{X}$
Truncated mean	$\bar{X}_\alpha$
SIGNAL PROCESSING: GENERAL NOTATIONS	
Carrier frequency	f_0
Continuous signals	$a(t)$, $b(t)$
Domain of definition of the signal $s(t)$	T_s
Discrete signal	$s(i)$, $s(t_i)$
Desired (reference) signal	$d(t)$
Error signal	$e(t)$
Energy per bit	E_b
Energy of signal	E
Envelope of signal $v(t)$	$E_v(t)$
Hyperspectral representation of discrete signal $a(i)$	$\mathbf{HST}[a(i)]$
Instantaneous values of signals	$a_t=a(t)$, $b_t=b(t)$
Impulse response function of digital filter	$h(i)$
Impulse response function of filter	$h(t)$
Linear signal space	$\mathcal{LS}(+)$
l_p-norm of signal $s(t)$	$\|s\|_p$
Metric signal space	$(\mathbf{\Gamma},\mu)$
Modulating function	$M[*,*]$
Negative part of signal $a(t)$	$a_-(t)$
White Gaussian noise power spectral density	N_0
Observation interval	T_{obs}

Continued on next page

Notion	**Notation**
Period of signal carrier oscillation	T_0
Positive part of signal $a(t)$	$a_+(t)$
Realization of signal $x(t)$	$x^*(t)$
Sampling interval	Δt
Samples of the observed signal $x(t)$	$\{x(t_j)\}$
Signal space with L-group properties	$\mathcal{L}(+, \vee, \wedge)$
Signals	$a(t), b(t), \ldots$
Signal-to-noise ratio (in power)	q^2
Time of arrival of signal	t_0
Upper frequency of noise (interference) PSD	$f_{n,\max}$
Variance of noise $n(t)$	D_n
Variance of signal $s(t)$	D_s
PARAMETER ESTIMATION AND SIGNAL FILTERING	
Autocorrelation matrix of vector signal $\mathbf{x}(t)$	$\mathbf{R}_{xx}(t)$
Cross-correlation vector of the input and the desired signals	$\mathbf{P}_{dx}(t)$
Cross-correlation coefficient between two signals	$r[a(t), b(t)]$
Estimator of the signal $s(t)$	$\hat{s}(t)$
Estimator of unknown scalar parameter λ	$\hat{\lambda}$
Estimator of measure of statistical interrelation	$\hat{N}(u; m)$
Estimator of mutual measure of statistical interrelation	$\hat{N}(u, v; m)$
Fisher information matrix	$\mathbf{\Phi} = \|\Phi_{ik}\|$
Hampel influence function	IF
Huber function	$HF(x, a)$
Vector estimator of unknown vector parameter $\boldsymbol{\lambda}$	$\hat{\boldsymbol{\lambda}}$
Vector of weight coefficients	$\mathbf{w}(t)$
Weight coefficient in k-th processing channel	$w_k(t)$
Weight matrix	$\mathbf{W}(t)$
DETECTION AND CLASSIFICATION	
Asymptotic relative efficiency (ARE) of algorithm α_A with respect to algorithm α_B	$e(\alpha_A, \alpha_B)$
ARE of sign algorithm with respect to linear correlation algorithm	$e_{SC,LC}$
ARE of Wilcoxon rank algorithm with respect to linear correlation algorithm	$e_{RW,LC}$
ARE of sign-rank algorithm with respect to linear correlation algorithm	$e_{SR,LC}$
ARE of detection algorithm, based on MSI in l_1-metrics space, with respect to linear matched filter algorithm	$e_{MSI(m),MF}$
ARE of detection algorithm, based on MSI in pseudometrics space, with respect to linear matched filter algorithm	$e_{MSI(pm),MF}$
ARE of detection algorithm, based on MSI in semimetrics space, with respect to linear matched filter algorithm	$e_{MSI(sm),MF}$
Conditional probability of correct detection	D
Conditional probability of false alarm	F
Conditional PDF of random variable y on H_i hypothesis	$p_y(x/H_i)$
Decision about signal presence	d_1

Continued on next page

Notion	Notation
Decision about signal absence	d_0
Decision about accepting the hypothesis H_k	d_k
Detection threshold	l_0
Domain of definition of the signal $s(t)$	T_s
Hypothesis about receiving the signal $s_k(t)$ in the observed stochastic process $x(t)$	H_k
Hypothesis concerning signal presence	H_1
Hypothesis concerning signal absence	H_0
Norm-factor	F_{nm}
Observation interval	T_{obs}
Probability of bit error	$\mathbf{P}_\mathrm{b}$
Probability of bit error (function)	$\mathbf{P}_\mathrm{b}(q_b)$
Probability of correct classification	$\mathbf{P}_{kk}$
Probability of symbol error	$\mathbf{P}_\mathrm{s} = 1 - \mathbf{P}_{kk}$
Probability of symbol error (function)	$\mathbf{P}_\mathrm{s}(q)$
Rank of the element x_k of the sample $\mathbf{x}$	R_k
Rank vector	$\mathbf{R} = [R_k]$
Signal duration	T
Time of signal arrival	t_0
Unknown detection parameter	$\theta, \theta \in \{0, 1\}$
RESOLUTION	
Ambiguity function	$\Psi(\tau, F)$
Frequency shift	F
Mismatching function	$w(\boldsymbol{\lambda}', \boldsymbol{\lambda})$
Mismatching in time	Δt
Mismatching in frequency	ΔF
Modulus of time-frequency mismatching function $w(i, k)$	$W(i, k) = |w(i, k)|$
Normalized mismatching function	$\rho(\boldsymbol{\lambda}', \boldsymbol{\lambda})$
Normalized time-frequency mismatching function	$\rho(\tau, F)$
Resolution measure in time delay	Δ_τ
Resolution measure in frequency shift	Δ_f
Time delay	τ
Time-frequency mismatching function of a discrete signal	$w(t_i, f_k), w(i, k)$
SPECTRAL ESTIMATION	
Autocorrelation function of the discrete stochastic signal $x(i)$	$r_x(m)$
Autocorrelation matrix of discrete signal $x(i)$	$\mathbf{R}_x$
Bartlett periodogram	$\hat{P}_B(f)$
Conjugate transpose of vector of complex harmonics	$\overline{\mathbf{e}(f)}^T$
Cross-correlation matrix estimate	$\hat{\mathbf{R}}_{uv}$
Eigenvector estimator of hyperspectral density	$\hat{H}_{EV}(f)$
Eigenvector estimator of power spectral density	$\hat{P}_{EV}(f)$
Estimator of autocorrelation function	$\hat{r}_x(m)$
Estimator of autocorrelation matrix	$\hat{\mathbf{R}}_x$
Estimator of hyperspectral density	$\hat{H}(f)$
Estimator of power spectral density	$\hat{P}_x(f)$

Continued on next page

Notion	Notation
Hyperspectral density estimator based on Bartlett's method	$\hat{H}_B(f)$
Hyperspectral density estimator based on Welch's method	$\hat{H}_W(f)$
Hyperspectral density of a discrete signal	$\hat{H}(f)$
Minimum norm estimator of hyperspectral density	$\hat{H}_{MN}(f)$
Minimum norm estimator of power spectral density	$\hat{P}_{MN}(f)$
Minimum variance estimator of hyperspectral density	$\hat{H}_{MV}(f)$
Minimum variance estimator of power spectral density	$\hat{P}_{MV}(f)$
MUSIC estimator of hyperpectral density	$\hat{H}_{MUSIC}(f)$
MUSIC estimator of power spectral density	$\hat{P}_{MUSIC}(f)$
Power spectral density of discrete signal $x(i)$	$P_x(f)$
Vector of complex harmonics	$\mathbf{e}(f)$
Welch periodogram	$\hat{P}_W(f)$
Window function	$w(i)$
ARRAY PROCESSING	
Autocorrelation matrix of received signals $a_i(t)$	$\mathbf{R}_a$
Carrier frequency of the signal $s(t)$	f_s
Conjugate transpose of vector of spatial harmonics	$\mathbf{e}(\theta_s, f_s)^T$
Cross-correlation coefficient between received signals $a_{i,j}(t)$	r_{ij}
Direction of arrival of the signal $s(t)$	θ_s
Eigenvector estimator of spatial hyperspectrum based on measure of statistical interrelationship	$\hat{H}_{EV}(\theta)$
Eigenvector estimator of spatial spectrum	$\hat{P}_{EV}(\theta)$
Interelement distance of antenna array	d
Minimum norm estimator of spatial hyperspectrum based on measure of statistical interrelationship	$\hat{H}_{MN}(\theta)$
Minimum norm estimator of spatial spectrum	$\hat{P}_{MN}(\theta)$
Minimum variance estimator of spatial hyperspectrum based on measure of statistical interrelationship	$\hat{H}_{MV}(\theta)$
Minimum variance estimator of spatial spectrum	$\hat{P}_{MV}(\theta)$
MUSIC estimator of spatial hyperspectrum based on measure of statistical interrelationship	$\hat{H}_{MUSIC}(\theta)$
Noise (interference) autocorrelation matrix estimate	$\hat{\mathbf{R}}_n$
MUSIC estimator of spatial spectrum	$\hat{P}_{MUSIC}(\theta)$
Propagation velocity of electromagnetic waves	c
Ratio of maximum to minimum frequencies	$f_{\max}/f_{\min}$
Received signal in the i-th receiving element of antenna array	$a_i(t)$
Variance of signal $a_i(t)$ in the i-th receiving element of antenna array	D_i
Vector of complex spatial harmonics	$\mathbf{e}(\theta_s, f_s)$
Vector of weight coefficients	$\mathbf{w}$
Weight coefficient in the i-th processing channel	w_i
WAVELETS AND IMAGE PROCESSING	
Additive noise	$n_+(i,j)$
Bivariate wavelet function	$g^{(i)}(x,y)$
Continuous wavelet basis function	$\psi(t; a, b)$

Continued on next page

Notion	Notation
Continuous wavelet transform of the signal $x(t)$	$CWT_x(a,b)$
Discrete wavelet transform of the signal $x(t_j)$	$DWT_x(u,s)$
Discrete wavelet basis function	$\psi(j;u,s)$
Discrete shift parameter (DWT)	u
Discrete scale parameter (DWT)	s
Sampling interval in time domain	Δt
Sampling interval in frequency domain	Δf
Discrete wavelet transform in l_1-metric sample space	$X_{l_1}(u,s)$
Discrete wavelet transform in pseudometric sample space	$X_p(u,s)$
Discrete wavelet transform in semimetric sample space	$X_s(u,s)$
Image array	$x(i,j)$
Impulse noise	$n_\oplus(i,j)$
Initial image signal	$s(i,j)$
Initial image signal estimator	$\hat{s}(i,j)$
Moving window	$W = \|w_{lk}\|$
Multiplicative noise	$n_\otimes(i,j)$
Set of basis functions	$\{\psi_{a,b}(t)\}$
Shift parameter (CWT)	a
Scale parameter (CWT)	b
GENERAL NOTATIONS	
Combination of k distinct elements from a set with n elements	C_n^k
Dirac delta function	$\delta(t)$
End of example	$\triangledown$
End of proof	$\square$
Euclidean metric	$\rho_E(x,y)$
Euclidean space (n-dimensional)	$\boldsymbol{R}^n$
Fourier transform	$\mathcal{F}[*]$
Heaviside unit step function	$1(t)$
Hilbert space	$\mathcal{HS}$
Hilbert transform	$\mathcal{H}[*]$
Identity matrix	$\mathbf{I}$
Indexed set	$\{A_t\}_{t \in T}$
Kronecker delta	δ_{ij}
Kronecker function	$\delta(a,b)$
Set of natural numbers	$\mathbf{N}, \mathbb{N}$
Sign function	$\mathrm{sgn}(x)$

1

Introduction to Signal Processing on L-groups

The chapter begins from considering the general questions concerning the application of L-group operations and the functions related with them and transforms to Signal Processing Theory. The goal of this chapter is acquainting the reader with those important parts of mathematical apparatus that have not yet found proper reflection in the university courses in mathematics intended for preparing specialists in signal processing for different branches of electronics. For these purposes the author uses the existing parallels between the signal spaces explored in the book and well-known linear spaces with the scalar (inner) product.

When studying the material of this chapter, it is important to understand the essence of main definitions, get accustomed to corresponding notations, learn performing operations with mathematical objects (signals and their samples) defined on L-groups. Reaching this goal is provided by examples exposed within every section whose consideration allows the reader to improve understanding of the stated material. In brief, this chapter contains information that is sufficient for (1) understanding the essence of general approach stated in the book; (2) understanding the main content of L-group signal processing; (3) helping the reader better understand robust and nonparametric statistical procedures; (4) drawing the parallels between known and suggested signal processing algorithms; and (5) providing a fresh view for further reading the literature on signal processing.

1.1   Operations of L-group

Any pair of signals $a(t)$, $b(t)$ can be considered as a partially ordered set $\mathcal{L}$, in which at each time instant $t \in T$ between two instantaneous values (samples) $a_t = a(t)$, $b_t = b(t)$ of signals $a(t)$, $b(t) \in \mathcal{L}$ one can define the relation of order $a_t \leq b_t$ (or $a_t \geq b_t$). Then partially ordered set $\mathcal{L}$ is a lattice with operations of join and meet, respectively: $a_t \vee b_t = \sup_{\mathcal{L}}\{a_t, b_t\}$, $a_t \wedge b_t = \inf_{\mathcal{L}}\{a_t, b_t\}$, and if $a_t \leq b_t$, then $a_t \wedge b_t = a_t$ and $a_t \vee b_t = b_t$ [29, 30]:

$$a_t \leq b_t \Leftrightarrow \begin{matrix} a_t \wedge b_t = a_t; \\ a_t \vee b_t = b_t. \end{matrix} \qquad (1.1.1)$$

In a natural way in partially ordered set $\mathcal{L}$, we define operation of addition $a(t)+b(t)$ between the signals $a(t)$, $b(t) \in \mathcal{L}$ at every time instant $t \in T$. Then partially ordered set $\mathcal{L}$ is *lattice-ordered group* $\mathcal{L}(+, \vee, \wedge)$ (L-group).

DOI: 10.1201/9781003275855-1

In any L-group the following statements hold [29, 31–35]:

(1) $\mathcal{L}(+)$ is an additive group, in which the following axiomatics holds:

$$a(t) + b(t) \in \mathcal{L}(+); \tag{1.1.2a}$$

$$a(t) + b(t) = b(t) + a(t); \tag{1.1.2b}$$

$$a(t) + (b(t) + c(t)) = (a(t) + b(t)) + c(t); \tag{1.1.2c}$$

$$\exists 0 \in \mathcal{L}(+): \ \forall a(t): \ a(t) + 0 = 0 + a(t); \tag{1.1.2d}$$

$$\forall a(t): \ \exists - a(t): \ a(t) + (-a(t)) = (-a(t)) + a(t) = 0, \tag{1.1.2e}$$

(2) $\mathcal{L}(\vee, \wedge)$ is a lattice, in which the following axiomatics holds:

$$a(t)\vee(b(t)\vee c(t)) = (a(t)\vee b(t))\vee c(t), \ a(t)\wedge(b(t)\wedge c(t)) = (a(t)\wedge b(t))\wedge c(t); \tag{1.1.3a}$$

$$a(t) \vee b(t) = b(t) \vee a(t), \ a(t) \wedge b(t) = b(t) \wedge a(t); \tag{1.1.3b}$$

$$a(t) \vee a(t) = a(t), \ a(t) \wedge a(t) = a(t); \tag{1.1.3c}$$

$$(a(t) \vee b(t)) \wedge a(t) = a(t), \ (a(t) \wedge b(t)) \vee a(t) = a(t), \tag{1.1.3d}$$

(3) for all the signals $a(t)$, $b(t)$, $c(t)$, $d(t) \in \mathcal{L}(+, \vee, \wedge)$, the following identities defining distributivity of operation of group addition $\mathcal{L}(+)$ with respect to lattice operations $\mathcal{L}(\vee, \wedge)$ of join and meet hold:

$$a(t) + (c(t) \wedge d(t)) + b(t) = (a(t) + c(t) + b(t)) \wedge (a(t) + d(t) + b(t)); \tag{1.1.4a}$$

$$a(t) + (c(t) \vee d(t)) + b(t) = (a(t) + c(t) + b(t)) \vee (a(t) + d(t) + b(t)). \tag{1.1.4b}$$

The relationships (1.1.2a,b,c,d,e) represent *axioms of closure, commutativity, associativity, neutral element, and inverse element* of additive group, correspondingly. The relationships (1.1.3a,b,c,d) are the *axioms of associativity, commutativity, idempotency, and absorption* of lattice, respectively.

Fig. 1.1.1a,b illustrates the initial signals $a(t)$ and $b(t)$ in the form of bell-shaped radiofrequency pulse and quasi-white normal noise, respectively. Fig. 1.1.2a,b depicts the results of interaction $x(t)$, $y(t)$ of signals $a(t)$ and $b(t)$ in the signal space with lattice properties $\mathcal{L}(\vee, \wedge)$ in the form of operations of join $x(t) = a(t)\vee b(t)$ and meet $y(t) = a(t) \wedge b(t)$, correspondingly. If we make a mental experiment to obtain the results of operations $x(t) \wedge a(t)$, $y(t) \vee a(t)$, we can easily convince ourselves that axioms of absorption really take place in lattice $\mathcal{L}(\vee, \wedge)$.

There is a simple and natural relation between operations of join $x(t) = a(t) \vee b(t)$ and meet $y(t) = a(t) \wedge b(t)$ and functions of minimum min and maximum max:

$$x(t) = a(t) \vee b(t) = \max_{t \in T}[a(t), b(t)]; \ y(t) = a(t) \wedge b(t) = \min_{t \in T}[a(t), b(t)]. \tag{1.1.5}$$

If $a(t) \in \mathcal{L}$, then the functions $a_+(t)$, $a_-(t)$ defined by relationships:

$$a_+(t) = a(t) \vee 0; \ a_-(t) = a(t) \wedge 0, \tag{1.1.6}$$

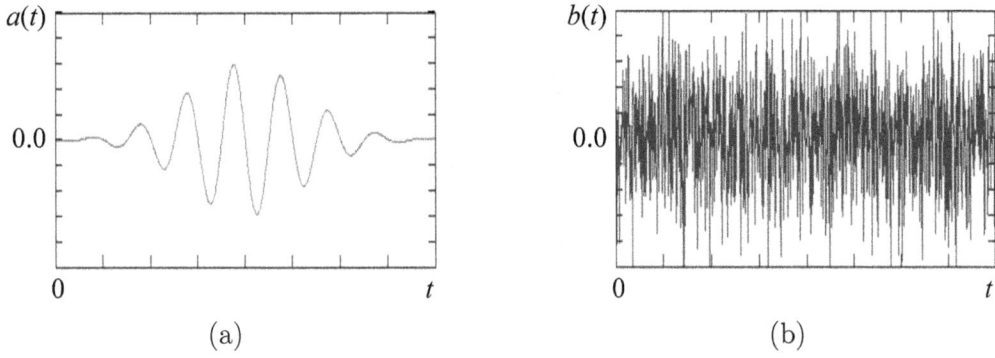

(a)　　　　　　　　　　　　(b)

FIGURE 1.1.1　The initial signals (a) $a(t)$; (b) $b(t)$

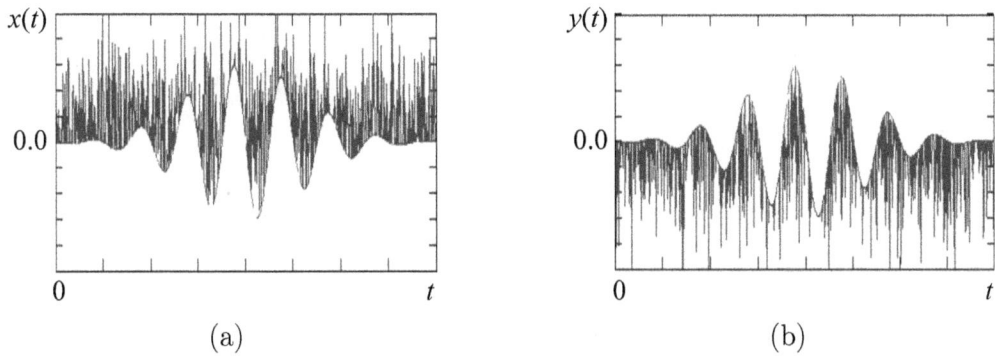

(a)　　　　　　　　　　　　(b)

FIGURE 1.1.2　Operations of (a) join $x(t) = a(t) \vee b(t)$; (b) meet $y(t) = a(t) \wedge b(t)$

are called *positive* and *negative parts* of a signal $a(t)$, respectively, so that the following identity holds:

$$a(t) = a_+(t) + a_-(t). \tag{1.1.7}$$

If $a(t) \in \mathcal{L}$, then the function $|a(t)|$ determined by the expression:

$$|a(t)| = a(t) \vee (-a(t)) = a_+(t) - a_-(t), \tag{1.1.8}$$

is called *modulus* of the signal $a(t)$.

Fig. 1.1.3a,b illustrates positive $a_+(t)$ and negative $a_-(t)$ parts of initial signal $a(t)$ (see Fig. 1.1.1a) obtained on the basis of relationships (1.1.6), and Fig. 1.1.3c depicts modulus $|a(t)|$ of initial signal $a(t)$ calculated by the expressions (1.1.8).

Modulus of difference of two initial signals $a(t)$ and $b(t)$ (see Fig. 1.1.1a,b) in signal space with lattice properties $\mathcal{L}(+, \vee, \wedge)$ and operations of join and meet $\vee, \wedge$ is determined by the difference of their operations of join and meet [29, § XIII.4;(22)] and shown in Fig. 1.1.3d:

$$|a(t) - b(t)| = a(t) \vee b(t) - a(t) \wedge b(t). \tag{1.1.9}$$

The signal space with lattice properties $\mathcal{L}(+, \vee, \wedge)$ can be realized by transforming the signals of linear space $\mathcal{LS}(+)$ in such a way that the results of interactions

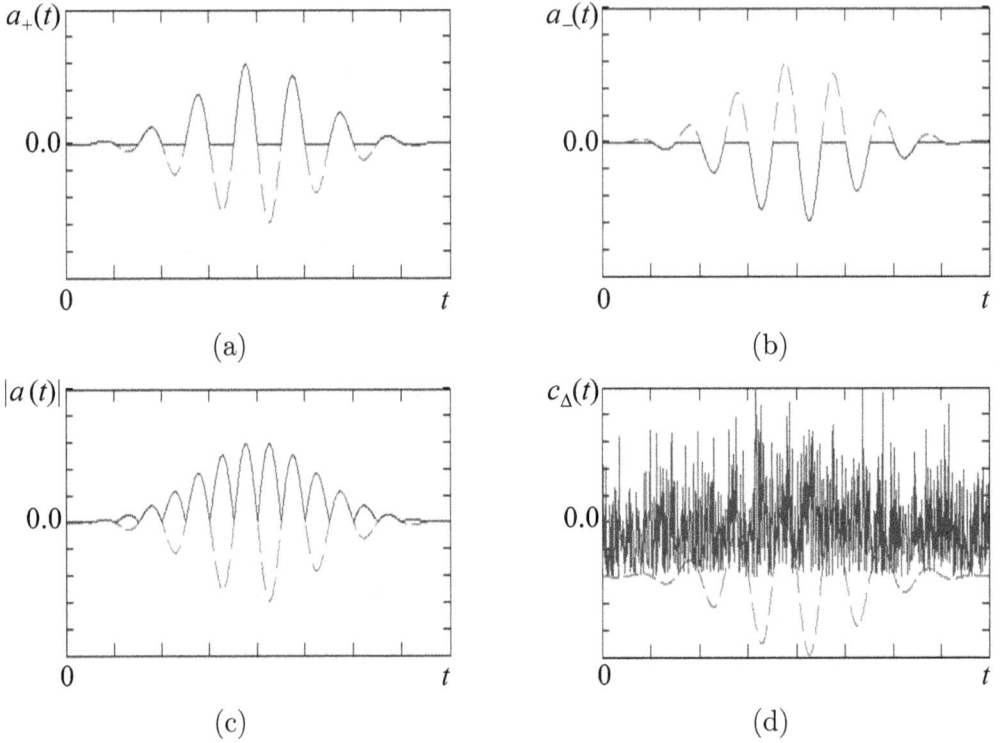

FIGURE 1.1.3   Illustrations of (a) positive part $a_+(t)$ of a signal $a(t)$; (b) negative part $a_-(t)$ of a signal $a(t)$; (c) modulus $|a(t)|$ of a signal $a(t)$; (d) modulus of the difference between two signals $a(t)$ and $b(t)$: $c_\Delta(t) = |a(t) - b(t)|$

$x(t) = a(t) \vee b(t)$, $y(t) = a(t) \wedge b(t)$ of the signals $a(t)$ and $b(t)$ in signal space $\mathcal{L}(+, \vee, \wedge)$ with operations of join and meet are realized according to the relationships:

$$x(t) = a(t) \vee b(t) = \{[a(t) + b(t)] + |a(t) - b(t)|\}/2; \tag{1.1.10a}$$

$$y(t) = a(t) \wedge b(t) = \{[a(t) + b(t)] - |a(t) - b(t)|\}/2, \tag{1.1.10b}$$

that are the consequences of known equations [29, § XIII.3;(14)], [29, § XIII.4;(22)].

The identities (1.1.10a,b) define the transform of linear signal space $\mathcal{LS}(+)$ into the signal space with lattice properties $\mathcal{L}(+, \vee, \wedge)$: T: $\mathcal{LS}(+) \to \mathcal{L}(+, \vee, \wedge)$. The mapping T^{-1} that is inverse with respect to a given one T: T^{-1}: $\mathcal{L}(+, \vee, \wedge) \to \mathcal{LS}(+)$ is defined by known identity [29, § XIII.3;(14)]:

$$a(t) + b(t) = a(t) \vee b(t) + a(t) \wedge b(t). \tag{1.1.11}$$

The existence of inverse transform T^{-1} (1.1.11) suggests that in signal space with lattice properties $\mathcal{L}(+, \vee, \wedge)$ one can provide quality indices of signal processing that are equivalent to quality indices obtained in linear space $\mathcal{LS}(+)$. It will be shown below when analyzing the corresponding algorithms. This statement is the direct consequence of Theorem 6.2.2 [24].

1.2 Interrelation between Operations of L-group and Known Nonlinear Functions

In this section we consider some nonlinear functions that are widely used in signal processing practice. Here, we pay main attention to some key questions of methodology of synthesizing nonlinear signal transformers based on operations of L-group $\mathcal{L}(+, \vee, \wedge)$.

1.2.1 Interrelation with Step Functions

The examples of step functions obtained the most prevalence in practice of signal processing are the following well-known ones: symmetric unit step function $u(x)$, Heaviside unit step function $u_-(x)$, unit step function $u_+(x)$ [36, (21.9-1)], sign function $\mathrm{sgn}(x)$, and Kronecker function $\delta(x,a)^*$:

$$u(x) = \begin{cases} 0, & x < 0; \\ 0.5, & x = 0; \\ 1, & x > 0, \end{cases} \tag{1.2.1a}$$

$$u_+(x) = \begin{cases} 0, & x \leqslant 0; \\ 1, & x > 0, \end{cases} \tag{1.2.1b}$$

$$u_-(x) = \begin{cases} 0, & x < 0; \\ 1, & x \geqslant 0, \end{cases} \tag{1.2.1c}$$

$$\mathrm{sgn}(x) = 2 \cdot u(x) - 1 = \begin{cases} -1, & x < 0; \\ 0, & x = 0; \\ 1, & x > 0, \end{cases} \tag{1.2.2}$$

$$\delta(x,a) = \begin{cases} 0, & x \neq a; \\ 1, & x = a. \end{cases} \tag{1.2.3}$$

In terms of L-group $\mathcal{L}(+, \vee, \wedge)$ with operations of join and meet $\vee, \wedge$, the formulas (1.2.1b,c), (1.2.2), and (1.2.3) can be written analytically by the following equations:

$$u_+(x) = 0 \vee [1 \wedge \alpha \cdot x] = 1 \wedge [0 \vee \alpha \cdot x]; \tag{1.2.4}$$

$$\mathrm{sgn}(x) = -1 \vee [1 \wedge \alpha \cdot x] = 1 \wedge [-1 \vee \alpha \cdot x]; \tag{1.2.5}$$

$$\delta(x,a) = 1 - (0 \vee [1 \wedge \alpha \cdot |x - a|]) = 1 - (1 \wedge [0 \vee \alpha \cdot |x - a|]); \tag{1.2.6}$$

$$u_-(x) = \delta(x,0) \vee [1 \wedge \alpha \cdot x] = 1 \wedge [\delta(x,0) \vee \alpha \cdot x], \tag{1.2.7}$$

where $\alpha = \mathrm{const}$, α is some large quantity: $\alpha \gg 1$.

When exploiting the relationships (1.2.4)...(1.2.7), one should take into account that if $\alpha = 10^n$; then, for instance, on small $x = 10^{-n-k} > 0$, $k = 1, 2, 3 \ldots$, sign function $\mathrm{sgn}(x)$ determined by the formula (1.2.5) is equal to $\mathrm{sgn}\left(10^{-n-k}\right) = 10^{-k}$

---

*Here we use a generalized version of the function $\delta(x,a)$ for real numbers

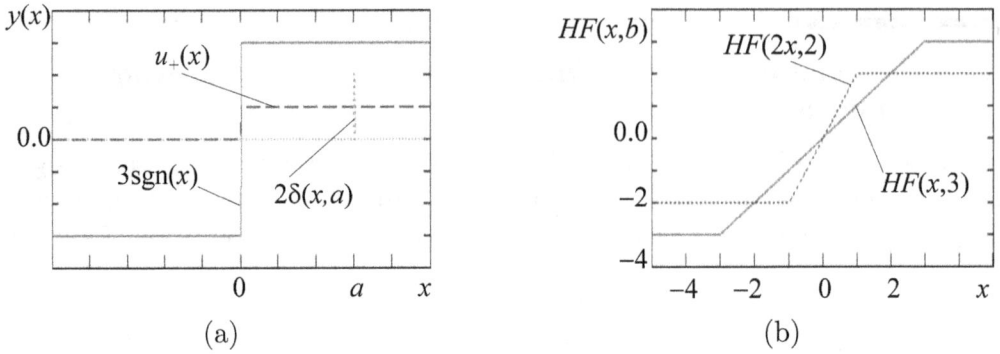

FIGURE 1.2.1   Functions: (a) $u_+(x)$, sgn(x), $\delta(x,a)$; (b) $HF(x,b)$

but not 1. Quite similarly, that is important for practical applications, if $\alpha = 2^n$, on small $x = 2^{-n-k} > 0$, $k = 1, 2, 3\ldots$, the function (1.2.5) is equal to sgn $\left(2^{-n-k}\right) = 2^{-k}$ but not 1. The last circumstance (using $\alpha = 2^n$) allows realizing the functions (1.2.4)...(1.2.7) for practical applications of digital signal processing without using operation of multiplication, since it can be replaced by operation of shift on the corresponding number of binary digits.

1.2.2   Interrelation with Estimation and Influence Functions

Solutions of some signal processing problems, including signal transformations, robust estimation, and so on, suppose using Huber function $HF(x,b)$ [11], [37, (2.3.15)]:

$$HF(x,b) = \min\left(b, \max\left(x, -b\right)\right), b = \text{const},$$

which can be written in terms of L-groups $\mathcal{L}(+, \vee, \wedge)$ in the following form:

$$HF(x,b) = -b \vee [x \wedge b] = b \wedge [x \vee -b]. \tag{1.2.8}$$

Fig. 1.2.1a shows unit step function $u_+(x)$ (dashed line), sign function sgn(x) (solid line), and also Kronecker function $\delta(x,a)$ determined by the formulas (1.2.4), (1.2.5), and (1.2.6), respectively.

Fig. 1.2.1b illustrates two different Huber functions $HF(x,b)$ determined by the formula (1.2.8).

In the monograph [37], the authors develop an approach to robust statistics based on influence functions. These functions are the basis for constructing robust procedures in various applications. Hampel influence functions could be represented in terms of L-group $\mathcal{L}(+, \vee, \wedge)$. Thus, for instance, influence functions corresponding to M-estimators IF_M, Tukey bisquare estimator IF_{bq}, and truncated median IF_{tmed} determined by the formulas [37, (2.6.2)], [37, (2.6.4)], [37, (2.6.5)] can be written in the following form, correspondingly:

$$IF_M(x) = c\left[HF(|x|, a) \wedge \left(a \cdot \frac{r - |x|}{r - b}\right)\right] \cdot (1 - u_+(|x| - r)) \cdot \text{sgn}(x) \tag{1.2.9a}$$

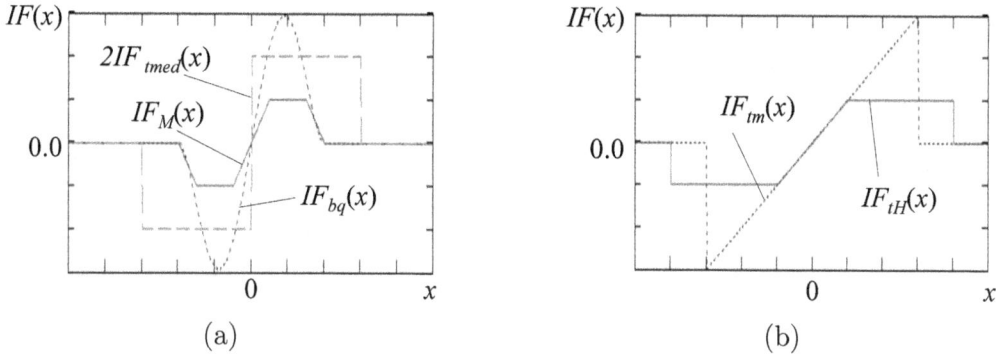

FIGURE 1.2.2   Functions: (a) IF_M, IF_{bq}, IF_{tmed}; (b) $HF(x, b)$

$$IF_{bq}(x) = c|x| \cdot \left(r^2 - x^2\right)^2 \cdot (1 - u_+(|x| - r)) \cdot \operatorname{sgn}(x) \qquad (1.2.9b)$$

$$IF_{tmed}(x) = c\,(1 - u_+(|x| - r)) \cdot \operatorname{sgn}(x) \qquad (1.2.9c)$$

where a, b, c, r = const; $HF(x, b)$ is Huber function (1.2.8), $u_+(x)$ is unit step function (1.2.4), $\operatorname{sgn}(x)$ is sign function (1.2.5); modulus function $|x|$ is calculated by the expression (1.1.7): $|x| = x \vee -x$.

Fig. 1.2.2a shows: influence function of M-estimator IF_M when $a = 1$, $c = 2$, $r = 4$ (1.2.9a) (solid line); influence function of Tukey bisquare estimator IF_{bq} when $c = 0.01$, $r = 4$ (1.2.9b) (dotted line); influence function of truncated median IF_{tmed} when $c = 1$, $r = 6$ (1.2.9c) (dashed line).

Influence functions, corresponding to truncated Huber estimator IF_{tH} and truncated mean estimator IF_{tm} and determined by formulas [37, (2.6.7)], [37, (2.6.8)], can be written in the following forms, respectively:

$$IF_{tH}(x) = HF(|x|, b) \cdot (1 - u_+(|x| - r)) \cdot \operatorname{sgn}(x); \qquad (1.2.10a)$$

$$IF_{tm}(x) = HF(|x|, r) \cdot (1 - u_+(|x| - r)) \cdot \operatorname{sgn}(x), \qquad (1.2.10b)$$

where b, r = const; $HF(x, b)$ is Huber function (1.2.8), $u_+(x)$ is unit step function (1.2.4), $\operatorname{sgn}(x)$ is sign function (1.2.5); modulus function $|x|$ is calculated by the expression (1.1.7): $|x| = x \vee -x$.

Fig. 1.2.2b depicts influence function of truncated Huber estimator IF_{tH} when $b = 2$, $r = 8$ (1.2.10a) (solid line) and influence function of truncated mean estimator IF_{tm} when $r = 6$ (1.2.10b) (dotted line).

1.2.3   Interrelation with Signal Limiters

Operations of L-group $\mathcal{L}(+, \vee, \wedge)$ can find their application in limiting amplifiers possessing amplitude characteristics with insensitivity zones described, for instance, by nonlinear odd functions $f(x)$, $g(x)$:

$$f(x) = (f_1(x) \vee f_2(x)) \cdot \operatorname{sgn}(x); \qquad (1.2.11)$$

$$f_1(x) = a_0 \cdot \ln\left(a_1 \cdot |x|\right), \quad f_2(x) = u_+(|x| - a_2); \qquad (1.2.11a)$$

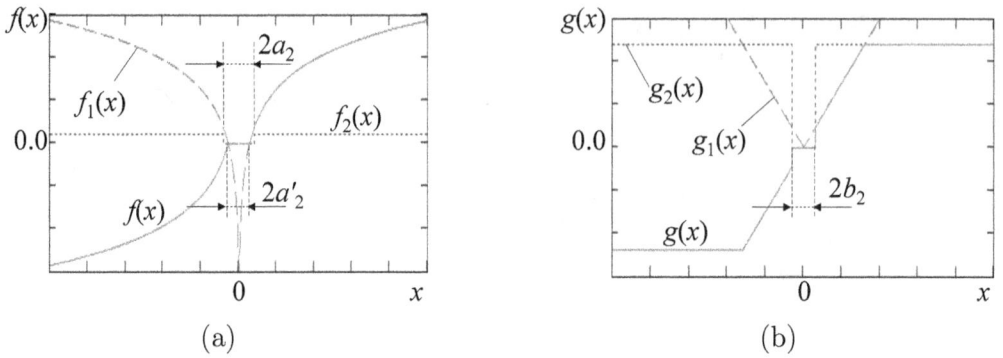

FIGURE 1.2.3   Functions: (a) $f(x)$, $f_1(x)$, $f_2(x)$; (b) $g(x)$, $g_1(x)$, $g_2(x)$

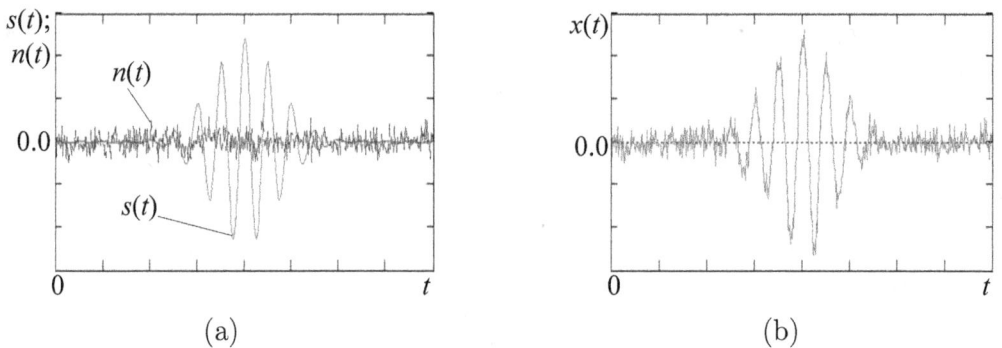

FIGURE 1.2.4   Illustrations of (a) useful signal $s(t)$ and interference (noise) $n(t)$; (b) additive sum $x(t) = s(t) + n(t)$

$$g(x) = (g_1(x) \wedge g_2(x)) \cdot \text{sgn}(x); \qquad (1.2.12)$$

$$g_1(x) = b_0 \cdot |x|, \ g_2(x) = b_1 \cdot u_+(|x| - b_2), \qquad (1.2.12a)$$

where a_0, a_1, a_2 = const; b_0, b_1, b_2 = const; a_2, b_2 are the constants defining the size of insensitivity zones; $u_+(x)$ is unit step function (1.2.4); $\text{sgn}(x)$ is sign function (1.2.5); modulus function $|x|$ is calculated by the expression (1.1.7): $|x| = x \vee -x$.

Plots of the functions $f(x)$ (1.2.11), $g(x)$ (1.2.12) are shown in the Fig. 1.2.3a,b by solid line, respectively. Also Fig. 1.2.3a,b illustrates the functions $f_1(x)$ (1.2.11a), $g_1(x)$ (1.2.12a) (dashed line) and the functions $f_2(x)$ (1.2.11a), $g_2(x)$ (1.2.12a) (dotted line), respectively.

Let $x(t) = s(t) + n(t)$ be additive sum of useful signal $s(t)$ and interference (noise) $n(t)$ acting in the input of limiting amplifiers with insensitivity zones whose amplitude characteristics are determined by the functions (1.2.11), (1.2.12) shown in Fig. 1.2.4a,b, correspondingly.

Then, in the outputs of these limiting amplifiers one can observe stochastic signals $y(t) = f[x(t)]$, $z(t) = g[x(t)]$ shown in Fig. 1.2.5a,b (solid line), respectively. Also both figures illustrate useful signal $s(t)$ (dotted line). Correlation coefficients between useful signal $s(t)$ and the signals in the outputs of the corresponding limiting amplifiers are approximately equal to each other in the case if the sizes of

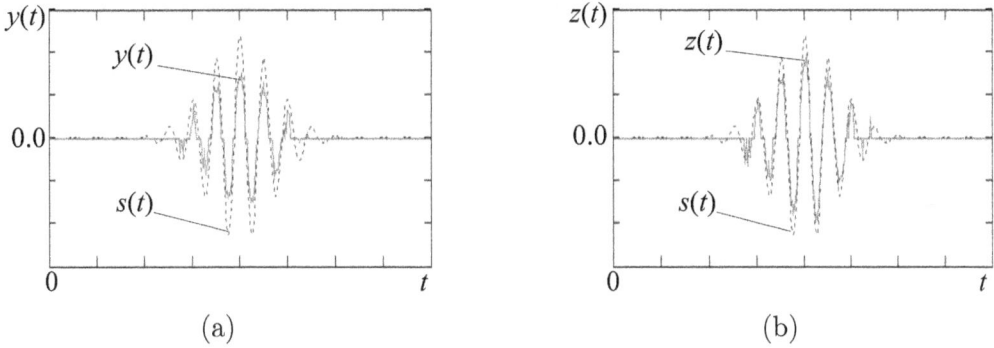

FIGURE 1.2.5   Illustrations of (a) useful signal $s(t)$ and the signal in the output of limiting amplifier $y(t)$; (b) useful signal $s(t)$ and the signal in the output of limiting amplifier $z(t)$

insensitivity zones are the same $a'_2 = b_2$:

$$r[y(t), s(t)] \approx r[z(t), s(t)],$$

while relative variance of noise in the output of the limiting amplifier with logarithmic amplitude characteristic (1.2.11) is approximately 1.5 times less than in the output of the limiting amplifier with linear part of amplitude characteristic:

$$\left(\frac{D\{z(t) - g[s(t)]\}}{D\{g[s(t)]\}} \right) \bigg/ \left(\frac{D\{y(t) - f[s(t)]\}}{D\{f[s(t)]\}} \right) \approx 1.5,$$

where $D[*]$ is a signal variance.

Analytical representation of piecewise functions (1.2.11), (1.2.12) by the operations of L-group $\mathcal{L}(+, \vee, \wedge)$ allow exploiting this approach to solving the signal processing problems, for instance, in radar detection systems for providing constant false alarm rate.

1.2.4   Interrelation with Quantizers

In this subsection we consider a simple example of constructing signal quantizer. Since the number of quantization levels can be arbitrary, we limit this number by several amplitude levels which will be enough representational. Amplitude characteristic $a(x)$ of a quantizer based on the operations of L-group $\mathcal{L}(+, \vee, \wedge)$ is constructed with help of basic even functions $\{h_i(x)\}$ of the following kind:

$$h_i(x) = i \cdot \Delta \cdot u_+(|x| - i \cdot \Delta), \qquad (1.2.13)$$

where $u_+(x)$ is unit step function (1.2.4) and Δ is some constant; $i = 1, 2, 3, \ldots, n$.

Amplitude characteristic $a(x)$ is determined by upper bound of basic even functions $\{h_i(x)\}$:

$$a(x) = \left(\bigvee_{i=1}^{n} h_i(x) \right) \cdot \operatorname{sgn}(x), \qquad (1.2.14)$$

where $\bigvee_{i=1}^{n} h_i(x) = h_1(x) \vee h_2(x) \vee \ldots \vee h_n(x)$; $\operatorname{sgn}(x)$ is sign function (1.2.5).

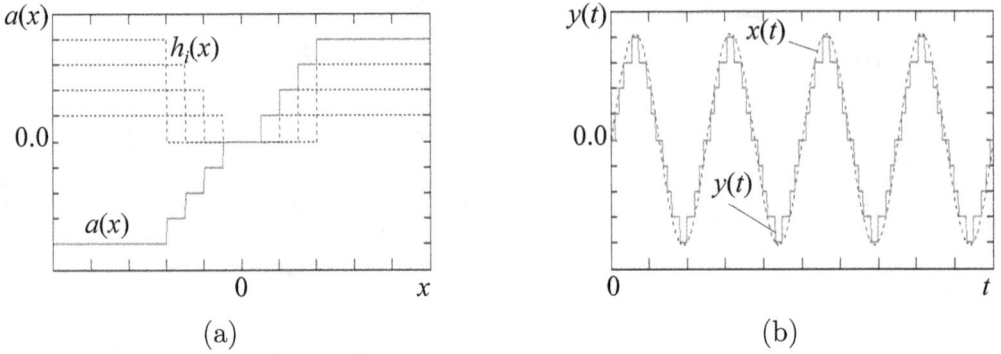

FIGURE 1.2.6   Illustrations of (a) four basic functions $\{h_i(x)\}$, $i = 1, 2, 3, 4$; (b) the signal $x(t)$ in the input of quantizer and the signal $y(t)$ in the output of quantizer

Fig. 1.2.6a demonstrates four basic functions $\{h_i(x)\}$, $i = 1, 2, 3, 4$ (dotted line) defining amplitude characteristic $a(x)$ of bipolar quantizer (1.2.14) on $\Delta = 1$. The latter is shown by the solid line. Fig. 1.2.6b illustrates the signal $x(t)$ affecting in the input of quantizer in the form of sinusoidal function (dotted line) and also the signal $y(t)$ in the output of the quantizer (solid line).

1.2.5   Interrelation with Companders

Some cases of signal processing require transformation of initial signal with some given distribution into a signal whose instantaneous values are uniformly distributed. These cases take place, for instance, when coding the source of messages, in particular, when coding voice signal. To realize such a transformation, one can use special devices allowing effectively change a distribution of input signal values. Such devices are called companders. Amplitude characteristic of this device can be determined by piecewise functions of the following type [38, (2.23)]:

$$y(x) = \begin{cases} \dfrac{A \cdot y_m \cdot |x|}{x_m(1 + \ln(A))}\mathrm{sgn}(x), & 0 < |x|/x_m \leqslant 1/A; \\ \dfrac{y_m \cdot [1 + \ln(A \cdot |x|/x_m)]}{1 + \ln(A)}\mathrm{sgn}(x), & 1/A < |x|/x_m < 1, \end{cases} \qquad (1.2.15)$$

where $A, x_m, y_m = \mathrm{const}$, $A > 0$, $x_m > 0$, $y_m > 0$.

Amplitude characteristic of compander (1.2.15) can be represented analytically in terms of L-group $\mathcal{L}(+, \vee, \wedge)$:

$$z(x) = [z_1(x) \vee z_2(x) \vee z_3(x)] \cdot \mathrm{sgn}(x); \qquad (1.2.16)$$

$$z_1(x) = [b_1(|x|) \vee b_2(|x|)] \cdot [1 - u_+(|x| - a)]; \qquad (1.2.16a)$$

$$z_2(x) = [b_1(|x|) \wedge b_2(|x|)] \cdot [u_+(|x| - a) - u_+(|x| - x_m)]; \qquad (1.2.16b)$$

$$z_3(x) = y_m \cdot u_+(|x| - x_m); \qquad (1.2.16c)$$

$$b_1(x) = \frac{A \cdot y_m \cdot |x|}{x_m(1 + \ln(A))}\mathrm{sgn}(x); \qquad (1.2.16d)$$

$$b_2(x) = \frac{y_m \cdot [1 + \ln(A \cdot |x|/x_m)]}{1 + \ln(A)} \text{sgn}(x), \qquad (1.2.16e)$$

where $z_1(x), z_2(x), z_3(x)$; $b_1(x), b_2(x)$ are auxiliary functions; a is a constant that is a root of equation $b_1(x) = b_2(x)$; $u_+(x)$ is unit step function (1.2.4); $\text{sgn}(x)$ is sign function (1.2.5); modulus function $|x|$ is calculated by the expression (1.1.7): $|x| = x \vee -x$.

Fig. 1.2.7a illustrates auxiliary functions $z_1(x), z_2(x), z_3(x)$ (1.2.16a,b,c); $b_1(x), b_2(x)$ (1.2.16d,e). Functions $z_1(x), z_2(x), z_3(x)$ are shown by solid, dashed, and dotted lines, respectively. Fig. 1.2.7b demonstrates amplitude characteristic of compander $z(x)$ (dashed line) (1.2.16), and also auxiliary functions $b_1(x), b_2(x)$ (solid and dotted lines, correspondingly) (1.2.16d,e).

Fig. 1.2.8a illustrates histogram $h_x(x)$ of amplitude values distribution of stochastic signal $x(t)$ acting in the input of compander. This distribution is rather well approximated by exponential distribution whose probability density function is shown by dotted line. Fig. 1.2.8b shows histogram $h_X(y)$ of amplitude values distribution of stochastic signal $X(t) = z[x(t)]$ in the output of compander. As shown in the figure, owing to limiting the input signal from above, amplitude values of the signal $X(t) = z[x(t)]$ in the output of compander are distributed in the interval $[0, y_m]$, and the resulting distribution becomes more close to uniform distribution (shown by dotted line).

1.3 Generalized Metric Spaces with *L*-group Properties

Metric spaces and other spaces of distances play an important role in mathematics and its applications, forming foundations of various applied methods intended for studying and analyzing the surround world. Fundamental regularities and properties of mathematical objects derive from mutual arrangement of their consisting elements. Such properties reflect peculiarities of interrelations between the elements

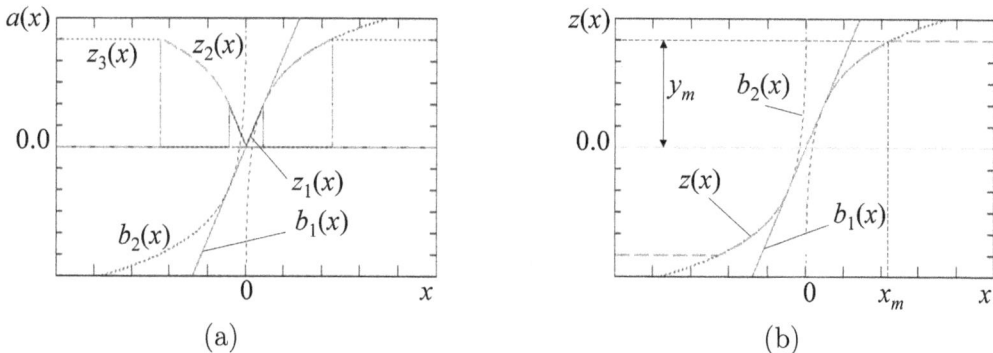

FIGURE 1.2.7 Illustrations of (a) auxiliary functions $z_1(x), z_2(x), z_3(x)$ (1.2.16a,b,c); $b_1(x), b_2(x)$ (1.2.16d,e); (b) the resulting amplitude characteristic of compander $z(x)$

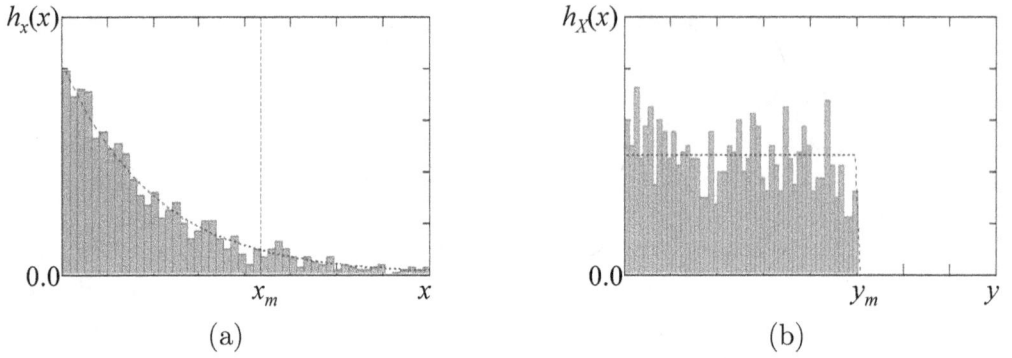

FIGURE 1.2.8   Illustrations of (a) histogram $h_x(x)$ of amplitude values distribution of the signal $x(t)$ in the input of compander; (b) histogram $h_X(y)$ of amplitude values distribution of the signal $X(t) = z[x(t)]$ in the output of compander

and also spatial structure of sets. One of the most important characteristic of structural relations between the elements of a set is the notion of closeness that is firmly coupled with the notion of distance between the elements. Any set, in which the notion of closeness is introduced by one or another way, is called a *space* and its elements are called *points of the space.*

In this Section, we briefly consider just those spaces that possess the lattice properties, or more precisely, *L*-group properties.

Definition 1.3.1. A set $\Gamma = \{a, b, c, \ldots\}$ is called *metric space* (Γ, ρ), if for any pair $\{a, b\}$ of the elements of this set $a, b \in \Gamma$ there exists such a function $\rho(a, b)$ that the following axioms hold:

    1. $\rho(a, b) \geqslant 0$ (non-negativity);

    2. $\rho(a, b) = 0 \Leftrightarrow a = b$ (identity of indiscernible elements);

    3. $\rho(a, b) = \rho(b, a)$ (symmetry);

    4. $\rho(a, b) \leqslant \rho(a, c) + \rho(c, b)$, $a, b, c \in \Gamma$ (triangle inequality).

Conditions 1,...,4 are called axioms of metric, and the function $\rho(a, b)$ is called *metric*. The set $\Gamma = \{a, b, c, \ldots\}$ is *metrizable*, if one can introduce some metric in this set.

For evaluating mutual arrangement of elements of the set $\Gamma = \{a, b, c, \ldots\}$, along with metric, other functions can be used that define spatial and structural relations between the elements of the set. These functions are based on incomplete axiomatics of metric space and called *generalized metrics* (see, for instance, [39–46]). A space based on incomplete axiomatics of metric space is called *generalized metric space.*

Definition 1.3.2. A set $\Gamma = \{a, b, c, \ldots\}$ is called *semimetric space* (Γ, ρ), if for any pair $\{a, b\}$ of the elements of this set $a, b \in \Gamma$ there exist such a function $\rho(a, b)$, that the following axioms hold:

    1. $\rho(a, b) \geqslant 0$ (non-negativity);

2. $\rho(a, b) = 0 \Leftrightarrow a = b$ (identity of indiscernible elements);

3. $\rho(a, b) = \rho(b, a)$ (symmetry).

The conditions 1, 2, and 3 are called axioms of semimetric and function $\rho(a, b)$ is called *semimetric*.

Definition 1.3.3. A set $\mathbf{\Gamma} = \{a, b, c, \ldots\}$ is called *pseudometric space* $(\mathbf{\Gamma}, \rho)$, if for any pair $\{a, b\}$ of the elements of this set $a, b \in \mathbf{\Gamma}$ there exist such a function $\rho(a, b)$, that the following axioms hold:

1. $\rho(a, b) \geqslant 0$ (non-negativity);

2. $\rho(a, a) = 0$ (indiscernibility of identical element);

3. $\rho(a, b) = \rho(b, a)$ (symmetry);

4. $\rho(a, b) \leqslant \rho(a, c) + \rho(c, b)$, $a, b, c \in \mathbf{\Gamma}$ (triangle inequality).

The conditions 1,...,4 are called axioms of pseudometric, and the function $\rho(a, b)$ is called *pseudometric*.

Different generalized metric spaces $(\mathbf{\Gamma}, \rho_F)$ can be formed, if various generalized metrics ρ_F are defined in the same set $\mathbf{\Gamma} = \{a, b, c, \ldots\}$. We can define new generalized metric $\rho_F(a, b)$ in the form of some functional $F(\rho)$ of the initial generalized metric $\rho(a, b)$:

$$\rho_F(a, b) = F[\rho(a, b)]. \tag{1.3.1}$$

Generalized metric space $(\mathbf{\Gamma}, \rho_F)$ is called *metrically transformed space*, and the functional $F(\rho)$ (1.3.1) of the initial generalized metric $\rho(a, b)$ is called *metric transformation of space* $(\mathbf{\Gamma}, \rho)$ [47, 48]. The most interesting transformations $F(\rho)$ (1.3.1) are those that preserve null $F(0) = 0$ of space $(\mathbf{\Gamma}, \rho)$ and also preserve these or those metric and topological properties of metrically transformed space $(\mathbf{\Gamma}, \rho_F)$ that are inherent to the initial generalized metric space $(\mathbf{\Gamma}, \rho)$.

Different transforms of generalized metrics (1.3.1) can be realized by convex upward functionals. Remind that real-valued function f defined in the interval D_f is called *convex upward function*, if for all $\forall x, y \in D_f$ the following condition holds

$$f(\alpha x + \beta y) \geqslant \alpha f(x) + \beta f(y);$$

$$0 \leqslant \alpha, \beta \leqslant 1, \ \alpha + \beta = 1,$$

and is called *convex downward function* on the inverse sign of the inequality. The sufficient condition preserving triangle inequality is defined by the following theorem (see Lemma 9.0.2 [47]).

Theorem 1.3.1. *Let $F(\rho)$ be monotone non-decreasing function, such that $F(0) = 0$, which realize metric transformation (1.3.1) of the space $(\mathbf{\Gamma}, \rho)$. Thus if $(\mathbf{\Gamma}, \rho)$ is pseudometric space, then $(\mathbf{\Gamma}, \rho_F)$ is also pseudometric space.*

The aforementioned theorem can be illustrated by the following example.

Example 1.3.1. The following convex upward functionals keep on metrically transformed pseudometric space to be pseudometric one:

$$\rho_F(a,b) = F(\rho(a,b)) = \frac{\rho(a,b)}{1 + \rho(a,b)}; \qquad (1.3.2a)$$

$$\rho_F(a,b) = F(\rho(a,b)) = [\rho(a,b)]^q, \ 0 < q \leqslant 1; \qquad (1.3.2b)$$

$$\rho_F(a,b) = F(\rho(a,b)) = \log[1 + \rho(a,b)]; \qquad (1.3.2c)$$

$$\rho_F(a,b) = F(\rho(a,b)) = 1 - \exp[-\lambda\rho(a,b)], \ \lambda > 0. \qquad (1.3.2d)$$

One can obtain some metric transformation of space $(\mathbf{\Gamma}, \rho)$ if two arbitrary points $a, b \in \mathbf{\Gamma}$ define a new distance $F(\rho)$ as a ratio of the initial distance $\rho(a,b)$ to perimeter of a triangle formed by these points a, b and some third point u:

$$\rho_F(a,b) = F(\rho(a,b)) = \frac{\rho(a,b)}{\rho(a,b) + \rho(a,u) + \rho(u,b)}. \qquad (1.3.3)$$

Metrically transformed space $(\mathbf{\Gamma}, \rho_F)$ stay pseudometric if the initial space is pseudometric.

In particularly, if we take as the point u null of space $(\mathbf{\Gamma}, \rho)$: $u \equiv 0$, then the last relationship takes the form:

$$\rho_F(a,b) = F(\rho(a,b)) = \frac{\rho(a,b)}{\rho(a,b) + \rho(a,0) + \rho(0,b)}. \qquad (1.3.3a)$$

$\triangledown$

Logic of stating this section is as follows. First, information involved in a message is transmitted by a stochastic signal that contains this message. Second, when processing the transmitted stochastic signal in a special way, information included in a message (and also in signal) is extracted from some finite set of samples of the processed stochastic signal. Such a set of samples is called *random sample*, so that each element of the sample is random variable. At first, with respect to a pair of arbitrary random variables ξ, η from partially ordered set $\mathbf{\Gamma}$; $\xi, \eta \in \mathbf{\Gamma}$ with properties of lattice-ordered group $\mathbf{\Gamma}(+, \vee, \wedge)$ (L-group), we consider main theorems formulated first in [24] for the samples of stochastic processes. Then these results will be extended on a pair of random samples $X = (X_1, \ldots, X_n)$, $Y = (Y_1, \ldots, Y_n)$, and also on their realizations (observations).

Any pair of random samples $X = (X_1, \ldots, X_n)$, $Y = (Y_1, \ldots, Y_n)$ can be considered as a partially ordered set $X \bigcup Y$, in which between two pairs of sample values $X_i, X_j \in X$, $X_i, Y_k \in X \bigcup Y$ there exist a relation of order $X_i \leqslant X_j$, $X_i \geqslant Y_k$ (or $X_i \geqslant X_j$, $X_i \leqslant Y_k$). Then partially ordered set $X \bigcup Y$ is a lattice with operations of join and meet, respectively:

$$X_i \vee X_j = \sup_{X \cup Y} \{X_i, X_j\}, \ X_i \wedge X_j = \inf_{X \cup Y} \{X_i, X_j\};$$

$$X_i \vee Y_k = \sup_{X \cup Y} \{X_i, Y_k\}, \ X_i \wedge Y_k = \inf_{X \cup Y} \{X_i, Y_k\},$$

and if $X_i \leqslant X_j$, then $X_i \wedge X_j = X_i$ and $X_i \vee X_j = X_j$, and if $X_i \geqslant Y_k$, then $X_i \vee Y_k = X_i$ and $X_i \wedge Y_k = Y_k$ [29–31, 49, 50]:

$$X_i \leqslant X_j \Leftrightarrow \left\{ \begin{array}{l} X_i \wedge X_j = X_i; \\ X_i \vee X_j = X_j, \end{array} \right. \quad X_i \geqslant Y_k \Leftrightarrow \left\{ \begin{array}{l} X_i \wedge Y_k = Y_k; \\ X_i \vee Y_k = X_i. \end{array} \right.$$

In natural way, in the sample space $\mathcal{L}(\mathcal{X}, \mathcal{B}_{\mathcal{X}}; +, \vee, \wedge)$, we define operation of addition $X_i + X_j$, $X_i + Y_k$ between pairs of elements $X_i, X_j \in X$, $X_i, Y_k \in X \bigcup Y$ of the samples $X, Y \in \mathcal{L}(\mathcal{X}, \mathcal{B}_{\mathcal{X}}; +, \vee, \wedge)$. Then the sample space $\mathcal{L}(\mathcal{X}, \mathcal{B}_{\mathcal{X}}; +, \vee, \wedge)$ is *lattice-ordered group* (L-group).

Then, in any L-group, the following statements holds [29, 31, 49, 50]:

1. $\mathcal{L}(\mathcal{X}, \mathcal{B}_{\mathcal{X}}; +)$ is a group;

2. $\mathcal{L}(\mathcal{X}, \mathcal{B}_{\mathcal{X}}; \vee, \wedge)$ is a lattice.

Thus, the sample space $\mathcal{L}(\mathcal{X}, \mathcal{B}_{\mathcal{X}}; +, \vee, \wedge)$ with L-group properties is defined as probabilistic space $(\mathcal{X}, \mathcal{B}_{\mathcal{X}})$, in which axioms of distributive lattice $(\mathcal{X}, \mathcal{B}_{\mathcal{X}}; \vee, \wedge)$ with operations of join and meet hold, respectively: $a \vee b = \sup_{\mathcal{L}} a, b$, $a \wedge b = \inf_{\mathcal{L}} a, b$; $a, b \in \mathcal{L}(\mathcal{X}, \mathcal{B}_{\mathcal{X}}; +, \vee, \wedge)$, and axioms of additive commutative group $(\mathcal{X}, \mathcal{B}_{\mathcal{X}}; +)$ also hold.

In most of practically important cases of signal processing, we mainly deal with stochastic signals (processes) $\xi(t)$, $\eta(t)$ with symmetric (even) univariate probability density functions (PDF) $p_\xi(x)$, $p_\eta(y)$: $p_\xi(x) = p_\xi(-x)$; $p_\eta(y) = p_\eta(-y)$. So, measures of closeness between a pair of instantaneous values (samples) $\xi_t = \xi$, $\eta_t = \eta$ of stochastic signals (processes) $\xi(t), \eta(t) \in \Gamma$ introduced in this section are predominantly oriented to the class of signals with even univariate PDFs $p_\xi(x)$, $p_\eta(y)$ that interact in partially ordered set Γ with the properties of lattice-ordered group $\Gamma(+, \vee, \wedge)$ (L-group).

Here one should notice the following. The main results describing metric relations between a pair of instantaneous values (samples) $\xi_t = \xi$, $\eta_t = \eta$ of stochastic signals (processes) $\xi(t), \eta(t) \in \Gamma$ were obtained in [24, Section 3.2]. The material stated there is intended to describe informational properties of stochastic signals (processes). Therefore, two of these signals related with each other by one-to-one transformation are considered to be identical within informational space, inasmuch as they carry the same information:

$$\xi'(t) = f[\xi(t)]; \quad \xi(t) = f^{-1}[\xi'(t)]; \tag{1.3.4a}$$

$$\xi(t) \equiv \xi'(t) \Leftrightarrow \mu[\xi(t), \xi'(t)] = 0, \tag{1.3.4b}$$

thus, metric $\mu[\xi(t), \xi'(t)]$ between them is equal to zero.

In this book, we distinguish stochastic signals $\xi(t)$, $\xi'(t)$ coupled with each other by one-to-one relationship (1.3.4a), so we repeat here main results from [24, Section 3.2] considering these signals to be related to pseudometric space, where symmetric implication (1.3.4b) does not hold.

Let ξ, η be random variables (with symmetric (even) univariate PDFs $p_\xi(x)$, $p_\eta(y)$: $p_\xi(x) = p_\xi(-x)$; $p_\eta(y) = p_\eta(-y)$, univariate cumulative distribution functions (CDF) $F_\xi(x)$, $F_\eta(y)$ and joint CDF $F_{\xi\eta}(x,y)$) from partially ordered set $\mathbf{\Gamma}$ with the properties of lattice-ordered group $\mathbf{\Gamma}(+, \vee, \wedge)$ (*L*-group): $\xi, \eta \in \mathbf{\Gamma}(+, \vee, \wedge)$.

Then the following theorem holds (see Theorem 3.2.1 [24]).

Theorem 1.3.2. *For a pair of random variables ξ, η from L-group $\mathbf{\Gamma}(+, \vee, \wedge)$, $\xi, \eta \in \mathbf{\Gamma}$, the functions $\mu_+(\xi, \eta)$, $\mu_-(\xi, \eta)$, $\mu(\xi, \eta)$ that are equal to:*

$$\mu_+(\xi, \eta) = 2(\mathbf{P}[\xi \vee \eta > 0] - \mathbf{P}[\xi \wedge \eta > 0]); \qquad (1.3.5a)$$

$$\mu_-(\xi, \eta) = 2(\mathbf{P}[\xi \wedge \eta < 0] - \mathbf{P}[\xi \vee \eta < 0]); \qquad (1.3.5b)$$

$$\mu(\xi, \eta) = [\mu_+(\xi, \eta) + \mu_-(\xi, \eta)]/2, \qquad (1.3.5c)$$

are pseudometrics.

For random variables ξ, η with symmetric (even) univariate PDFs $p_\xi(x)$, $p_\eta(y)$, pseudometrics $\mu_+(\xi, \eta)$, $\mu_-(\xi, \eta)$, $\mu(\xi, \eta)$ are identically equal and determined by the relationship [24, (3.2.5)], [24, (3.2.6)]:

$$\mu_+(\xi, \eta) = \mu_-(\xi, \eta) = \mu(\xi, \eta) = 2[F_\xi(0) + F_\eta(0)] - 4F_{\xi\eta}(0, 0). \qquad (1.3.6)$$

Definition 1.3.4. The function $\mu(\xi, \eta)$ defined by the relationship (1.3.5c) is called *pseudometric* between two random variables ξ, η from partially ordered set $\mathbf{\Gamma}$ with the properties of lattice-ordered group $\mathbf{\Gamma}(+, \vee, \wedge)$ (*L*-group): $\xi, \eta \in \mathbf{\Gamma}(+, \vee, \wedge)$.

Thus, partially ordered set $\mathbf{\Gamma}$ with operations $\chi_+ = \xi + \eta$, $\chi_\vee = \xi \vee \eta$, $\chi_\wedge = \xi \wedge \eta$ is pseudometric space $(\mathbf{\Gamma}, \mu)$ with respect to pseudometric μ (1.3.5c) introduced here.

The following theorem defines invariant of a group of continuous mappings of a pair of random variables ξ, η in *L*-group $\mathbf{\Gamma}$ (see Theorem 3.2.2 [24]):

Theorem 1.3.3. *For a pair of random variables ξ, η in L-group $\mathbf{\Gamma}$: $\xi, \eta \in \mathbf{\Gamma}$, pseudometric $\mu(\xi, \eta)$ between them is an invariant of a group H of their continuous mappings $\{h_{\alpha,\beta}\}$, $h_{\alpha,\beta} \in H$; $\alpha, \beta \in A$ preserving null 0 of L-group $\mathbf{\Gamma}$: $h_{\alpha,\beta}(0) = 0$:*

$$\mu(\xi, \eta) = \mu(\xi', \eta'); \qquad (1.3.7)$$

$$h_\alpha : \xi \to \xi', \ h_\beta : \eta \to \eta'; \qquad (1.3.7a)$$

$$h_\alpha^{-1} : \xi' \to \xi, \ h_\beta^{-1} : \eta' \to \eta; \qquad (1.3.7b)$$

where ξ', η' are the results of mappings of random variables ξ, η in L-group $\mathbf{\Gamma}'$: $h_{\alpha,\beta}: \mathbf{\Gamma} \to \mathbf{\Gamma}'$.

Exploiting the formulas (1.3.5) and (1.3.6) for solving problems of mathematical statistics, in general, and problems of signal processing, in particular, need some refinement.

Let, as before, $X = (X_1, \ldots, X_n)$, $Y = (Y_1, \ldots, Y_n)$ be random samples in the sample space $\mathcal{L}(\mathcal{X}, \mathcal{B}_\mathcal{X}; +, \vee, \wedge)$ with properties of *L*-group: $X, Y \in$

$\mathcal{L}(\mathcal{X}, \mathcal{B}_{\mathcal{X}}; +, \vee, \wedge)$. Apparently, for the pair of statistically dependent elements X_i, Y_i of the samples X, Y considered as a pair of random variables, the results of Theorems 1.3.2 and 1.3.3 with respect to pseudometric $\mu(X_i, Y_i)$ (1.3.5) stay true. Further, we consider the features of using the obtained results with respect to realizations (observations) of random samples X, Y.

Let, also, $x = (x_1, x_2, \ldots, x_n)$, $y = (y_1, y_2, \ldots, y_n)$ be realizations (observations) of random samples X, Y; $\{x_i\}$, $\{y_i\}$ be the values of realizations of random variables $\{X_i\}$, $\{Y_i\}$ of the samples X, Y; $i = 1, 2, \ldots, n$; $n \in \mathbf{N}$.

We introduce a function $I_z(A)$ denoting distribution concentrated in a point z [51]:

$$I_z(A) = \begin{cases} 1, & z \in A; \\ 0, & z \notin A, \end{cases}$$

where A is some set.

Then, according to the formulas (1.3.5a), (1.3.5b), (1.3.5c), pseudometrics $\mu_+(x, y)$, $\mu_-(x, y)$, $\mu(x, y)$ between realizations (observations) $x = (x_1, x_2, \ldots, x_n)$, $y = (y_1, y_2, \ldots, y_n)$ of the samples X, Y are defined by the following relationships:

$$\mu_+(x, y) = \frac{2}{n} \left(\sum_{i=1}^{n} I_{v_i}(v_i > 0) - \sum_{i=1}^{n} I_{u_i}(u_i > 0) \right); \qquad (1.3.8a)$$

$$\mu_-(x, y) = \frac{2}{n} \left(\sum_{i=1}^{n} I_{u_i}(u_i < 0) - \sum_{i=1}^{n} I_{v_i}(v_i < 0) \right); \qquad (1.3.8b)$$

$$\mu(x, y) = [\mu_+(x, y) + \mu_-(x, y)]/2, \qquad (1.3.8c)$$

where $v_i = x_i \vee y_i$, $u_i = x_i \wedge y_i$.

Substituting the values of pseudometrics (1.3.8a) and (1.3.8b) into the formula (1.3.8c), we obtain the expression that is convenient for practical calculating pseudometric $\mu(x, y)$ between realizations (observations) $x = (x_1, x_2, \ldots, x_n)$, $y = (y_1, y_2, \ldots, y_n)$ of the samples X, Y in the following form:

$$\mu_p(x, y) = \frac{1}{n} \sum_{i=1}^{n} [\text{sgn}(x_i \vee y_i) - \text{sgn}(x_i \wedge y_i)], \qquad (1.3.9)$$

where $\text{sgn}(x)$ is sign function determined by the formulas (1.2.5) and (1.2.2).

Thus, the sample space $\mathcal{L}(\mathcal{X}, \mathcal{B}_{\mathcal{X}}; +, \vee, \wedge)$ with L-group properties is metrizable by introducing pseudometric (1.3.9) and becomes pseudometric space $(\mathcal{X}, \mu)$. According to the Theorem 1.3.3, pseudometric (1.3.9) is invariant of a group of continuous mappings of a pair of random samples X, Y in the sample space $\mathcal{L}(\mathcal{X}, \mathcal{B}_{\mathcal{X}}; +, \vee, \wedge)$ with L-group properties $X, Y \in \mathcal{L}(\mathcal{X}, \mathcal{B}_{\mathcal{X}}; +, \vee, \wedge)$.

The following theorem illustrates properties of pseudometric space $(\mathcal{X}, \mu)$ with pseudometric (1.3.9).

Theorem 1.3.4. *In sample space $\mathcal{L}(\mathcal{X}, \mathcal{B}_{\mathcal{X}}; +, \vee, \wedge)$ with L-group properties and pseudometric (1.3.9) between the realizations (observations) $x = (x_1, x_2, \ldots, x_n)$,*

$y = (y_1, y_2, \ldots, y_n)$ of random samples X, Y: $X = (X_1, X_2, \ldots, X_n)$, $Y = (Y_1, Y_2, \ldots, Y_n)$: $X, Y \in \mathcal{L}(\mathcal{X}, \mathcal{B}_\mathcal{X}; +, \vee, \wedge)$, the identity of parallelogram holds:

$$\mu(x, y) + \mu(x, -y) = \mu(x, \mathbf{0}) + \mu(\mathbf{0}, y), \tag{1.3.10}$$

where $\mathbf{0}$ is null sample: $\mathbf{0} = (0, 0, \ldots, 0)$.

Proof. It is obvious that probabilities of events $X_i = 0$, $Y_i = 0$; $i = 1, 2, \ldots, n$; $n \in \mathbf{N}$ are equal to: $\mathbf{P}[X_i = 0] = 0$, $\mathbf{P}[Y_i = 0] = 0$. Hence, the values $\{x_i\}$, $\{y_i\}$ of realizations of random variables $\{X_i\}$, $\{Y_i\}$ of the samples X, Y are considered to be not equal to zero:

$$x_i \neq 0, \; y_i \neq 0. \tag{1.3.11}$$

Evidently, for $\forall a \neq 0$, the following identity holds:

$$\mathrm{sgn}(a \vee 0) - \mathrm{sgn}(a \wedge 0) = 1. \tag{1.3.12}$$

Besides, for $\forall a \neq 0$, $\forall b \neq 0$, the following implications hold:

$$\mathrm{sgn}(a) = \mathrm{sgn}(b) \Rightarrow \begin{cases} \mathrm{sgn}(a \vee b) - \mathrm{sgn}(a \wedge b) = 0; \\ \mathrm{sgn}(a \vee -b) - \mathrm{sgn}(a \wedge -b) = 2; \end{cases} \tag{1.3.13a}$$

$$\mathrm{sgn}(a) = -\mathrm{sgn}(b) \Rightarrow \begin{cases} \mathrm{sgn}(a \vee b) - \mathrm{sgn}(a \wedge b) = 2; \\ \mathrm{sgn}(a \vee -b) - \mathrm{sgn}(a \wedge -b) = 0. \end{cases} \tag{1.3.13b}$$

For $\forall a \neq 0$, $\forall b \neq 0$, joint fulfillment of (1.3.13a,b) implies the identity:

$$[\mathrm{sgn}(a \vee b) - \mathrm{sgn}(a \wedge b)] + [\mathrm{sgn}(a \vee -b) - \mathrm{sgn}(a \wedge -b)] = 2. \tag{1.3.14}$$

For $\forall a \neq 0$, $\forall b \neq 0$, the following identity follows from the equality (1.3.12):

$$[\mathrm{sgn}(a \vee 0) - \mathrm{sgn}(a \wedge 0)] + [\mathrm{sgn}(b \vee 0) - \mathrm{sgn}(b \wedge 0)] = 2. \tag{1.3.15}$$

Joint fulfillment of identities (1.3.14) and (1.3.15) implies the relationship (1.3.10). $\square$

In the sample space $\mathcal{L}(\mathcal{X}, \mathcal{B}_\mathcal{X}; +, \vee, \wedge)$ with the properties of L-group, one can also introduce known l_1-metric [36, 47] between realizations (observations) $x = (x_1, x_2, \ldots, x_n)$, $y = (y_1, y_2, \ldots, y_n)$ of random samples X, Y:

$$\rho(x, y) = \sum_{i=1}^{n} |x_i - y_i|. \tag{1.3.16}$$

Taking into account the relationship (1.1.9), metric (1.3.16) can be written in another form using operations of lattice of sample space $\mathcal{L}(\mathcal{X}, \mathcal{B}_\mathcal{X}; +, \vee, \wedge)$ with L-group properties:

$$\rho(x, y) = \sum_{i=1}^{n} (x_i \vee y_i - x_i \wedge y_i). \tag{1.3.17}$$

For most practical applications of Signal Processing Theory, generalized metrics

(1.3.9), (1.3.16), and (1.3.17) could be more preferable than metric $\rho_E(x, y)$ of n-dimensional Euclidean space

$$\rho_E(x, y) = \sqrt{\sum_{i=1}^{n} (x_i - y_i)^2},$$

induced by scalar product (x, y) between realizations (observations) $x = (x_1, x_2, \ldots, x_n)$, $y = (y_1, y_2, \ldots, y_n)$ of the samples X, Y:

$$(x, y) = \sum_{i=1}^{n} x_i \cdot y_i.$$

This is due to the fact that calculating the generalized metrics (1.3.9), (1.3.16), and (1.3.17) does not require performing operation of multiplication unlike Euclidean metric (or scalar product), that can provide essential advantage in computational rate within those applications where it is really necessary.

When solving some signal processing problems, it is convenient to use those generalized metrics whose values do not exceed some given quantity. For this purpose one can use an idea of metric transformation of a space (see the formula (1.3.1) and Theorem 1.3.1).

Thus, for instance, substituting metric (1.3.16) into the formula (1.3.3a), one can obtain metric of the following kind:

$$\rho_F(x, y) = \frac{\sum_{i=1}^{n} |x_i - y_i|}{\sum_{i=1}^{n} |x_i - y_i| + \sum_{i=1}^{n} |x_i| + \sum_{i=1}^{n} |y_i|}. \tag{1.3.18}$$

Substituting metric (1.3.16) into the formulas (1.3.2a) and (1.3.2b), we obtain metrics of the following kinds, respectively:

$$\rho_F(x, y) = \frac{\sum_{i=1}^{n} |x_i - y_i|}{1 + \sum_{i=1}^{n} |x_i - y_i|}; \tag{1.3.19a}$$

$$\rho_F(x, y) = \left[\sum_{i=1}^{n} |x_i - y_i| \right]^q, \quad 0 < q \leqslant 1. \tag{1.3.19b}$$

However, to solve signal processing problems related with an idea of exploiting orthogonal functions (signals) when transmitting opposite signals (suppose, from a set $\{-1, 1\}$), the boundedness of metric provided by the functions (1.3.2a,b,d), (1.3.3), and (1.3.3a) can be insufficient. Therefore, in sample space $\mathcal{L}(\mathcal{X}, \mathcal{B}_{\mathcal{X}}; +, \vee, \wedge)$ with L-group properties, it is desirable to require a generalized metric $\rho(x, y)$ between realization (observations) $x = (x_1, x_2, \ldots, x_n)$,

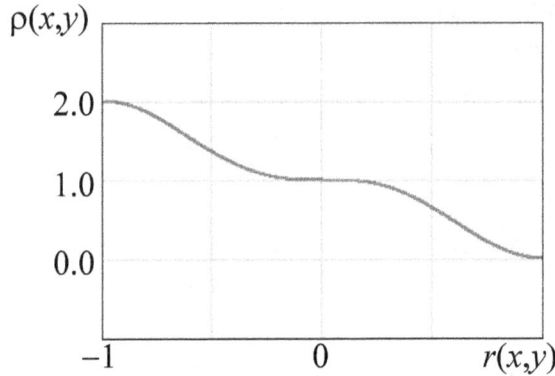

FIGURE 1.3.1   The dependence (1.3.20)

$y = (y_1, y_2, \ldots, y_n)$ of the samples X, Y to be such a function g of correlation coefficient $r(x, y)$ between these samples

$$\rho(x, y) = g[r(x, y)], \tag{1.3.20}$$

that the condition of oddness of the function $\rho(x, y) - 1$ holds:

$$g[r(x, y)] - 1 = -(g[-r(x, y)] - 1), \tag{1.3.21}$$

besides, the following relations hold:

$$g[r(x, y) = -1] = 2 = \max[\rho(x, y)]; \tag{1.3.22a}$$

$$g[r(x, y) = 0] = 1; \tag{1.3.22b}$$

$$g[r(x, y) = 1] = 0 = \min[\rho(x, y)]. \tag{1.3.22c}$$

An appearance of dependence that is defined by (1.3.20) and meets the requirements (1.3.21), (1.3.22) is shown in Fig. 1.3.1.

Pseudometric $\mu_p(x, y)$ (1.3.9) between realizations (observations) $x = (x_1, x_2, \ldots, x_n)$, $y = (y_1, y_2, \ldots, y_n)$ of the samples X, Y, introduced in the sample space $\mathcal{L}(\mathcal{X}, \mathcal{B}_\mathcal{X}; +, \vee, \wedge)$ with L-group properties, satisfies the conditions (1.3.21), (1.3.22) listed above. Nevertheless, taking into account statistical interrelations between the samples X, Y, pseudometric $\mu(x, y)$ (1.3.9) ignores amplitude differences between the elements $\{X_i\}$, $\{Y_i\}$ of the samples X, Y that can negatively affect on the signal processing efficiency.

For the purpose of taking into consideration both statistical interrelations and amplitude differences between the elements $\{X_i\}$, $\{Y_i\}$ of the samples X, Y, as a generalized metric that satisfies the requirements (1.3.21), (1.3.22) one can suggest semimetric $\mu_s(x, y)$:

$$\mu_s(x, y) = \frac{1}{\|x\| + \|y\|} \sum_{i=1}^{n} |x_i - y_i| \cdot [\operatorname{sgn}(x_i \vee y_i) - \operatorname{sgn}(x_i \wedge y_i)]; \tag{1.3.23}$$

$$\|x\| = \sum_{i=1}^{n} |x_i|; \|y\| = \sum_{i=1}^{n} |y_i|,$$

where $\|x\|$, $\|y\|$ are l_1-norms of realizations (observations) $x = (x_1, x_2, \ldots, x_n)$, $y = (y_1, y_2, \ldots, y_n)$ of the samples X, Y, so that semimetric is bounded above by the sum of norms that figures in denominator of fraction to meet the property (1.3.22).

Taking into account the relationships (1.1.8), (1.1.9), semimetric (1.3.23) can be written in another form using operations of lattice of the sample space $\mathcal{L}(\mathcal{X}, \mathcal{B}_{\mathcal{X}}; +, \vee, \wedge)$ with L-group properties:

$$\mu_s(x, y) = \frac{\sum\limits_{i=1}^{n} [x_i \vee y_i - x_i \wedge y_i] \cdot [\text{sgn}(x_i \vee y_i) - \text{sgn}(x_i \wedge y_i)]}{\sum\limits_{i=1}^{n} [(x_i \vee (-x_i)) + (y_i \vee (-y_i))]}, \tag{1.3.23a}$$

so that we assume that $\text{sgn}(x)$ is sign function defined by (1.2.5).

To take into consideration both statistical interrelations and amplitude differences between the elements $\{X_i\}$, $\{Y_i\}$ of the samples X, Y, as a generalized metric that meets the requirements (1.3.21), (1.3.22), one can suggest a normalized l_1-metric $\mu_{l_1}(x, y)$:

$$\mu_{l_1}(x, y) = \frac{2}{\|x + y\| + \|x - y\|} \sum_{i=1}^{n} |x_i - y_i|, \tag{1.3.24}$$

where $\|x + y\|$, $\|x - y\|$ are l_1-norms of sum and difference of realizations (observations) $x = (x_1, x_2, \ldots, x_n)$, $y = (y_1, y_2, \ldots, y_n)$ of the samples X, Y, respectively.

Taking into account the relations (1.1.8), (1.1.9), metric (1.3.24) can be written in another form, using operations of lattice of the sample space $\mathcal{L}(\mathcal{X}, \mathcal{B}_{\mathcal{X}}; +, \vee, \wedge)$ with L-group properties:

$$\mu_{l_1}(x, y) = \frac{2 \sum\limits_{i=1}^{n} [x_i \vee y_i - x_i \wedge y_i]}{\sum\limits_{i=1}^{n} [((x_i + y_i) \vee -(x_i + y_i)) + ((x_i - y_i) \vee -(x_i - y_i))]}. \tag{1.3.24a}$$

Thus, in sample space $\mathcal{L}(\mathcal{X}, \mathcal{B}_{\mathcal{X}}; +, \vee, \wedge)$ with L-group properties, in addition to normalized generalized metrics $\mu_p(x, y)$ (1.3.9), $\mu_s(x, y)$ (1.3.23), and $\mu_{l_1}(x, y)$ (1.3.24) between realizations (observations) $x = (x_1, x_2, \ldots, x_n)$, $y = (y_1, y_2, \ldots, y_n)$ of the samples X, Y, we consider their non-normalized analogues determined by the following relationships:

$$M_p(x, y) = \sum_{i=1}^{n} [\text{sgn}(x_i \vee y_i) - \text{sgn}(x_i \wedge y_i)]; \tag{1.3.25}$$

$$M_s(x, y) = \sum_{i=1}^{n} [x_i \vee y_i - x_i \wedge y_i] \cdot [\text{sgn}(x_i \vee y_i) - \text{sgn}(x_i \wedge y_i)]. \tag{1.3.26}$$

Also we consider l_1-metric (1.3.16) that is written in terms of the sample space $\mathcal{L}(\mathcal{X}, \mathcal{B}_{\mathcal{X}}; +, \vee, \wedge)$ with L-group properties in the following form:

$$M_{l_1}(x, y) = \sum_{i=1}^{n} (x_i \vee y_i - x_i \wedge y_i). \tag{1.3.27}$$

Calculating metrics (1.3.25...1.3.27) can be performed without using operation of multiplication that is doubtless advantage when organizing calculations, first, in applications that do not assume exploiting considerable computational resources, second, in applications requiring high computational rate. The same consideration could be referred to normalized generalized metrics (1.3.9), (1.3.23), and (1.3.24).

1.4 Measures of Statistical Interrelation between Signals in Space with L-group Properties

1.4.1 Normalized Measure of Statistical Interrelation between Stochastic Signals in Space with L-group Properties

As before, we consider the stochastic signals (processes) $\xi(t)$, $\eta(t)$ with symmetric (even) univariant PDFs $p_\xi(x)$, $p_\eta(y)$ in the form: $p_\xi(x) = p_\xi(-x)$; $p_\eta(y) = p_\eta(-y)$. So, introduced in this section measures of closeness between a pair of instantaneous values (samples) $\xi_t = \xi$, $\eta_t = \eta$ of stochastic signals (processes) $\xi(t), \eta(t) \in \Gamma$ are oriented to the class of signals with even univariate PDFs $p_\xi(x)$, $p_\eta(y)$ that interact in a partially ordered set Γ with the properties of lattice-ordered group $\Gamma(+, \vee, \wedge)$ (L-group).

Thus, let ξ, η be random variables (with symmetric (even) univariate PDFs $p_\xi(x)$, $p_\eta(y)$: $p_\xi(x) = p_\xi(-x)$; $p_\eta(y) = p_\eta(-y)$, univariate CDFs $F_\xi(x)$, $F_\eta(y)$, and joint CDF $F_{\xi\eta}(x, y)$) from partially ordered set Γ with properties of lattice-ordered group $\Gamma(+, \vee, \wedge)$ (L-group): $\xi, \eta \in \Gamma(+, \vee, \wedge)$.

Then any pair of stochastic processes $\xi(t), \eta(t) \in \Gamma$ with even univariate PDFs can be associated with the following measure between random variables (signal samples) $\xi_t = \xi$, $\eta_t = \eta$.

Definition 1.4.1. *Normalized measure of statistical interrelation* (NMSI) between random variables ξ, η is the quantity $\nu(\xi, \eta)$ determined by normalized generalized metric $\mu(\xi, \eta)$:

$$\nu(\xi, \eta) = 1 - \mu(\xi, \eta). \tag{1.4.1}$$

The last relationship and formula (1.3.6) imply that NMSI $\nu(\xi, \eta)$ is defined over joint CDF $F_{\xi\eta}(x, y)$ and univariate CDFs $F_\xi(x)$, $F_\eta(y)$ of random variables ξ, η:

$$\nu(\xi, \eta) = 1 + 4F_{\xi\eta}(0, 0) - 2(F_\xi(0) + F_\eta(0)). \tag{1.4.2}$$

For random variables ξ, η the following theorem holds (see Corollary 3.2.1 [24]), which is a consequence from Theorem 1.3.3 and relationship (1.4.1).

Theorem 1.4.1. *For a pair of random variables ξ, η (with even univariate PDFs) in L-group Γ: $\xi, \eta \in \Gamma$, NMSI $\nu(\xi, \eta)$ is invariant of a group H of their continuous mappings $\{h_{\alpha,\beta}\}$, $h_{\alpha,\beta} \in H$; $\alpha, \beta \in A$ preserving null 0 of L-group Γ: $h_{\alpha,\beta}(0) = 0$:*

$$\nu(\xi, \eta) = \nu(\xi', \eta'); \tag{1.4.3}$$

$$h_\alpha : \xi \to \xi', \ h_\beta : \eta \to \eta'; \tag{1.4.3a}$$

$$h_\alpha^{-1} : \xi' \to \xi, \ h_\beta^{-1} : \eta' \to \eta; \tag{1.4.3b}$$

where ξ', η' are results of mapping of random variables ξ, η in L-group Γ': $h_{\alpha,\beta}$: $\Gamma \to \Gamma'$.

This theorem defines invariant of a group H of continuous mappings $\{h_{\alpha,\beta}\}$, $h_{\alpha,\beta} \in H$ of a pair of random variables ξ, η in L-group Γ.

To determine a dependence between NMSI $\nu(\xi, \eta)$ (1.4.1) and pseudometric $\mu(\xi, \eta)$ (1.3.6) on correlation coefficient $r_{\xi\eta}$ for a pair of Gaussian random variables ξ, η with zero expectations, we use the following theorem (see Theorem 3.2.3 [24]).

Theorem 1.4.2. *For Gaussian random variables ξ, η with even univariate PDFs and correlation coefficient $r_{\xi\eta}$, NMSI $\nu(\xi, \eta)$ is equal to:*

$$\nu(\xi, \eta) = \frac{2}{\pi} \arcsin(r_{\xi\eta}). \tag{1.4.4}$$

Pointed above Theorem is illustrated by the following example.

Example 1.4.1. Exploiting the result (1.4.4) and formula (1.4.1), we obtain that for a pair of Gaussian random variables ξ, η with zero expectations and correlation coefficient $r_{\xi\eta}$, pseudometric $\mu(\xi, \eta)$ (1.3.6) between them is determined by the quantity:

$$\mu(\xi, \eta) = 1 - \frac{2}{\pi} \arcsin(r_{\xi\eta}). \tag{1.4.5}$$

It is easy to verify that the function (1.4.5) meets the listed above requirements (1.3.21), (1.3.22). The plots of dependences (1.4.4) and (1.4.5) for a pair of Gaussian random variables ξ, η with zero expectations are shown in Fig. 1.4.1 by dashed and solid lines, respectively. Here we notice that the dependence (1.4.4) is an odd function that is a consequence from more general properties of pseudometric (1.4.5) meeting the formulated above requirements (1.3.21), (1.3.22). $\triangledown$

1.4.2 Sample Normalized Measures of Statistical Interrelation between Signals in Space with *L*-group Properties

Using the formulas (1.4.1) and (1.4.2) for mathematical statistics problems, in general, and for signal processing problems, in particular, needs some specification.

Let, as before, $X = (X_1, \ldots, X_n)$, $Y = (Y_1, \ldots, Y_n)$ be random samples in the sample space $\mathcal{L}(\mathcal{X}, \mathcal{B}_\mathcal{X}; +, \vee, \wedge)$ with L-group properties: $X, Y \in \mathcal{L}(\mathcal{X}, \mathcal{B}_\mathcal{X}; +, \vee, \wedge)$. Further we consider the features of exploiting the obtained results with respect to realizations (observations) of random samples X, Y. Let, also, $x = (x_1, x_2, \ldots, x_n)$,

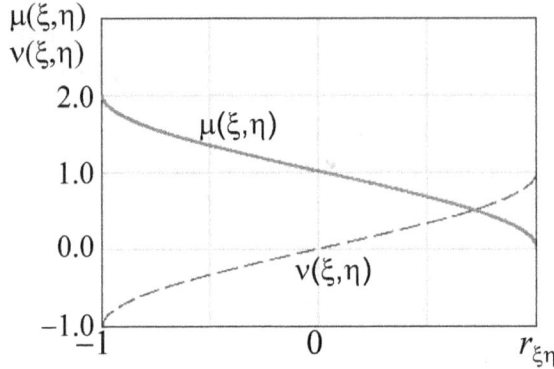

FIGURE 1.4.1   Plots of dependences (1.4.4) and (1.4.5)

$y = (y_1, y_2, \ldots, y_n)$ be realizations (observations) of random samples X, Y; $\{x_i\}$, $\{y_i\}$ be the values of realizations of random variables $\{X_i\}$, $\{Y_i\}$ of the samples X, Y; $i = 1, 2, \ldots, n$; $n \in \mathbf{N}$.

Substituting the expression for pseudometric (1.3.9) into the general relationship between NMSI and pseudometric (1.4.1), we obtain expression that is necessary for practical calculations of NMSI between realizations (observations) $x = (x_1, x_2, \ldots, x_n)$, $y = (y_1, y_2, \ldots, y_n)$ of the samples X, Y in the sample space $\mathcal{L}(\mathcal{X}, \mathcal{B}_{\mathcal{X}}; +, \vee, \wedge)$ with L-group properties:

$$\nu_p(x, y) = 1 - \frac{1}{n} \sum_{i=1}^{n} [\mathrm{sgn}(x_i \vee y_i) - \mathrm{sgn}(x_i \wedge y_i)], \qquad (1.4.6)$$

where $\mathrm{sgn}(x)$ is sign function determined by the formulas (1.2.5), (1.2.2).

Here we notice that the formulas for pseudometrics $\mu(\xi, \eta)$ (1.3.5c), (1.3.6) and NMSI $\nu(\xi, \eta)$ (1.4.1), (1.4.2) define probabilistic description of distance and measure of closeness between random variables ξ, η, whereas the expressions for pseudometric $\mu(x, y)$ (1.3.9) and NMSI $\nu(x, y)$ (1.4.6) are statistical description of these characteristics for random samples X, Y.

Quite similarly, substituting the formula for semimetric $\mu_s(x, y)$ (1.3.23) into general relationship between NMSI and normalized generalized metric (1.4.1), we obtain an expression that is necessary for practical calculations of NMSI $\nu_s(x, y)$ between realizations (observations) $x = (x_1, x_2, \ldots, x_n)$, $y = (y_1, y_2, \ldots, y_n)$ of random samples X, Y in the sample space $\mathcal{L}(\mathcal{X}, \mathcal{B}_{\mathcal{X}}; +, \vee, \wedge)$ with L-group properties:

$$\nu_s(x, y) = 1 - \frac{1}{\|x\| + \|y\|} \sum_{i=1}^{n} |x_i - y_i| \cdot [\mathrm{sgn}(x_i \vee y_i) - \mathrm{sgn}(x_i \wedge y_i)]; \qquad (1.4.7)$$

$$\|x\| = \sum_{i=1}^{n} |x_i|; \ \|y\| = \sum_{i=1}^{n} |y_i|,$$

where $\|x\|$, $\|y\|$ are l_1-norms of realizations (observations) $x = (x_1, x_2, \ldots, x_n)$, $y = (y_1, y_2, \ldots, y_n)$ of random samples X, Y in normed sample space.

Taking into account the equations (1.1.8), (1.1.9), NMSI (1.4.7) can be written in another form, using operations of lattice of the sample space $\mathcal{L}(\mathcal{X}, \mathcal{B}_{\mathcal{X}}; +, \vee, \wedge)$ with L-group properties:

$$\nu_s(x, y) = 1 - \frac{\displaystyle\sum_{i=1}^{n} [x_i \vee y_i - x_i \wedge y_i] \cdot [\operatorname{sgn}(x_i \vee y_i) - \operatorname{sgn}(x_i \wedge y_i)]}{\displaystyle\sum_{i=1}^{n} [(x_i \vee (-x_i)) + (y_i \vee (-y_i))]}, \qquad (1.4.7a)$$

where we assume that $\operatorname{sgn}(x)$ is sign function determined by the formula (1.2.5).

To determine statistical interrelation between realizations (observations) $x = (x_1, x_2, \ldots, x_n)$, $y = (y_1, y_2, \ldots, y_n)$ of random samples X, Y in the sample space $\mathcal{L}(\mathcal{X}, \mathcal{B}_{\mathcal{X}}; +, \vee, \wedge)$ with L-group properties, one can also propose NMSI based on normalized generalized metric (1.3.24):

$$\nu_{l_1}(x, y) = \frac{1}{\|x + y\| + \|x - y\|} \sum_{i=1}^{n} [|x_i + y_i| - |x_i - y_i|], \qquad (1.4.8)$$

where $\|x + y\|$, $\|x - y\|$ are l_1-norms of sums and differences of realizations (observations) $x = (x_1, x_2, \ldots, x_n)$, $y = (y_1, y_2, \ldots, y_n)$ of random samples X, Y, respectively.

NMSI determined by the formulas (1.4.6), (1.4.7), and (1.4.8) allow introducing the notion of *orthogonality* for signals in spaces with L-group properties, which, in general case, differs from the corresponding notion in linear space with scalar product.

Definition 1.4.2. Two discrete signals $x(t)$ and $y(t)$, $t \in T$:

$$T = \{0, \Delta t, \ldots, \Delta t \cdot (i - 1), \ldots, \Delta t \cdot (n - 1)\},$$

represented by their own samples $x = (x_1, x_2, \ldots, x_n)$ and $y = (y_1, y_2, \ldots, y_n)$, respectively, $x(t_i) = x_i$, $y(t_i) = y_i$ are said to be *orthogonal* in the sample space $\mathcal{L}(\mathcal{X}, \mathcal{B}_{\mathcal{X}}; +, \vee, \wedge)$ with pseudometric (1.3.25), if NMSI $\nu_p(x, y)$ (1.4.6) between them is equal to zero:

$$\nu_p(x, y) = 0.$$

Definition 1.4.3. Two discrete signals $x(t)$ and $y(t)$, $t \in T$:

$$T = \{0, \Delta t, \ldots, \Delta t \cdot (i - 1), \ldots, \Delta t \cdot (n - 1)\},$$

represented by their own samples $x = (x_1, x_2, \ldots, x_n)$ and $y = (y_1, y_2, \ldots, y_n)$, respectively, $x(t_i) = x_i$, $y(t_i) = y_i$ are said to be *orthogonal* in the sample space $\mathcal{L}(\mathcal{X}, \mathcal{B}_{\mathcal{X}}; +, \vee, \wedge)$ with semimetric (1.3.26), if NMSI $\nu_s(x, y)$ (1.4.7) between them is equal to zero:

$$\nu_s(x, y) = 0.$$

Definition 1.4.4. Two discrete signals $x(t)$ and $y(t)$, $t \in T$:

$$T = \{0, \Delta t, \dots, \Delta t \cdot (i-1), \dots, \Delta t \cdot (n-1)\},$$

represented by their own samples $x = (x_1, x_2, \dots, x_n)$ and $y = (y_1, y_2, \dots, y_n)$, respectively, $x(t_i) = x_i$, $y(t_i) = y_i$ are said to be *orthogonal* in the sample space $\mathcal{L}(\mathcal{X}, \mathcal{B}_{\mathcal{X}}; +, \vee, \wedge)$ with metric (1.3.27), if NMSI $\nu_{l_1}(x, y)$ (1.4.8) between them is equal to zero:

$$\nu_{l_1}(x, y) = 0.$$

1.4.3 Sample Measures of Statistical Interrelation between Signals in Space with L-group Properties

Along with the notion of generalized metric by the following definition, we introduce the notion that generalizes the presence of statistical interrelation within a pair of random samples X, Y.

Definition 1.4.5. *Generalized measure of statistical interrelation* $N(x, y)$ between realizations (observations) $x = (x_1, x_2, \dots, x_n)$ and $y = (y_1, y_2, \dots, y_n)$ of random samples X, Y in the sample space $\mathcal{L}(\mathcal{X}, \mathcal{B}_{\mathcal{X}}; +, \vee, \wedge)$ is said to be some function of generalized metric $M(x, y)$ of this space:

$$N(x, y) = F[M(x, y)], \tag{1.4.9}$$

that possesses the following properties:

  1. $N(x, y) = -N(\pm x, \mp y)$ (oddness);

  2. $N(x, y) = 0 \Rightarrow x \perp y$ for $x \neq 0$, $y \neq 0$ (orthogonality of random samples X, Y);

  3. $N(x, x) \geqslant N(x, y)$ (boundedness).

Further we also use measures of statistical interrelation (MSIs) between realizations (observations) $x = (x_1, x_2, \dots, x_n)$, $y = (y_1, y_2, \dots, y_n)$ of random samples X, Y in the sample space $\mathcal{L}(\mathcal{X}, \mathcal{B}_{\mathcal{X}}; +, \vee, \wedge)$ with L-group properties that are the functions of NMSIs (1.4.6), (1.4.7), (1.4.8), respectively:

$$N_p(x, y) = n - \sum_{i=1}^{n} [\operatorname{sgn}(x_i \vee y_i) - \operatorname{sgn}(x_i \wedge y_i)]; \tag{1.4.10}$$

$$N_s(x, y) = \|x\| + \|y\| - \sum_{i=1}^{n} |x_i - y_i| \cdot [\operatorname{sgn}(x_i \vee y_i) - \operatorname{sgn}(x_i \wedge y_i)]; \tag{1.4.11}$$

$$N_{l_1}(x, y) = \sum_{i=1}^{n} [|x_i + y_i| - |x_i - y_i|], \tag{1.4.12}$$

where $\|x\|$, $\|y\|$ are l_1-norms of realizations (observations) $x = (x_1, x_2, \dots, x_n)$, $y = (y_1, y_2, \dots, y_n)$ of random samples X, Y.

Taking into account the fact that the differences of sign functions figuring in the relationships (1.4.7), (1.4.11) take the values in the set $\{0, 1, 2\}$, calculating NMSIs (1.4.6), (1.4.7), (1.4.8) and MSIs (1.4.10), (1.4.11) (1.4.12) can be organized without using operation of multiplication (if do not take into account the sole division by denominator in (1.4.7) and (1.4.8)), that is doubtless advantage when organizing calculations, first, in applications that do not assume exploiting the considerable computational resources, and second, in applications requiring high computational rate.

1.4.4 Estimating Sample Measures of Statistical Interrelation between Signals in Space with *L*-group Properties

Let $x_i = u_i + j \cdot v_i$ be complex discrete stochastic signal, $i = 1, 2, \ldots, n$; where $u_i = \mathrm{Re}[x_i]$, $v_i = \mathrm{Im}[x_i]$ are real and imaginary components. Then estimator of measure of statistical interrelation (MSI) (1.4.10) $\hat{N}_p(u; m)$, $(\hat{N}_p(v; m))$ of its real $u_i = \mathrm{Re}[x_i]$ (imaginary $v_i = \mathrm{Im}[x_i]$) components is defined by relationship:

$$\hat{N}_p(u; m) = n - \frac{n}{n-m} \sum_{i=1}^{n-m} [\mathrm{sgn}(u_{i+m} \vee u_i) - \mathrm{sgn}(u_{i+m} \wedge u_i)], \qquad (1.4.13)$$

where m is a delay parameter, $m = 1, 2, \ldots, L$; $L \ll n$.

Estimators of mutual MSIs (1.4.10) $\hat{N}_p(u, v; m)$, $\hat{N}_p(v, u; m)$ of real u_i and imaginary v_i components of a signal x_i are defined by the following relationships:

$$\hat{N}_p(u, v; m) = n - \frac{n}{n-m} \sum_{i=1}^{n-m} [\mathrm{sgn}(u_{i+m} \vee v_i) - \mathrm{sgn}(u_{i+m} \wedge v_i)]; \qquad (1.4.14a)$$

$$\hat{N}_p(v, u; m) = n - \frac{n}{n-m} \sum_{i=1}^{n-m} [\mathrm{sgn}(v_{i+m} \vee u_i) - \mathrm{sgn}(v_{i+m} \wedge u_i)]. \qquad (1.4.14b)$$

Quite similarly, estimator of MSI (1.4.11) $\hat{N}_s(u; m)$, $(\hat{N}_s(v; m))$ of its real $u_i = \mathrm{Re}[x_i]$ (imaginary $v_i = \mathrm{Im}[x_i]$) component is determined by the following relationship:

$$\hat{N}_s(u; m) = \frac{1}{n-m} \left(\sum_{i=1}^{n-m} [|u_{i+m}| + |u_i|] \right.$$
$$\left. + \sum_{i=1}^{n-m} |u_{i+m} - u_i| [\mathrm{sgn}(u_{i+m} \vee u_i) - \mathrm{sgn}(u_{i+m} \wedge u_i)] \right), \qquad (1.4.15)$$

where m is delay parameter, $m = 1, 2, \ldots, L$; $L \ll n$.

Estimators of mutual MSIs (1.4.11) $\hat{N}_s(u, v; m)$, $\hat{N}_s(v, u; m)$ of real u_i and imaginary v_i components of a signal x_i are determined by the following relationships:

$$\hat{N}_s(u,v;m) = \frac{1}{n-m}\left(\sum_{i=1}^{n-m}[|u_{i+m}| + |v_i|]\right.$$

$$\left. + \sum_{i=1}^{n-m}|u_{i+m} - v_i|\left[\mathrm{sgn}(u_{i+m} \vee v_i) - \mathrm{sgn}(u_{i+m} \wedge v_i)\right]\right); \quad (1.4.16a)$$

$$\hat{N}_s(v,u;m) = \frac{1}{n-m}\left(\sum_{i=1}^{n-m}[|v_{i+m}| + |u_i|]\right.$$

$$\left. + \sum_{i=1}^{n-m}|v_{i+m} - u_i|\left[\mathrm{sgn}(v_{i+m} \vee u_i) - \mathrm{sgn}(v_{i+m} \wedge u_i)\right]\right). \quad (1.4.16b)$$

Estimator of MSI (1.4.12) $\hat{N}_{l_1}(u;m)$, $\hat{N}_{l_1}(v;m)$ of its real $u_i = \mathrm{Re}[x_i]$ (imaginary $v_i = \mathrm{Im}[x_i]$) components is defined by the relationship:

$$\hat{N}_{l_1}(u;m) = \frac{1}{n-m}\left(\sum_{i=1}^{n-m}[|u_{i+m} + u_i| - |u_{i+m} - u_i|]\right), \quad (1.4.17)$$

where m is delay parameter, $m = 1, 2, \ldots, L$; $L << n$, and estimators of mutual MSIs (1.4.11) $\hat{N}_{l_1}(u,v;m)$, $\hat{N}_{l_1}(v,u;m)$ of real u_i and imaginary v_i components of signal are defined by the following relationships:

$$\hat{N}_{l_1}(u,v;m) = \frac{1}{n-m}\sum_{i=1}^{n-m}[|u_{i+m} + v_i| - |u_{i+m} - v_i|]; \quad (1.4.18a)$$

$$\hat{N}_{l_1}(v,u;m) = \frac{1}{n-m}\sum_{i=1}^{n-m}[|v_{i+m} + u_i| - |v_{i+m} - u_i|]. \quad (1.4.18b)$$

Here, as before, notations "p, s" figuring in subscript index in the formulas (1.4.13)...(1.4.16) point at relation to pseudometrics and semimetrics, respectively, and notation "l_1" figuring in subscript index in the formulas (1.4.17), (1.4.18) point at relation to l_1-metric. MSIs (1.4.13), (1.4.15), (1.4.17) and mutual MSIs (1.4.14a,b), (1.4.16a,b), (1.4.18a,b) are substantive analogues of autocorrelation function (ACF) and cross-correlation function (CCF), correspondingly, with a difference that they, in general case, take into account both linear and nonlinear statistical relations between signal samples.

Then, taking into consideration the relationships (1.4.13)...(1.4.18), the resulting MSI estimators $\hat{N}(x;m)$ of complex discrete stochastic signal x_i are defined by the following expressions, respectively:

$$\hat{N}_p(x;m) = \hat{N}_p(u;m) + \hat{N}_p(v;m) - j[\hat{N}_p(u,v;m) - \hat{N}_p(v,u;m)]; \quad (1.4.19a)$$

$$\hat{N}_s(x;m) = \hat{N}_s(u;m) + \hat{N}_s(v;m) - j[\hat{N}_s(u,v;m) - \hat{N}_s(v,u;m)]; \quad (1.4.19b)$$

$$\hat{N}_{l_1}(x;m) = \hat{N}_{l_1}(u;m) + \hat{N}_{l_1}(v;m) - j[\hat{N}_{l_1}(u,v;m) - \hat{N}_{l_1}(v,u;m)]. \quad (1.4.19c)$$

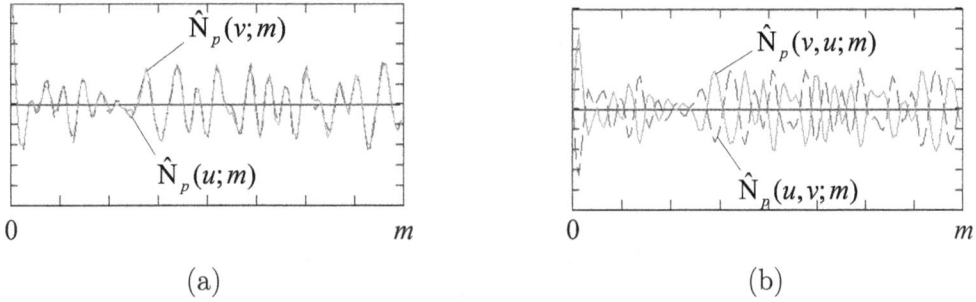

FIGURE 1.4.2   Estimates of (a) MSIs $\hat{N}_p(u; m)$ (dotted line), $\hat{N}_p(v; m)$ (solid line) (1.4.13); (b) mutual MSIs $\hat{N}_p(u, v; m)$ (dotted line), $\hat{N}_p(v, u; m)$ (solid line) (1.4.14a,b)

Example 1.4.2. Let $x(i)$ be an additive mixture of M complex discrete harmonic signals that are observed in the presence of complex quasi-white Gaussian noise $n(i)$ with zero mean:

$$x(i) = \sum_{l=1}^{M} s_l(i) + n(i) = \sum_{l=1}^{M} A_l \exp[j2\pi f_l \cdot i + \varphi_l] + n(i), \qquad (1.4.20)$$

where $i = 1, 2, \ldots, n$; $A_l = $ const, $f_l = $ const, φ_l is random initial phase uniformly distributed in the interval $\varphi_l \in [0, 2\pi]$; $n(i) = n_c(i) + j \cdot n_s(i)$; $\mathbf{M}\{n_c^2(i)\} = \mathbf{M}\{n_s^2(i)\} = D_n/2$, $\mathbf{M}\{*\}$ is a symbol of mathematical expectation; D_n is a variance of quasi-white Gaussian noise $n(i)$ with zero mean.

In the considered example, amplitudes of harmonics are chosen to be equal $A_l = A$, so that the following relationship holds: $A^2/2 = D_n$, D_n is a variance of noise $n(i)$; $F_n > 2\max_{l}\{f_l\}$, where F_n is an upper bound frequency of power spectral density of quasi-white Gaussian noise $n(i)$; $M = 6$. Number of samples n of stochastic processes $x(i)$ used when forming MSI estimator is equal to $n = 1024$. Maximal index L of a shift m of MSI figuring in the estimators (1.4.13)...(1.4.18) is chosen to be equal to $L = 128$.

Fig.1.4.2a and 1.4.2b show estimates of MSIs $\hat{N}_p(u; m)$, $\hat{N}_p(v; m)$ (1.4.13) and mutual MSIs $\hat{N}_p(u, v; m)$, $\hat{N}_p(v, u; m)$ (1.4.14a,b), respectively. Correlation coefficient $r[\hat{N}_p(u; m), \hat{N}_p(v; m)]$ between MSI estimates $\hat{N}_p(u; m)$, $\hat{N}_p(v; m)$ (1.4.13), correlation coefficient $r[\hat{N}_p(u, v; m), \hat{N}_p(v, u; m)]$ between mutual MSI estimates $\hat{N}_p(u, v; m)$, $\hat{N}_p(v, u; m)$ (1.4.14a,b), and also correlation coefficient $r[\hat{r}_x(m), \hat{N}_p(x; m)]$ between estimate $\hat{r}_x(m)$ of ACF of the signal $x(i)$ (1.4.19) and MSI estimate $\hat{N}_p(x; m)$ (1.4.17), correspondingly, are equal to:

$$r[\hat{N}_p(u; m), \hat{N}_p(v; m)] = 0.985;$$

$$r[\hat{N}_p(u, v; m), \hat{N}_p(v, u; m)] = -0.979;$$

$$r[\hat{r}_x(m), \hat{N}_p(x; m)] = 0.99.$$

As can be seen from the last formula, MSI estimators $\hat{N}_p(x; m)$, $\hat{N}_s(x; m)$,

$\hat{N}_{l_1}(x; m)$ defined by the relationships (1.4.19a,b,c), respectively, can be used instead of ACF estimator $\hat{r}_x(m)$ when solving some problems of statistical and spectral analysis. $\triangledown$

1.5 Metric Characteristics of Random Samples with Arbitrary Distributions

In most practically important cases of signal processing we mainly deal with stochastic signals (processes) $\xi(t)$, $\eta(t)$ with symmetric (even) univariate PDFs $p_\xi(x)$, $p_\eta(y)$: $p_\xi(x) = p_\xi(-x)$; $p_\eta(y) = p_\eta(-y)$. Introduced in the Section 1.3 characteristics of statistical interrelation of a pair of instantaneous values (samples) ξ_t, η_t of stochastic signals (processes) $\xi(t), \eta(t) \in \mathbf{\Gamma}$, and also the results formulated in the form of theorems, are predominately oriented to the class of signals with even univariate PDFs $p_\xi(x)$, $p_\eta(y)$ that interact in partially ordered set $\mathbf{\Gamma}$ with the properties of lattice-ordered group $\mathbf{\Gamma}(+, \vee, \wedge)$ (*L*-group). However, within rather important applications of signal processing (for instance, in image processing, etc) we deal with random samples whose elements possess rather arbitrary distributions.

In this section we generalize main results of the Section 1.3 concerning random samples $X = (X_1, \dots, X_n)$, $Y = (Y_1, \dots, Y_n)$ with arbitrary distributions. We assume that the elements $\{X_i\}$, $\{Y_j\}$ of the samples X, Y of sample space $\mathcal{L}(\mathcal{X}, \mathcal{B}_\mathcal{X}; +, \vee, \wedge)$ with *L*-group properties: $X, Y \in \mathcal{L}(\mathcal{X}, \mathcal{B}_\mathcal{X}; +, \vee, \wedge)$ are statistically independent with arbitrary identical univariate cumulative distribution functions (CDFs): $F_X(x_i) \equiv F_X(x)$, $F_Y(y_j) \equiv F_Y(y)$, so that a pair of elements X_i, Y_i are statistically dependent with joint CDF $F_{XY}(x_i, y_i) \equiv F_{XY}(x, y)$.

Logic of stating the main material of the section is as follows. At first, as applied to a pair of arbitrary random variables ξ, η from partially ordered set $\mathbf{\Gamma}$: $\xi, \eta \in \mathbf{\Gamma}$ with properties of lattice-ordered group $\mathbf{\Gamma}(+, \vee, \wedge)$ (*L*-group), we obtain main results formulated in the form of theorems. Then we expand these results with respect to a pair of random samples $X = (X_1, \dots, X_n)$, $Y = (Y_1, \dots, Y_n)$ and their realizations (observations).

Theorem 1.5.1. *For a pair of random variables ξ, η with medians of distributions $m_\xi = \mathrm{med}(\xi)$, $m_\eta = \mathrm{med}(\eta)$, $\xi, \eta \in \mathbf{\Gamma}$, the functions $\mu_+(\xi, \eta)$, $\mu_-(\xi, \eta)$, $\mu(\xi, \eta)$ that are equal to:*

$$\mu_+(\xi, \eta) = 2(\mathbf{P}[(\xi - m_\xi) \vee (\eta - m_\eta) > 0] - \mathbf{P}[(\xi - m_\xi) \wedge (\eta - m_\eta) > 0]); \quad (1.5.1a)$$

$$\mu_-(\xi, \eta) = 2(\mathbf{P}[(\xi - m_\xi) \wedge (\eta - m_\eta) < 0] - \mathbf{P}[(\xi - m_\xi) \vee (\eta - m_\eta) < 0]); \quad (1.5.1b)$$

$$\mu(\xi, \eta) = [\mu_+(\xi, \eta) + \mu_-(\xi, \eta)]/2, \quad (1.5.1c)$$

are pseudometrics.

Proof. Consider probabilities $\mathbf{P}[(\xi - m_\xi) \wedge (\eta - m_\eta) > 0]$, $\mathbf{P}[(\xi - m_\xi) \vee (\eta - m_\eta) > 0]$,

$\mathbf{P}[(\xi - m_\xi) \wedge (\eta - m_\eta) < 0]$, $\mathbf{P}[(\xi - m_\xi) \vee (\eta - m_\eta) < 0]$ that, according to the formulas [52, (3.2.80)], [52, (3.2.85)], are equal to:

$$\mathbf{P}[(\xi - m_\xi) \wedge (\eta - m_\eta) > 0] = 1 - (F_\xi(m_\xi) + F_\eta(m_\eta) - F_{\xi\eta}(m_\xi, m_\eta)); \quad (1.5.2a)$$

$$\mathbf{P}[(\xi - m_\xi) \vee (\eta - m_\eta) > 0] = 1 - F_{\xi\eta}(m_\xi, m_\eta); \quad (1.5.2b)$$

$$\mathbf{P}[(\xi - m_\xi) \wedge (\eta - m_\eta) < 0 = F_\xi(m_\xi) + F_\eta(m_\eta) - F_{\xi\eta}(m_\xi, m_\eta); \quad (1.5.2c)$$

$$\mathbf{P}[(\xi - m_\xi) \vee (\eta - m_\eta) < 0] = F_{\xi\eta}(m_\xi, m_\eta), \quad (1.5.2d)$$

where $F_{\xi\eta}(m_\xi, m_\eta)$ is joint CDF of random variables ξ, η; $F_\xi(m_\xi)$, $F_\eta(m_\eta)$ are univariate CDFs of random variables ξ, η.

Then the function $\mathbf{P}[(\xi - m_\xi) > 0]$, $\mathbf{P}[(\eta - m_\eta) > 0]$ on lattice $\mathbf{\Gamma}$ is *valuation* for which the following identity holds [29, § X.1 (V1)]:

$$\mathbf{P}[(\xi - m_\xi) > 0] + \mathbf{P}[(\eta - m_\eta) > 0]$$

$$= \mathbf{P}[(\xi - m_\xi) \vee (\eta - m_\eta) > 0] + \mathbf{P}[(\xi - m_\xi) \wedge (\eta - m_\eta) > 0]. \quad (1.5.3)$$

Besides, valuation $\mathbf{P}[(\xi - m_\xi) > 0]$ is isotonic, since the following implication holds [29, § X.1 (V2)]:

$$\xi - m_\xi \geq \xi' - m'_\xi \Rightarrow \mathbf{P}[(\xi - m_\xi) > 0] \geq \mathbf{P}[(\xi' - m'_\xi) > 0]. \quad (1.5.4)$$

From joint fulfillment of relationships (1.5.3) and (1.5.4), according to Theorem 1 [29, § X.1], the quantity $\mu_+(\xi, \eta)$ (1.5.10) that is equal to:

$$\mu_+(\xi, \eta) = 2(\mathbf{P}[(\xi - m_\xi) \vee (\eta - m_\eta) > 0] - \mathbf{P}[(\xi - m_\xi) \wedge (\eta - m_\eta) > 0])$$

$$= 2[F_\xi(m_\xi) + F_\eta(m_\eta)] - 4F_{\xi\eta}(m_\xi, m_\eta) = 2 - 4F_{\xi\eta}(m_\xi, m_\eta), \quad (1.5.5)$$

is pseudometric.

Substituting the relationships (1.5.2c), (1.5.2d) into the formula (1.5.1b), we obtain the expression for the function $\mu_-(\xi, \eta)$:

$$\mu_-(\xi, \eta) = 2[F_\xi(m_\xi) + F_\eta(m_\eta)] - 4F_{\xi\eta}(m_\xi, m_\eta) = 2 - 4F_{\xi\eta}(m_\xi, m_\eta), \quad (1.5.6)$$

i.e. functions $\mu_+(\xi, \eta)$, $\mu_-(\xi, \eta)$ determined by the relationships (1.5.1a), (1.5.1b) are identically equal:

$$\mu_+(\xi, \eta) \equiv \mu_-(\xi, \eta),$$

and function $\mu_-(\xi, \eta)$ is also pseudometric.

Obviously, sum of two identical pseudometrics (1.5.1c) $\mu(\xi, \eta)$ is also pseudometric that is determined by relationship:

$$\mu(\xi, \eta) = 2[F_\xi(m_\xi) + F_\eta(m_\eta)] - 4F_{\xi\eta}(m_\xi, m_\eta) = 2 - 4F_{\xi\eta}(m_\xi, m_\eta). \quad (1.5.7)$$

$\square$

Definition 1.5.1. Quantity $\mu(\xi, \eta)$ defined by the relationships (1.5.1c), (1.5.7) is called *pseudometrics* between two random variables ξ, η from partially ordered set $\mathbf{\Gamma}$ with properties of lattice-ordered group $\mathbf{\Gamma}(+, \vee, \wedge)$ (*L*-group): $\xi, \eta \in \mathbf{\Gamma}(+, \vee, \wedge)$.

Thus, partially ordered set $\mathbf{\Gamma}$ with operations $\chi_+ = \xi + \eta$, $\chi_\vee = \xi \vee \eta$, $\chi_\wedge = \xi \wedge \eta$ is pseudometric space $(\mathbf{\Gamma}, \mu)$, in which generalized metric (pseudometric) $\mu(\xi, \eta)$ is defined by the formula (1.5.7).

Then any pair of random variables $\xi, \eta \in \mathbf{\Gamma}$ can be associated with the following normalized measure between them.

Definition 1.5.2. *Normalized measure of statistical interrelation* (NMSI) between random variables ξ, η is the quantity $\nu(\xi, \eta)$ defined by normalized generalized metric $\mu(\xi, \eta)$ between them:

$$\nu(\xi, \eta) = 1 - \mu(\xi, \eta). \tag{1.5.8}$$

The last relationship and the formula (1.5.7) imply that NMSI $\nu(\xi, \eta)$ is determined over joint CDF $F_{\xi\eta}(m_\xi, m_\eta)$ of random variables ξ, η:

$$\nu(\xi, \eta) = 4F_{\xi\eta}(m_\xi, m_\eta) - 1. \tag{1.5.9}$$

Theorem 1.5.2. *For a pair of random variables ξ, η in L-group $\mathbf{\Gamma}$: $\xi, \eta \in \mathbf{\Gamma}$, pseudometric $\mu(\xi, \eta)$ is invariant of a group H of their continuous mappings $\{h_{\alpha,\beta}\}$, $h_{\alpha,\beta} \in H$; $\alpha, \beta \in A$:*

$$\mu(\xi, \eta) = \mu(\xi', \eta'); \tag{1.5.10}$$

$$h_\alpha : \xi \to \xi', \ h_\beta : \eta \to \eta'; \tag{1.5.10a}$$

$$h_\alpha^{-1} : \xi' \to \xi, \ h_\beta^{-1} : \eta' \to \eta; \tag{1.5.10b}$$

where ξ', η' are results of mapping of random variables ξ, η in L-group $\mathbf{\Gamma}'$: $h_{\alpha,\beta}: \mathbf{\Gamma} \to \mathbf{\Gamma}'$.

Proof. When realizing one-to-one mappings $\{h_{\alpha,\beta}\}$, $h_{\alpha,\beta} \in H$ (1.5.10a,b), the property of differential probability invariance holds that implies the identity of joint CDFs $F_{\xi\eta}(m_\xi, m_\eta)$, $F_{\xi'\eta'}(m'_\xi, m'_\eta)$ and univariate CDFs $F_\xi(x)$, $F'_\xi(x')$; $F_\eta(y)$, $F'_\eta(y')$ of the pairs of random variables ξ, η; ξ', η', respectively:

$$F_{\xi'\eta'}(m'_\xi, m'_\eta) = F_{\xi\eta}(m_\xi, m_\eta); \tag{1.5.11}$$

$$F'_\xi(x') = F_\xi(x); \tag{1.5.11a}$$

$$F'_\eta(y') = F_\eta(y). \tag{1.5.11b}$$

Thus, taking into account the equality (1.5.7), the relationships (1.5.11), (1.5.11a,b) imply the equalities $F_{\xi\eta}(m_\xi, m_\eta) = F_{\xi'\eta'}(m'_\xi, m'_\eta)$, $F_\xi(m_\xi) = F'_\xi(m'_\xi)$, $F_\eta(m_\eta) = F'_\eta(m'_\eta)$, (where $m_{\xi'} = \text{med}(\xi')$, $m_{\eta'} = \text{med}(\eta')$ are medians of random variables ξ', η', respectively), and hence, the identity (1.5.10). $\square$

Theorem 1.5.2 has the following corollary:

Corollary 1.5.1. *For a pair of random variables ξ, η in L-group Γ: $\xi, \eta \in \Gamma$, NMSI $\nu(\xi, \eta)$ is an invariant of a group H of their continuous mappings $\{h_{\alpha,\beta}\}$, $h_{\alpha,\beta} \in H$; $\alpha, \beta \in A$:*

$$\nu(\xi, \eta) = \nu(\xi', \eta'); \tag{1.5.12}$$

$$h_\alpha : \xi \to \xi', \ h_\beta : \eta \to \eta'; \tag{1.5.12a}$$

$$h_\alpha^{-1} : \xi' \to \xi, \ h_\beta^{-1} : \eta' \to \eta; \tag{1.5.12b}$$

where ξ', η' are the results of mappings of random variables ξ, η in L-group Γ': $h_{\alpha,\beta} : \Gamma \to \Gamma'$.

Example 1.5.1. Define NMSI $\nu(\xi, \eta)$ for two jointly Gaussian random variables ξ, η with joint CDF $F_{\xi\eta}(x, y)$ and correlation coefficient ρ. Find an expression for joint CDF $F_{\xi\eta}(x, y)$ of Gaussian random variables ξ, η in a point $x = m_\xi$, $y = m_\eta$, which, according to the formula (1.5.12) from application II of work [53], is determined over double integral $K_{00}(\alpha)$:

$$F_{\xi\eta}(m_\xi, m_\eta) = \left(\frac{\sqrt{1 - \rho^2}}{\pi} \right) K_{00}(\alpha),$$

where $\alpha = \pi - \arccos(\rho)$, $K_{00}(\alpha) = \alpha/(2 \sin \alpha)$ (see formula (1.5.14) in the same place), $\sin \alpha = \sqrt{1 - \rho^2}$. Then, after performing necessary transformations we obtain the resulting expression for CDF $F_{\xi\eta}(m_\xi, m_\eta)$:

$$F_{\xi\eta}(m_\xi, m_\eta) = \left(1 + \frac{2}{\pi} \arcsin[\rho] \right) / 4. \tag{1.5.13}$$

Having substituted the last expression $F_{\xi\eta}(m_\xi, m_\eta)$ together with the values of univariate CDFs $F_\xi(m_\xi) = 0.5$, $F_\eta(m_\eta) = 0.5$ into the formula (1.5.9), we obtain the desired expression for NMSI $\nu(\xi, \eta)$:

$$\nu(\xi, \eta) = \nu(\rho) = \frac{2}{\pi} \arcsin[\rho]. \tag{1.5.14}$$

On arbitrary one-to-one transformations of two jointly Gaussian random variables ξ, η, according to Corollary 1.5.1 of Theorem 1.5.2, their NMSI $\nu(\xi, \eta)$ stays invariable and is determined just over correlation coefficient ρ by the dependence (1.5.14), whose general form and also its appearance on large values of ρ in logarithmic scale are shown in Fig. 1.5.1a and 1.5.1b, respectively.

Correlation coefficient $\rho_{\xi\eta}$ of a pair of random variables ξ, η with arbitrary joint CDF $F_{\xi\eta}(x, y)$ characterizes just linear statistical relation between these random variables. Hence, correlation coefficient $\rho_{\xi'\eta'}$ of a pair of random variables ξ', η' can essentially decrease $\rho_{\xi\eta} > \rho_{\xi'\eta'}$, if random variables ξ, η are exposed to nonlinear one-to-one transformation (1.5.12a,b). *Correlation ratio $\theta_{\xi\eta}$ (normalized variance function)* is considered to be more complete characteristic of statistical interrelation between two random variables [54, 55], that is defined by the following expression:

$$\theta_{\xi\eta}^2 = \frac{1}{D_\xi} \mathbf{M}_\eta \{ [\mathbf{M}\{\xi/\eta\} - \mathbf{M}\{\xi\}]^2 \},$$

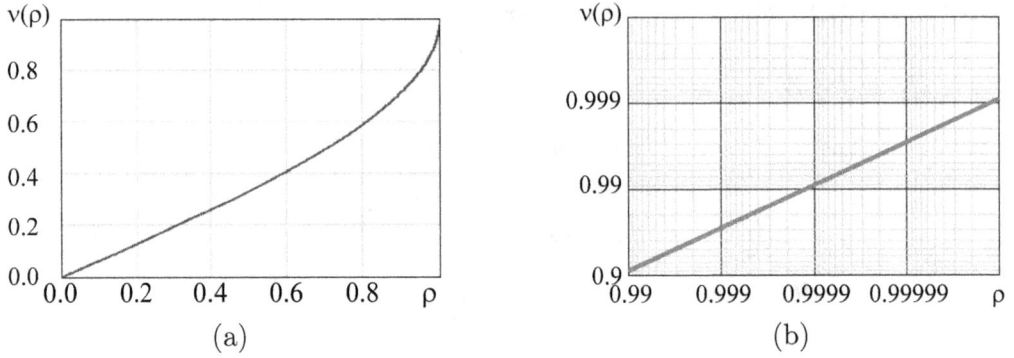

FIGURE 1.5.1   Illustrations of (a) dependence (1.5.14); (b) dependence (1.5.14) in logarithmic scale

where D_ξ is a variance of random variable ξ; $\mathbf{M}(*)$ is a symbol of mathematical expectation; $\mathbf{M}\{\xi\}$ is mathematical expectation of random variable ξ; $\mathbf{M}\{\xi/\eta\}$ is conditional mathematical expectation;

$$\mathbf{M}_\eta\{[\mathbf{M}\{\xi/\eta\} - \mathbf{M}\{\xi\}]^2\} = \int_{-\infty}^{\infty} [\mathbf{M}\{\xi/\eta\} - \mathbf{M}\{\xi\}]^2 p_\eta(y)dy,$$

where $p_\eta(y)$ is probability density function (PDF) of random variable η.

Notice, that correlation ratio, first, is nonsymmetric: $\theta_{\xi\eta} \neq \theta_{\eta\xi}$, and second, it is not invariant with respect to one-to-one mappings of random variables inasmuch as it is not preserved when realizing the last ones. In signal processing problems, as a rule, there is no need to solve problems that are similar to studying dependences of a quantity of grown vegetables and fruits on a quantity of atmospheric precipitation and vice versa, but it is important that a measure of statistical interrelation would possess the property of symmetry. Therefore, in signal processing practice, correlation coefficient is used more frequently than correlation ratio. However, correlation coefficient between a pair of random variables ξ, η preserves only within a group of their linear one-to-one transformations.

Unlike correlation coefficient $\rho_{\xi\eta}$ and correlation ratio $\theta_{\xi\eta}$ between random variables ξ, η with an arbitrary distribution, their NMSI $\nu(\xi, \eta)$ stays invariable on mappings $\{h_{\alpha,\beta}\}$, $\alpha, \beta \in A$ of these random variables ξ, η from an arbitrary group H: $h_{\alpha,\beta} \in H$ (1.5.12a), (1.5.12b).

Exploiting the formulas (1.5.1c), (1.5.7), (1.5.8), and (1.5.9) to solve the problems of mathematical statistics, in general, and signal processing, in particular, needs some refinement.

Let, as before, $X = (X_1, X_2, \ldots, X_n)$, $Y = (Y_1, Y_2, \ldots, Y_n)$ be random samples in sample space $\mathcal{L}(\mathcal{X}, \mathcal{B}_\mathcal{X}; +, \vee, \wedge)$ with L-group properties: $X, Y \in \mathcal{L}(\mathcal{X}, \mathcal{B}_\mathcal{X}; +, \vee, \wedge)$. Obviously, for a pair of statistically dependent elements X_i, Y_i of random samples X, Y, as well as for a pair of random variables, the results of Theorems 1.5.1, 1.5.2 concerning generalized metric $\mu(X_i, Y_i)$ (1.5.7) and NMSI $\nu(X_i, Y_i)$

(1.5.8) between them stay true. Further we consider exploiting the obtained results with respect to realizations (observations) of random samples X, Y.

Let, as before, $x = (x_1, x_2, \ldots, x_n)$, $y = (y_1, y_2, \ldots, y_n)$ be realizations (observations) of random samples X, Y; $\{x_i\}$, $\{y_i\}$ are values of realizations of random variables $\{X_i\}$, $\{Y_i\}$ of random samples X, Y, correspondingly; $i = 1, 2, \ldots, n$; $n \in \mathbf{N}$.

Introduce a function $I_z(A)$, denoting a distribution concentrated in a point z [51]:

$$I_z(A) = \begin{cases} 1, & z \in A; \\ 0, & z \notin A, \end{cases}$$

where A is some set.

Then, according to the formulas (1.5.1a,b,c), pseudometrics $\mu_+(\xi, \eta)$, $\mu_-(\xi, \eta)$, $\mu(\xi, \eta)$ between realizations (observations) $x = (x_1, x_2, \ldots, x_n)$, $y = (y_1, y_2, \ldots, y_n)$ of random samples X, Y are determined by the following relationships:

$$\mu_+(x, y) = \frac{2}{n} \left(\sum_{i=1}^{n} I_{v_i}(v_i > 0) - \sum_{i=1}^{n} I_{u_i}(u_i > 0) \right); \qquad (1.5.15a)$$

$$\mu_-(x, y) = \frac{2}{n} \left(\sum_{i=1}^{n} I_{u_i}(u_i < 0) - \sum_{i=1}^{n} I_{v_i}(v_i < 0) \right); \qquad (1.5.15b)$$

$$\mu(x, y) = [\mu_+(x, y) + \mu_-(x, y)]/2, \qquad (1.5.15c)$$

where $v_i = (x_i - m_x) \vee (y_i - m_y)$, $u_i = (x_i - m_x) \wedge (y_i - m_y)$; m_x, m_y are sample medians of realizations $x = (x_1, x_2, \ldots, x_n)$, $y = (y_1, y_2, \ldots, y_n)$ of random samples X, Y, respectively:

$$\sum_{i=1}^{n} I_{x_i}(x_i < m_x) = \sum_{i=1}^{n} I_{x_i}(x_i > m_x);$$

$$\sum_{i=1}^{n} I_{y_i}(y_i < m_y) = \sum_{i=1}^{n} I_{y_i}(y_i > m_y).$$

Substituting the relationships for pseudometrics (1.5.15a) and (1.5.15b) into the formula (1.5.15c), we obtain the expression that is necessary for practical calculations of pseudometric between realizations (observations) $x = (x_1, x_2, \ldots, x_n)$, $y = (y_1, y_2, \ldots, y_n)$ of random samples X, Y of the following kind:

$$\mu_p(x, y)$$
$$= \frac{1}{n} \sum_{i=1}^{n} [\mathrm{sgn}((x_i - m_x) \vee (y_i - m_y)) - \mathrm{sgn}((x_i - m_x) \wedge (y_i - m_y))], \quad (1.5.16)$$

where $\mathrm{sgn}(x)$ is sign function determined by the formulas (1.2.5), (1.2.2); m_x, m_y are sample medians of realizations $x = (x_1, x_2, \ldots, x_n)$, $y = (y_1, y_2, \ldots, y_n)$ of random samples X, Y, respectively.

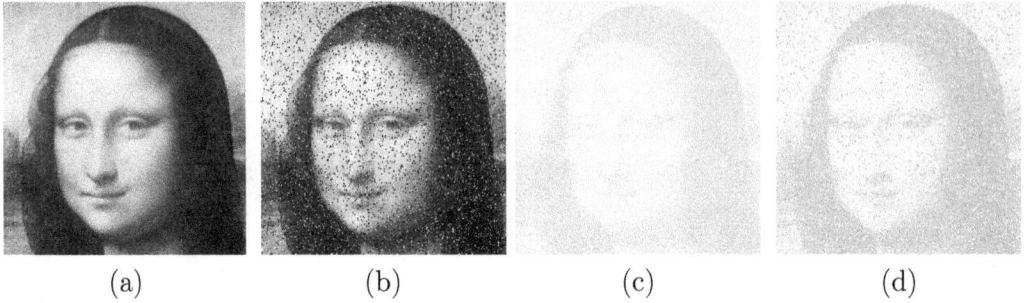

(a)   (b)   (c)   (d)

FIGURE 1.5.2   Illustrations of (a) initial image; (b) initial image corrupted by "salt" and "pepper" impulse noise; (c) image obtained as a result of transformation (1.5.18a); (d) image obtained as a result of transformation (1.5.18b). From [56], with permission

Thus, the sample space $\mathcal{L}(\mathcal{X}, \mathcal{B}_{\mathcal{X}}; +, \vee, \wedge)$ with L-group properties is metricized by introducing pseudometric (1.5.1c), (1.5.7), becoming pseudometric space $(\mathcal{X}, \mu)$. Remind, Theorem 1.5.2 asserts that pseudometric (1.5.7) is an invariant of a group of continuous mappings of random samples X, Y in the sample space $\mathcal{L}(\mathcal{X}, \mathcal{B}_{\mathcal{X}}; +, \vee, \wedge)$ with L-group properties: $X, Y \in \mathcal{L}(\mathcal{X}, \mathcal{B}_{\mathcal{X}}; +, \vee, \wedge)$.

Substituting the expression for pseudometric (1.5.16) into general relationship between NMSI and normalized generalized metric (1.5.8), we obtain the expression that is necessary for practical calculating NMSI $\nu(x, y)$ between realizations (observations) $x = (x_1, x_2, \ldots, x_n)$, $y = (y_1, y_2, \ldots, y_n)$ of random samples X, Y in the sample space $\mathcal{L}(\mathcal{X}, \mathcal{B}_{\mathcal{X}}; +, \vee, \wedge)$ with L-group properties:

$$\nu_p(x, y)$$
$$= 1 - \frac{1}{n} \sum_{i=1}^{n} [\operatorname{sgn}((x_i - m_x) \vee (y_i - m_y)) - \operatorname{sgn}((x_i - m_x) \wedge (y_i - m_y))], \quad (1.5.17)$$

where $\operatorname{sgn}(x)$ is sign function determined by the formulas (1.2.5), (1.2.2)

Here we notice that formulas defining pseudometric $\mu(\xi, \eta)$ (1.5.1c), (1.5.7), and NMSI $\nu(\xi, \eta)$ (1.5.8), (1.5.9) establish probabilistic description of a distance and measure of closeness between random variables ξ, η, whereas the expressions for pseudometric (1.5.16) and NMSI (1.5.17) provide statistical description of these characteristics for random samples X, Y.

The following example helps to understand usefulness of invariance property of the introduced probabilistic-statistical characteristic that is based on metric relations between random variables.

Example 1.5.2. Fig. 1.5.2a demonstrates an initial image and Fig. 1.5.2b depicts the same image corrupted by simultaneous impact of weak "salt" and "pepper" impulse noise with probabilities of appearance that are equal to 0.05 each.

To make the following explanations more comprehensible, we assume that the images shown in Fig. 1.5.2a, b are given by square matrices of data samples M, M' concerning the values of image intensities m_{ij}, m'_{ij} for corresponding pixels (image elements), respectively: $M = \|m_{ij}\|$; $M' = \|m'_{ij}\|$; $i = 1, 2, \ldots, n$; $j = 1, 2, \ldots, n$.

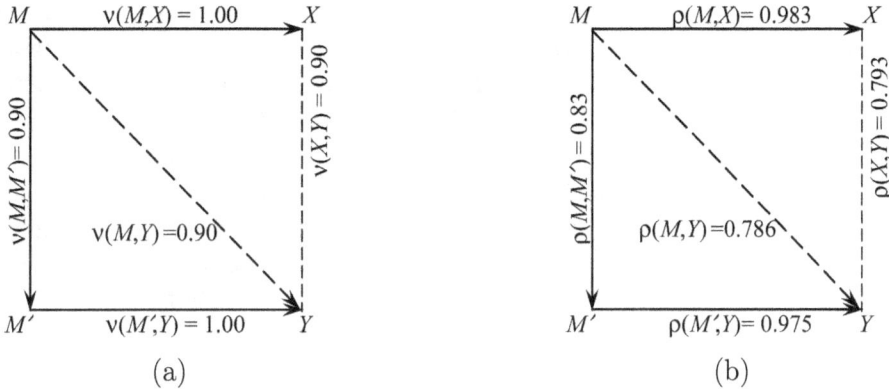

FIGURE 1.5.3   Values of (a) NMSIs ν; (b) correlation coefficients ρ between images given by matrices M,M', X, Y, respectively. From [56], with permission

When making an experiment, a size of images (dimensionality of matrices) $n \times n$ is equal to $n \times n = 256 \times 256$ pixels. Further the images M, M' are transformed by one-to-one mapping of the following type:

$$X = \|x_{ij}\| = \left\|10\sqrt[3]{m_{ij}} + 180\right\|; \tag{1.5.18a}$$

$$Y = \|y_{ij}\| = \left\|6\sqrt[2]{m'_{ij}} + 150\right\|. \tag{1.5.18b}$$

The images, given by square matrices of data samples X, Y concerning the intensities x_{ij}, y_{ij} of pixels $X = \|x_{ij}\|$ (1.5.18a); $Y = \|y_{ij}\|$ (1.5.18b), are shown in Fig. 1.5.2c, d, respectively. The choice of listed above image transformations represented for the example is realized to provide, first, their nonlinearity, and second, clearness of images obtained in the result of their transformation that is sufficient for visual perception.

In the result of statistical modeling, we obtain values of NMSIs ν and correlation coefficients ρ between the images given by matrices M, M', X, Y, that are shown in the diagrams below (see Fig. 1.5.3a, b, respectively).

The values of sample NMSIs ν calculated on the basis of the formula (1.5.17) are equal to: for a pair of images shown in Fig. 1.5.2a and 1.5.2c (matrices M, X), $\nu(M, X)= 1.00$; for a pair of images shown in Fig. 1.5.2b and 1.5.2d (matrices M', Y), $\nu(M', Y) = 1.00$ (see Fig. 1.5.3a). In this case, values of sample NMSIs are equal to 1, inasmuch as the pointed pairs of images are related with each other by one-to-one transformations (1.5.18a,b). In the same time, values of sample correlation coefficients for the same pairs of images are equal to $\rho(M, X) = 0.983$ and $\rho(M', Y) = 0.975$, respectively (see Fig. 1.5.3b). In this case, correlation coefficients are less than 1 due to nonlinearity of the pair of mappings (1.5.18a,b). For a pair of images shown in Fig. 1.5.2a and 1.5.2b (matrices M, M'), Fig. 1.5.2a and 1.5.2d (matrices M, Y), and Fig. 1.5.2c and 1.5.2d (matrices X, Y), the values of sample NMSIs ν are equal to $\nu(M, M') = \nu(M,Y) = \nu(X,Y) = 0.90$. The same obtained value of sample NMSIs in all three pairs of images is explained by the fact that value of generalized metric determined by the formula (1.5.16) is equal to $\mu = 0.10$. $\triangledown$

The aforementioned example shows that in some signal processing problems, for the purpose of assessing statistical relations between random samples (and also data samples), it is expedient to use NMSI introduced by Definition 1.5.2 and also pseudometric introduced by Definition 1.5.1 that are related with each other by the coupling equation (1.5.8). Practical significance of obtained theoretical results lies in the possibility of exploiting introduced NMSI with respect to random samples with arbitrary properties to characterize their informational distinctions, and also to reveal group properties of their mappings. NMSI based on pseudometric of a sample space with lattice properties possesses the property of invariance with respect to a group of mappings of random samples, and corresponds to that very part of information quantity which is contained in the transformed signal with respect to initial one.

1.6 Hyperspectral Representation of Signals in Space with L-group Properties

Theory and practice of signal processing widely use methods based on vector representation of signals, so that they are considered to be the elements of some metric space; the properties of signals are considered as the properties of this space; and a transformation of signals is considered as a mapping of one space into another. Signal Theory use a linear space $\mathcal{LS}$ with scalar product as a signal space [57].

1.6.1 Linear Space with Scalar Product

Introducing the notion of linear space in Signal Theory was the first step on the way to geometric interpretation of a signal space. The constraints imposed by the axioms of linear space are rather tough. Not every set of signals can be adequately described in terms of linear space [24]. Meanwhile, the notions of signal space, scalar product, basis, norm, and metric introduced within a concept of linear space allow formalizing a description of processes related to signal transmitting, receiving, and processing.

Definition 1.6.1. A set $\mathcal{LS} = \{a, b, c, \ldots\}$ is called real *linear space* over a field of real numbers (scalars), if: each pair of elements (vectors) $a, b \in \mathcal{LS}$ can be associated with a sole element of the space $\mathcal{LS}$ that is called *sum* and denoted by $a + b$; each element $a \in \mathcal{LS}$ and each real number λ, $\lambda \in \mathbf{R}$ can be associated with a sole element of the space $\mathcal{LS}$ that is called *scalar multiplication* of element a by scalar λ and denoted by $\lambda \cdot a$.

The following groups of axioms hold [42–44, 58–60]:

1. axioms of additive commutative group:

$$a + b \in \mathcal{LS}; \tag{1.6.1a}$$

$$a + b = b + a; \tag{1.6.1b}$$

$$a + (b + c) = (a + b) + c; \tag{1.6.1c}$$

$$\exists 0 \in \mathcal{LS} : \forall a \in \mathcal{LS} : a + 0 = 0 + a; \tag{1.6.1d}$$

$$\forall a \in \mathcal{LS} : \exists -a : a + (-a) = (-a) + a = 0; \tag{1.6.1e}$$

2. axioms of scalar multiplication:

$$1 \cdot a = a; \tag{1.6.2a}$$

$$\lambda(\mu a) = (\lambda\mu)a; \tag{1.6.2b}$$

$$(\lambda + \mu)a = \lambda a + \mu a; \tag{1.6.2c}$$

$$\lambda(a + b) = \lambda a + \lambda b, \tag{1.6.2d}$$

where $a, b \in \mathcal{LS}$, $\lambda, \mu, 1 \in \mathbf{R}$.

The relationships (1.6.1a...e) represent axioms of closure, commutativity, associativity, null element, and inverse element, respectively. The relationships (1.6.2a...d) represent the axioms: (1.6.2a)—identity element of scalar multiplication; (1.6.2b)—associativity of scalar multiplication; (1.6.2c)—distributivity of scalar multiplication with respect to operation of addition of a field (scalar addition); (1.6.2d)—distributivity with respect to operation of addition of a group (vector addition).

Definition 1.6.2. Real function defined on a set of ordered pairs of elements of real linear space $\mathcal{LS}$ denoted by (a, b), $a, b \in \mathcal{LS}$ is called a *scalar product*, if for all elements $a, b, c \in \mathcal{LS}$ and all scalars $\lambda, \mu \in \mathbf{R}$ it satisfies the following conditions:

$$(a, b) = (b, a); \tag{1.6.3a}$$

$$(\lambda a + \mu b, c) = \lambda(a, c) + \mu(b, c); \tag{1.6.3b}$$

$$(a, a) \geq 0; \tag{1.6.3c}$$

$$(a, a) = 0 \Rightarrow a = 0. \tag{1.6.3d}$$

The relationships (1.6.3a...d) represent axioms of commutativity, linearity, non-negativity, and non-degeneracy, respectively.

For complex linear space, the notion of scalar product is introduced analogously taking into account complex conjugation of elements (see, for instance, [59, 61–64]).

Scalar product (a, b) defined in linear space $\mathcal{LS}$ induces a *norm* $\|a\|$ of an element a and metric $\rho(a, b)$ between the elements $a, b \in \mathcal{LS}$:

$$\|a\| = \sqrt{(a, a)};$$

$$\rho(a, b) = \sqrt{(a - b, a - b)}.$$

Thus, every linear space with scalar product is, first, normed space, and second, metric space.

Let $\{e_1, \ldots, e_n\}$ be the system of n linearly independent elements in linear space $\mathcal{LS}$ with scalar space and a be some element of this space, $a \in \mathcal{LS}$. It is known (see, for instance, [58–60, 65, 66]), that there exist a sole representation of element a of linear space $\mathcal{LS}$, $a \in \mathcal{LS}$ with scalar product in the form of linear combination of linearly independent basis elements $\{e_1, \ldots, e_n\}$:

$$a = \sum_{i=1}^{n} A_i e_i; \tag{1.6.4}$$

$$A_k = (a, e_k)/(e_k, e_k), \tag{1.6.4a}$$

where $A = \{A_k\}$ are coefficients of decomposition of the element a over basis elements $\{e_1, \ldots, e_n\}$; (x, y) is scalar product of the elements x, $y \in \mathcal{LS}$.

Representation (1.6.4) ensures the best approximation of the element a in linear space $\mathcal{LS}$ providing minimization of the function:

$$\left(a - \sum_{i=1}^{n} A_i e_i, a - \sum_{i=1}^{n} A_i e_i\right) \to \min. \tag{1.6.5}$$

If the space $\mathcal{LS}$ is n-dimensional and the set of elements $\{e_1, \ldots, e_n\}$ forms a basis, then one can always choose such coefficients $\{A_i\}$, $i = 1, 2, \ldots, n$ that the equality (1.6.4) holds, and hence, the expression (1.6.5) tends to zero.

The set $\{e_1, \ldots, e_n\}$ can be always replaced by orthogonal system of nonzero vectors. So further we assume that the system $\{e_1, \ldots, e_n\}$ is orthogonal, i.e. the following relationship holds:

$$(e_i, e_k) = \begin{cases} (e_i, e_i), & i = k; \\ 0, & i \neq k. \end{cases}$$

If the system is $\{e_1, \ldots, e_n\}$ *orthonormalized*, i.e., $(e_i, e_i) = 1$, then formula (1.6.4a) takes a simpler form:

$$A_k = (a, e_k). \tag{1.6.6}$$

Representation (1.6.4) of the element a in linear space $\mathcal{LS}$, $a \in \mathcal{LS}$ with scalar product in the form of linear combination of basis elements $\{e_1, \ldots, e_n\}$ is called *generalized Fourier series* decomposition [59, 67].

Example 1.6.1. Within signal processing problems, as a base transform that provide transferring from time domain (space X) into frequency one (space X'), a pair of discrete cosine transform (DCT) is often used [68–70]:

$$A(u) = \frac{2}{\sqrt{2n}} \sum_{i=0}^{n-1} C(u) a(i) \cos\left(\frac{(2i+1)\pi u}{2n}\right) \Leftrightarrow A = DCT(a); \tag{1.6.7a}$$

$$a(i) = \frac{2}{\sqrt{2n}} \sum_{u=0}^{n-1} C(u) A(u) \cos\left(\frac{(2i+1)\pi u}{2n}\right) \Leftrightarrow a = DCT^{-1}(A), \tag{1.6.7b}$$

where $C(u) = \begin{cases} 1/\sqrt{2}, & u = 0; \\ 1, & u \neq 0. \end{cases}$

Transform (1.6.7) possesses the property of isomorphism between two coupled linear spaces X, X' with scalar product and preserves scalar product between the elements $a, b \in X$; $A, B \in X'$:

$$(a, b) = (A, B); \tag{1.6.8}$$

$$X \underset{DCT^{-1}}{\overset{PCT}{\longleftrightarrow}} X', \ a \underset{DCT^{-1}}{\overset{PCT}{\longleftrightarrow}} A, \ b \underset{DCT^{-1}}{\overset{PCT}{\longleftrightarrow}} B. \tag{1.6.8a}$$

Analogously, the property of preserving scalar product (1.6.8) between the elements $a, b \in X$ coupled by the mapping (1.6.7), (1.6.8a) implies also preservation of norms $\|a\|$, $\|b\|$ of the elements $a, b \in X$:

$$\|a\|^2 = (a, a) = \|A\|^2 = (A, A); \tag{1.6.9a}$$

$$\|b\|^2 = (b, b) = \|B\|^2 = (B, B), \tag{1.6.9b}$$

and metric $\|a - b\|$ between the elements $a, b \in X$:

$$\|a - b\|^2 = (a - b, a - b) = \|A - B\|^2 = (A - B, A - B). \tag{1.6.10}$$

$\triangledown$

1.6.2 Hyperspectral Representation of Deterministic Signals Based on Generalized Metrics in *L*-group

Introducing linear space $\mathcal{LS}$ with scalar product in a set of signals, there appears a possibility of using such a characteristic as an angle φ_{ab} between two signals $a, b \in \mathcal{LS}$:

$$\cos(\varphi_{ab}) = \frac{(a, b)}{\sqrt{(a, a) \cdot (b, b)}}. \tag{1.6.11}$$

From the very beginning of the chapter, we agreed on considering the signals only with even univariate PDFs (and hence, with zero mean), unless otherwise stated. In this case, as applied to realizations of stochastic signals $x(t)$, $y(t)$, cosine of an angle between them (1.6.11) coincide with the notion of sample correlation coefficient $r(x, y)$:

$$r(x, y) = \frac{\sum\limits_{i=1}^{n} x_i y_i}{\sqrt{\left(\sum\limits_{i=1}^{n} x_i^2\right) \cdot \left(\sum\limits_{i=1}^{n} y_i^2\right)}}, \tag{1.6.11a}$$

where $x = (x_1, x_2, \ldots, x_n)$, $y = (y_1, y_2, \ldots, y_n)$ are realizations (observations) of random sample X, Y of stochastic signals $x(t)$, $y(t)$: $x(t_i) = x_i$, $y(t_i) = y_i$; $i = 1, 2, \ldots, n$.

Here we use known signal representation (1.6.6) in orthonormalized basis, and on the other hand, formulated idea on MSIs oddness $N_p(x, y)$ (1.4.10), $N_s(x, y)$ (1.4.11), $N_{l_1}(x, y)$ (1.4.12) to introduce the notion of *hyperspectral representation of signals*.

Definition 1.6.3. *Hyperspectral representation* $A(u)$ of discrete signal $a(i)$, $i = 0, 1, \ldots, n-1$ in sample space $\mathcal{L}(\mathcal{X}, \mathcal{B}_\mathcal{X}; +, \vee, \wedge)$ with L-group properties: $a(i) \in \mathcal{L}(\mathcal{X}, \mathcal{B}_\mathcal{X}; +, \vee, \wedge)$ is a mapping **HST** based on generalized MSI $\mathrm{N}(x, y)$ (1.4.9) and defined by the corresponding relationship:

$$A(u) = \mathbf{HST}[a(i)] = \mathrm{N}(a(i), \mathrm{cosF}(i, u)), \tag{1.6.12}$$

$$\mathrm{cosF}(i, u) = \cos\left(\frac{(2i+1)\pi u}{2n}\right). \tag{1.6.12a}$$

Then, depending on a type of sample space, we differ the following kinds of hyperspectral representation of discrete signals:

$$A_p(u) = n - \sum_{i=0}^{n-1} [\mathrm{sgn}(a(i) \vee \mathrm{cosF}(i, u)) - \mathrm{sgn}(a(i) \wedge \mathrm{cosF}(i, u))]; \tag{1.6.13a}$$

$$A_s(u) = \sum_{i=0}^{n-1} (|a(i)| + |\mathrm{cosF}(i, u)|) - \sum_{i=0}^{n-1} \{|a(i) - \mathrm{cosF}(i, u)|$$
$$\times [\mathrm{sgn}(a(i) \vee \mathrm{cosF}(i, u)) - \mathrm{sgn}(a(i) \wedge \mathrm{cosF}(i, u))]\}; \tag{1.6.13b}$$

$$A_{l_1}(u) = \sum_{i=0}^{n-1} (|a(i) + \mathrm{cosF}(i, u)| - |a(i) - \mathrm{cosF}(i, u)|). \tag{1.6.13c}$$

More exact term for the mappings (1.6.13a,b,c) is discrete hyperspectral cosine representation, however, hereinafter, we use more short term that figures in the definition.

Here, the result of hyperspectral representation $A_p(u)$, $A_s(u)$, $A_{l_1}(u)$ (1.6.13a,b,c) is called *hyperspectrum of signal* $a(i)$, $i = 0, 1, \ldots, n-1$, bearing in mind that hyperspectral representation is a nonlinear mapping, unlike, for instance, DCT and a lot of other linear mappings that allow obtaining signal spectrum for its following analysis and/or processing.

Hereinafter, subscript indexes "p, s, l_1" mean that corresponding notation relates to pseudometric (1.3.25), semimetric (1.3.26), and metric (1.3.27), respectively. Calculating hyperspectral representations (1.6.13a,b,c) is realized without exploiting operation of multiplication that is doubtless advantage when organizing calculations, first, in applications that do not assume exploiting considerable computational resources, and second, in applications requiring high computational rate.

The relationships (1.6.13) defining hyperspectral representation $A_p(u)$, $A_s(u)$, $A_{l_1}(u)$ of discrete signal $a(i)$, $i = 0, 1, \ldots, n-1$ in sample space $\mathcal{L}(\mathcal{X}, \mathcal{B}_\mathcal{X}; +, \vee, \wedge)$ with L-group properties $a(i) \in \mathcal{L}(\mathcal{X}, \mathcal{B}_\mathcal{X}; +, \vee, \wedge)$ can be written in brief form:

$$A_p(u) = \mathrm{N}_p(a(i), \mathrm{cosF}(i, u)) \Leftrightarrow A_p = \mathbf{HST}_p(a); \tag{1.6.14a}$$

$$A_s(u) = \mathrm{N}_s(a(i), \mathrm{cosF}(i, u)) \Leftrightarrow A_s = \mathbf{HST}_s(a); \tag{1.6.14b}$$

$$A_{l_1}(u) = \mathrm{N}_{l_1}(a(i), \mathrm{cosF}(i, u)) \Leftrightarrow A_{l_1} = \mathbf{HST}_{l_1}(a), \tag{1.6.14c}$$

where $N_p(x, y)$, $N_s(x, y)$, $N_{l_1}(x, y)$ are MSIs defined by the relationships (1.4.10), (1.4.11), and (1.4.12), respectively; $\cos F(i, u)$ is the function determined by (1.6.12a).

The sense of relationships (1.6.14a,b,c) lies in the fact that hyperspectral representation of signals $\mathbf{HST}_p$, $\mathbf{HST}_s$, $\mathbf{HST}_{l_1}$ supposes decomposition of an initial signal $a(i)$ over a series of basis functions (1.6.13), so that decomposition coefficients $A_p(u)$, $A_s(u)$, $A_{l_1}(u)$ are calculated on the basis of corresponding statistics (1.6.14a,b,c) in the form of MSI $N_p(x, y)$ (1.4.10), $N_s(x, y)$ (1.4.11), $N_{l_1}(x, y)$ (1.4.12) between a sample of signal and basis function (1.6.12a).

Here we should discuss the following feature. Basis function $\cos F(i, u)$ (1.6.12a), being orthogonal in linear sample space

$$(\cos F(i, u_l), \cos F(i, u_m)) = 0, \ u_l \neq u_m,$$

is not orthogonal in terms of sample space $\mathcal{L}(\mathcal{X}, \mathcal{B}_{\mathcal{X}}; +, \vee, \wedge)$ with L-group properties and generalized metrics $M_p(x, y)$ (1.3.25), $M_s(x, y)$ (1.3.26), $M_{l_1}(x, y)$ (1.3.27), i.e., in general case, the following relationships hold:

$$\nu_p \left[\cos F(i, u_l), \cos F(i, u_m)\right] \neq 0, \ u_l \neq u_m;$$

$$\nu_s \left[\cos F(i, u_l), \cos F(i, u_m)\right] \neq 0, \ u_l \neq u_m;$$

$$\nu_{l_1} \left[\cos F(i, u_l), \cos F(i, u_m)\right] \neq 0, \ u_l \neq u_m.$$

Taking into account the fact that, according to Theorem 1.4.1, NMSI (1.4.1) is invariant of a group of continuous mappings, this theorem has a corollary with respect to hyperspectral representation of signals (1.6.14a).

Corollary 1.6.1. *Hyperspectral representation $A_p(u)$ (1.6.14a) of discrete signal $a(i)$ in the sample space $\mathcal{L}(\mathcal{X}, \mathcal{B}_{\mathcal{X}}; +, \vee, \wedge)$ with L-group properties: $a(i) \in \mathcal{L}(\mathcal{X}, \mathcal{B}_{\mathcal{X}}; +, \vee, \wedge)$ is an invariant of a group $H = \{h_\alpha\}$ of odd mappings:*

$$A_p(u) = N_p(a(i), \cos F(i, u)) = N_p(h_\alpha[a(i)], \cos F(i, u)) = A_\alpha(u); \qquad (1.6.15)$$

$$A_p(u) \underset{h_\alpha^{-1}}{\overset{h_\alpha}{\longleftrightarrow}} A_\alpha(u), \ h_\alpha[z] = -h_\alpha[-z].$$

Elucidate the introduced above notion of hyperspectral representation by the following example.

Example 1.6.2. Consider two radiofrequency (RF) signals with rectangular envelopes $s_1(i)$, $s_2(i)$ shown in Fig. 1.6.1a by solid and dashed lines, respectively; the first one is RF biphase-modulated signal in the form of 7-element Barker code signal [71], and the second one is a harmonic signal (RF pulse). Durations of signals are chosen to be equal, and their frequencies are related as 2:1. Fig. 1.6.1b illustrates the results of DCTs $S_1(u)$, $S_2(u)$ of these signals $s_1(i)$, $s_2(i)$ defined by the formula (1.6.7a), so that $i = 0, 1, \ldots, n - 1$, $u = 0, 1, \ldots, n - 1$, $n = 1024$.

Fig. 1.6.2a and 1.6.2b depict the results of hyperspectral representations $A_{1,p}(u)$, $A_{2,p}(u)$ and $A_{1,s}(u)$, $A_{2,s}(u)$ of these signals $s_1(i)$, $s_2(i)$ determined by the

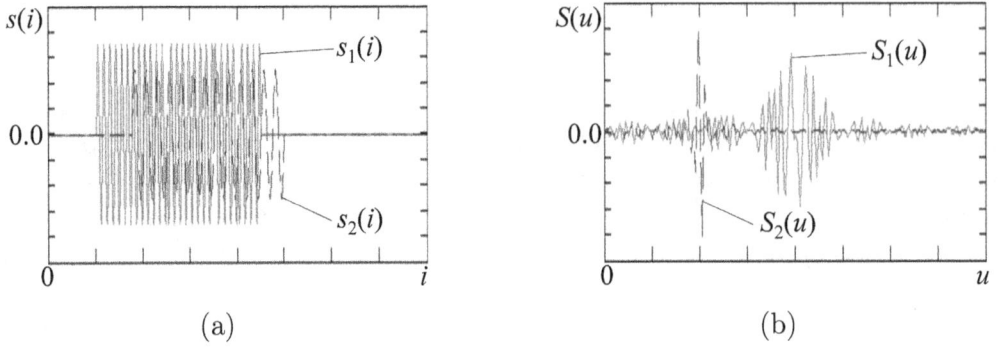

FIGURE 1.6.1   Illustrations of (a) Barker code RF signal $s_1(i)$ and RF pulse $s_2(i)$; (b) DCTs $S_1(u)$, $S_2(u)$ of the signals $s_1(i)$, $s_2(i)$

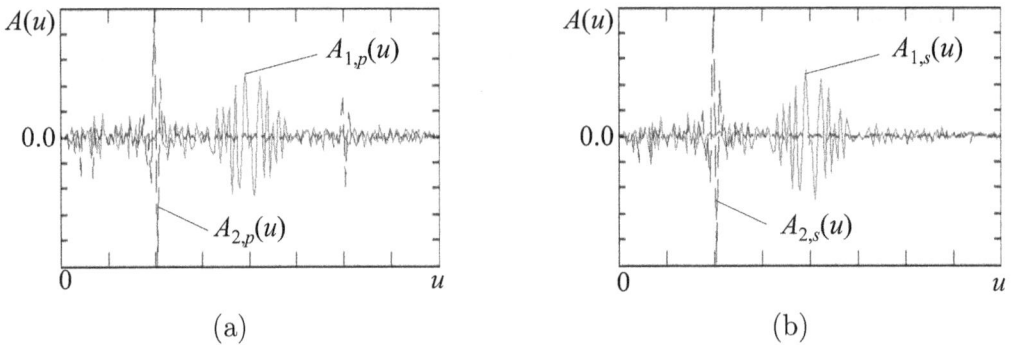

FIGURE 1.6.2   Hyperspectra (a) $A_{1,p}(u)$, $A_{2,p}(u)$ of the signals $s_1(i)$, $s_2(i)$; (b) $A_{1,s}(u)$, $A_{2,s}(u)$ of the signals $s_1(i)$, $s_2(i)$

formulas (1.6.14a) and (1.6.14b), respectively, so that, similarly, $i = 0, 1, \ldots, n-1$, $u = 0, 1, \ldots, n-1$, $n = 1024$. Mentally matching the results shown in Fig. 1.6.2a and 1.6.2b and also the results of DCTs depicted in Fig. 1.6.1b, one can notice the following features. Ratio of central frequencies of hyperspectra of the signals $s_1(i)$, $s_2(i)$ is equal to 2:1. Hyperspectrum $A_{2,p}(u)$ of harmonic signal $s_2(i)$ contains noticeable third harmonic due to odd nonlinear mappings (1.6.14). Correlation coefficients between spectra $S_1(u)$, $S_2(u)$ and hyperspectra $A_{1,p}(u)$, $A_{2,p}(u)$; $A_{1,s}(u)$, $A_{2,s}(u)$ of the signals $s_1(i)$, $s_2(i)$ are equal to: $r[S_1(u), A_{1,p}(u)] = 0.8$; $r[S_2(u), A_{2,p}(u)] = 0.81$; $r[S_1(u), A_{1,s}(u)] = 0.92$; $r[S_2(u), A_{2,s}(u)] = 0.925$. Higher values of correlation coefficients between spectra and hyperspectra that take place on the mapping (1.6.14b), and also absence of noticeable third harmonic in hyperspectrum $A_{1,s}(u)$ of the signal $s_1(i)$ (Fig. 1.6.2b), allow telling on lesser degree of nonlinearity of this mapping as against the mapping (1.6.14a). Corollary 1.6.1 defines an interesting property of hyperspectral mapping (1.6.14a), according to which arbitrary odd functions $h_1[s_1(i)]$ and $h_2[s_2(i)]$ of initial signals $s_1(i)$ and $s_2(i)$ have hyperspectra that are identical to hyperspectra $A_{1,p}(u)$ and $A_{2,p}(u)$, respectively (see Fig. 1.6.2a).

Fig. 1.6.3a and Fig. 1.6.3b show the results of hyperspectral representation $X_p(u)$ and $X_s(u)$ of additive sum $x(i) = s_1(i) + s_2(i)$ of initial signals $s_1(i)$, $s_2(i)$ defined by the formulas (1.6.14a) and (1.6.14b), respectively.

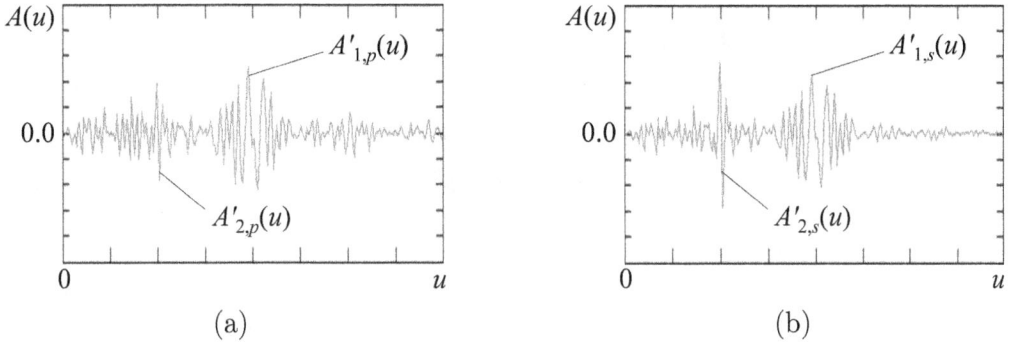

FIGURE 1.6.3 Hyperspectra of additive sum $x(i) = s_1(i) + s_2(i)$ of initial signals $s_1(i)$, $s_2(i)$: (a) $X_p(u)$; (b) $X_s(u)$

As follows from the figures, despite nonlinearity of the mappings (1.6.14a), (1.6.14b), partial hyperspectra $A'_{1,p}(u)$, $A'_{2,p}(u)$ and $A'_{1,s}(u)$, $A'_{2,s}(u)$ of the signals $s_1(i)$, $s_2(i)$ are clearly distinguishable within the corresponding hyperspectra $X_p(u)$ and $X_s(u)$ of additive sum $x(i) = s_1(i) + s_2(i)$. Correlation coefficients between hyperspectra $X_p(u)$ and $X_s(u)$ and sum of linear spectra $S_1(u)$, $S_2(u)$ obtained by DCT, respectively, are equal to: $r[S_1(u) + S_2(u), X_p(u)] = 0.78$; $r[S_1(u) + S_2(u), X_s(u)] = 0.916$.

Here a phenomenon of *quasi-linearity* takes place, which is expressed by the following relationships:

$$\mathbf{HST}_p(s_1 + s_2) \approx \mathbf{HST}_p(s_1) + \mathbf{HST}_p(s_2);$$

$$\mathbf{HST}_s(s_1 + s_2) \approx \mathbf{HST}_s(s_1) + \mathbf{HST}_s(s_2).$$

In more detail, quasi-linearity will be discussed in the following section. ▽

Taking into account, first, quite satisfactory (in adequacy degree) representation of discrete signals in frequency domain based on nonlinear mappings (1.6.14a), (1.6.14b), (1.6.14c), and second, a simplicity of calculating the corresponding MSIs that do not require performing operation of multiplication, unlike known linear transforms that allow transferring from time to frequency form of signal representation, one can conclude that algorithms of hyperspectral representation of discrete signals can be used in various applications of signal processing despite their nonlinearity.

1.6.3 Estimating Hyperspectral Density of Stochastic Signal on L-groups

Development of spectral methods of signal processing related to stochastic signals requires introducing a special notion, which would correspond to a representation of a signal in frequency domain and describe statistical properties of a signal in time domain. Such a characteristic of stochastic signal was introduced in Section 1.4. It creates a basis on which we introduce a function that is a substantial analogue of power spectral density (PSD) of a stochastic process.

Definition 1.6.4. *Estimator of hyperspectral density* $\hat{H}(f)$ of complex discrete stochastic signal $x(i)$ is a function of MSI estimator $\hat{N}(x; m)$ (1.4.9) that is defined by the following relationship:

$$\hat{H}(f) = N(\text{Re}[\hat{N}(x; m)], \text{cosF}(f, m)) + N(\text{Im}[\hat{N}(x; m)], \text{sinF}(f, m)); \qquad (1.6.16)$$

$$\text{cosF}(f, m) = \cos(2\pi f \cdot mT/N); \ \text{sinF}(f, m) = \sin(2\pi f \cdot mT/N); \qquad (1.6.16a)$$

where N(a, b) is MSI of signals a, b; T is a duration of complex-valued stochastic process $x(i)$; N is a number of samples $\{x(i)\}$ of stochastic process $x(t)$ taken in the interval $[0, T]$; $T/(N-1) = \Delta t$; Δt is a sampling interval; f is discrete frequency parameter.

Then, depending on a type of sample space, we distinguish the following kinds of hyperspectral density estimators:

$$\hat{H}_p(f) = N_p(\text{Re}[\hat{N}_p(x; m)], \text{cosF}(f, m)) + N_p(\text{Im}[\hat{N}_p(x; m)], \text{sinF}(f, m)); \quad (1.6.17)$$

$$N_p[a(m), b(m)] = L - \sum_{m=0}^{L-1} [\text{sgn}(a(m) \vee b(m)) - \text{sgn}(a(m) \wedge b(m))]; \qquad (1.6.17a)$$

$$\hat{H}_s(f) = N_s(\text{Re}[\hat{N}_s(x; m)], \text{cosF}(f, m)) + N_s(\text{Im}[\hat{N}_s(x; m)], \text{sinF}(f, m)); \quad (1.6.18)$$

$$N_s[a(m), b(m)] = \sum_{m=0}^{L-1} |a(m)| + \sum_{m=0}^{L-1} |b(m)|$$

$$- \sum_{m=0}^{L-1} |a(m) - b(m)| \cdot [\text{sgn}(a(m) \vee b(m)) - \text{sgn}(a(m) \wedge b(m))]; \qquad (1.6.18a)$$

$$\hat{H}_{l_1}(f) = N_{l_1}(\text{Re}[\hat{N}_{l_1}(x; m)], \text{cosF}(f, m)) + N_{l_1}(\text{Im}[\hat{N}_{l_1}(x; m)], \text{sinF}(f, m));$$
$$(1.6.19)$$

$$N_{l_1}[a(m), b(m)] = \sum_{m=0}^{L-1} (|a(m) + b(m)| - |a(m) - b(m)|); \qquad (1.6.19a)$$

where L is maximal index of parameter m, $L << N$; $a(m)$, $b(m)$ are arbitrary function.

Usefulness of the introduced characteristic is illustrated by the following example.

Example 1.6.3. Let $x(i)$ be additive mixture of M complex discrete harmonic signals observed in the presence of complex quasi-white Gaussian noise $n(i)$ with zero mean (1.4.20). In the considered example, amplitudes of harmonics are chosen to be equal $A_l = A$, so that the relationship holds: $A^2/2 = D_n$, D_n is a variance of noise $n(i)$; $F_n > 2\max_{l}\{f_l\}$, F_n is upper bound frequency of PSD of Gaussian noise $n(i)$; $M = 6$. The number n of samples of stochastic process $x(i)$ used when

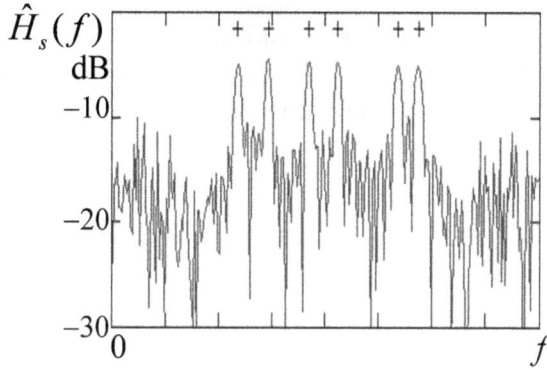

FIGURE 1.6.4   Hyperspectral density estimate $\hat{H}_s(f)$ (1.6.18) of the signal $x(i)$ (1.4.20), $M = 6$

forming MSI estimator $\hat{N}_s(x; m)$ (1.4.15) is equal to $n = 1024$. Maximal index L of parameter m figuring in the formula (1.6.17) is chosen to be equal to $L = 128$.

Fig. 1.6.4 depicts hyperspectral density estimate $\hat{H}_s(f)$ (1.6.18) of the signal $x(i)$ (1.4.20) that corresponds to the situation described above. True locations of harmonic signal frequencies are denoted by "+". $\triangledown$

In more detail spectral estimation methods based on L-group algorithms are considered in Chapter 5.

1.7   Digital Filtering Based on L-group Operations

Theory of linear digital filtering extends all main results of theory of linear systems related to continuous signals to the case of discrete signals. Linear stationary system transforms input continuous signal $x(t)$ in such a way that on its output there appears a signal $y(t)$ defined by operation of convolution between input signal $x(t)$ and impulse response $h(t)$ of the filter:

$$y(t) = F[x(t); h(t)] = \int_{-\infty}^{\infty} x(\tau)h(t - \tau)d\tau, \qquad (1.7.1)$$

so that the superposition principle and condition of time shift invariance hold, respectively:

$$F[\alpha_1 x_1(t) + \alpha_2 x_2(t); h(t)] = \alpha_1 F[x_1(t); h(t)] + \alpha_2 F[x_2(t); h(t)]; \qquad (1.7.2a)$$

$$F[\delta(t - t_0); h(t)] = h(t - t_0), \qquad (1.7.2b)$$

where $\delta(t)$ is Dirac delta function.

According to the definition, *linear digital filter* is a discrete system (either physical device or algorithm) which transforms a sequence of the samples $x(i) = (x(0), x(1), \ldots, x(n-1))$, $i = 0, 1, \ldots, n-1$ of discrete input signal into the sequence of samples of discrete output signal $y(i)$, thus, this filter is considered to be a linear and stationary system:

$$y(i) = F[x(i); h(i)] = \sum_{k=-\infty}^{\infty} x(k)h(i-k), \qquad (1.7.3)$$

where impulse response $h(i)$ is defined as a reaction of a filter to unit impulse:

$$F[\delta(i - i_0); h(i)] = h(i - i_0), \qquad (1.7.4)$$

$$\delta(i - i_0) = \begin{cases} 1, & i = i_0; \\ 0, & i \neq i_0. \end{cases}$$

The expression (1.7.3) is called *discrete linear convolution*. For physically realizable system $h(i) = 0$ if $i < 0$, hence the upper limit of summation in the formula (1.7.3) can be replaced by i and the lower limit by zero:

$$y(i) = F[x(i); h(i)] = \sum_{k=0}^{i} x(k)h(i-k). \qquad (1.7.5)$$

This means that when calculating the next sample the system can operate only with the last values of the input signal and knows nothing concerning future samples of the signal.

Formula (1.7.5), playing a leading role in theory of linear digital filtering, shows that output sequence $y(i)$ is a discrete convolution between the input signal $x(i)$ and impulse response $h(i)$ of the filter.

The subjects of further consideration are stationary nonlinear digital filters performing an operation that corresponds to the convolution (1.7.5) but is based on generalized metrics $M_p(x, y)$ (1.3.25), $M_s(x, y)$ (1.3.26), $M_{l_1}(x, y)$ (1.3.27).

In order to introduce the notions of *digital hyperfilter* and *hyperconvolution* of signals, we use the known representation of a signal in the output of linear digital filter (1.7.5), and, on the other hand, formulated idea on oddness of MSIs $N_p(x, y)$ (1.4.10), $N_s(x, y)$ (1.4.11), $N_{l_1}(x, y)$ (1.4.12). Notice, that normalized generalized metrics $\mu_p(x, y)$ (1.3.9), $\mu_s(x, y)$ (1.3.23), $\mu_{l_1}(x, y)$ (1.3.24) related to aforementioned MSIs meet the requirements (1.3.21), (1.3.22) (see Figs. 1.3.1, 1.4.1).

Definition 1.7.1. *Digital hyperfilter* $F_p[x(i); h(i)]$ $(F_s[x(i); h(i)], F_{l_1}[x(i); h(i)])$ of a discrete signal $x(i)$, $i = 0, 1, \ldots, n - 1$ in sample space $\mathcal{L}(\mathcal{X}, \mathcal{B}_{\mathcal{X}}; +, \vee, \wedge)$ with L-group properties: $x(i), h(i) \in \mathcal{L}(\mathcal{X}, \mathcal{B}_{\mathcal{X}}; +, \vee, \wedge)$ is a mapping based on one of MSIs $N_p(x, y)$ (1.4.10), $(N_s(x, y)$ (1.4.11), $N_{l_1}(x, y)$ (1.4.12)) and defined by the corresponding relationship:

$$z_p(i) = F_p[x(i); h(i)]$$

$$= \sum_{k=0}^{i} [1 - (\text{sgn}(x(k) \vee h(i-k)) - \text{sgn}(x(k) \wedge h(i-k)))] \cdot |\text{sgn}(h(i-k))|;$$

(1.7.6a)

$$z_s(i) = F_s[x(i); h(i)] = \sum_{k=0}^{i} (|x(k)| + |h(i-k)|)$$

$$- \sum_{k=0}^{i} (\text{sgn}(x(k) \vee h(i-k)) - \text{sgn}(x(k) \wedge h(i-k))) \cdot |x(k) - h(i-k)|; \quad (1.7.6b)$$

$$z_{l_1}(i) = F_{l_1}[x(i); h(i)] = \sum_{k=0}^{i} [|x(k) + h(i-k)| - |x(k) - h(i-k)|], \quad (1.7.6c)$$

where, in general case, $h(i)$ is an arbitrary function (and it is not obligatory an impulse response), and a multiplier $|\text{sgn}(h(i-k))|$ in (1.7.6a) is used to provide physical realizability of a filter.

The subscript indices "p, s, l_1" in notations, as before, mean relation to generalized metrics (1.3.25), (1.3.26), (1.3.27), respectively.

The results of digital filtering $z_p(i)$ (1.7.6a), $z_s(i)$ (1.7.6b), $z_{l_1}(i)$ (1.7.6c) are called *hyperconvolutions* between the functions $x(i)$ and $h(i)$ bearing in mind that $F_p[x(i); h(i)]$, $F_s[x(i); h(i)]$, $F_{l_1}[x(i); h(i)]$ are nonlinear mappings unlike linear convolution (1.7.5), hence superposition principle (1.7.2a) does not hold.

Meanwhile, as for some nonlinear filters, superposition principle holds approximately that allow us to introduce the following useful notion.

Definition 1.7.2. Digital hyperfilter $F[x(i); h(i)]$ is said to be *quasi-linear*, if within a given application the relationship

$$F[x_1(i) + x_2(i); h(i)] \approx F[x_1(i); h(i)] + F[x_2(i); h(i)] \quad (1.7.7)$$

holds with a required accuracy.

It should be noted, that, in general case, the stationarity condition (1.7.2b) for digital hyperfilter (1.7.6) does not hold and is written in less rough formulation:

$$z(i) = F[x(i); h(i)] \Rightarrow z(i - i_0) = F[x(i - i_0); h(i)], \quad (1.7.8)$$

so that the following implication and inequality hold:

$$F[\delta(i); h(i)] = h'(i) \Rightarrow F[\delta(i - i_0); h(i)] = h'(i - i_0), \ h(i) \neq h'(i),$$

where, for instance, as regards hyperfilter (1.7.6a) $h'(i) = \text{sgn}(h(i))$ and for hyperfilter (1.7.6b) $h'(i) \approx h(i)$.

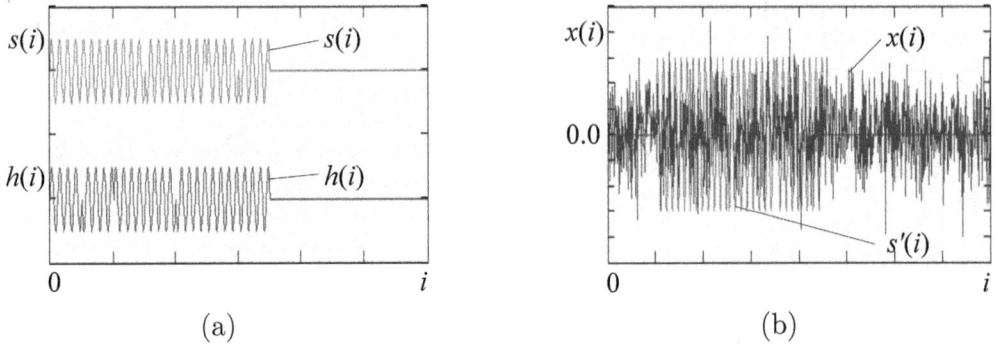

FIGURE 1.7.1   Illustrations of (a) RF Barker code signal $s(i)$ and impulse response $h(i)$ of matched filter (1.7.9); (b) delayed copy $s'(i) = s(i - i_0)$ of signal $s(i)$ and additive mixture $x(i)$

Taking into account the fact that the difference of sign functions figuring in the formulas (1.7.6a,b) takes the values in the set $\{0, 1, 2\}$, and multiplier $|\operatorname{sgn}(h(i-k))|$ in (1.7.6a) takes the values in the set $\{0, 1\}$, nonlinear digital filtering can be performed without using operation of multiplication, that is a doubtless advantage when organizing calculations, first, in applications that do not assume using considerable computational resources, and second, in applications that require high computational rate.

The notion of digital hyperfilter introduced by the Definition 1.7.1 is illustrated by the following example.

Example 1.7.1. Consider RF biphase-modulated signal $s(i)$ with a rectangular envelope in the form of 7-element Barker code signal [71] shown in Fig. 1.7.1a (in the upper part of the figure), and also impulse response $h(i)$ of a matched filter (in the lower part of the figure), which corresponds to this signal and related to the latter by known relationship (see, for instance, [72, (5.69)]):

$$h(i) = s(T - i), \tag{1.7.9}$$

where T is a signal duration, $T=448$; $i = 0, 1, \ldots, n - 1$; $n=1024$.

Fig. 1.7.1b depicts a delayed copy $s'(i) = s(i - i_0)$ of the signal $s(i)$ and also additive mixture $x(i)$ of the signal $s'(i)$ and quasi-white Gaussian noise $n(i)$: $x(i) = s'(i) + n(i)$. Here, for convenience of visual perception, the signals $x(i)$ and $s'(i)$ are shown in different amplitude scale. Signal-to-noise ratio (SNR) $q^2 = E/N_0$ is equal to $q^2 = 20$, where E, N_0 are a signal energy and noise PSD, respectively.

Fig. 1.7.2a illustrates the signal $y(i)$ in the output of digital matched filter performing operation of linear discrete convolution $y(i) = F[x(i); h(i)]$ (1.7.5) between additive mixture $x(i)$ and impulse response $h(i)$ (1.7.9), and also delayed useful signal $s'(i)$. Here, similarly, for convenience of visual perception, the signals $y(i)$ and $s'(i)$ are shown in different amplitude scale. Fig. 1.7.2b demonstrates the signal $y_0(i)$ in the output of digital matched filter performing operation of linear discrete convolution $y_0(i) = F[s'(i); h(i)]$ (1.7.5) between useful signal $s'(i)$ and impulse response $h(i)$ (1.7.9).

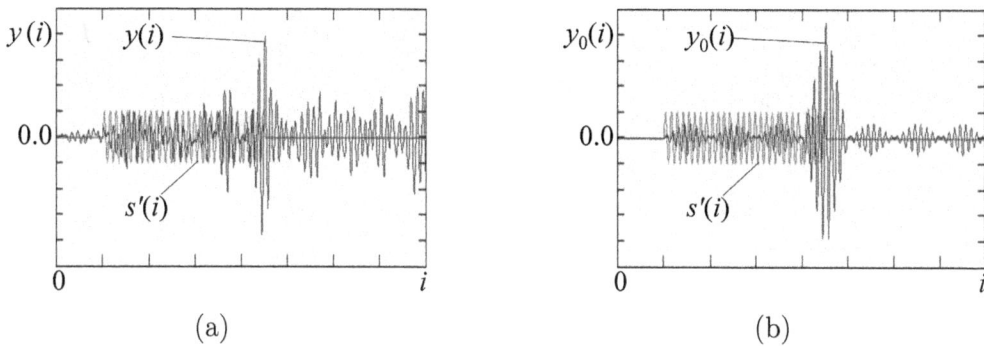

FIGURE 1.7.2   Illustrations of the signals in the output of linear digital matched filter (a) $y(i)$ in the presence of noise; (b) $y_0(i)$ in the absence of noise

Fig. 1.7.3a depicts the signal $z_p(i)$ in the output of digital hyperfilter performing operation of hyperconvolution $z_p(i) = F_p[x(i); h(i)]$ (1.7.6a) between additive mixture $x(i) = s'(i) + n(i)$ and impulse response $h(i)$ (1.7.9), and also the useful signal $s'(i)$. Here, analogously, for convenience of visual perception, the signals $z_p(i)$ and $s'(i)$ are shown in different amplitude scales. Fig. 1.7.3b depicts the signal $z_{p,0}(i)$ in the output of digital hyperfilter $z_{p,0}(i) = F_p[s'(i); h(i)]$ (1.7.6a) performing operation of hyperconvolution between delayed useful signal $s'(i)$ and impulse response $h(i)$ (1.7.9).

Fig. 1.7.4a illustrates the signal $z_s(i)$ in the output of digital hyperfilter performing operation of hyperconvolution $z_s(i) = F_s[x(i); h(i)]$ (1.7.6b) between additive mixture $x(i) = s'(i) + n(i)$ and impulse response $h(i)$ (1.7.9), and also delayed useful signal $s'(i)$. Here, similarly, the signals $z_s(i)$ and $s'(i)$ are shown in different amplitude scales. Fig. 1.7.4b demonstrates the signal $z_{s,0}(i)$ in the output of digital hyperfilter performing operation of hyperconvolution $z_{s,0}(i) = F_s[s'(i); h(i)]$ (1.7.6b) between delayed useful signal $s'(i)$ and impulse response $h(i)$ (1.7.9).

$\triangledown$

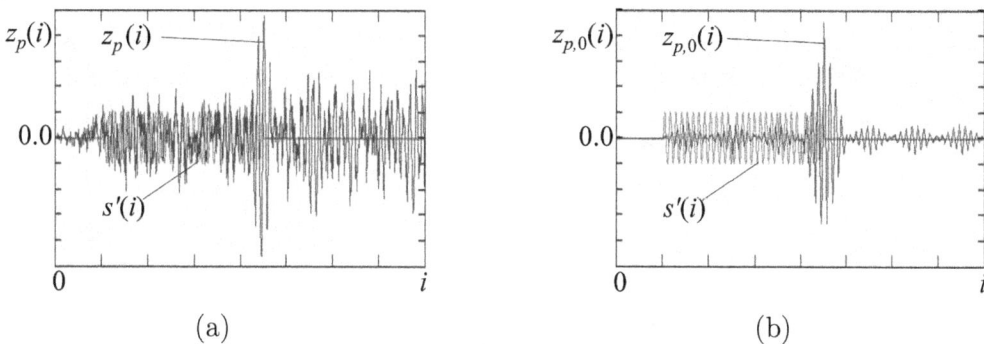

FIGURE 1.7.3   Illustrations of the signals in the output of digital hyperfilter (a) $z_p(i)$ in the presence of noise; (b) $z_{p,0}(i)$ in the absence of noise

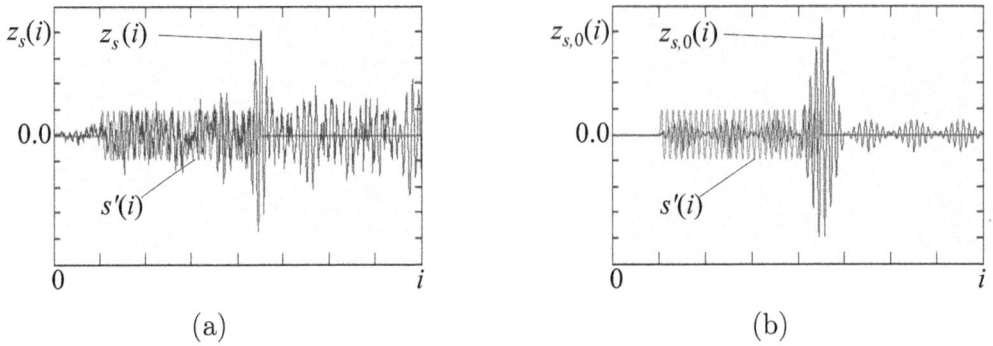

FIGURE 1.7.4 Illustrations of the signals in the output of digital hyperfilter (a) $z_s(i)$ in the presence of noise; (b) $z_{s,0}(i)$ in the absence of noise

Comparing the results of digital filtering shown in Figs. 1.7.3a,b and 1.7.4a,b with well known classic results shown in Fig. 1.7.2a,b obtained under the same initial conditions, one can conclude that digital hyperfilters based on algorithms of signal processing in spaces with L-group properties (1.7.6a,b) allow us effectively to solve the problems that for a long time considered to be a prerogative of linear filters. L-group algorithms (1.7.6) do not require performing operation of multiplication that makes these algorithms attractive when organizing calculations, first, in applications that do not assume exploiting considerable computational resources, and second, in applications requiring high computational rate. Questions, related with efficiency of digital hyperfilters for solving signal processing problems, such as, for instance, signal detection, signal extraction in the presence of interference (noise), and other, require corresponding research.

1.8   Statistical Demultiplexing Based on L-group Operations

Let $\{e_1(t), \ldots, e_N(t)\}$ be a system of N orthogonal elements in linear space $\mathcal{LS}$ with scalar product and $a(t)$ be some element of this space, $a(t) \in \mathcal{LS}$. Return to the representation of an element $a(t)$ in linear space $\mathcal{LS}$, $a(t) \in \mathcal{LS}$ with scalar product that is based on the system of orthogonal basis elements $\{e_1(t), \ldots, e_N(t)\}$ (1.6.4):

$$a(t) = \sum_{k=1}^{N} A_k(t)e_k(t); \tag{1.8.1}$$

$$t \in T_s = [0, T];$$

$$A_k(t) = (a(t), e_k(t))/(e_k(t), e_k(t)); \tag{1.8.2}$$

$$(e_l(t), e_k(t)) = \begin{cases} (e_l(t), e_l(t)), & l = k; \\ 0, & l \neq k. \end{cases}$$

where $\{A_k(t)\}$ are coefficients of decomposition of the element $a(t)$ over orthogonal basis elements $\{e_1(t),\ldots,e_N(t)\}$; $(x(t),y(t)) = \int\limits_{T_s} x(t) \cdot y(t)dt$ is scalar product of the elements $x(t)$, $y(t) \in \mathcal{LS}$.

Fundamental relationship (1.8.1) allows using it both in spectral analysis of signals and in various other applications of Signal Processing Theory. In this section we discuss the questions related with exploiting a pair of mappings (1.8.1), (1.8.2) for communication channel demultiplexing. Really, if, for instance, we interpret a set $\{A_k(t)\}$, $k = 1,\ldots,N$ as the elements of a message (symbols) that are formed in time interval $T_s = [0, T]$ simultaneously in N channels and united into a group signal $a(t)$ according to the expression (1.8.1) for a following transmitting. Then, in the receiving side (ignoring for a while the presence of interference) in the k-th receiving channel, according to the formula (1.8.2), one can extract the corresponding element of the message $A_k(t)$. If in such multichannel system with channel multiplexing/demultiplexing, the aforementioned orthonormalized basis $\{e_1(t),\ldots,e_N(t)\}$ is used, that is $(e_k(t),e_k(t)) = 1$, then the formula (1.8.2) takes the simpler form:

$$A_k(t) = (a(t), e_k(t)). \tag{1.8.3}$$

To formulate an approach to *statistical demultiplexing*[†] in data receiving channels, we use, in the one hand, known signal representation (1.8.3) in orthonormalized basis, and on the other hand, formulated idea on oddness of MSIs $N_p(x,y)$ (1.4.10), $N_s(x,y)$ (1.4.11), $N_{l_1}(x,y)$ (1.4.12) (see Definition 1.4.5).

Taking into account the fact that MSIs $N_p(x,y)$ (1.4.10), $N_s(x,y)$ (1.4.11), $N_{l_1}(x,y)$ (1.4.12) are nonlinear functions that can negatively affect on extracting arbitrary symbols $\{A_k(t)\}$, we introduce additional constraints on the elements of a transmitted message. Thus, we assume that the symbols of transmitted message $\{A_k(t)\}$, first, are not functions of time $A_k(t) = A_k$, and second, take the values in the set $A_k \in \{-1, 1\}$, thus, these symbols take discrete values.

By the analogy with the approach for linear space with scalar product which is defined by the relationships (1.8.1), (1.8.2), we formulate an idea on representation of discrete group signal $a(i)$, $i = 0, 1,\ldots,n-1$ for information transmitting system with channel multiplexing/demultiplexing in the sample space $\mathcal{L}(\mathcal{X},\mathcal{B}_{\mathcal{X}}; +,\vee,\wedge)$ with L-group properties: $a(i) \in \mathcal{L}(\mathcal{X},\mathcal{B}_{\mathcal{X}}; +,\vee,\wedge)$ in the form of linear combination of orthogonal (in the sense of Definitions 1.4.2, 1.4.3, or 1.4.4) basis functions $\{\psi_1(i),\ldots,\psi_N(i)\}$:

$$a(i) = \sum_{k=1}^{N} a_k(i) = \sum_{k=1}^{N} A_k\psi_k(i); \tag{1.8.4}$$

$$\nu_p(\psi_l(i),\psi_k(i)) = \begin{cases} 1, & l = k; \\ 0, & l \neq k, \end{cases} \tag{1.8.5a}$$

$$\nu_s(\psi_l(i),\psi_k(i)) = \begin{cases} 1, & l = k; \\ 0, & l \neq k, \end{cases} \tag{1.8.5b}$$

[†]This term has nothing to do with known notion of "statistical multiplexing"

$$\nu_{l_1}(\psi_l(i), \psi_k(i)) = \begin{cases} 1, & l = k; \\ 0, & l \neq k, \end{cases} \qquad (1.8.5c)$$

where $\{A_k\}$ are the symbols of a message that are transmitted at the same time in the channels of the system by a channel signal $a_k(i)$ and take their values in the set $A_k \in \{-1, 1\}$; $a_k(i)$ is a channel signal that is formed as a result of multiplication between a symbol A_k and corresponding basis function $\psi_k(i)$: $a_k(i) = A_k\psi_k(i)$; $\nu_p(x(i), y(i))$ is NMSI (1.4.6); $\nu_s(x(i), y(i))$ is NMSI (1.4.7); $\nu_{l_1}(x(i), y(i))$ is NMSI (1.4.8), $x(i), y(i) \in \mathcal{LS}$.

Definition 1.8.1. *Statistical demultiplexing* of additive mixture $x(i) = a(i) + n(i)$ between discrete group signal $a(i)$, $i = 0, 1, \ldots, n - 1$ (1.8.4) and noise $n(i)$ in the sample space $\mathcal{L}(\mathcal{X}, \mathcal{B_X}; +, \vee, \wedge)$ with L-group properties: $a(i), n(i), x(i) \in \mathcal{L}(\mathcal{X}, \mathcal{B_X}; +, \vee, \wedge)$ is a mapping that forms an estimator $\hat{A}_k$ of a symbol A_k that is based on one of MSIs $N_p(x, y)$ (1.4.10), $N_s(x, y)$ (1.4.11), $N_{l_1}(x, y)$ (1.4.12) and defined by the following relationship:

$$\hat{A}_{p,k} = \mathrm{sgn}\left(N_p[x(i), \psi_k(i)]\right); \qquad (1.8.6a)$$

$$\hat{A}_{s,k} = \mathrm{sgn}\left(N_s[x(i), \psi_k(i)]\right). \qquad (1.8.6b)$$

$$\hat{A}_{l_1,k} = \mathrm{sgn}\left(N_{l_1}[x(i), \psi_k(i)]\right). \qquad (1.8.6c)$$

where $\hat{A}_{p,k}$, $\hat{A}_{s,k}$, $\hat{A}_{l_1,k}$ are estimators of a symbol A_k of a message which is transmitted by channel signal $a_k(i)$ and take values in the set $A_k \in \{-1, 1\}$; $\psi_k(i)$ is basis function.

The term «statistical» in Definition 1.8.1 underline the feature that a decision on receiving one or another symbol is made on the basis of one or another statistics (1.8.6a,b,c). The kind of channel multiplexing/demultiplexing (FDM, TDM, CDM, ...) is defined by concrete type of basis functions $\{\psi_k(i)\}$.

Hereinafter the subscript indices "p, s, l_1" in notations corresponds to the proper generalized metrics (1.3.25...1.3.27). Calculating MSIs (1.4.10...1.4.12) can be organized without using operation of multiplication, that is a doubtless advantage when performing calculations, first, in applications that do not assume exploiting considerable computational resources, and second, in applications requiring high computational rate.

The relationships (1.8.6) defining statistical demultiplexing of additive mixture $x(i) = a(i) + n(i)$ between discrete group signal $a(i)$, $i = 0, 1, \ldots, n - 1$, (1.8.4) and interference $n(i)$ in the sample space $\mathcal{L}(\mathcal{X}, \mathcal{B_X}; +, \vee, \wedge)$ with L-group properties can be written in the following expanded forms:

$$\hat{A}_{p,k} = \mathrm{sgn}\left(N_p[x(i), \psi_k(i)]\right)$$
$$= \mathrm{sgn}\left(n - \sum_{i=0}^{n-1} [\mathrm{sgn}(x(i) \vee \psi_k(i)) - \mathrm{sgn}(x(i) \wedge \psi_k(i))]\right); \quad (1.8.7a)$$

$$\hat{A}_{s,k} = \text{sgn}\left(N_s[x(i), \psi_k(i)]\right) = \text{sgn}\left(\sum_{i=0}^{n-1} (|x(i)| + |\psi_k(i)|)\right.$$

$$\left. - \sum_{i=0}^{n-1} |x(i) - \psi_k(i)| \cdot [\text{sgn}(x(i) \vee \psi_k(i)) - \text{sgn}(x(i) \wedge \psi_k(i))]\right); \quad (1.8.7\text{b})$$

$$\hat{A}_{l_1,k} = \text{sgn}\left(N_{l_1}[x(i), \psi_k(i)]\right)$$

$$= \text{sgn}\left(\sum_{i=0}^{n-1} (|x(i) + \psi_k(i)| - |x(i) - \psi_k(i)|)\right). \quad (1.8.7\text{c})$$

The sense of relationships (1.8.7a,b,c) can be elucidated by the fact that *statistical demultiplexing* supposes, first, representation of discrete group signal $a(i)$ in the form of linear combination of basis functions (1.8.4), and second, in the receiving side of information transmitting system, the estimators $\hat{A}_{p,k}$, $\hat{A}_{s,k}$, $\hat{A}_{l_1,k}$ of the elements A_k of messages in k-th receiving channel are obtained by using one of the corresponding statistics (1.8.6a,b,c) in the form of sign function of MSIs $N_p(x(i), \psi_k(i))$, $N_s(x(i), \psi_k(i))$, and $N_{l_1}(x(i), \psi_k(i))$ between the signal and basis function with the property defined by the formulas (1.8.5a,b,c).

Elucidate the aforementioned by the following example.

Example 1.8.1. As an example we use two systems $\{\psi_k^c(t)\}$, $\{\psi_k^s(t)\}$, $k = 1, \ldots, N$ of orthogonal (in the sense of Definitions 1.4.2, 1.4.3, or 1.4.4) basis functions determined by Walsh functions:

$$\psi_k^c(t) = wal_k^N(t) \cdot \cos\left(2\pi t/T_0 + \pi/4\right); \quad (1.8.8\text{a})$$

$$\psi_k^s(t) = wal_k^N(t) \cdot \sin\left(2\pi t/T_0 + \pi/4\right); \quad (1.8.8\text{b})$$

$$wal_k^N(t) = H^N_{\left[\frac{t}{Tch}\right],k}; \quad (1.8.9)$$

$$t \in T = [0, T_s[, \ t = i \cdot \Delta t, \ i = 0, 1, \ldots, n-1,$$

where $wal_k^N(t)$ is k-th Walsh function of order N; time parameter t takes discrete values in the interval $t \in T$: $t = i \cdot \Delta t$, $i = 0, 1, \ldots, n-1$; $[x]$ is an integer part of x; T_{ch} is a duration of elementary signal, $T_{ch} = T_s/N$; T_s is a duration of a transmitted symbol A_k (see formula (1.8.4)), $T_s = n \cdot \Delta t$; H^N is Hadamard matrix of order N: $H^N = \|H_{r,c}\|$, r, c are rows and columns indexes of a matrix, so that Hadamard matrix H^{2n} of order $2n$ is formed iteratively basing on the relationship:

$$H^{2n} = \left\| \begin{array}{cc} H^n & H^n \\ H^n & -H^n \end{array} \right\|, \ H^1 = \|1\|. \quad (1.8.10)$$

We consider information transmitting system with DS-CDM(A)-QPSK based on two systems of orthogonal functions $\{\psi_k^c(t)\}$ (1.8.8a), $\{\psi_k^s(t)\}$ (1.8.8b), so that

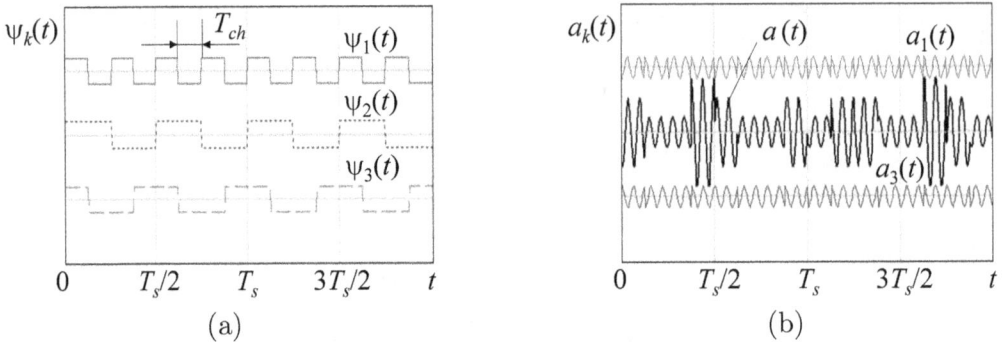

FIGURE 1.8.1   Illustrations of (a) Walsh function (1.8.9) of order $N=8$; (b) channel signals $a_k(t)$ and a group signal $a(t)$

a channel signal $a_k(t)$ is formed according to the relationships [38, (9.4.4)], [73, (12.22)] which are written here in more compact form:

$$a_k(t) = \frac{1}{\sqrt{2}} \left[A_k^c \cdot \psi_k^c(t) + A_k^s \cdot \psi_k^s(t) \right], \ t \in T, \tag{1.8.11}$$

where A_k^c, A_k^s are the symbols of messages of in-phase and quadrature components of QPSK signal, respectively, that are transmitted by k-th channel signal $a_k(t)$.

According to Definition 1.8.1, we consider that group signal $a(t)$ (1.8.4) additively interacts with interference:

$$x(t) = a(t) + n(t); \tag{1.8.12}$$

$$a(t) = \sum_{k=1}^{N} a_k(t). \tag{1.8.12a}$$

According to general algorithms of processing (1.8.7a,b,c), estimators $\hat{A}_k^c$, $\hat{A}_k^s$ of symbols A_k^c, A_k^s of in-phase and quadrature components of QPSK signal are determined by the relationships:

$$\hat{A}_k^c = \mathrm{sgn}\,(u_k), \ \hat{A}_k^s = \mathrm{sgn}\,(v_k); \tag{1.8.13}$$

$$u_k = \mathrm{N}[x(t), \psi_k^c(t)]; \tag{1.8.13a}$$

$$v_k = \mathrm{N}[x(t), \psi_k^s(t)], \tag{1.8.13b}$$

where $\mathrm{N}(x, y)$ is a generalized MSI of a kind (1.4.10...1.4.12).

Fig. 1.8.1a illustrates Walsh functions (1.8.9) of order $N=8$ formed in the 1st, 2nd, 3rd channels, respectively. Fig. 1.8.1b depicts channel signals $a_k(t)$ formed according to (1.8.11) in the 1st and the 3rd channels, and also a group signal $a(t)$ which is formed basing on the relationship (1.8.12a) as a sum of channel signals $\{a_k(t)\}$. For convenience of visual perception, the corresponding signals in these figures are shown shifted along the axis of ordinates.

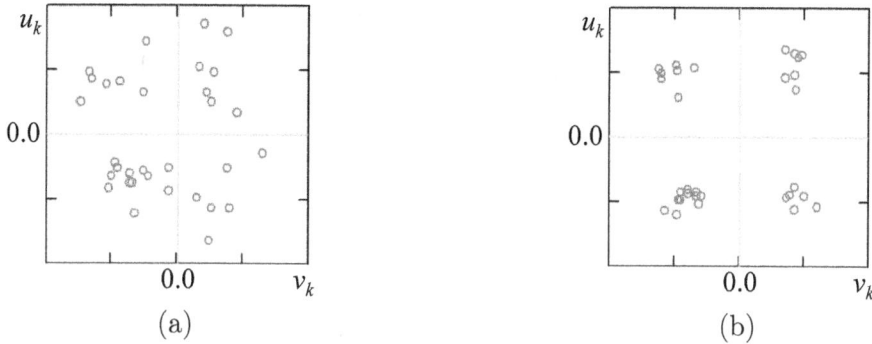

FIGURE 1.8.2   I-Q diagrams of the results of demodulating QPSK channel signals based on *L*-group algorithms and corresponding MSIs (a) N_p (1.4.10); (b) N_s (1.4.11)

Consider results of simulating the algorithm (1.8.13) of processing of a group signal $a(t)$ in the presence of quasi-white Gaussian noise $n(t)$ (1.8.12) obtained by statistical modelling based on Monte Carlo method. Statistical modelling was realized under the following constraint conditions.

1. Symbol duration T_s of QPSK signal is equal to $T_s=256$; number of elementary signals N_{ch} transmitted within symbol duration of QPSK signal is equal to $N_{ch}=8$; duration of elementary signal $T_{ch} = T_s/N_{ch}$ is equal to $T_{ch}=32$; number N of orthogonal Walsh functions from the system $\{wal_k^N(t)\}$, $k = 1, \ldots, N$ determined by the relationship (1.8.9) is equal to $N = N_{ch}=8$; oscillation period T_0 of a carrier is equal to $T_0=16$.

2. Interference $n(t)$ is quasi-white Gaussian noise with PSD $N(f) = N_0 \cdot [1(f) - 1(f - f_{\max})]$, N_0=const, $f_{\max} = 8/T_0$; SNR per bit $E_b/N_0=16$, (where E_b is a signal energy per one bit of information (bit energy)).

Fig. 1.8.2a, b illustrates I-Q diagrams of the results of demodulating QPSK channel signals $\{a_k(t)\}$ obtained within simulating *L*-group algorithms of signal demultiplexing and demodulation in signal space with *L*-group properties (1.8.13) based on MSIs N_p (1.4.10) and N_s (1.4.11), respectively. $\triangledown$

Inasmuch as practical interest of exploiting *L*-group algorithms (1.8.7a,b,c) with respect to concrete information transmitting systems with TDM, FDM, and CDM requires carrying out comparative analysis of their efficiency with known algorithms in the presence of interference (noise), we return to this question within Chapter 7 devoted to exploring *L*-group algorithms based on MSIs (1.4.10...1.4.12) with applications to multichannel communication systems and multi-station multiple access networks.

2

Estimation of Signal Parameters in Sample Spaces with L-group Properties

Estimation of signals and their parameters is the most general problem of signal processing in the presence of interference (noise). Some other problems of signal processing relate to estimation, for instance, signal detection, signal classification, and signal resolution. In known literature, the problems of signal processing in the presence of interference (noise) are formulated in terms of linear signal space, where the result of interaction x between the signal s and interference (noise) n is described by operation of addition of an additive commutative group: $x = s + n$. Quite similarly (i.e., by operation of addition), the literature describes the results of interaction between unknown nonrandom parameters of the signal and estimation errors (measurement errors) caused by the influence of interference (noise). Characteristics and behavior of estimators under additive (in terms of linear sample space) interaction between the estimated parameter and the measurement errors from some arbitrary family of distributions are discussed in the corresponding literature [54, 55, 74, 75].

In most works on point estimation, the model of indirect measurement of an unknown nonrandom scalar location parameter λ is described by its additive interaction with statistically independent measurement errors in *linear sample space* $\mathcal{LS}(\mathcal{X}, \mathcal{B}_\mathcal{X}; +)$:

$$X_i = f(\lambda) + N_i, \tag{2.0.1}$$

where $f(\lambda)$ is some known one-to-one function of a measured parameter; $\{N_i\}$ are the independent measurement errors represented by the sample $N = (N_1, \ldots, N_n)$, $N_i \in N$, $N \in \mathcal{LS}(\mathcal{X}, \mathcal{B}_\mathcal{X}; +)$, with a distribution from a distribution class with symmetric (even) probability density function (PDF) $p_N(z) = p_N(-z)$; $\{X_i\}$ are the observations represented by the sample $X = (X_1, \ldots, X_n)$, $X_i \in X$: $X \in \mathcal{LS}(\mathcal{X}, \mathcal{B}_\mathcal{X}; +)$; "+" is operation of addition of linear sample space $\mathcal{LS}(\mathcal{X}, \mathcal{B}_\mathcal{X}; +)$; $i = 1, \ldots, n$ is the index of elements of statistical collections $\{N_i\}$, $\{X_i\}$; n is a size of the samples $N = (N_1, \ldots, N_n)$, $X = (X_1, \ldots, X_n)$.

The subject of further consideration is estimation of unknown nonrandom parameter in a sample space with lattice-ordered group (L-group) properties. In existing algebraic literature L-groups are known for a long time and are well explored [29, 30, 32–35, 49, 50, 76–78]. Since in this chapter we consider estimation algorithms based on L-group operations, our further statement is relied on the notion introduced by the following definition.

Definition 2.0.1. *Sample space* $\mathcal{L}(\mathcal{X}, \mathcal{B}_\mathcal{X}; +, \vee, \wedge)$ *with L-group properties* is a probabilistic space $(\mathcal{X}, \mathcal{B}_\mathcal{X})$ in which axioms of distributive lattice $\mathcal{L}(\mathcal{X}; \vee, \wedge)$ hold (where $a \vee b = \sup_\mathcal{L}\{a, b\}$, $a \wedge b = \inf_\mathcal{L}\{a, b\}$ are operations of join and meet of

DOI: 10.1201/9781003275855-2

lattice, respectively; $a, b \in \mathcal{L}(\mathcal{X}; \vee, \wedge)$), and axioms of additive commutative group $\mathcal{L}(\mathcal{X}, \mathcal{B}_{\mathcal{X}}; +)$ also hold.

The approach based on exploiting sample space with L-group properties allows essentially expanding algebraic properties of usual sample space, and on the other hand, describing signal processing algorithms in terms of L-groups as a more complete algebraic system than additive group of linear space, using operations of both a lattice and an additive group.

The estimators obtained on the basis of least squares method (LSM) and least moduli method (LMM), according to the criteria of minimum of sums of squares and moduli of measurement errors, respectively, were the first and simplest estimators [54, 79]:

$$\hat{\lambda}_{\text{LSM}} = \arg \min_{\lambda} \left\{ \sum_i (X_i - f(\lambda))^2 \right\}; \qquad (2.0.2a)$$

$$\hat{\lambda}_{\text{LMM}} = \arg \min_{\lambda} \left\{ \sum_i |X_i - f(\lambda)| \right\}. \qquad (2.0.2b)$$

Extrema of the functions $\sum_i (X_i - f(\lambda))^2$ and $\sum_i |X_i - f(\lambda)|$ determined by criteria (2.0.2a) and (2.0.2b), respectively, are found as the roots of the equations:

$$d \sum_i (X_i - f(\hat{\lambda}))^2 / d\hat{\lambda} = 0; \qquad (2.0.3a)$$

$$d \sum_i |X_i - f(\hat{\lambda})| / d\hat{\lambda} = 0. \qquad (2.0.3b)$$

The estimators $\hat{\lambda}_{\text{LSM}}$ and $\hat{\lambda}_{\text{LMM}}$ in the form of the function $f^{-1}[*]$ of sample mean and sample median $\text{med}\{*\}$ of the observations $\{X_i\}$ are the solutions of the equations (2.0.3a) and (2.0.3b), respectively:

$$\hat{\lambda}_{\text{LSM}} = f^{-1} \left(\frac{1}{n} \sum_{i=1}^{n} X_i \right); \qquad (2.0.4a)$$

$$\hat{\lambda}_{\text{LMM}} = f^{-1}[\underset{i \in \mathbf{N} \cap [1,n]}{\text{med}} \{X_i\}], \qquad (2.0.4b)$$

where $f^{-1}[x]$ is an inverse function of $f(x)$; $\mathbf{N}$ is set of natural numbers.

The estimators (2.0.4a) and (2.0.4b) are asymptotically efficient in the case of Gaussian and Laplace distributions of measurement errors, respectively.

Poisson S.D. was the first who showed that the sample mean of initial random sample $X = (X_1, X_2, \ldots, X_n)$, whose elements possess PDF in the form $p(x) = a/(\pi(x^2 + a^2))$, $a > 0$, has the same distribution. This convincing example demonstrates that least mean squares estimator, to which Gauss appealed "basing on a simplicity", is not pertinent when estimating parameters of distributions with "heavy tails".

Besides, as it was quite fairly noticed by J.W. Tukey [80] and P.J. Huber [81], even in the simple case when the elements of a random sample $X = (X_1, X_2, \ldots, X_n)$ are independent and distributed with a CDF $F(x)$ in the form of Tukey ε-contaminated model $T(\varepsilon, \tau)$:

$$F(x; \varepsilon, \tau) = (1 - \varepsilon)\Phi(x) + \varepsilon\Phi(x/\tau), \tag{2.0.5}$$

where $\Phi(x) = \frac{1}{2\pi} \int\limits_{-\infty}^{x} \exp\left(\frac{-y^2}{2}\right) dy$ is CDF of standard normal distribution when a parameter ε takes the values in the interval $\varepsilon \in [0.002, 0.5]$, and $\tau = 3$, the estimator $\hat{\lambda}_{\text{LMM}}$ of a location parameter (2.0.4b) possess the higher absolute efficiency as against the estimator $\hat{\lambda}_{\text{LMS}}$ (2.0.4a).

Instead of supposing a distribution F of random sample to be known, P. Huber suggested to use a class of "ε-contaminated" distributions [11]:

$$F(x) = (1 - \varepsilon)G(x) + \varepsilon H(x), \tag{2.0.6}$$

where G is known distribution and H is an arbitrary unknown distribution, so that both distributions G and H are symmetrical with respect to zero; $\varepsilon = \text{const}$.

In practical applications, well-grounded assumptions on a known prior distribution take place extremely rarely. In this regard, some mathematical statisticians formulate a couple of questions [82].

1. What is the behavior of optimal (according to some criterion) estimators obtained for concrete distributions, in the case of their exploiting with respect to random samples with other distributions?

2. How one can construct estimators that are robust with respect to various distributions from a given class, including a class of "ε-contaminated" distributions?

Within the following discussion, we consider possible approaches providing responses to these questions.

Any pair of random samples $X = (X_1, X_2, \ldots, X_n)$, $Y = (Y_1, Y_2, \ldots, Y_n)$ from $\mathcal{L}(\mathcal{X}, \mathcal{B}_{\mathcal{X}}; +, \vee, \wedge)$ can be considered as a partially ordered set $X \bigcup Y$, in which for two pairs of sample values $X_i, X_j \in X$, $X_i, Y_k \in X \bigcup Y$ there exist a *relation of order* $X_i \leq X_j$, $X_i \geq Y_k$ (or $X_i \geq X_j$, $X_i \leq Y_k$). Partially ordered set $X \bigcup Y$ is a lattice with operations of join and meet, respectively:

$$X_i \vee X_j = \sup_{X \bigcup Y} \{X_i, X_j\}, \quad X_i \wedge X_j = \inf_{X \bigcup Y} \{X_i, X_j\};$$

$$X_i \vee Y_k = \sup_{X \bigcup Y} \{X_i, Y_k\}, \quad X_i \wedge Y_k = \inf_{X \bigcup Y} \{X_i, Y_k\},$$

and if $X_i \leq X_j$, then $X_i \wedge X_j = X_i$ and $X_i \vee X_j = X_j$, and if $X_i \geq Y_k$, then $X_i \vee Y_k = X_i$ and $X_i \wedge Y_k = Y_k$ [29, 31–34, 50, 76]:

$$X_i \leq X_j \Leftrightarrow \begin{cases} X_i \wedge X_j = X_i; \\ X_i \vee X_j = X_j, \end{cases} \quad X_i \geq Y_k \Leftrightarrow \begin{cases} X_i \wedge Y_k = Y_k; \\ X_i \vee Y_k = X_i. \end{cases}$$

In natural way in the sample space $\mathcal{L}(\mathcal{X}, \mathcal{B}_\mathcal{X}; +, \vee, \wedge)$, we define operation of addition $X_i + X_j$, $X_i + Y_k$ between pairs of elements $X_i, X_j \in X$, $X_i, Y_k \in X \bigcup Y$ of the samples $X, Y \subset \mathcal{L}(\mathcal{X}, \mathcal{B}_\mathcal{X}; +, \vee, \wedge)$. Then the sample space $\mathcal{L}(\mathcal{X}, \mathcal{B}_\mathcal{X}; +, \vee, \wedge)$ is *lattice-ordered group* (*L-group*).

In any *L*-group the following statements hold [29, 31–34, 50, 76]:
(1) $\mathcal{L}(\mathcal{X}, \mathcal{B}_\mathcal{X}; +)$ is a group; (2) $\mathcal{L}(\mathcal{X}, \mathcal{B}_\mathcal{X}; \vee, \wedge)$ is a lattice; (3) for arbitrary elements X_i, Y_j, A_k, B_l from $\mathcal{L}(\mathcal{X}, \mathcal{B}_\mathcal{X}; +, \vee, \wedge)$ the following identities hold:

$$A_k + (X_i \wedge Y_j) + B_l = (A_k + X_i + B_l) \wedge (A_k + Y_j + B_l);$$

$$A_k + (X_i \vee Y_j) + B_l = (A_k + X_i + B_l) \vee (A_k + Y_j + B_l).$$

2.1   Sample Ordering Algorithms Based on Lattice Operations

Often it is necessary to sort the elements of initial sample $X = (X_1, X_2, \ldots, X_n)$ in increasing (or decreasing) order, i.e. by ordering one can obtain a new sample $X' = (X_{(1)}, X_{(2)}, \ldots, X_{(n)})$ in which the elements are related by a *relation of order*:

$$X_{(1)} \leq X_{(2)} \leq \ldots \leq X_{(n-1)} \leq X_{(n)}. \tag{2.1.1}$$

Hereinafter, under a *sample ordering* we will mean arranging the elements of a sample in a sequence ordered by some criterion. The sample $X' = (X_{(1)}, X_{(2)}, \ldots, X_{(n)})$ is called *variational series*, and $X_{(i)}$ is called *i-th order statistic*, $X_{(i)} \in X'$ [51, 82, 83].

The following methods of sample ordering are known [83, 84].

2.1.1   Method of Ordering by a Choice

Let $X = (X_1, X_2, \ldots, X_n)$ be initial sample that must be ordered. Method of ordering by a choice is realized by the following algorithm.

Step 1. Choose the greatest element $X_{(n)}$ from $X = (X_1, X_2, \ldots, X_n)$:

$$X_{(n)} = \max(X_1, X_2, \ldots, X_n) = X_1 \vee X_2 \vee \ldots \vee X_n.$$

Step 2. Change the places of the greatest element $X_{(n)}$ and the element X_n.

Step 3. Choose the greatest element $X_{(n-1)}$ from a sequence $X_1, X_2, \ldots, X_{n-1}$:

$$X_{(n-1)} = \max(X_1, X_2, \ldots, X_{n-1}) = X_1 \vee X_2 \vee \ldots \vee X_{n-1}.$$

Step 4. Change the places of the greatest element $X_{(n-1)}$ and the element X_{n-1}.

Then the procedure is repeated. This method requires $(n-1) + (n-2) + \ldots + 1 = n(n-1)/2$ operations of join between the elements.

2.1.2 Method of Ordering by Pairwise Successive Permutation

Method of ordering by pairwise successive permutation is realized with the help of the following algorithm.

Step 1. At the first pairwise arranging the elements of initial sample $X = (X_1, X_2, \ldots, X_n)$ that should be ordered, calculate operation of join (meet) between the elements $X_{(i)}$ and $X_{(i+1)}$, so that if $X_i \vee X_{i+1} = X_i$ ($X_i \wedge X_{i+1} = X_{i+1}$) change the places of the elements, if, on the contrary, the identities hold $X_i \vee X_{i+1} = X_{i+1}$ ($X_i \wedge X_{i+1} = X_i$) then a pair of elements $X_{(i)}$ and $X_{(i+1)}$ stay on their own places. When finishing the procedure, the greatest element $X_{(n)}$ is moved into position of X_n and does not take part in the following arrangements.

Step 2. At the second (and, in general, i-th) pairwise arrangement, the aforementioned procedure is repeated only for the elements $X_1, X_2, \ldots, X_{n-(i-1)}$.

The required number of binary operations is approximately the same as in the first method, however, it is determined by a degree of initial ordering of initial sample.

2.1.3 Method of Ordering by Serial-parallel 2^m-union

Method of ordering by serial-parallel 2^m-union is realized by means of the following algorithm.

Step 1. At first arranging the elements of initial sample $X = (X_1, X_2, \ldots, X_n)$ that should be ordered and in which $n = 2^k$, $k \in \mathbf{N}$, the neighbor elements are united in pairs $((X_1, X_2); (X_3, X_4), \ldots, (X_{n-1}, X_n))$ and for each pair operation of join (meet) between the elements $X_{(i)}$ and $X_{(i+1)}$ is calculated, so that if $X_i \vee X_{i+1} = X_i$ ($X_i \wedge X_{i+1} = X_{i+1}$) then the elements are inverted, if, on the contrary, the identities hold $X_i \vee X_{i+1} = X_{i+1}$ ($X_i \wedge X_{i+1} = X_i$), then a pair of elements $X_{(i)}$ and $X_{(i+1)}$ stay on their own places. Thus, the sample $X' = (X'_1, X'_2, \ldots, X'_n)$ is formed.

Step 2. At the second arrangement within the sample $X' = (X'_1, X'_2, \ldots, X'_n)$, the adjacent elements are united into quadruples $((X'_1, X'_2, X'_3, X'_4), \ldots, (X'_{n-3}, X'_{n-2}, X'_{n-1}, X'_n))$. Then, for each quadruple, the previous method is applied. As a result, the sample $X'' = (X''_1, X''_2, \ldots, X''_n)$ is formed.

Step 3. At the third arranging the elements of the sample $X'' = (X''_1, X''_2, \ldots, X''_n)$, the adjacent elements are united into double quadruples, whereupon for each double quadruple the previous method is used.

If necessary, within the following steps the procedure is repeated with preliminary uniting the neighbor elements from a set of 2^m elements, $m = 4, 5, \ldots$. Necessary number of steps is equal to $\log_2 n$, and the number of binary operations does not exceed $n \log_2 n$.

2.1.4 Method of Ordering by Series-parallel Pairwise Union with Pairwise Permutation

Method of ordering by series-parallel pairwise union with pairwise permutation is realized by the following algorithm.

Step 1. At first arranging the elements of initial sample $X = (X_1, X_2, \ldots, X_n)$ that should be ordered and $n = 2k$, $k \in \mathbf{N}$, the adjacent elements are united in pairs $((X_1, X_2); (X_3, X_4), \ldots, (X_{n-1}, X_n))$ and for each pair operations of join $X_i \vee X_{i+1} = X_{(i+1)}$ and meet $X_i \wedge X_{i+1} = X_{(i)}$ between the elements $X_{(i)}$ and $X_{(i+1)}$ are calculated, so that within each pair of elements, the results of operations $X_{(i)}, X_{(i+1)}$ are arranged in increasing order: $X_{(i)} \leq X_{(i+1)}$. Thus, the sample $X' = (X'_1, X'_2, \ldots, X'_n)$ is formed.

Step 2. At the second (or, in general, even j-th, $j = 2m$, $m \in \mathbf{N}$) arrangement of the sample $X' = (X'_1, X'_2, \ldots, X'_n)$, the neighbor elements, starting from the second one and finishing by penultimate element, are united in pairs $((X'_2, X'_3), (X'_4, X'_5), \ldots, (X'_{n-2}, X'_{n-1}))$ and for each pair operations of join $X'_i \vee X'_{i+1} = X'_{(i+1)}$ and meet $X'_i \wedge X'_{i+1} = X'_{(i)}$ between the elements $X'_{(i)}$ and $X'_{(i+1)}$ are calculated, so that within each pair of elements, the results of operations $X'_{(i)}, X'_{(i+1)}$ are arranged in increasing order: $X'_{(i)} \leq X'_{(i+1)}$. Thus, the sample $X'' = (X''_1, X''_2, \ldots, X''_n)$ is formed.

Step 3. At the third (or, in general, odd j-th $j = 2m + 1$, $m \in \mathbf{N}$) arrangement of the sample $X'' = (X''_1, X''_2, \ldots, X''_n)$, the adjacent element, starting from the first one and finishing by the last element, are united in pairs $((X''_1, X''_2); (X''_3, X''_4), \ldots, (X''_{n-1}, X''_n))$ and for each pair operations of join $X''_i \vee X''_{i+1} = X''_{(i+1)}$ and meet $X''_i \wedge X''_{i+1} = X''_{(i)}$ between the elements $X''_{(i)}$ and $X''_{(i+1)}$ are calculated, so that within every pair of elements, the results of operations $X''_{(i)}, X''_{(i+1)}$ are arranged in increasing order: $X''_{(i)} \leq X''_{(i+1)}$.

As in previous method, necessary number of binary operations does not exceed $n \log_2 n$.

Block diagram of a unit called *systolic processor* [85–87] that realizes a method of sample ordering by series-parallel pairwise union with pairwise permutation (for the case of sample size equal to 7 and non-negativity of sample elements) is shown in Fig. 2.1.1. Hereinafter, we call such a device *systolic processor with binary elements* intended for sample ordering.

An example of operating of systolic processor depicted in Fig. 2.1.1 is elucidated in the Table 2.1.1, where $y_1, y_2, \ldots, y_8$ are state vectors describing the signals in the

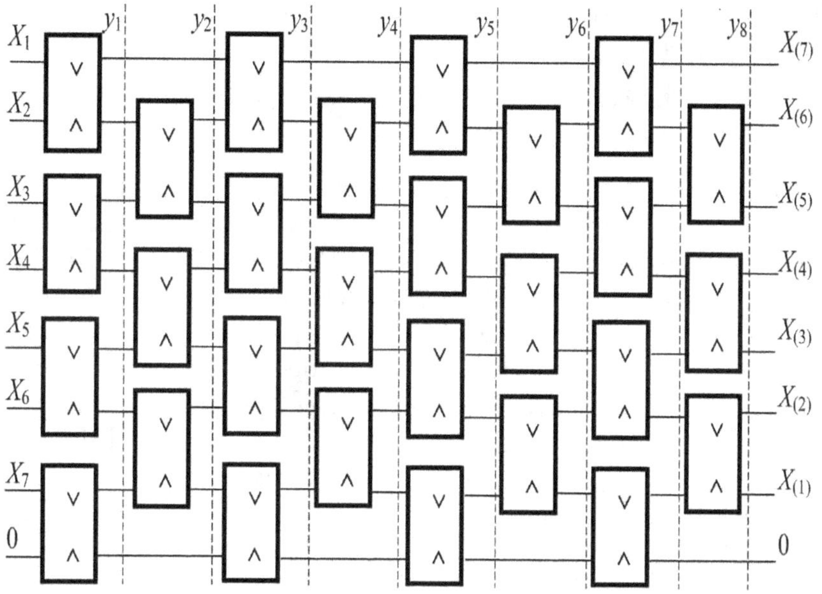

FIGURE 2.1.1   Systolic processor with binary elements

outputs of the corresponding cascade of data processing. According to aforementioned processing algorithm, a number of two-input elements, performing operations of join/meet at their two outputs and situated in an odd processing cascade, is one more than the number of these elements in an even cascade. The values of elements of initial sample $X = (X_1, X_2, \ldots, X_n)$ and null element 0 of the sample space $\mathcal{L}(\mathcal{X}, \mathcal{B}_{\mathcal{X}}; +, \vee, \wedge)$ are fed in the inputs of systolic processor, and in its outputs the values of ordered sample $X' = (X_{(1)}, X_{(2)}, \ldots, X_{(n)})$ are formed.

TABLE 2.1.1   An example of operation of systolic processor with binary elements

Elements of initial sample X	Values of the elements of initial sample	State vectors							
		y_1	y_2	y_3	y_4	y_5	y_6	y_7	y_8
X_1	17	17	17	17	17	17	22	22	24
X_2	9	9	9	9	9	22	17	24	22
X_3	1	1	5	5	22	9	24	17	17
X_4	5	5	1	22	5	24	9	12	12
X_5	12	12	22	1	24	5	12	9	9
X_6	22	22	12	24	1	12	5	5	5
X_7	24	24	24	12	12	1	1	1	1
0	0	0	0	0	0	0	0	0	0

2.1.5   Method of Ordering by Series-parallel Ternary Union with Ternary Permutation

Method of ordering by series-parallel ternary union with ternary permutation is realized by the following algorithm.

Step 1.   At first arranging the elements of initial sample $X = (X_1, X_2, \ldots, X_n)$ that should be ordered in which $n = 3 \cdot k$, $k = 2m + 1$, $m \in \mathbf{N}$, the neighbor elements are united in triplets $((X_1, X_2, X_3), (X_4, X_5, X_6), \ldots,$ $(X_{n-2}, X_{n-1}, X_n))$ and for each triplets operations of join $X_{(i+2)} = X_i \vee X_{i+1} \vee X_{i+2}$ and meet $X_{(i)} = X_i \wedge X_{i+1} \wedge X_{i+2}$ between the elements X_i, X_{i+1}, X_{i+2} are calculated, and also operation of a sample median $X_{(i+1)} = \mathrm{med}[X_i, X_{i+1}, X_{i+2}]$ is performed, so that within each triplet the results of operations $X_{(i)}, X_{(i+1)}, X_{(i+2)}$ are arranged in increasing order: $X_{(i)} \leq X_{(i+1)} \leq X_{(i+2)}$. Thus, the sample $X' = (X'_1, X'_2, \ldots, X'_n)$ is formed.

Step 2.   At the second (or, in general, even j-th $j = 2 \cdot m, m \in \mathbf{N}$) arrangement of the sample $X' = (X'_1, X'_2, \ldots, X'_n)$, the adjacent elements, starting from the second one and finishing by penultimate element and excluding the central element $X'_{[n+1]/2}$, are united in triplets $((X'_2, X'_3, X'_4), \ldots, (X'_{n-3}, X'_{n-2}, X'_{n-1}))$ and for each ternary operations of join $X'_{(i+2)} = X'_i \vee X'_{i+1} \vee X'_{i+2}$ and meet $X'_{(i)} = X'_i \wedge X'_{i+1} \wedge X'_{i+2}$ between the elements X'_i, X'_{i+1}, X'_{i+2} are calculated, and also operation of a sample median is performed $X'_{(i+1)} = \mathrm{med}[X'_i, X'_{i+1}, X'_{i+2}]$, so that within every triplet the results of operations $X'_{(i)}, X'_{(i+1)}, X'_{(i+2)}$ are arranged in increasing order: $X'_{(i)} \leq X'_{(i+1)} \leq X'_{(i+2)}$. Thus, the sample $X'' = (X''_1, X''_2, \ldots, X''_n)$ is formed.

Step 3.   At the third (or, in general, odd j-th $j = 2 \cdot m + 1$, $m \in \mathbf{N}$) arrangement of the sample $X'' = (X''_1, X''_2, \ldots, X''_n)$, the neighbor elements, starting from the first one and finishing by the last element, are united in triplets $((X''_1, X''_2, X''_3), \ldots, (X''_{n-2}, X''_{n-1}, X''_n))$ and for each triplet operations of join $X''_{(i+2)} = X''_i \vee X''_{i+1} \vee X''_{i+2}$ and meet $X''_{(i)} = X''_i \wedge X''_{i+1} \wedge X''_{i+2}$ between the elements $X''_i, X''_{i+1}, X''_{i+2}$ are calculated, and also operation of a sample median $X''_{(i+1)} = \mathrm{med}[X''_i, X''_{i+1}, X''_{i+2}]$ is performed, so that within each triplet the results of operations are arranged in increasing order: $X''_{(i)} \leq X''_{(i+1)} \leq X''_{(i+2)}$.

As in the previous method, necessary number of ternary operations does not exceed $n \log_3 n$.

Block diagram of systolic processor that realizes a method of sample ordering by series-parallel ternary union with ternary permutation (for the case of sample size equal to 9) is shown in Fig. 2.1.2. Hereinafter such a device we call *systolic processor with ternary elements* for sample ordering.

An example of operation of systolic processor depicted in Fig. 2.1.2 is elucidated in the Table 2.1.2, where $y_1, y_2, \ldots, y_8$ are the state vectors describing the

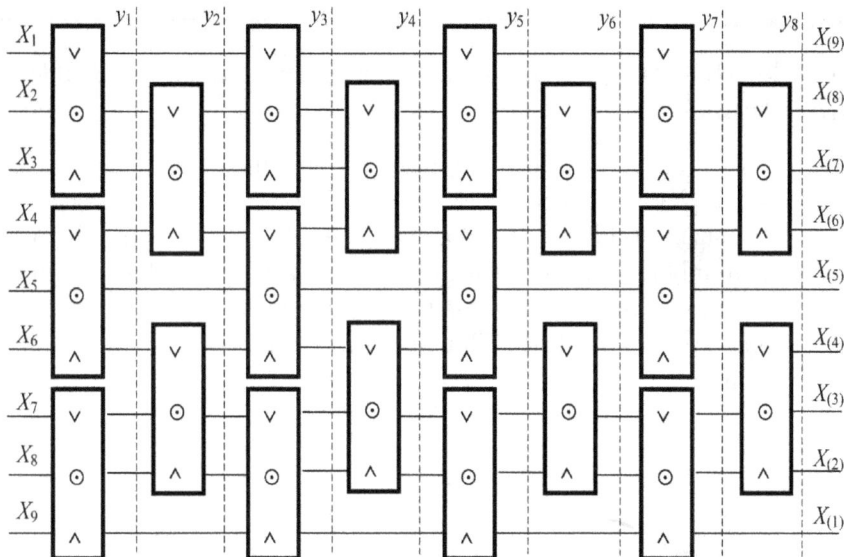

FIGURE 2.1.2 Systolic processor with ternary elements

signals in the output of the corresponding processing cascade. According to afore-mentioned data processing algorithm, a number of three-input elements, performing in their own outputs operations of join/meet and also operation of a sample median (denoted by the symbol $\odot$) in odd processing cascade, is one more than a number of these elements in even cascade. The values of elements of initial sample $X = (X_1, X_2, \ldots, X_n)$ (if necessary and also null element 0 of sample space $\mathcal{L}(\mathcal{X}, \mathcal{B}_{\mathcal{X}}; +, \vee, \wedge)$) are fed in the inputs of systolic processor; in its output the values of the elements of ordered sample $X' = (X_{(1)}, X_{(2)}, \ldots, X_{(n)})$ are formed.

TABLE 2.1.2 Example of operation of systolic processor with ternary elements

Elements of initial sample X	Values of elements of initial sample	State vectors							
		y_1	y_2	y_3	y_4	y_5	y_6	y_7	y_8
X_1	21	47	47	69	69	89	89	89	89
X_2	17	21	69	47	89	69	81	81	81
X_3	47	17	21	21	47	47	69	69	75
X_4	54	69	17	89	21	81	47	75	69
X_5	49	54	54	54	54	54	54	54	54
X_6	69	49	89	17	81	21	75	47	49
X_7	81	89	81	81	75	75	49	49	47
X_8	75	81	49	75	17	49	21	21	21
X_9	89	75	75	49	49	17	17	17	17

2.2 Sample Median Calculation Algorithms Based on Lattice Operations

Let $X = (X_1, X_2, \ldots, X_n)$ be initial sample whose sample median is calculated: $\widetilde{X} = \text{med}[X] = \text{med}[X_1, X_2, \ldots, X_n]$.

If the sample size is odd, i.e. $n = 2k + 1$, $k \in \mathbf{N}$, median estimate $\widetilde{X}$ takes a position with a number $(n + 1)/2$ within ordered sample. Median estimate $\widetilde{X}$ is larger than all previous values and is less than all next values, whose number is the same and is equal to $k = (n - 1)/2$. Thus, the sample median $\widetilde{X}$ can be found as a function of the kind:

$$\widetilde{X} = \text{med}[X] = \bigvee_{i=1}^{I} v_i = v_1 \vee \ldots \vee v_I; \tag{2.2.1a}$$

$$\widetilde{X} = \text{med}[X] = \bigwedge_{i=1}^{I} u_i = u_1 \wedge \ldots \wedge u_I, \tag{2.2.1b}$$

where $I = C_n^{(n+1)/2}$ is a number of $(n + 1)/2$ combinations from n elements; $v_i = \bigwedge_j X_j \big|_{X_j \in X/C_n^{(n+1)/2}}$ is meet of some $(n + 1)/2$ combination from n elements of initial sample $X = (X_1, X_2, \ldots, X_n)$; $u_i = \bigvee_j X_j \big|_{X_j \in X/C_n^{(n+1)/2}}$ is join of some $(n + 1)/2$ combination from n elements of initial sample $X = (X_1, X_2, \ldots, X_n)$.

Forming the sample median $\widetilde{X}$ according to the formulas (2.2.1a), (2.2.1b) for the case $n = 3$, $C_n^{(n+1)/2} = 3$, $X = (X_1, X_2, X_3)$ is based on the following relationships, respectively:

$$\widetilde{X} = \text{med}[X] = (X_1 \wedge X_2) \vee (X_2 \wedge X_3) \vee (X_3 \wedge X_1); \tag{2.2.2a}$$

$$\widetilde{X} = \text{med}[X] = (X_1 \vee X_2) \wedge (X_2 \vee X_3) \wedge (X_3 \vee X_1). \tag{2.2.2b}$$

Block diagrams of processing unit calculating the sample median $\widetilde{X}$ according to the formulas (2.2.2a), (2.2.2b) (for the case $n = 3$, $C_n^{(n+1)/2} = 3$, $X = (X_1, X_2, X_3)$) are shown in Fig. 2.2.1a, 2.2.1b, respectively. Processing units calculating the sample median $\widetilde{X}$ according to formulas (2.2.2a), (2.2.2b) consist of three two-input elements performing meet/join operations whose outputs are united by three-input element performing join/meet operations, respectively. Forming the sample median $\widetilde{X}$ according to (2.2.1a), (2.2.1b) for the case $n = 5$, $C_n^{(n+1)/2} = 10$, $X = (X_1, X_2, X_3, X_4, X_5)$ is based on the following relationships, correspondingly:

$$\widetilde{X} = \text{med}[X] = (X_1 \wedge X_2 \wedge X_3) \vee (X_1 \wedge X_2 \wedge X_4) \vee (X_1 \wedge X_2 \wedge X_5) \vee$$
$$\vee (X_1 \wedge X_3 \wedge X_4) \vee (X_1 \wedge X_3 \wedge X_5) \vee (X_1 \wedge X_4 \wedge X_5) \vee (X_2 \wedge X_3 \wedge X_4) \vee$$
$$\vee (X_2 \wedge X_3 \wedge X_5) \vee (X_2 \wedge X_4 \wedge X_5) \vee (X_3 \wedge X_4 \wedge X_5); \tag{2.2.3a}$$

$$\widetilde{X} = \mathrm{med}[X] = (X_1 \vee X_2 \vee X_3) \wedge (X_1 \vee X_2 \vee X_4) \wedge (X_1 \vee X_2 \vee X_5) \wedge$$
$$\wedge (X_1 \vee X_3 \vee X_4) \wedge (X_1 \vee X_3 \vee X_5) \wedge (X_1 \vee X_4 \vee X_5) \wedge (X_2 \vee X_3 \vee X_4) \wedge$$
$$\wedge (X_2 \vee X_3 \vee X_5) \wedge (X_2 \vee X_4 \vee X_5) \wedge (X_3 \vee X_4 \vee X_5). \quad (2.2.3b)$$

The examples of calculating the sample median $\widetilde{X}$ according to the formulas (2.2.3a), (2.2.3b) for the case $n = 5$, $C_n^{(n+1)/2} = 10$, $X = (X_1, X_2, X_3, X_4, X_5)$ are shown in the Tables 2.2.1a, b, respectively.

TABLE 2.2.1a   Example of calculating the sample median $\widetilde{X}$ according to the formula (2.2.3a)

Element of a sample	Value of element	Combination function	Value of function	Result
$X_1 =$	14	$X_1 \wedge X_2 \wedge X_3$	4	**18**
$X_2 =$	18	$X_1 \wedge X_2 \wedge X_4$	14	
$X_3 =$	4	$X_1 \wedge X_2 \wedge X_5$	14	
$X_4 =$	56	$X_1 \wedge X_3 \wedge X_4$	4	
$X_5 =$	45	$X_1 \wedge X_3 \wedge X_5$	4	
		$X_1 \wedge X_4 \wedge X_5$	14	
		$X_2 \wedge X_3 \wedge X_4$	4	
		$X_2 \wedge X_3 \wedge X_5$	4	
		$X_2 \wedge X_4 \wedge X_5$	18	
		$X_3 \wedge X_4 \wedge X_5$	4	

TABLE 2.2.1b   Example of calculating the sample median $\widetilde{X}$ according to the formula (2.2.3b)

Element of a sample	Value of element	Combination function	Value of function	Result
$X_1 =$	14	$X_1 \vee X_2 \vee X_3$	18	**18**
$X_2 =$	18	$X_1 \vee X_2 \vee X_4$	56	
$X_3 =$	4	$X_1 \vee X_2 \vee X_5$	45	
$X_4 =$	56	$X_1 \vee X_3 \vee X_4$	56	
$X_5 =$	45	$X_1 \vee X_3 \vee X_5$	45	
		$X_1 \vee X_4 \vee X_5$	56	
		$X_2 \vee X_3 \vee X_4$	56	
		$X_2 \vee X_3 \vee X_5$	45	
		$X_2 \vee X_4 \vee X_5$	56	
		$X_3 \vee X_4 \vee X_5$	56	

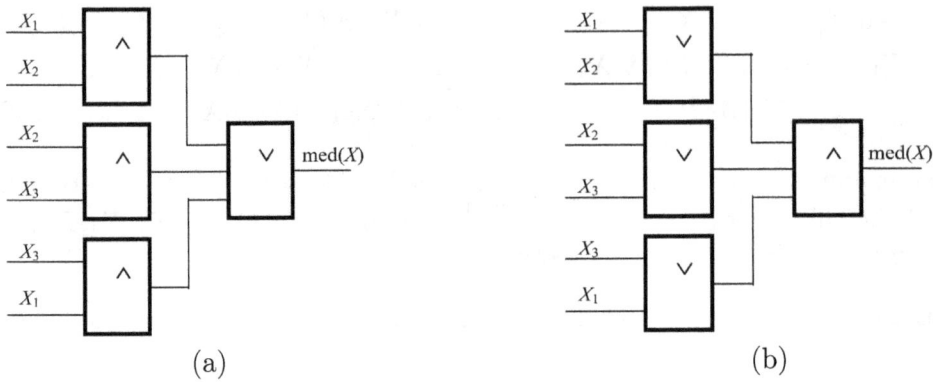

FIGURE 2.2.1　Processing units calculating the sample median $\widetilde{X}$ according to the formulas: (a)—(2.2.2a), (b)—(2.2.2b)

2.3　Algorithm of Forming M-estimator Based on L-group Operations

Some compromise between the estimators in the form of sample mean $\bar{X}$ and sample median $\widetilde{X}$ was suggested by P. Huber [81]. This approach is based on the fact that sample mean $\bar{X}$ and sample median $\widetilde{X}$ are the estimators (2.0.2a), (2.0.2b) minimizing the functions $\sum_i (X_i - m)^2$ and $\sum_i |X_i - m|$, respectively, where m is a location parameter, P. Huber suggested minimizing the following function instead mentioned above [11]:

$$\sum_i \rho(X_i - m) \to \min; \tag{2.3.1}$$

$$\rho(x) = \begin{cases} \frac{1}{2}x^2, & |x| \le k; \\ k|x| - \frac{1}{2}k^2, & |x| \ge k, \end{cases} \tag{2.3.1a}$$

where $k > 0$.

Definition 2.3.1. *Huber estimators* are the estimators minimizing (2.3.1), where ρ is given by the relationship (2.3.1a).

Huber estimators form a subset of the class of M-estimators which are defined as follows.

Definition 2.3.2. *M-estimators* are the estimators minimizing (2.3.1) on an arbitrary kind of a function ρ.

If ρ has a derivative $\rho' = \psi$, then M-estimators are defined as a solution of the equation:

$$\sum_i \psi(X_i - m) = 0 \tag{2.3.2}$$

In a special case, when G in the distribution F (2.0.6) is standard normal distribution, the solution of the problem of minimizing an upper bound variance of the distribution F is Huber M-estimator that correspond to (2.3.1a), so that minimax function ψ is determined by the relationship:

$$\psi(x) = \begin{cases} -k, & x \leq -k; \\ x, & |x| < k; \\ k, & x \geq k, \end{cases} \tag{2.3.3}$$

or, more compactly:

$$\psi(x) = k \wedge (x \vee -k) = -k \vee (x \wedge k), \tag{2.3.4}$$

where ε and k are coupled by the identity [75, § 5.6 (14)]:

$$\frac{1}{1-\varepsilon} = \int_{-k}^{k} \varphi(x)dx + \frac{2}{k}\varphi(k), \quad \varphi(x) = \frac{1}{\sqrt{2\pi}}\exp[-x^2/2].$$

Taking into account (2.3.4) and (2.3.2), Huber M-estimator is determined as a solution of the equation:

$$\sum_i k \wedge [(X_i - m) \vee -k] = 0. \tag{2.3.5}$$

Here we notice that the equation (2.3.5) is the relationship (2.3.2) written in terms of the sample space $\mathcal{L}(\mathcal{X}; +, \vee, \wedge)$ with L-group properties.

To solve the equation (2.3.5), one can use some numerical method, for instance, Newton method. As a result, all the procedure of finding a parameter m reduces to iterative calculating the expression:

$$\hat{m}_j = \hat{m}_{j-1} + \hat{\sigma} \cdot \frac{\sum_{i=1}^{n} k \wedge [(X_i - \hat{m}_{j-1}) \vee -k]}{\sum_{i=1}^{n} [1(X_i + k - \hat{m}_{j-1}) - 1(X_i - k - \hat{m}_{j-1})]}, \tag{2.3.6}$$

where $\hat{m}_j$, $\hat{m}_{j-1}$ is an estimator of location parameter m on the current j-th and on the previous $(j-1)$-th steps of iterations; $\hat{\sigma}$ is some estimator of scale parameter of a distribution; $1(t)$ is Heaviside unit step function.

As an initial value $\hat{m}_{j=0}$ of location parameter m, one can accept a median value $\mathrm{med}(X)$ of the random sample $X = (X_1, X_2, \ldots, X_n)$: $\hat{m}_{j=0} = \mathrm{med}(X)$, mean value of the sample $\bar{X} = (X_1 + \ldots + X_n)/n$: $\hat{m}_{j=0} = \bar{X}$, or a value that is equal to zero: $\hat{m}_{j=0} = 0$.

Notice that a denominator of fraction in the formula (2.3.6) is a derivative of the function (2.3.5) that figures in numerator. For practical applications this denominator can be substituted by a sample size n. Taking into account this replacement, the formula (2.3.6) takes the simpler form:

$$\hat{m}_j = \hat{m}_{j-1} + \frac{\hat{\sigma}}{n} \cdot \sum_{i=1}^{n} k \wedge [(X_i - \hat{m}_{j-1}) \vee -k]. \tag{2.3.7}$$

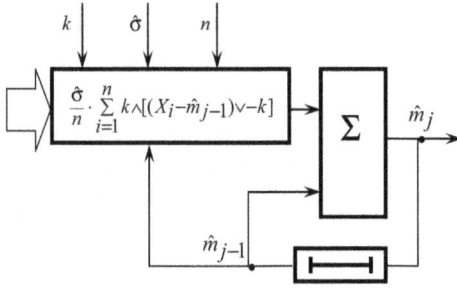

FIGURE 2.3.1   Block diagram of processing unit of calculating a location parameter m

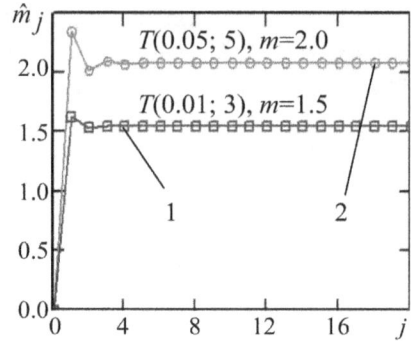

FIGURE 2.3.2   The results of calculating M-estimator according to formula (2.3.7)

Block diagram of processing unit providing recursive calculating estimated location parameter m is shown in Fig. 2.3.1. Fig. 2.3.2 demonstrates the results of calculating M-estimator by this device according to the formula (2.3.7). Iterative curves 1, 2 correspond to the conditions of estimating location parameters $m = 2.0,\ 1.5$ for Tukey model $T(0.05;\ 5)$, $T(0.01;\ 3)$ (2.0.5), respectively. In both cases, the size of a sample n is equal to 20. As follows from the curves 1, 2, used Newton method provides fast convergence of the results $\hat{m}_j$ to estimated parameter m in the case of both a strong contamination ($T(0.05;\ 5)$) and relatively weak contamination ($T(0.01;\ 3)$) of normally distributed estimation results.

To disadvantages of the method one should refer, first, the need in prior information concerning a distribution of sample elements to form the values of a coefficient k and the estimator $\hat{\sigma}$ of scale parameter, and second, a possibility of iterative procedure divergence in the case of insufficient accuracy of the estimate $\hat{\sigma}$.

2.4   Algorithms of Forming L-estimators Based on L-group Operations

Despite the fact, that M-estimators possess quite satisfactory robust properties, in many practical applications L-estimators could be more preferable from the standpoint of providing the simplicity of calculations.

Analysis of interrelation between M-estimators and L-estimators is discussed, for instance, in [75, 88].

Definition 2.4.1. L-estimators $T_{n,L}(X)$ are the estimators in the form of linear combination of order statistics [37, 82]:

$$T_{n,L}(X) = \sum_{i=1}^{n} a_i X_{(i)}, \qquad (2.4.1)$$

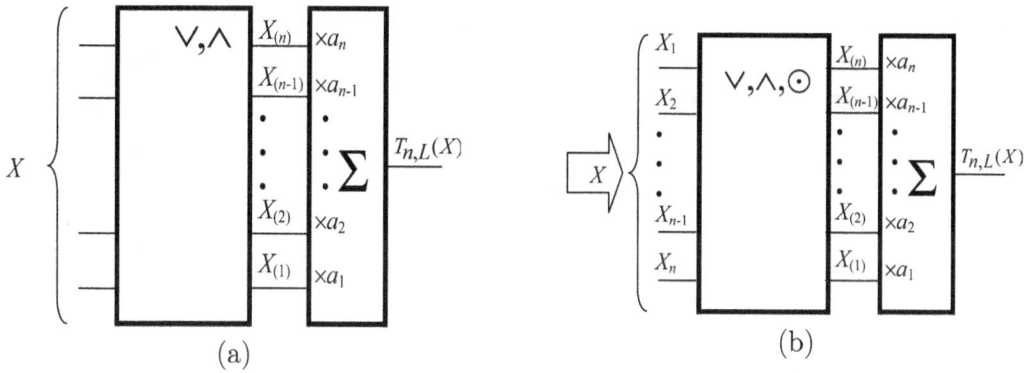

FIGURE 2.4.1 Processing units forming L-estimators based on systolic processors with: (a) binary elements; (b) ternary elements

where $X = (X_1, X_2, \ldots, X_n)$ is initial sample; $X' = (X_{(1)}, X_{(2)}, \ldots, X_{(n)})$ is the initial ordered sample; a_i are some coefficients: $a_{n-i+1} = a_i$, $\sum_{i=1}^{n} a_i = 1$.

Generalized forms of block diagrams of the processing units forming L-estimators based on systolic processors with binary/ternary elements are shown in Fig. 2.4.1a, b, respectively. Hereinafter, systolic processors with binary/ternary elements are denoted by the symbols of lattice binary operations $\vee, \wedge$ of the sample space $\mathcal{L}(\mathcal{X}; +, \vee, \wedge)$ with L-group properties and also by the symbol of calculating a median $\odot$, respectively. Notice, that for processing the sample $X = (X_1, X_2, \ldots, X_n)$ with even number of elements $n = 2 \cdot k$, $k \in \mathbf{N}$, it is expedient to use systolic processor with binary elements (see Fig. 2.1.1), whereas for processing the sample with odd number of elements, so that $n = 3 \cdot k$, $k \in \mathbf{N}$, it is better to exploit systolic processor with ternary elements (see Fig. 2.1.2).

Among L-estimators, there are winsorized mean estimators, truncated mean estimators, and linearly weighted mean estimators.

Definition 2.4.2. *Winsorized mean estimators* $W_n(X, p)$ *are L-estimators of the following form:*

$$W_n(X, p) = \frac{1}{n}\left[(r+1)(X_{(r+1)} + X_{(n-r)}) + \sum_{i=r+2}^{n-r-1} X_{(i)}\right]; \qquad (2.4.2)$$

$$W_n(X, \tfrac{1}{2n}) = X_{((n+1)/2)}, \qquad (2.4.2a)$$

where in formula (2.4.2) $p = \frac{1}{2} - \frac{r}{n}$, $0 < r < \frac{1}{2}(n-1)$, and in formula (2.4.2a) $r = \frac{1}{2}(n-1)$, n is odd.

Block diagrams of processing units forming winsorized mean estimators $W_n(X, p)$ based on systolic processors with binary/ternary elements that calculate estimators according to the formulas (2.4.2), (2.4.2a) are shown in Fig. 2.4.2a,b, respectively. Notice, that for processing the sample $X = (X_1, X_2, \ldots, X_n)$ according to the formula (2.4.2), one can use systolic processor with binary (ternary) elements depending on evenness (oddness) of sample size n, whereas for processing the sample with odd number of elements according to the formula (2.4.2a), it is better to exploit systolic processor with ternary elements.

Definition 2.4.3. *Truncated mean estimators* $L_n(X, \alpha)$, *that also will be denoted* $\bar{X}_\alpha$, *are L-estimators of the following form:*

$$L_n(X, \alpha) = \frac{1}{n - 2r} \sum_{i=r+1}^{n-r} X_{(i)} \tag{2.4.3}$$

where $\alpha = r/n < 1/2$; $r = [\alpha n]$, $[t]$ is the integer part of t.

Truncated mean estimators $\bar{X}_\alpha$ compile an important segment in the class of L-estimators. Besides, sample mean $\bar{X}$ and sample median $\widetilde{X}$ are two extreme variants of the estimator (2.4.3), when no one sample element ($\alpha = 0$) is rejected, or all the elements are rejected with the exception of either the element $X_{((n+1)/2)}$ if n is odd, or the pair of elements $X_{(n/2)}$, $X_{(n/2+1)}$ if n is even. Block diagrams of processing units forming truncated mean estimators $\bar{X}_\alpha$ based on systolic processors with binary/ternary elements are shown in Fig. 2.4.3a, b, respectively.

Definition 2.4.4. *Linearly weighted mean estimators* $L_n(X, p)$ *are L-estimators of the following form:*

$$L_n(X, p)$$
$$= \begin{cases} \displaystyle\sum_{j=1}^{n/2-r} \frac{(2j-1)(X_{(r+j)} + X_{(n-r+1-j)})}{2(n/2-r)}, & \mathrm{mod}_2 n = 0; \\[4mm] \displaystyle\sum_{j=1}^{\frac{n-1}{2}-r} \frac{(2j-1)(X_{(r+j)} + X_{(n-r+1-j)}) + (n-2r)X_{((n+1)/2)}}{\left[\frac{n-1}{2}-r\right]^2 + \left[\frac{n+1}{2}-r\right]^2}, & \mathrm{mod}_2 n = 1, \end{cases}$$
$$\tag{2.4.4}$$

where $\mathrm{mod}_2 n$ is n modulo 2.

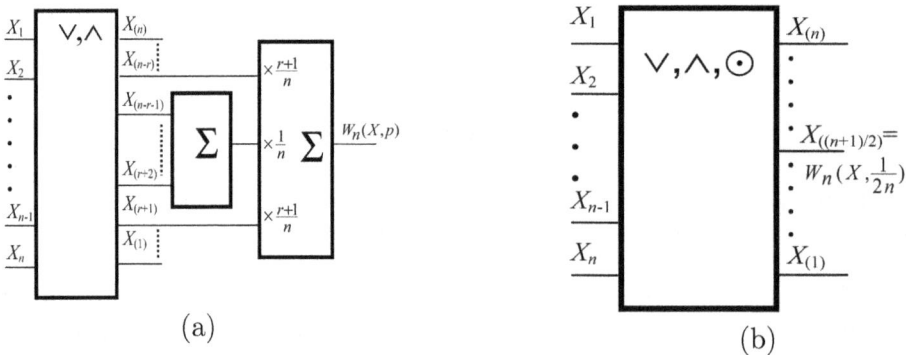

FIGURE 2.4.2   Processing units forming winsorized mean estimators $W_n(X, p)$ based on systolic processors with: (a) binary elements; (b) ternary elements

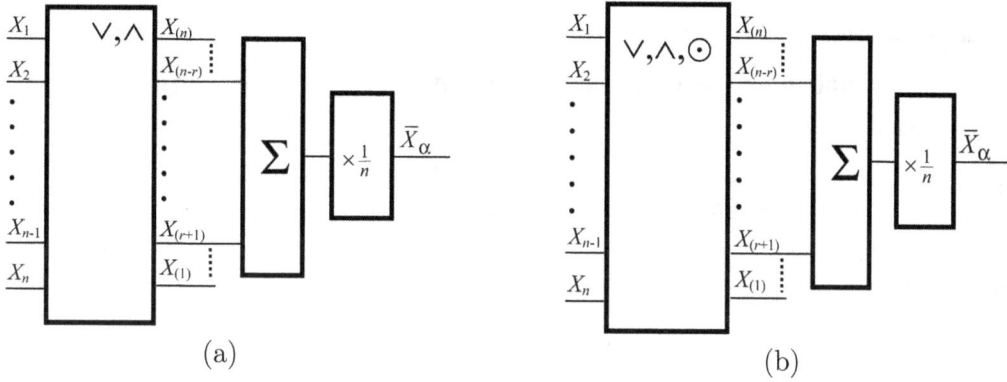

FIGURE 2.4.3   Processing units forming truncated mean estimators $\bar{X}_\alpha$ based on systolic processors with: (a) binary elements; (b) ternary elements

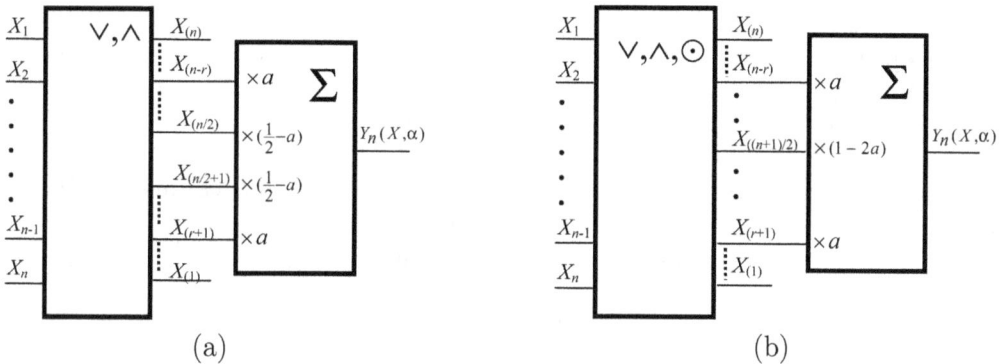

FIGURE 2.4.4   Processing units forming estimators of weighted sum $Y_n(X, \alpha)$ based on systolic processors with: (a) binary elements; (b) ternary elements

Definition 2.4.5. *Estimators of weighted sum* $Y_n(X, \alpha)$ *of a pair of symmetrical order statistics are L-estimators of the following form:*

$$Y_n(X, \alpha)$$
$$= \begin{cases} \alpha(X_{(r+1)} + X_{(n-r)}) + (\frac{1}{2} - \alpha)(X_{(n/2)} + X_{(n/2+1)}), & \mathrm{mod}_2 n = 0; \\ \alpha(X_{(r+1)} + X_{(n-r)}) + (1 - 2\alpha)X_{((n+1)/2)}, & \mathrm{mod}_2 n = 1. \end{cases} \quad (2.4.5)$$

Block diagrams of processing units forming estimators of weighted sum of a pair of symmetrical order statistics $Y_n(X, \alpha)$ based on systolic processors with binary/ternary elements are shown in Fig. 2.4.4a, b, respectively. In the case of even sample size ($\mathrm{mod}_2 n = 0$), calculating the estimator $Y_n(X, \alpha)$ by the upper formula of the equation system (2.4.5) is based on systolic processor with binary elements (Fig. 2.4.4a). In the case of odd sample size ($\mathrm{mod}_2 n = 1$), calculating the estimator $Y_n(X, \alpha)$ by the lower formula of the system (2.4.5) is based on systolic processor with ternary elements (Fig. 2.4.4b).

2.5 Algorithms of Forming R-estimators Based on L-group Operations

Briefly consider one more class of estimators of a center of symmetric distribution, namely, R-estimators. R-estimators are based on rank criteria for hypothesis testing on symmetry center of a distribution [89]. Here we cite the definition of R-estimators from [90] which is written in terms on the sample space $\mathcal{L}(\mathcal{X}, \mathcal{B}_{\mathcal{X}}; +, \vee, \wedge)$ with L-group properties.

Definition 2.5.1. R-*estimator* is a solution of the following equation with respect to parameter m:

$$W(m) = \sum_{i=1}^{2n} e_i(m) K\left(\frac{i}{2n+1}\right) = 0 \qquad (2.5.1)$$

where $e_i(m) = 1[(X_i - m) \wedge -(X_i - m)]$; $1(t)$ is Heaviside unit step function; $X = (X_1, X_2, \ldots, X_n)$ is initial sample; $K(t)$ is nondecreasing function defined in the interval $]0,1[$ which satisfies the condition: $K(1 - t) = -K(t)$.

Among R-estimators, one can consider Hodges-Lehmann estimators $W_n(X)$, $W_n'(X)$ that are medians of $n(n + 1)/2$ $(n(n - 1)/2)$ pairwise averages $m_{i,j} = \frac{1}{2}(X_{(i)} + X_{(j)})$ including (excluding) the observations themselves, respectively:

$$W_n(X) = \operatorname*{med}_{i \leq j}\{m_{i,j}\}; \qquad (2.5.2a)$$

$$W_n'(X) = \operatorname*{med}_{i < j}\{m_{i,j}\}, \qquad (2.5.2b)$$

where $\operatorname{med}\{a_{i,j}\}$ is a sample median of some sample set of the elements $\{a_{i,j}\}$, $i \in I$, $j \in J$.

Block diagrams of processing units forming Hodges-Lehmann estimators $W_n'(X)$ based on systolic processors with binary/ternary elements are shown in Fig. 2.5.1a,b, respectively. In the case of even number $K = n(n-1)/2$ ($\operatorname{mod}_2 K = 0$) of pairwise averages $m_{i,j} = Y_k$, $k = 1, \ldots, K$, calculating the estimator $W_n'(X)$ according to the formula (2.5.2b) is based on systolic processor with binary elements (Fig. 2.5.1a). In the case of odd number $K = n(n - 1)/2$ ($\operatorname{mod}_2 K = 1$) of pairwise averages $m_{i,j} = Y_k$, $k = 1, \ldots, K$, forming the estimator $W_n'(X)$ according to the formula (2.5.2b) is based on systolic processor with ternary elements (Fig. 2.5.1b).

Hodges estimator [91] is also related to R-estimators:

$$R_n(X) = \operatorname*{med}_{1 \leq i \leq [\frac{n}{2}]}\{m_{i,n-i+1}\}, \qquad (2.5.3)$$

where $[t]$ is an integer part of t.

Statistic $R_n(X)$ is appeared to be much simpler calculated than Hodges-Lehmann estimators $W_n(X)$, $W_n'(X)$ that, however, are a little bit more efficient than $R_n(X)$.

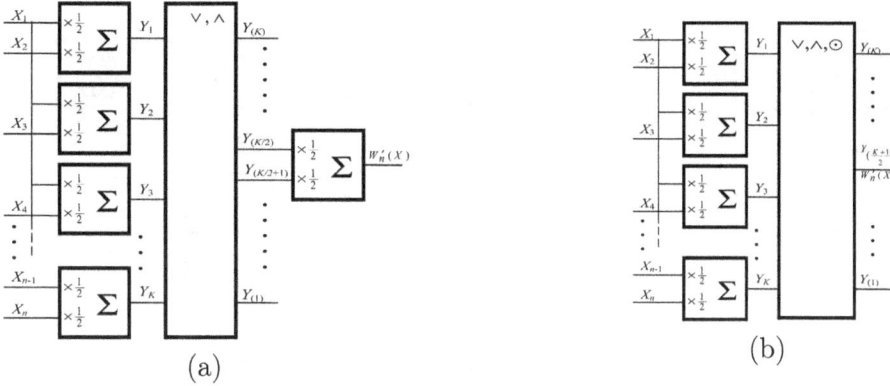

FIGURE 2.5.1 Processing unit forming Hodges-Lehmann estimator $W'_n(X)$ based on systolic processor with: (a) binary elements; (b) ternary elements

Block diagrams of processing units forming Hodges estimators $R_n(X)$ based on systolic processors with binary/ternary elements are shown in Fig. 2.5.2a, b, respectively. In the case of even number $K = n(n-1)/2$ ($\mathrm{mod}_2 K = 0$) of pairwise averages $m_{i,j} = Y_k$, $k = 1, \ldots, K$, calculating the estimator $R_n(X)$ according to the formula (2.5.3) is based on systolic processor with binary elements (Fig. 2.5.1a). In the case of odd number $K = n(n-1)/2$ ($\mathrm{mod}_2 K = 1$) of pairwise averages $m_{i,j} = Y_k$, $k = 1, \ldots, K$, forming the estimator $R_n(X)$ according to the formula (2.5.3) is realized by systolic processor with ternary elements (Fig. 2.5.1b).

2.6 Comparative Efficiency of Some Estimators

One of the sides of estimator quality is its asymptotic relative efficiency (ARE) with respect to sample mean estimator $\bar{X}$ that shows how much better (or worse) is an explored estimator with respect to such a widely used estimator as the estimator $\bar{X}$ is.

Asymptotic relative efficiency $e_{\bar{X}_\alpha, \bar{X}}$ of truncated mean estimator $L_n(X, \alpha)$ with respect to sample mean estimator $\bar{X}$ is defined according to the formula [75]:

$$e_{\bar{X}_\alpha, \bar{X}} = D/D_\alpha, \qquad (2.6.1)$$

where D is a variance of a distribution F; D_α is a variance of truncated mean estimator $L_n(X, \alpha) = \bar{X}_\alpha$ that, according to the formula [75, § 5.4 (2)], is equal to:

$$D_\alpha = \frac{2}{(1 - 2\alpha)^2} \left[\int_0^{\xi(1-\alpha)} t^2 f(t)dt + \alpha \xi^2 (1 - \alpha) \right] ; \quad F[\xi(\beta)] = \beta, \qquad (2.6.1a)$$

where $f(x)$ is PDF of a sample $X = (X_1, X_2, \ldots, X_n)$ with a distribution F.

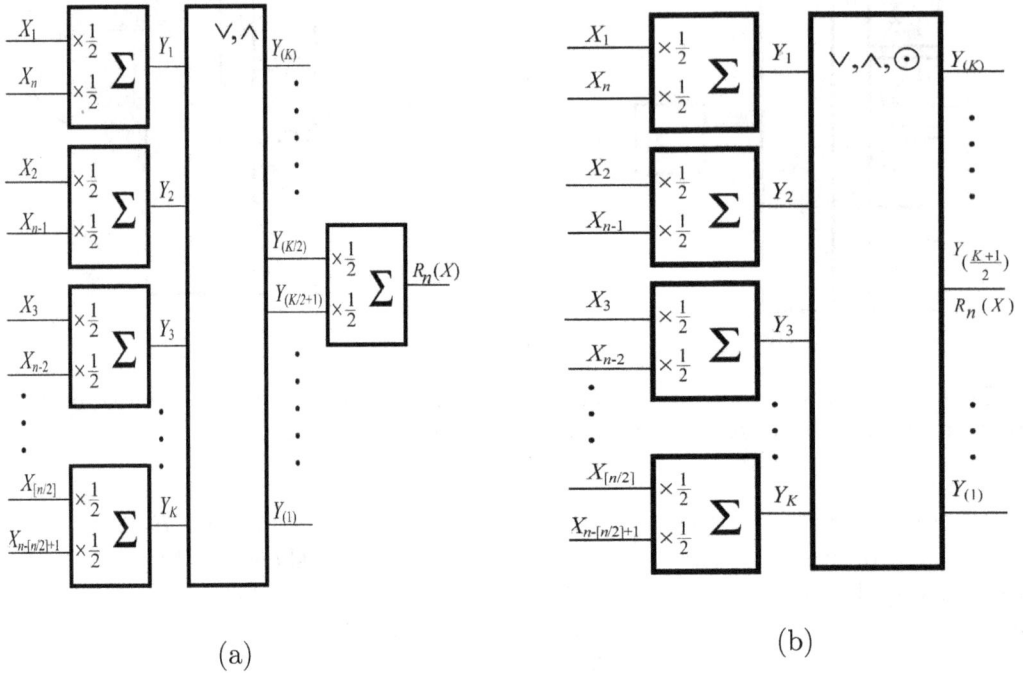

(a)   (b)

FIGURE 2.5.2   Processing units forming Hodges estimator $R_n(X)$ based on systolic processors with: (a) binary elements; (b) ternary elements

For symmetric distributions F, ARE of Hodges-Lehmann estimator $W_n(X)$ with respect to sample mean estimator $\bar{X}$, according to the formula [75, § 5.6 (29)], is equal to:

$$e_W = 12 \left(\int f^2(x)dx \right)^2 D, \qquad (2.6.2)$$

where D is a variance of distribution F.

Values of AREs of truncated mean estimator $L_n(X, \alpha)$ and Hodges-Lehmann estimator $W_n(X)$ with respect to sample mean estimator $\bar{X}$ obtained by the formulas (2.6.1) and (2.6.2) for some symmetrical distributions are shown in the Table 2.6.2.

Notations of distributions from the Table 2.6.2 together with their name and PDFs are shown in the Table 2.6.1.

TABLE 2.6.1   Notations of distributions from the Table 2.6.2

Notation	Distribution	PDF
$N(a,b)$	Normal	$\dfrac{1}{\sqrt{2\pi}b} \exp\left[-(x-a)^2/(2b^2)\right]$
$T(a,\varepsilon,\tau)$	Tukey	$(1-\varepsilon)N(a,1) + \varepsilon N(a,\tau)$

continued on next page

TABLE 2.6.1   Notations of distributions from the Table 2.6.2

Notation	Distribution	PDF
$St(a,b,\nu)$	Student	$\dfrac{\Gamma[(\nu+1)/2]}{\sqrt{\pi\nu b^2}\Gamma(\nu/2)}\dfrac{1}{(1+(x-a)^2/(\nu b^2))^{(\nu+1)/2}}$
$L(a,b)$	Logistic	$\dfrac{1}{b}\dfrac{\exp[-(x-a)/b]}{[1+\exp[-(x-a)/b]]^2}$
$C(a,b)$	Cauchy	$\dfrac{1}{\pi b}\left[1+\left(\dfrac{x-a}{b}\right)^2\right]^{-1}$
$DE(a,b)$	Double exponential (Laplace)	$\dfrac{1}{2b}\exp[-|x-a|/b]$
$Tr(a,b)$	Triangular	$\dfrac{1}{b^2}(b-|a-x|)\times$ $\times[1(x-(a-b))-1(x-(a+b))]$
$U(a,b)$	Uniform	$\dfrac{1}{2b}[1(x-(a-b))-1(x-(a+b))]$

Another side of estimator quality is absolute efficiency, i.e. asymptotic efficiency with respect to the best estimator. Absolute efficiency $e_{\bar{X}_\alpha}$ of truncated mean estimator $L_n(X,\alpha)$ is equal to [75, § 5.4 (7)]:

$$e_{\bar{X}_\alpha} = 1/(D_\alpha I_f), \qquad (2.6.3)$$

where D_α is a variance of truncated mean estimator $L_n(X,\alpha){=}\bar{X}_\alpha$ determined by the function (2.6.1a); I_f is Fisher information.

TABLE 2.6.2   Asymptotic relative efficiency of truncated mean estimator $L_n(X,\alpha)$ and Hodges-Lehmann estimator $W_n(X)$ with respect to sample mean estimator $\bar{X}$

Distribution	$L_n(X,\alpha)$ $\alpha=0$	$1/20$	$1/8$	$1/4$	$3/8$	$1/2$	$W_n(X)$
$N(a,b)$	1	0.974	0.927	0.835	0.743	0.637	0.955
$T(a,0.01,3)$	1	1.027	0.987	0.886	0.787	0.678	1.009
$T(a,0.05,3)$	1	1.216	1.180	1.089	0.959	0.833	1.196
$L(a,b)$	1	1.072	1.088	1.039	0.943	0.822	1.097
$DE(a,b)$	1	1.201	1.387	1.635	1.841	2.00	1.50
$St(a,b,5)$	1	1.20	1.249	1.206	1.112	0.961	1.241
$St(a,b,3)$	1	1.688	1.907	1.964	1.836	1.618	1.896
$C(a,b)$	1	∞	∞	∞	∞	∞	∞
$Tr(a,b)$	1	0.898	0.822	0.744	0.699	0.667	0.889
$U(a,b)$	1	0.833	0.667	0.500	0.400	0.333	1.000

Absolute efficiency e_W of Hodges-Lehmann estimator $W_n(X)$ is defined by the formula [75, § 5.6 (35)]:

$$e_W = 12 \left(\int f^2(x)dx \right)^2 / I_f, \qquad (2.6.4)$$

where $f(x)$ is PDF of the elements of random sample $X = (X_1, X_2, \ldots, X_n)$ with a distribution F.

Values of absolute efficiency of truncated mean estimator $L_n(X, \alpha)$ and Hodges-Lehmann estimator $W_n(X)$ calculated on the basis of the formulas (2.6.3) and (2.6.4), respectively, for some symmetric distributions are listed in the Table 2.6.3.

TABLE 2.6.3   Absolute efficiency of truncated mean estimator $L_n(X, \alpha)$ and Hodges-Lehmann estimator $W_n(X)$

| Distribution | $L_n(X, \alpha)$ | | | | | | $W_n(X)$ |
	$\alpha = 0$	$1/20$	$1/8$	$1/4$	$3/8$	$1/2$	
$N(a, b)$	1	0.974	0.927	0.835	0.743	0.637	0.955
$T(a, 0.01, 3)$	0.953	0.979	0.941	0.845	0.75	0.647	0.962
$T(a, 0.05, 3)$	0.783	0.953	0.924	0.853	0.751	0.652	0.937
$L(a, b)$	0.912	0.978	0.993	0.947	0.86	0.75	1.00
$DE(a, b)$	0.50	0.601	0.693	0.817	0.921	1.00	0.75
$St(a, b, 5)$	0.80	0.96	0.999	0.965	0.89	0.769	0.993
$St(a, b, 3)$	0.501	0.846	0.956	0.984	0.92	0.811	0.95
$C(a, b)$	0	0.226	0.499	0.785	0.875	0.811	0.608

As for efficiency of Huber M-estimator, taking into consideration the identity [75, § 5.6 (16)], it is equivalent to efficiency of truncated mean estimator $L_n(X, \alpha)$.

2.7   Signal Parameter Estimation Algorithms in Spaces with L-group Properties

Let $s(t)$ be useful signal that is determined by a deterministic one-to-one function $M[*, *]$ of a signal-carrier $c(t)$ and informational parameter $\boldsymbol{\lambda}$ which is invariable in a domain T_s of the signal $s(t)$:

$$s(t) = M[c(t), \boldsymbol{\lambda}],$$

where $\boldsymbol{\lambda} = [\lambda_1, \ldots, \lambda_m]$, $\boldsymbol{\lambda} \in \Lambda$ is a vector of unknown informational parameters of the signal $s(t)$, $\lambda_1 = \text{const}, \ldots, \lambda_m = \text{const}$.

Then *estimation* of unknown vector parameter $\boldsymbol{\lambda} = [\lambda_1, \ldots, \lambda_m]$ of the signal $s(t)$ in the presence of interference (noise) $n(t)$ lies in forming (according to a certain criterion) a vector estimator $\hat{\boldsymbol{\lambda}}$ in the form of vector functional $F_{\hat{\lambda}}[x(t)]$ of the observed realization $x(t)$:

$$\hat{\boldsymbol{\lambda}} = F_{\hat{\boldsymbol{\lambda}}}[x(t)], \ t \in T_s, \ \hat{\boldsymbol{\lambda}} \in \boldsymbol{\Lambda}.$$

Within this section, we consider signal parameter estimation algorithms in the sample space $\mathcal{L}(\mathcal{X}, \mathcal{B}_{\mathcal{X}}; +, \vee, \wedge)$ with L-group properties (see Definition 2.0.1).

Most used model of direct measuring unknown nonrandom scalar parameter is defined by additive interaction between a location parameter λ and independent measurement errors in linear sample space $\mathcal{LS}(\mathcal{X}, \mathcal{B}_{\mathcal{X}}; +)$:

$$X_i = \lambda + N_i, \tag{2.7.1}$$

where $\{N_i\}$ are statistically independent identically distributed estimation (measurement) errors with a distribution from a class of distributions with symmetric PDF $p_N(z) = p_N(-z)$ represented by the sample $N = (N_1, \ldots, N_n)$, $N_i \in N$, so that $N \in \mathcal{L}(\mathcal{X}, \mathcal{B}_{\mathcal{X}}; +, \vee, \wedge)$ and $N \in \mathcal{LS}(\mathcal{X}, \mathcal{B}_{\mathcal{X}}; +)$;
$\{X_i\}$ are the results of observations represented by the sample $X = (X_1, \ldots, X_n)$;
$X_i \in X$, $X \in \mathcal{L}(\mathcal{X}, \mathcal{B}_{\mathcal{X}}; +, \vee, \wedge)$ and $X \in \mathcal{LS}(\mathcal{X}, \mathcal{B}_{\mathcal{X}}; +)$;
"+" is operation of addition of the sample space $\mathcal{L}(\mathcal{X}, \mathcal{B}_{\mathcal{X}}; +, \vee, \wedge)$ with L-group properties $\mathcal{L}(\mathcal{X}; +, \vee, \wedge)$ that corresponds to operation of addition of linear sample space $\mathcal{LS}(\mathcal{X}, \mathcal{B}_{\mathcal{X}}; +)$;
$i = 1, \ldots, n$ is the index of the elements of statistical totalities $\{N_i\}$, $\{X_i\}$;
n is a sample size of the samples $N = (N_1, \ldots, N_n)$, $X = (X_1, \ldots, X_n)$.

In the sample space $\mathcal{L}(\mathcal{X}, \mathcal{B}_{\mathcal{X}}; +, \vee, \wedge)$ with L-group properties, the problem of estimating a location parameter λ in the presence of measurement errors distributed on a class with symmetric PDF $p_N(z) = p_N(-z)$ is also described by the model (2.7.1). Such an analogy is explained by a commonality of the properties of such sample space, in particular, due to the properties of additive commutative group $\mathcal{L}(\mathcal{X}, \mathcal{B}_{\mathcal{X}}; +)$, $\mathcal{L}(\mathcal{X}, \mathcal{B}_{\mathcal{X}}; +) \subset \mathcal{L}(\mathcal{X}, \mathcal{B}_{\mathcal{X}}; +, \vee, \wedge)$.

In signal processing, statistical model of direct measurement of unknown nonrandom scalar parameter λ has a form that is similar to (2.7.1):

$$x(t_j) = \lambda + n(t_j), \tag{2.7.2}$$

where $x(t_j) \equiv X_i$, $n(t_j) \equiv N_i$; $t_j = t - \Delta t \cdot j$, $j = 0, 1, \ldots, n-1$, $i = 1, \ldots, n$, $j = i - 1$; $\{n(t_j)\}$ are independent measurement errors with a distribution from a class of symmetric PDFs $p_n(z) = p_n(-z)$ $(p_n(z) \equiv p_N(z))$ due to affecting noise $n(t)$ with zero mean; $\{x(t_j)\}$ are results of observations (samples) of continuous observed stochastic process $x(t)$; t is time parameter; Δt is a sample interval determining independence of the samples $\{n(t_j)\}$; $j = 0, 1, \ldots, n-1$ is an index of samples $\{n(t_j)\}$, $\{x(t_j)\}$.

Generalized block diagram of processing unit forming the estimator of location parameter λ (using one or another criterion) based on the model (2.7.2) is shown in Fig. 2.7.1. As follows from the figure, with help of time-delay line, a sample of independent observations $\{x(t_j)\}$ of stochastic process $x(t)$ is formed. Filter, realizing algorithm of processing the observations $\{x(t_j)\}$ according to a given criterion, forms the estimator $\hat{\lambda}$ of a parameter λ.

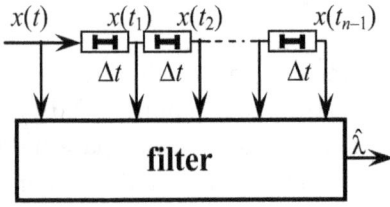

FIGURE 2.7.1 Generalized block diagram of location parameter estimator

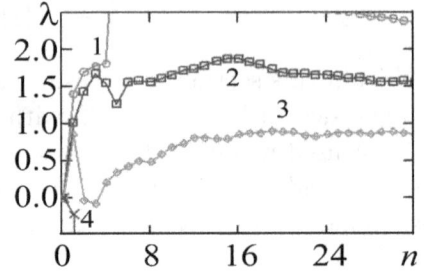

FIGURE 2.7.2 Convergence rate of estimate in the output of homogeneous filter

The subject of further consideration is comparative analysis of location parameter estimation by homogeneous filter (moving average filter) forming the estimator according to well known linear algorithm [92, (15.1)]:

$$\hat{\lambda}_{\text{hmF}} = \sum_{j=0}^{n-1} x(t - \Delta t \cdot j), \qquad (2.7.3)$$

and also by four filters realizing estimation in the sample space $\mathcal{L}(\mathcal{X}, \mathcal{B}_{\mathcal{X}}; +, \vee, \wedge)$ with L-group properties and forming the estimators (2.2.1a,b), (2.3.7), (2.4.3), (2.5.2a) according to the following algorithms, respectively:

$$\hat{\lambda}_{\text{MF}} = \operatorname*{med}_{j=0,\dots,n-1} \{x(t - \Delta t \cdot j)\}; \qquad (2.7.4a)$$

$$\hat{\lambda}_{\text{HF}} = \hat{\lambda}(t) = \hat{\lambda}(t - \Delta t) + \frac{\hat{\sigma}}{n} \cdot \sum_{j=0}^{n-1} k \wedge [(x(t - \Delta t \cdot j) - \hat{\lambda}(t - \Delta t)) \vee -k]; \quad (2.7.4b)$$

$$\hat{\lambda}_{\text{tmF}} = \frac{1}{n - 2r} \sum_{i=r+1}^{n-r} X_{(i)}; \qquad (2.7.4c)$$

$$\hat{\lambda}_{\text{HL}} = \operatorname*{med}_{i \le k} \{m_{i,k}\}, \qquad (2.7.4d)$$

where $\hat{\lambda}_{\text{MF}}$, $\hat{\lambda}_{\text{HF}}$, $\hat{\lambda}_{\text{tmF}}$, $\hat{\lambda}_{\text{HL}}$ are the estimators formed by median filter, Huber filter, truncated mean filter, and Hodges-Lehmann filter, respectively;
$\alpha = r/n < 1/2$, $r = [\alpha n]$, $[t]$ is the integer part of t;
$X_{(i)}$ is i-th order statistic of variational series $X' = (X_{(1)}, X_{(2)}, \dots, X_{(n)})$; $X_{(i)} \in X'$, $X_i \equiv x(t - \Delta t \cdot j)$, $j = i - 1$;
$m_{i,k} = \frac{1}{2}(X_{(i)} + X_{(k)})$ are $n(n+1)/2$ pairwise averages;
$\text{med}\{a_{i,k}\}$ is a sample median of a set of the elements $\{a_{i,k}\}$.

Fig. 2.7.2, 2.7.3a,b,c,d illustrate the results of statistical modeling of convergence rate of estimate in the output of the filters forming the estimators according

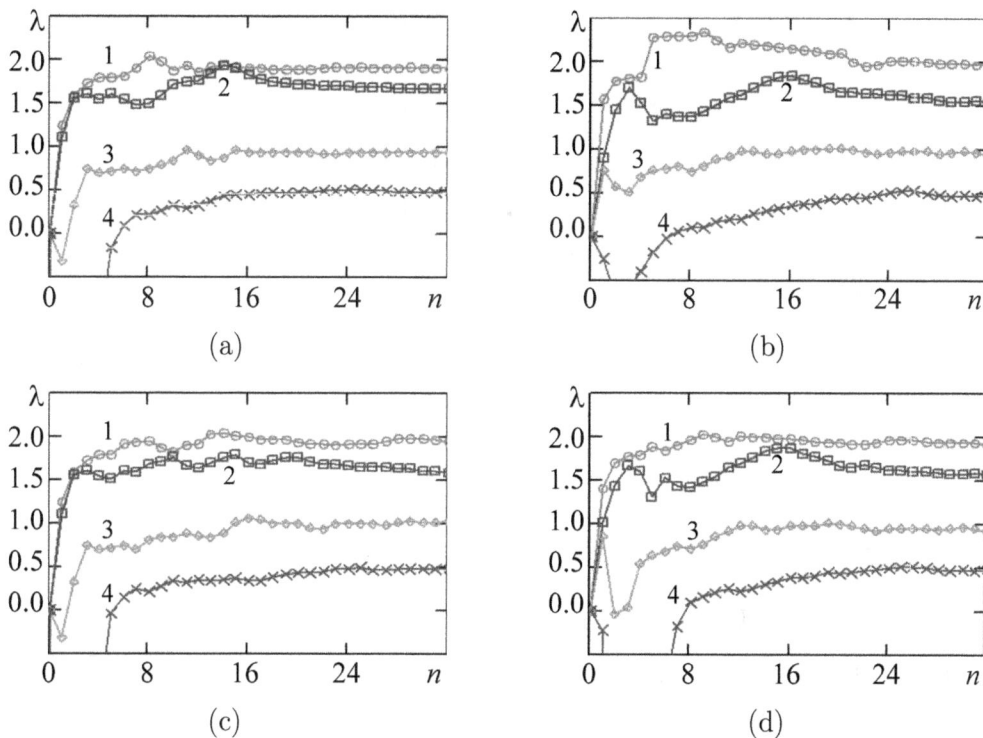

FIGURE 2.7.3 Convergence rate of estimate in the outputs of the filters forming the estimators according to algorithms (2.7.4a...2.7.4d), respectively: (a) median filter; (b) Huber filter; (c) truncated mean filter; (d) Hodges-Lehmann filter

to algorithms (2.7.3), (2.7.4a...2.7.4d), respectively. Simulation is carried out under the same conditions used for the estimation model (2.7.2), so that we consider independent measurement errors $\{n(t_j)\}$ with the following four distributions: 1—ε-contaminated Tukey distribution $T(0.05, 5)$ with a sample variance 1.0 and location parameter λ=2.0 (curve denoted by circles); 2—logistic distribution with a sample variance 1.0 and location parameter λ=1.5 (curve denoted by squares); 3—Laplace (double exponential) distribution with a sample variance 1.0 and location parameter λ=1.0 (curve denoted by rhombuses); 4—Cauchy distribution with a sample variance and location parameter λ=0.5 (curve denoted by crosses). Estimate convergence to the estimated parameter λ (within the model (2.7.2)) is shown depending on a number n of processed values $\{x(t_j)\}$. Sample size (filter window) used in processing is chosen to be equal to 33.

Realizations of discrete observations $\{x(t_j)\}$ of stochastic process $x(t)$ in the inputs of filters with four mentioned distributions: Tukey, logistic, Laplace, and Cauchy for the first 200 samples are shown in Fig. 2.7.4a,b,c,d, respectively.

As follows from Fig. 2.7.3a,b,c,d, median filter, Huber filter, truncated mean filter, and Hodges-Lehmann filter possess quite good convergence to estimated parameter for all four pointed distributions of observations $\{x(t_j)\}$. Homogeneous filter behaves itself similarly for first three distributions, that, however, is explained

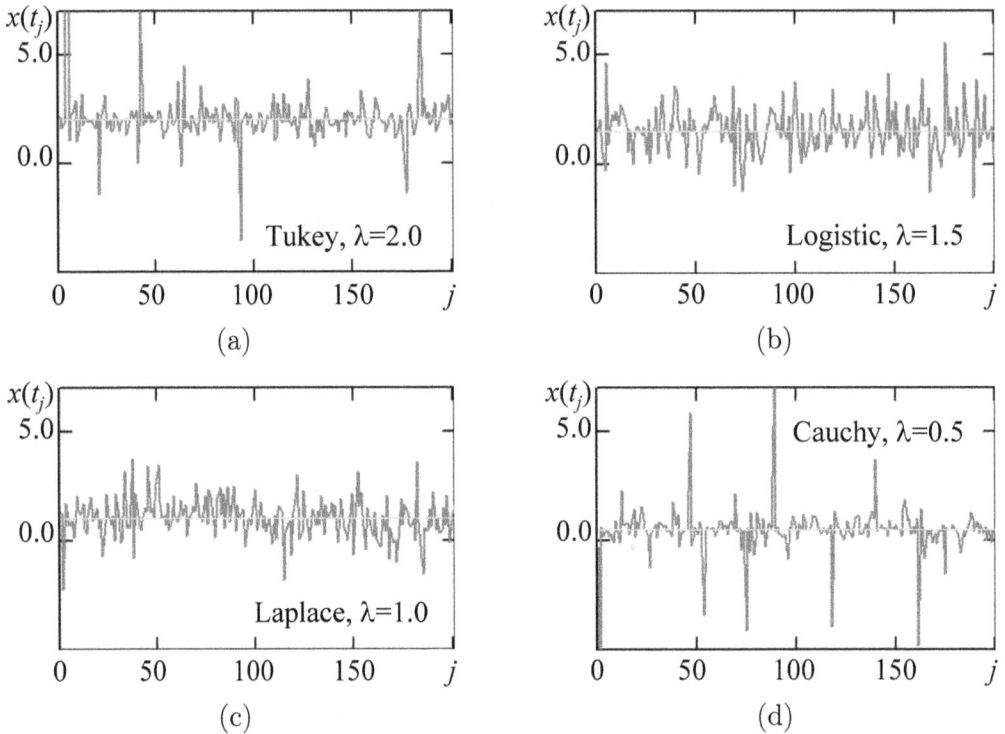

FIGURE 2.7.4　Realizations of observations $\{x(t_j)\}$ of stochastic process $x(t)$ in the inputs of filters with distributions: (a) Tukey; (b) logistic; (c) Laplace; (d) Cauchy

by rather small sample variance of observations $\{x(t_j)\}$. On the contrary, as for Cauchy distribution, in the case of rather large outliers among the observations $\{x(t_j)\}$ (see Fig. 2.7.4d), homogeneous filter can not adequately form the estimate whose values tend to arbitrarily large negative value on the 2nd sample.

Figs. 2.7.5a,b,c, 2.7.6a,b,c illustrate the results of statistical modeling of a dependence of relative sample variance $\delta_{\hat\lambda} = D_{\hat\lambda}/D_x$ (D_x is a sample variance of observations $\{x(t_j)\}$) of an estimate $\hat\lambda$ of location parameter λ within the estimation model (2.7.2) in the outputs of filters operating according to algorithms (2.7.3), (2.7.4a...2.7.4d) on the number j of samples used when forming the estimate $\{\hat\lambda(t_j)\}$, $\hat\lambda = \hat\lambda(t_j)$.

Independent measurement errors $\{n(t_j)\}$ are described by the following six symmetric distributions: Fig. 2.7.5a — normal $N(0, b)$; Fig. 2.7.5b — "ε-contaminated" Tukey $T(\varepsilon, \tau)$ with $\varepsilon = 0.05$, $\tau = 5$; Fig. 2.7.5c — logistic $L(0, b)$; Fig. 2.7.6a — Student $St(3)$; Fig. 2.7.6b — Laplace (double exponential) $DE(0, b)$; Fig. 2.7.6c — Cauchy $C(0, b)$ with sample variance 50.0.

Random variables with aforementioned distributions, excepting Tukey and Laplace distributions, are simulated by standard built-in MathCAD's random numbers generator. Random variable ξ_T with Tukey distribution $T(\varepsilon, \tau)$ is simulated basing on linear combination of two independent random variables ξ_1, ξ_2 with standard normal distribution $N(0, 1)$ and random variable with Bernoulli distribution

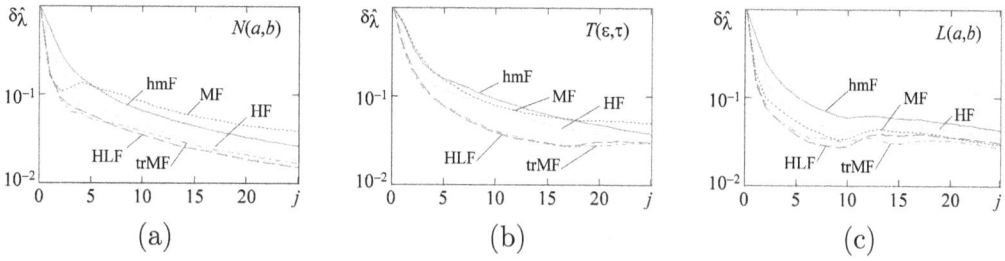

FIGURE 2.7.5   Dependences of relative sample variance $\delta_{\hat{\lambda}}$ in the outputs of filters on the number j of samples in the presence of measurement errors with the following distributions: (a) normal; (b) Tukey; (c) logistic

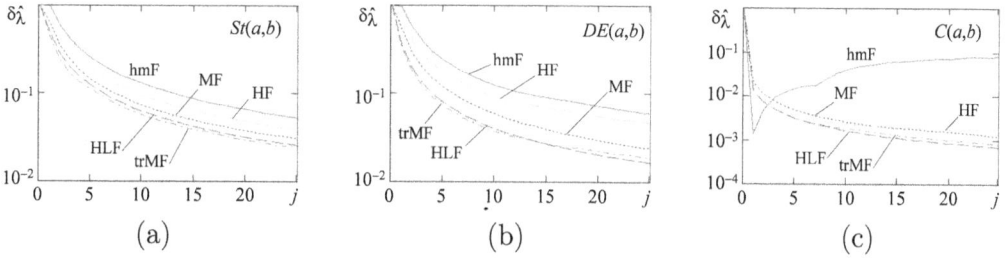

FIGURE 2.7.6   Dependences of relative sample variance $\delta_{\hat{\lambda}}$ in the outputs of filters on a number j of samples in the presence of measurement errors with the following distributions: (a) Student; (b) Laplace (double exponential); (c) Cauchy

$\xi_B(\varepsilon)$ with parameter ε:

$$\xi_T(\varepsilon, \tau) = (1 - \xi_B(\varepsilon)) \cdot \xi_1 + \xi_B(\varepsilon) \cdot \tau \cdot \xi_2.$$

Random variable with Laplace distribution is obtained basing on the difference of two independent random variables with exponential distribution.

The size of filter window n used when processing is equal to 33. Notations shown in Figs. 2.7.5a,b,c, 2.7.6a,b,c correspond to the following devices: hmF—homogeneous filter; HF—Huber filter; MF—median filter; trMF—truncated mean filter; HLF—Hodges-Lehmann filter.

As follows from the simulation results, Hodges-Lehmann filter and truncated mean filter, as compared to other filters, possess the higher accuracy of estimating location parameter achieved at all mentioned distributions excepting Laplace one. Median filter, as expected, shows the higher efficiency on Laplace distribution of measurement errors. On the contrary, in normal and logistic distribution of errors, median filter provides the worst accuracy of estimation. Huber filter, that must have an efficiency rather close to efficiency of truncated mean filter, behaves itself in this way for all mentioned distributions excepting Tukey model.

In normal distribution of measurement errors $\{n(t_j)\}$, the least efficiency is related with homogeneous filter, that, however, contradicts to known theoretical results. This contradiction could be explained by some mismatch between normality and real statistical properties of small samples that are normally distributed.

In the case of processing the observations $\{x(t_j)\}$ with Cauchy distribution, homogeneous filter is inefficient. On the contrary, relative sample variance $\delta_{\hat{\lambda}}$ of Huber filter, truncated mean filter, Hodges-Lehmann filter, and median filter is two orders less than that of homogeneous filter.

As for rather essential "contradictions" with theoretical results, describing asymptotic relative efficiency of estimators (see, for instance, table 2.6.3), they could be explained by rather small number (up to 32) of processed observations $\{x(t_j)\}$. For such values of n, statistical properties of a sample could noticeably differ from declared properties of random number generators, so that some correspondence become clear for sample size $n \geq 1000$.

As follows from the simulation results shown in Figs. 2.7.5a,b,c, 2.7.6a,b,c, filters realizing estimation in the sample space $\mathcal{L}(\mathcal{X}, \mathcal{B}_\mathcal{X}; +, \vee, \wedge)$ with L-group properties, i.e., those that form the estimators (2.7.4a,b,c,d) and possess robust property with respect to distribution of measurement error $\{n(t_j)\}$.

Fig. 2.7.7 illustrates the simulation results of the dependences of relative sample variance $\delta_{\hat{\lambda}} = D_{\hat{\lambda}}/D_x$ of estimate $\hat{\lambda}$ of a location parameter λ within the estimation model (2.7.2) for the filters described by algorithms (2.7.3), (2.7.4a...2.7.4d) on the ratio m_1^2/m_2 of squared expectation m_1^2 to a second moment m_2 of envelopes of measurement errors $\{n(t_j)\}$. Independent measurement errors $\{n(t_j)\}$ are described by six symmetric distributions: normal $N(0, b)$, $m_1^2/m_2 = \pi/4$; logistic $L(0, b)$, $m_1^2/m_2 \approx 0.75$; Laplace (double exponential) $DE(0, b)$, $m_1^2/m_2 \approx 0.71$; Student $St(3)$, $m_1^2/m_2 \approx 0.67$; ε-contaminated Tukey $T(\varepsilon, \tau)$ with $\varepsilon = 0.05$, $\tau = 5$, $m_1^2/m_2 \approx 0.48$; Cauchy $C(0, b)$, $m_1^2/m_2 \approx 0.15$. Ratios m_1^2/m_2 for envelopes of measurement errors $\{n(t_j)\}$ with all distributions, excepting normal one, were determined experimentally. Filter window size n used in processing is equal to 32. Notations shown in Fig. 2.7.8 correspond to the following devices: hmF – homogeneous filter, solid line denoted by crosses $\times - \times - \times$; HF – Huber filter, solid line denoted by squares $\square - \square - \square$; MF – median filter, dotted line denoted by rhombuses $\diamondsuit - \diamondsuit - \diamondsuit$; trMF – truncated mean filter, dashed line denoted by circles $\circ - \circ - \circ$; HLF – Hodges-Lehmann filter, dashed-dotted line denoted by pluses $+ - + - +$.

Conditionally, interference whose envelopes have the ratio m_1^2/m_2 within double inequality $0 < m_1^2/m_2 \leq 0.725$ are related to noises of impulse type, while interference whose envelopes have the ratio m_1^2/m_2 within double inequality $0.725 < m_1^2/m_2 \leq 1$ are related to noises of harmonic type [93]. Remind that stochastic process with Rayleigh distribution corresponds to an envelope of Gaussian stochastic process with $m_1^2/m_2 = \pi/4 \approx 0.785$.

As follows from the results of simulating the dependences of relative sample variance $\delta_{\hat{\lambda}} = D_{\hat{\lambda}}/D_x$ of estimate $\hat{\lambda}$ of a location parameter λ on the ratio between two first moments m_1^2/m_2 of envelope of measurement errors $\{n(t_j)\}$ shown in Fig. 2.7.7, at a qualitative level, one can underline two main features. First, for homogeneous filter (curve hmF) relative sample variance $\delta_{\hat{\lambda}}$ weakly depends on a distribution, so that the difference between the largest $\delta_{\hat{\lambda},\max}$ and the smallest $\delta_{\hat{\lambda},\min}$ values of relative sample variance $\delta_{\hat{\lambda}}$ does not exceeds $\sqrt{3}$: $\delta_{\hat{\lambda},\max}/\delta_{\hat{\lambda},\min} < \sqrt{3}$. Second, for median filter (MF), Huber filter (HF), truncated mean filter (trMF), and

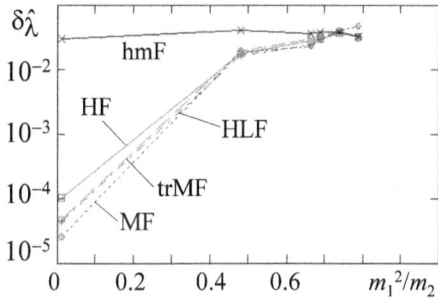

FIGURE 2.7.7 Dependences of relative sample variance $\delta_{\hat{\lambda}}$ on the ratio m_1^2/m_2

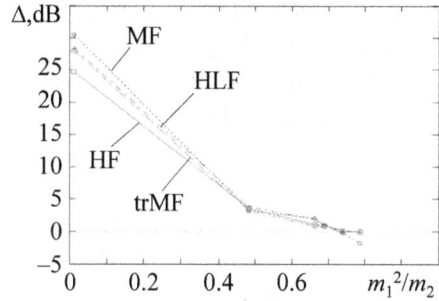

FIGURE 2.7.8 Dependences of gain (loss) $\Delta = 10 \lg(\delta_{\hat{\lambda},\text{hmF}}/\delta_{\hat{\lambda}})$ on the ratio m_1^2/m_2

Hodges-Lehmann filter (HLF), on the contrary, relative sample variance $\delta_{\hat{\lambda}}$ rather essentially depends on a distribution, so that the difference between the largest $\delta_{\hat{\lambda},\text{max}}$ and the smallest $\delta_{\hat{\lambda},\text{min}}$ values of relative sample variance $\delta_{\hat{\lambda}}$ is greater than 2.5 orders: $\delta_{\hat{\lambda},\text{max}}/\delta_{\hat{\lambda},\text{min}} > 10^{2.5}$, and the curves illustrating estimation accuracy of these filters are rather close to each other, especially in the interval $0.45 \leq m_1^2/m_2 \leq 0.71$.

The behavior of dependence $\delta_{\hat{\lambda}}$ for homogeneous filter is quite understandable: sample variance $D_{\hat{\lambda}}$ of the estimator in the output of the filter is equal to $D_{\hat{\lambda}} = D_x/n$ (D_x is a sample variance of observations $\{x(t_j)\}$), therefore $\delta_{\hat{\lambda}} = D_{\hat{\lambda}}/D_x = 1/n$, that for a pointed sample size $n=33$ is equal to $\delta_{\hat{\lambda}} \approx 0.03$ regardless of distribution of errors.

For the filters forming the estimators according to algorithms (2.7.4a...2.7.4d), the relationship $\delta_{\hat{\lambda}} = D_{\hat{\lambda}}/D_x = 1/n$ does not hold, since these filters, unlike homogeneous one, exclude from the processing the observation values $\{x(t_j)\}$ that are considered to be extremal (from the standpoint of own processing algorithm). As a consequence, in the presence of measurement errors $\{n(t_j)\}$ distributed with «tails» that are heavier than those of normal distribution (with a ratio m_1^2/m_2 from the interval $0 < m_1^2/m_2 \leq 0.725$), filters realizing estimation in the sample space $\mathcal{L}(\mathcal{X}, \mathcal{B}_{\mathcal{X}}; +, \vee, \wedge)$ with L-group properties can provide higher estimation accuracies than linear filters operating in linear sample space $\mathcal{LS}(\mathcal{X}, \mathcal{B}_{\mathcal{X}}; +)$.

Fig. 2.7.8 depicts dependences $\Delta = 10 \lg(\delta_{\hat{\lambda},\text{hmF}}/\delta_{\hat{\lambda}})$ of gain (loss) obtained when processing the observations in one of four filters forming the estimators according to algorithms (2.7.4a...2.7.4d) with respect to homogeneous filter hmF on the ratio m_1^2/m_2 of squared expectation m_1^2 to the second moment m_2 of envelope of measurement errors $\{n(t_j)\}$, where $\delta_{\hat{\lambda},\text{hmF}}$, $\delta_{\hat{\lambda}}$ are relative sample variance $\delta_{\hat{\lambda}} = D_{\hat{\lambda}}/D_x$ of estimate $\hat{\lambda}$ of a location parameter λ for homogeneous filter and other filters, respectively. Initial data used for calculating correspond to dependencies shown in Fig. 2.7.7. Notations shown in Fig. 2.7.8 correspond to the following devices: HF – Huber filter, solid line denoted by squares $\square - \square - \square$; MF – median filter, dotted line denoted by rhombuses $\Diamond - \Diamond - \Diamond$; trMF – truncated mean filter, dashed line denoted by circles o — o — o; HLF – Hodges-Lehmann filter, dashed-dotted line denoted by pluses + − + − +.

As follows from simulation results shown in Fig. 2.7.8, in normal distribution of measurement errors $\{n(t_j)\}$, median filter is inferior to homogeneous one in 1.61 dB; truncated mean filter — in 0.7 dB; Hodges-Lehmann filter — in 0.28 dB, and Huber filter provides the same relative sample variance as homogeneous one.

In logistic distribution of measurement errors $\{n(t_j)\}$, median filter loses 0.85 dB to homogeneous one, the rest of the filters provide approximately the same efficiency.

In double exponential distribution of measurement errors $\{n(t_j)\}$, all filters provide a gain in processing: median filter – in 2.1 dB; Hodges-Lehmann filter – in 1.2 dB; truncated mean filter – in 0.7 dB; and Huber filter – in 0.93 dB.

The same situation, but with a larger gain, is appeared in the case when measurement errors $\{n(t_j)\}$ possess Student distribution $St(3)$: all the filters provide gain in 2.8 dB.

In ε-contaminated Tukey distribution $T(\varepsilon, \tau)$ with $\varepsilon = 0.05$, $\tau = 5$, median filter provides gain in 4.5 dB; truncated mean filter and Huber filter — 3.5 dB; Hodges-Lehmann filter — 5.0 dB.

More essential gain take place in the case of Cauchy distributed measurement results $\{n(t_j)\}$: median filter provides gain in 28.3 dB; truncated mean filter and Hodges-Lehmann filter in 27.3 dB, and Huber filter in 22.4 dB.

On the whole, the obtained results rather well describe the known theoretical results, listed, for instance, in the table 2.6.3, whose meaning lies in follows. Filters based on estimation algorithms in the sample space $\mathcal{L}(\mathcal{X}, \mathcal{B}_{\mathcal{X}}; +, \vee, \wedge)$ with L-group properties, i.e., those that form the estimators (2.7.4a...2.7.4d) provide a gain as against homogeneous filter on a wide class of distributions that describe the behavior of interference (noise) of impulse type, so that the less is the ratio m_1^2/m_2 characterizing a ratio m_1^2/m_2 between first two moments of envelope of measurements errors the greater the obtained gain is.

It is desirable to check the last statement for the case of measurement errors $\{n(t_j)\}$ with other statistical properties. To obtain the results that are similar to those shown in Figs. 2.7.7, 2.7.8, but for other distributions, we use the following classes of distributions of envelopes of measurement errors:
Weibull $W(c, \alpha)$, $m_1^2/m_2 = \Gamma^2(1 + \frac{1}{\alpha})/\Gamma(1 + \frac{2}{\alpha})$:

$$p_W(x) = c\alpha x^{\alpha-1} \exp\{-cx^\alpha\}, \ x \geq 0, \ c > 0, \ \alpha > 0; \qquad (2.7.5)$$

Gamma $\Gamma(\beta, \alpha)$, $m_1^2/m_2 = \alpha/(1 + \alpha)$:

$$p_\Gamma(x) = \frac{1}{\beta^\alpha \Gamma(\alpha)} x^{\alpha-1} \exp\{-x/\beta\}, \ x \geq 0, \ \beta > 0, \ \alpha > 0; \qquad (2.7.6)$$

Lognormal $Ln(m, \sigma)$, $m_1^2/m_2 = \exp(-\sigma^2)$:

$$p_{Ln}(x) = \frac{1}{x\sigma\sqrt{2\pi}} \exp\left\{-\frac{(\ln x - m)^2}{2\sigma^2}\right\}, \ x \geq 0, \ m \geq 0, \ \sigma > 0. \qquad (2.7.7)$$

An appearance of PDFs of Weibull $W(c, \alpha)$, lognormal $Ln(0, \sigma)$ and gamma

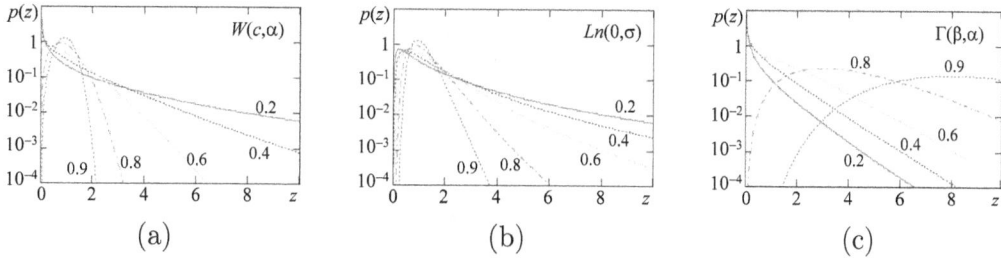

FIGURE 2.7.9   PDFs of distributions: (a) Weibull $W(c, \alpha)$; (b) lognormal $Ln(0, \sigma)$; (c) gamma $\Gamma(\beta, \alpha)$

$\Gamma(\beta, \alpha)$ distributions for the values $m_1^2/m_2 = 0.2$, 0.4, 0.6, 0.8, 0.9 are shown in logarithmic scale in Fig. 2.7.9a,b,c, respectively.

In general, a relation between the ratio m_1^2/m_2 of envelope of noise (measurement errors) and distribution "tails" can be revealed by the following empiric relationship:

$$\int_{\sqrt{m_2}/2}^{\infty} p(z)\,\mathrm{d}\,z \in \left[\frac{5}{6}\frac{m_1^2}{m_2}, \frac{5}{6}\frac{m_1^2}{m_2} + \frac{7}{32} \right] \tag{2.7.8}$$

where $p(z)$ is PDF of envelope of noise (measurement errors).

Some values of integral (2.7.8) for PDFs (2.7.5)...(2.7.7) are shown in the Table 2.7.1.

TABLE 2.7.1   Values of integral (2.7.8)

m_1^2/m_2	\multicolumn{3}{c}{Distribution, value of integral $\int_{\sqrt{m_2}/2}^{\infty} p(z)\,\mathrm{d}\,z$}		
	Weibull, $W(c, \alpha)$	lognormal, $Ln(0, \sigma)$	Gamma, $\Gamma(\beta, \alpha)$
0.2	0.238	0.235	0.239
0.4	0.406	0.407	0.407
0.6	0.584	0.599	0.585
0.8	0.795	0.839	0.811
0.9	0.919	0.965	0.947

Within the estimation model (2.7.2), Fig. 2.7.10a,b,c,d illustrates the results of simulating dependences of relative sample variance $\delta_{\hat{\lambda}} = D_{\hat{\lambda}}/D_x$ of estimate $\hat{\lambda}$ of a location parameter λ in the outputs of the filters forming the estimators according to algorithms (2.7.4a...2.7.4d), respectively, on the ratio m_1^2/m_2 of squared expectation m_1^2 to the second moment m_2 of envelopes of measurement errors $\{n(t_j)\}$ distributed on three PDF classes that are described by (2.7.5)...(2.7.7). Parameter m of lognormal distribution $Ln(m, \sigma)$ (2.7.7) is chosen to be equal to zero: $m = 0$. In each of three classes of distributions, envelopes of independent measurement errors $\{n(t_j)\}$ are characterized by the ratio m_1^2/m_2 that takes the following values: $m_1^2/m_2 = 0.2$; 0.4; 0.6; 0.7; 0.8; 0.9. Notations shown in Fig. 2.7.10a,b,c,d correspond to the following devices: MF – median filter, HF – Huber filter, trMF – truncated mean filter, HLF – Hodges-Lehmann filter. Envelope distributions of measurements errors described by PDFs (2.7.5)...(2.7.7) are denoted by the

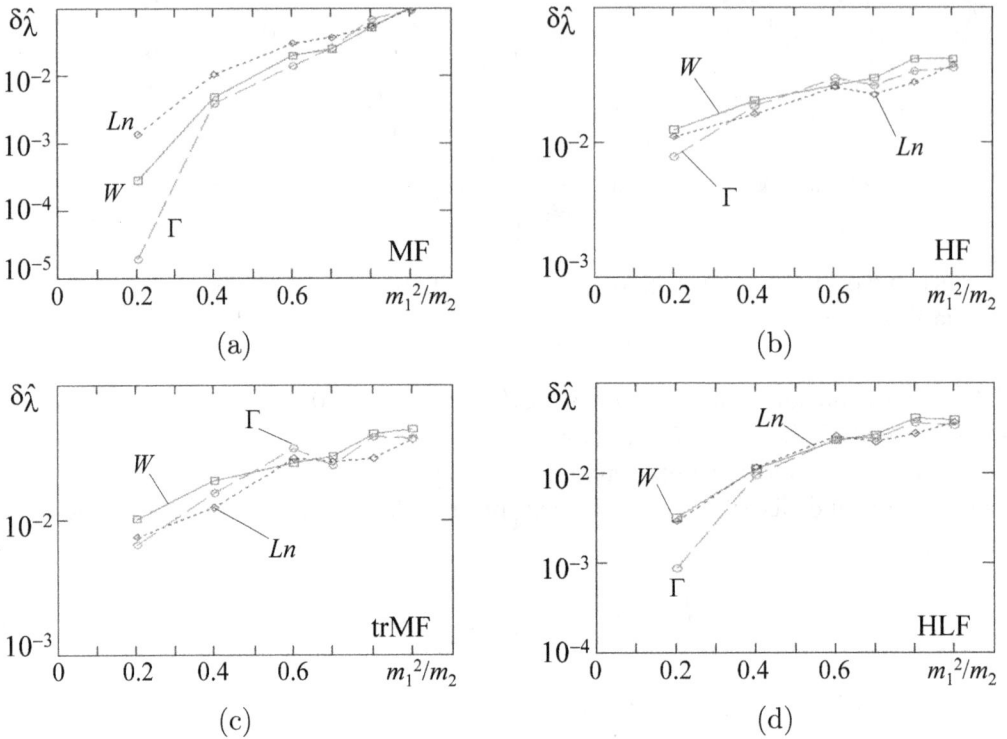

FIGURE 2.7.10   Dependences of relative sample variance $\delta_{\hat{\lambda}} = D_{\hat{\lambda}}/D_x$ on the ratio m_1^2/m_2 for: (a) median filter; (b) Huber filter; (c) truncated mean filter; (d) Hodges-Lehmann filter

following way: Weibull (W) – solid line denoted by squares $\square - \square - \square$; gamma ($\Gamma$) – dashed line denoted by circles $\circ - \circ - \circ$; lognormal (Ln) – dotted line denoted by rhombuses $\diamond - \diamond - \diamond$.

As follows from the dependences of relative sample variance $\delta_{\hat{\lambda}} = D_{\hat{\lambda}}/D_x$ of estimate $\hat{\lambda}$ of a location parameter λ on the ratio m_1^2/m_2 shown in Fig. 2.7.10a,b,c,d, at a qualitative level, one can underline two main features that have been already mentioned above. First, for homogeneous filter (results are not depicted in figures) relative sample variance $\delta_{\hat{\lambda}}$ faintly depends on a distribution from the considered three classes, so that the difference between the largest $\delta_{\hat{\lambda},\max}$ and the least $\delta_{\hat{\lambda},\min}$ values of relative sample variance $\delta_{\hat{\lambda}}$ does not exceed $\sqrt{3}$: $\delta_{\hat{\lambda},\max}/\delta_{\hat{\lambda},\min} < \sqrt{3}$. Second, for median filter (MF), Huber filter (HF), truncated mean filter (trMF), and Hodges-Lehmann filter (HLF), on the contrary, relative sample variance $\delta_{\hat{\lambda}}$ essentially depends on a distribution, so that a difference between the largest $\delta_{\hat{\lambda},\max}$ and the least $\delta_{\hat{\lambda},\min}$ value of relative sample variance $\delta_{\hat{\lambda}}$ is greater than $3\ldots5$ orders for median filter: $\delta_{\hat{\lambda},\max}/\delta_{\hat{\lambda},\min} > 10^3\ldots10^5$; is about $1\ldots1.5$ orders for Hodges-Lehmann filter: $10 < \delta_{\hat{\lambda},\max}/\delta_{\hat{\lambda},\min} < 10^{1.5}$; and is within $3.3 < \delta_{\hat{\lambda},\max}/\delta_{\hat{\lambda},\min} < 7.5$ for Huber filter and truncated mean filters. Quite similarly, curves characterizing estimation accuracy of these filters are rather close to each other in the interval $0.4 \leq m_1^2/m_2 \leq 0.7$.

Fig. 2.7.11a,b,c,d depict dependencies of gain (loss) $\Delta = 10\lg(\delta_{\hat{\lambda},\mathrm{hmF}}/\delta_{\hat{\lambda}})$, obtained when processing the observations in one of four filters forming the estimators according to algorithms (2.7.4a...2.7.4d) with respect to homogeneous filter, on the ratio m_1^2/m_2. Here $\delta_{\hat{\lambda},\mathrm{hmF}}$, $\delta_{\hat{\lambda}}$ are relative sample variances $\delta_{\hat{\lambda}} = D_{\hat{\lambda}}/D_x$ of estimates $\hat{\lambda}$ of a location parameter λ for homogeneous filter hmF and other filters forming the estimators according to equations (2.7.4a...2.7.4d), respectively. Initial data used for calculating correspond to the dependencies depicted in Fig. 2.7.10a,b,c,d. Notations shown in Fig. 2.7.11a,b,c,d correspond to the following devices: MF – median filter, HF – Huber filter, trMF – truncated mean filter, HLF – Hodges-Lehmann filter. Envelope distributions of measurement errors $\{n(t_j)\}$ defined on three PDF classes (2.7.5)...(2.7.7) are denoted in the following way: Weibull (W) – solid line denoted by squares $\square - \square - \square$; gamma ($\Gamma$) – dashed line denoted by circles $\circ - \circ - \circ$; lognormal (Ln) – dotted line denoted by rhombuses $\diamond - \diamond - \diamond$.

As follows from simulation results shown in Fig. 2.7.11a,b,c,d, when the ratio m_1^2/m_2 takes a value $m_1^2/m_2=0.9$, median filter is inferior to homogeneous one in about 5.0 dB; Huber filter is inferior to the latter in about 1.0 dB; truncated mean filter loses 1.25...1.62 dB, and Hodges-Lehmann filter is inferior to homogeneous one in 0.6...0.8 dB.

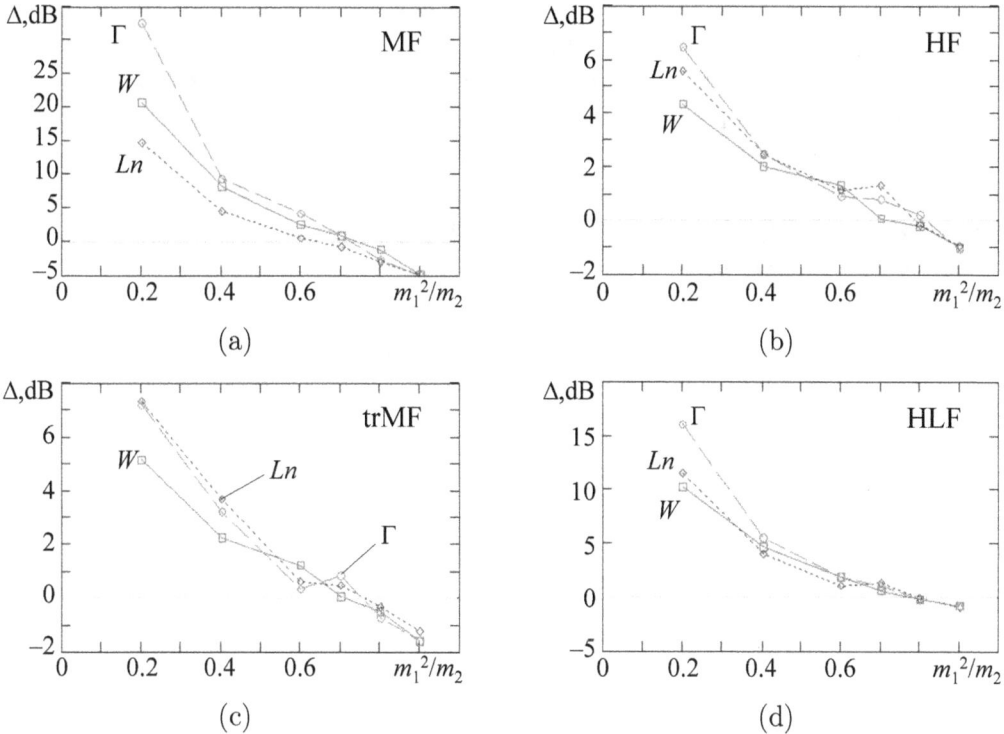

FIGURE 2.7.11   Dependence of gain (loss) $\Delta = 10\lg(\delta_{\hat{\lambda},\mathrm{hmF}}/\delta_{\hat{\lambda}})$ on the ratio m_1^2/m_2 for: (a) median filter; (b) Huber filter; (c) truncated mean filter; (d) Hodges-Lehmann filter

When the ratio m_1^2/m_2 takes a value $m_1^2/m_2=0.8$, median filter is inferior to homogeneous one in 1.15...3.0 dB; Huber filter loses up to 0.15...0.2 dB;

truncated mean filter loses $0.25\ldots0.73$ dB, and Hodges-Lehmann filter is inferior to homogeneous one in $0.10\ldots0.15$ dB.

When the ratio m_1^2/m_2 takes a value $m_1^2/m_2=0.7$, almost all filters provide gain in estimation with some exception: median filter – $0.75\ldots0.8$ dB (in lognormal distribution this filter still loses 0.6 dB); Huber filter – $0.05\ldots1.4$ dB; truncated mean filter – $0.05\ldots0.95$ dB, and Hodges-Lehmann filter – $0.6\ldots1.3$ dB.

When the ratio m_1^2/m_2 takes a value $m_1^2/m_2=0.6$, all filters provide gain in estimation: median filter – $0.45\ldots4.3$ dB; Huber filter – $0.9\ldots1.43$ dB; truncated mean filter – $0.32\ldots1.3$ dB, and Hodges-Lehmann filter – $1.1\ldots1.8$ dB.

The same situation, but with a greater gain, is observed when the ratio m_1^2/m_2 takes a value $m_1^2/m_2=0.4$: median filter provides gain $4.1\ldots9.1$ dB (with respect to homogeneous filter); Huber filter – $1.8\ldots2.37$ dB; truncated mean filter – $2.2\ldots3.7$ dB, and Hodges-Lehmann filter – $3.7\ldots5.5$ dB.

More considerable gain take place when the ratio m_1^2/m_2 takes a value $m_1^2/m_2=0.2$: median filter provides gain $14.7\ldots28.5$ dB; Huber filter – $4.1\ldots6.25$ dB; truncated mean filter – $5.1\ldots7.3$ dB, and Hodges-Lehmann filter – $10.2\ldots16$ dB.

Summarizing the aforementioned, notice the following. Filters based on estimation algorithms formulated in sample space $\mathcal{L}(\mathcal{X},\mathcal{B}_{\mathcal{X}};+,\vee,\wedge)$ with L-group properties, i.e., those that form the estimators (2.7.4a,b,c,d), provide gain in estimation accuracy as against homogeneous filter on a wide class of distributions describing the behavior of interference (noise) of impulse type for which the ratio m_1^2/m_2 of first two moments of measurement errors envelope takes the values in the interval $0 < m_1^2/m_2 \le 0.725$, so that the less is the ratio m_1^2/m_2 the greater the obtained gain is. Apparently, L-group estimation algorithms formulated in sample space $\mathcal{L}(\mathcal{X},\mathcal{B}_{\mathcal{X}};+,\vee,\wedge)$ with L-group properties possess mentioned advantages as against all estimation algorithms constructed in linear sample space $\mathcal{LS}(\mathcal{X},\mathcal{B}_{\mathcal{X}};+)$.

3

Signal Filtering Algorithms in Spaces with L–group Properties

Let $s(t)$ be useful signal that additively interacts with interference (noise) $n(t)$ in signal space with L-group properties $\mathcal{L}(+, \vee, \wedge)$:

$$x(t) = s(t) + n(t); \ t \in T^*, \tag{3.0.1}$$

where "+" is a binary operation of addition of L-group $\mathcal{L}(+, \vee, \wedge)$;
T^* is an interval of processing, $T^* \subset T_s$;
$T_s = [t_0, t_0 + T]$ is a domain of useful signal $s(t)$;
T is a signal duration $s(t)$.

Probabilistic-statistical properties of stochastic signal $s(t)$ and interference $n(t)$ can be known in a certain extent and arbitrary simultaneously.

Obtaining an *estimator* $\hat{s}(t + \tau)$ of a signal $s(t)$ as a functional $F_{\hat{s}}[x(t)]$ of an observed stochastic process $x(t)$, $t \in T^*$ when $\tau = 0$ is called the problem of signal *filtering* (*extracting*), and when $\tau < 0$, it is called the problem of signal *smoothing* (*interpolation*) [94–97]:

$$\hat{s}(t + \tau) = F_{\hat{s}}[x(t)], \ t \in T^*. \tag{3.0.2}$$

In the case, when power spectral density (PSD) of a signal is exactly known, as a processing unit forming signal estimator can be used Wiener filter [94–96, 98]. In the case, when energetic relations between useful and interference signals and also dynamic model of the system are known, then Kalman filter can be used to solve a problem of filtering [94–96, 99].

In the case, when it is known that useful stochastic signal is a band-pass one (i.e. when ratio $\Delta F_s / f_0$ of its effective PSD bandwidth ΔF_s to the central frequency of PSD f_0 is much less than 1: $\Delta F / f_0 << 1$), and it is known, that effective PSD bandwidth ΔF_n of interference is much greater than effective PSD of the signal ΔF_s ($\Delta F_n >> \Delta F_s$), then signal filtering algorithms can exploit any algorithms of point estimation, such, for instance, as (2.7.3), (2.7.4a,b,c,d).

Depending on prior information concerning spectral characteristics of useful and interference signals, other known filters can be used for signal filtering.

As a rule, the greater prior information quantity on useful signal is used for its filtering (extracting), the better quality of signal processing is provided during signal filtering. When filtering, signal processing quality is assessed by filtering mean squared error (or a variance of filtering error). Within this section, however, filtering efficiency is also evaluated by correlation coefficient between the signal $s(t)$ and its estimator $\hat{s}(t)$.

DOI: 10.1201/9781003275855-3

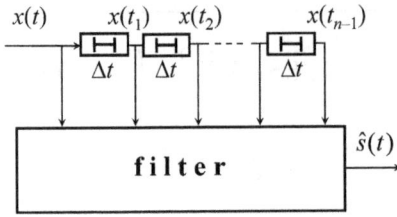

FIGURE 3.1.1   Generalized block diagram
of signal filtering unit

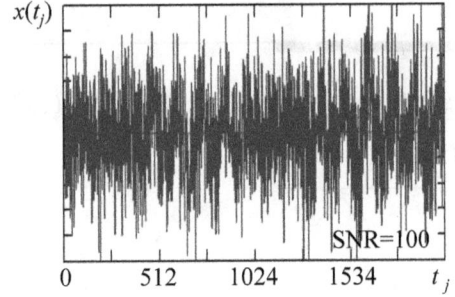

FIGURE 3.1.2   Additive mixture of signal
and noise in the input of a filter

The subject of the following consideration is a comparative efficiency of signal filtering (extracting) algorithms that realize signal processing in the presence of interference (noise) in signal spaces with L-group properties.

At first, we consider algorithms forming the estimators in the form of (2.7.3), (2.7.4a,b,c,d) that operate under conditions with scanty prior information on the processed signals, and therefore, providing satisfactory (with respect to applied problems) quality indices of signal processing when signal-to-noise ratio (SNR) is sufficiently large. Then we consider filtering algorithms requiring larger quantity of prior information concerning useful signal spectral characteristics.

3.1   Robust Signal Filtering Algorithms Based on L-group Operations

In signal processing problems, statistical model of signal filtering has usually the form that is similar to (3.0.1):

$$x(t_j) = s(t_j) + n(t_j); \ t_j \in T^*, \tag{3.1.1}$$

where $t_j = t - j\Delta t$, $j = 0, 1, \ldots, n - 1$, $t_j \in T^*$, $T^* \subset T_s$.
T^* is a processing interval: $T^* = [t - (n-1)\Delta t, t]$; $n \in \mathbf{N}$, $\mathbf{N}$ is the set of natural numbers;
$T_s = [t_0, t_0 + T]$ is a domain of useful signal $s(t)$;
T is a duration of useful signal $s(t)$;
$\{n(t_j)\}$ are independent samples of interference (noise) distributed on a class of distributions with symmetric probability density function (PDF) $p_n(z) = p_n(-z)$;
$\{s(t_j)\}$ are the samples of useful signal $s(t)$;
$\{x(t_j)\}$ are the samples of the observed stochastic process $x(t)$;
t is time parameter;
Δt is a sampling interval that provides independence of the samples $\{n(t_j)\}$;
$j = 0, 1, \ldots, n - 1$ is an index of samples $\{n(t_j)\}$, $\{x(t_j)\}$.

Instantaneous values (samples) of the signal $\{s(t_j)\}$ and interference $\{n(t_j)\}$ are the elements of the sample space: $s(t_j), n(t_j) \in \mathcal{L}(+, \vee, \wedge)$. Temporal samples of interference $n(t_j)$ are considered to be independent, taken over the sampling interval Δt, so that $\Delta t << 1/f_0$, where f_0 is an unknown carrier frequency of the signal $s(t)$.

Generalized block diagram of processing unit realizing signal filtering based on additive model of signal interaction (3.1.1), when using one or another criterion of efficiency, is shown in Fig. 3.1.1. As follows from the figure, the set of samples $\{x(t_j)\}$ of the observations is formed by series time-delay lines. Filter, performing algorithm of processing the observations $\{x(t_j)\}$ according to a given criterion of signal filtering efficiency, forms the estimator $\hat{s}(t)$ of useful signal $s(t)$.

In this section, we consider comparative analysis of efficiency of homogeneous filter (moving average filter) forming the estimator [92, (15.1)]:

$$\hat{s}_{\text{hmF}}(t) = \sum_{j=0}^{n-1} x(t - \Delta t \cdot j), \tag{3.1.2}$$

and also four filters realizing signal processing in the space $\mathcal{L}(+, \vee, \wedge)$ with L-group properties and forming the estimators (2.7.4a,b,c,d), respectively:

$$\hat{s}_{\text{MF}}(t) = \underset{j=0,\ldots,n-1}{\text{med}} \{x(t - \Delta t \cdot j)\}; \tag{3.1.3a}$$

$$\hat{s}_{\text{HF}}(t) = \hat{s}(t - \Delta t) + \frac{\hat{\sigma}}{n} \cdot \sum_{j=0}^{n-1} k \wedge [(x(t - \Delta t \cdot j) - \hat{s}(t - \Delta t)) \vee -k]; \tag{3.1.3b}$$

$$\hat{s}_{\text{tmF}}(t) = \frac{1}{n - 2r} \sum_{i=r+1}^{n-r} X_{(i)}; \tag{3.1.3c}$$

$$\hat{s}_{\text{HL}}(t) = \underset{i \leq k}{\text{med}} \{m_{i,k}\}, \tag{3.1.3d}$$

where $\hat{s}_{\text{MF}}(t)$, $\hat{s}_{\text{HF}}(t)$, $\hat{s}_{\text{tmF}}(t)$, $\hat{s}_{\text{HL}}(t)$ are the estimators formed by median filter, Huber filter, truncated mean filter, and Hodges-Lehmann filter, respectively; $\alpha = r/n < 1/2$, $r = [\alpha n]$, $[u]$ is an integer part of u; $X_{(i)}$ is i-th order statistic of variational series $X' = (X_{(1)}, X_{(2)}, \ldots, X_{(n)})$; $X_{(i)} \in X'$, $X_i \equiv x(t - \Delta t \cdot j)$, $j = i - 1$; $m_{i,k} = \frac{1}{2}(X_{(i)} + X_{(k)})$ are $n(n + 1)/2$ pairwise averages including observations themselves; $\text{med}\{a_{i,k}\}$ is a sample median of some set of elements $\{a_{i,k}\}$.

Fig. 3.1.2 illustrates an example of realization of additive mixture between useful amplitude-modulated signal and Gaussian noise in the input of explored filter on signal-to-noise ratio (SNR) that is equal to 100. Filters defined by relations (3.1.3a)...(3.1.3d) will be investigated on SNR that is equal to 30, 100, and 300 (in units of power) under various noise distributions.

Fig. 3.1.3a,b,c depict the dependences of relative filtering error δ on the ratio m_1^2/m_2 of squared expectation m_1^2 to the second moment m_2 of interference (noise) envelope for SNRs that are equal to SNR=30, 100, 300, respectively:

$$\delta = \mathbf{M}\{(\hat{s}(t) - s(t))^2\}/2D_s, \tag{3.1.4}$$

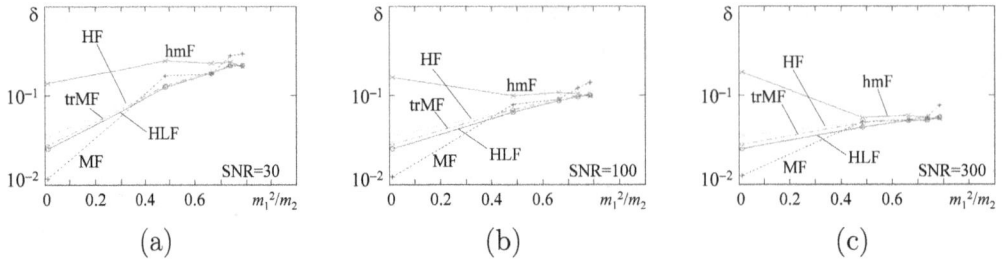

FIGURE 3.1.3   Dependences of relative filtering error δ on the ratio m_1^2/m_2 for the following SNRs: (a) SNR=30; (b) SNR=100; (c) SNR=300

where D_s is a signal variance; $\hat{s}(t)$ is an estimator of the signal $s(t)$.

These dependences are obtained by simulation within the model of additive interaction between the signal and interference (noise) (3.1.1) in the outputs of filters forming the estimators according to the relationships (3.1.2), (3.1.3a,b,c,d).

Interference (noise) $n(t)$ is characterized by six symmetric distributions: normal $N(0, b)$, $m_1^2/m_2 = \pi/4$; logistic $L(0, b)$, $m_1^2/m_2 \approx 0.75$; Laplace (double exponential) $DE(0, b)$, $m_1^2/m_2 \approx 0.71$; Student $St(3)$, $m_1^2/m_2 \approx 0.67$; ε-contaminated Tukey $T(\varepsilon, \tau)$ with $\varepsilon = 0.05$, $\tau = 5$, $m_1^2/m_2 \approx 0.48$; Cauchy $C(0, b)$, $m_1^2/m_2 \approx 0.15$.

Ratios m_1^2/m_2 for all kinds of distributions, excepting normal, were determined experimentally.

Random variables simulated according to aforementioned distributions, excepting Tukey and Laplace ones, are obtained by built-in MathCAD random number generators. Random variable ξ_T with Tukey $T(\varepsilon, \tau)$ distribution is obtained on the basis of linear combination of two independent random variables ξ_1, ξ_2 with standard normal distribution $N(0, 1)$ and random variable with Bernoulli distribution $\xi_B(\varepsilon)$ with parameter ε:

$$\xi_T(\varepsilon, \tau) = (1 - \xi_B(\varepsilon)) \cdot \xi_1 + \xi_B(\varepsilon) \cdot \tau \cdot \xi_2.$$

Laplace distributed random variable is obtained on the basis of a difference between two independent exponentially distributed random variables.

Window size n of filters, constructed according to block diagram shown in Fig. 3.1.1, is equal to 51. Notations depicted in Fig. 3.1.3a,b,c and also in other figures correspond to the following processing units: hmF – homogeneous filter; MF – median filter; HF – Huber filter; trMF – truncated mean filter; HLF – Hodges-Lehmann filter.

Useful signal $s(t)$ used in simulating is an amplitude-modulated harmonic one with oscillation period that is equal to $T_0=128$. Interference (noise) has a normal distribution with zero mean. The signal $s(t)$ (dotted line) and its estimates $\hat{s}(t)$ (solid lines) in the outputs of the corresponding filters (on SNR that is equal to 100) are shown in Fig. 3.1.4a,b,c.

The obtained results shown in Fig. 3.1.3a,b,c allows revealing the following qualitative features of signal filtering. For all processing units, in the presence of interference (noise) with Cauchy distribution, relative filtering error δ does not practically

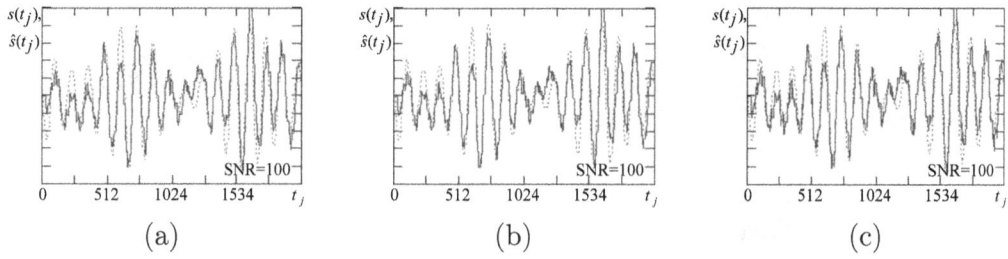

FIGURE 3.1.4   Signal $s(t)$ (dotted line) and its estimates $\hat{s}(t)$ (solid line) in the outputs of the corresponding filters: (a) Huber filter; (b) truncated mean filter; (c) Hodges-Lehmann filter

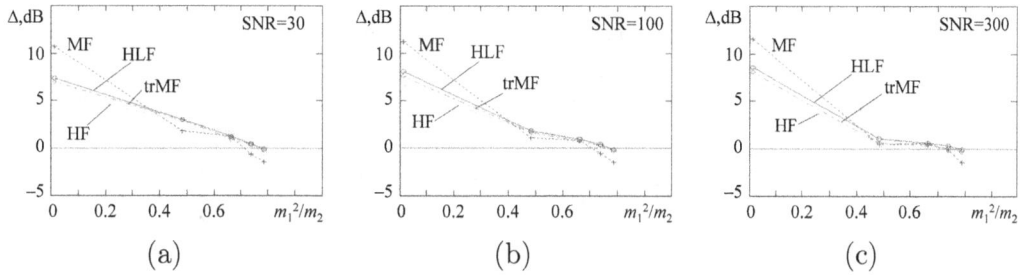

FIGURE 3.1.5   Dependences $\Delta = 10\lg(\delta_{\mathrm{hmF}}/\delta)$ of gain (loss) on the ratio m_1^2/m_2 for the following SNRs: (a) SNR=30; (b) SNR=100; (c) SNR=300

depend on SNR and is only determined by window size n used in processing. When interference (noise) has other kinds of distribution (i.e., if $m_1^2/m_2 \in [0.48, \pi/4]$), relative filtering error δ depends on both window size n (number of instantaneous values (samples) of the observed process $\{x(t_j)\}$ used in processing) and SNR. As for the filters performing signal processing in the space $\mathcal{L}(+, \vee, \wedge)$ with L-group properties, dependence of relative filtering error on the ratio m_1^2/m_2 is characterized by the following feature: the less m_1^2/m_2 is, the better quality index of signal filtering δ is. However, taking into account the fact that the intervals $m_1^2/m_2 \in [0.2, 0.6] \cup [0.8, 0.9]$ were not seized by experiment, the mentioned feature needs an additional research that will be realized below. Huber, Hodges-Lehmann, and truncated mean filters are characterized by rather close values of relative filtering error δ.

Fig. 3.1.5a,b,c illustrate the dependences $\Delta = 10\lg(\delta_{\mathrm{hmF}}/\delta)$ of gain (loss) on the ratio m_1^2/m_2 of squared expectation m_1^2 to the second moment m_2 of interference (noise) envelope for SNRs equal to SNR=30, 100, 300, respectively, obtained as a result of processing additive mixture $x(t)$ between signal and interference (noise) in one of four filters forming signal estimators according to the equations (3.1.3a,b,c,d) with respect to homogeneous filter hmF (3.1.2), where δ, δ_{hmF} are the values of relative filtering error of an arbitrary and homogeneous hmF filters. For the purpose of plotting these dependences we used the simulation results shown in Fig. 3.1.3a,b,c.

Dependences of gain (loss) $\Delta = 10\lg(\delta_{\mathrm{hmF}}/\delta)$, obtained in processing the observed stochastic process $x(t)$ in one of these filters shown in Fig. 3.1.5a,b,c, allow pointing the following qualitative features of signal filtering.

First, Huber, Hodges-Lehmann, truncated mean, and median filters are more efficient than homogeneous one for all used interference (noise) distributions, excepting normal and logistic ones, i.e., in the interval of values $m_1^2/m_2 \in [0.15, 0.71]$, in other words, for Cauchy, Tukey, Student, and Laplace distributions, providing rather close values of quality indices of signal processing. On the contrary, they lose in both normal and logistic distributions. Third, gain (loss) $\Delta = 10\lg(\delta_{\mathrm{hmF}}/\delta)$, obtained by the filters realizing signal processing in the space $\mathcal{L}(+, \vee, \wedge)$ with L-group properties, faintly depends on SNR when it is less than or equal to SNR ≤ 100.

We itemize concrete values of gain (loss) $\Delta = 10\lg(\delta_{\mathrm{hmF}}/\delta)$ on SNR equal to 100. Thus, in normal distribution of interference (noise), median filter is inferior to homogeneous one in 2.1 dB; the rest of the filters lose no more than 0.4 dB. In logistic distribution of interference (noise) all filters lose no more than 0.2 dB. In Laplace distribution of interference (noise) all filters provide gain when processing, but this gain does not exceed 0.7 dB. In Student distribution $St(3)$ and ε-contaminated Tukey $T(\varepsilon, \tau)$ distribution with $\varepsilon = 0.05$, $\tau = 5$, all filters provide gain in $1.7\ldots2.3$ dB. More considerable gain takes place in Cauchy distributed interference (noise): median filter provides gain in 8 dB; Huber filter – in 4.2 dB; truncated mean filter – in 5.1 dB; Hodges-Lehmann filter – in 5.4 dB.

In general, the obtained results sufficiently well describe known theoretical results shown, for instance, in the Table 2.6.3, whose essence is in follows. Filters based on processing algorithms operating in the space $\mathcal{L}(+, \vee, \wedge)$ with L-group properties, i.e. those that forms the estimators determined by (3.1.3a,b,c,d), provide gain with respect to homogeneous filter in a wide class of distributions describing the behavior of interference (noise) of impulse type, so that the less the ratio m_1^2/m_2 is, the greater the obtained gain is.

It is desirable to check the last statement in the presence of interference (noise) with other probabilistic-statistical properties. During a simulation, in order to obtain the results that are similar to those shown in Figs. 3.1.3, 3.1.5, but for other distributions, we use the following classes of interference (noise) envelope distributions: Weibull $W(c, \alpha)$ (2.7.5), gamma $\Gamma(\beta, \alpha)$ (2.7.6), lognormal $Ln(m, \sigma)$ (2.7.7).

Appearance of PDFs for Weibull $W(c, \alpha)$, lognormal $Ln(m, \sigma)$, and gamma $\Gamma(\beta, \alpha)$ distributions on m_1^2/m_2=0.2, 0.4, 0.6, 0.8, 0.9 were shown in Fig. 2.7.9a,b,c, respectively.

Figs. 3.1.6a,b,c, 3.1.7a,b,c, 3.1.8a,b,c illustrate the dependences of relative filtering error $\delta = \mathbf{M}\{(\hat{s}(t) - s(t))^2\}/2D_s$ (D_s is a signal variance) on the ratio m_1^2/m_2 of squared expectation m_1^2 to the second moment m_2 of interference (noise) envelope. These dependences are obtained by simulation within the model of additive interaction between the signal and interference (noise) (3.1.1). Estimators $\hat{s}(t)$ are formed in the outputs of the filters according to the relationships (3.1.2), (3.1.3a,b,c,d). Envelopes of interference (noise) $n(t)$ have Weibull W, lognormal Ln, and gamma Γ distributions for SNRs that are equal to SNR=30, 100, 300, respectively.

Window size n of the filters constructed according to the block diagram shown in Fig. 3.1.1 is equal to 51. Notations shown in Figs. 3.1.6, 3.1.7, 3.1.8 and also in other figures correspond to the following processing units: hmF – homogeneous

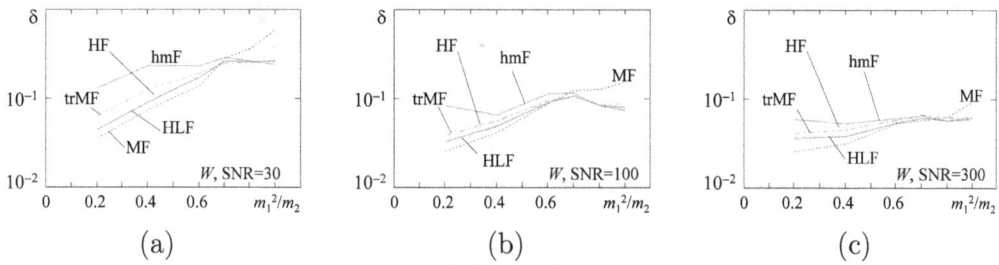

FIGURE 3.1.6   Dependences of relative filtering error δ on the ratio m_1^2/m_2 for Weibull distribution W of interference (noise) envelope and SNRs: (a) SNR=30; (b) SNR=100; (c) SNR=300

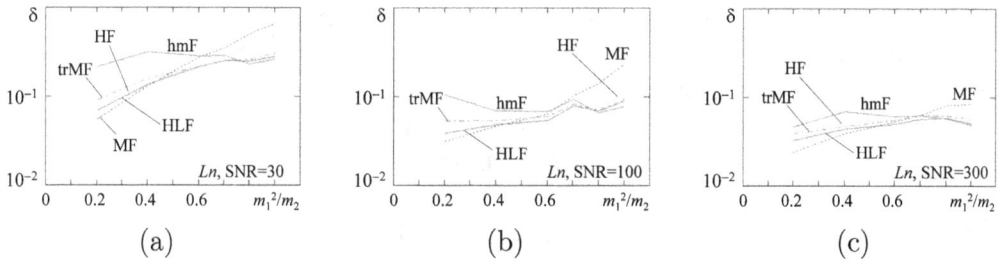

FIGURE 3.1.7   Dependences of relative filtering error δ on the ratio m_1^2/m_2 for lognormal distribution Ln of interference (noise) envelope and SNRs: (a) SNR=30; (b) SNR=100; (c) SNR=300

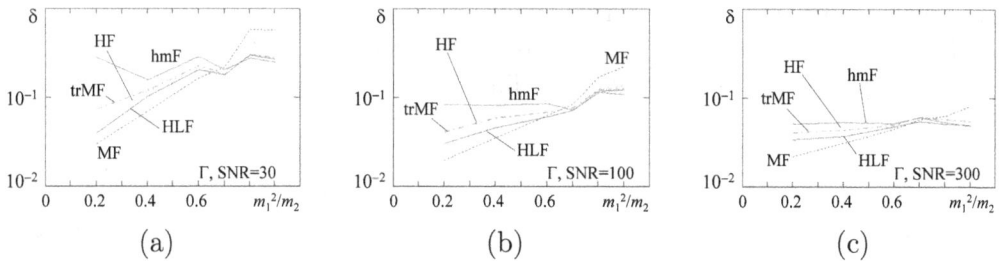

FIGURE 3.1.8   Dependences of relative filtering error δ on the ratio m_1^2/m_2 for gamma distribution Γ of interference (noise) envelope and SNRs: (a) SNR=30; (b) SNR=100; (c) SNR=300

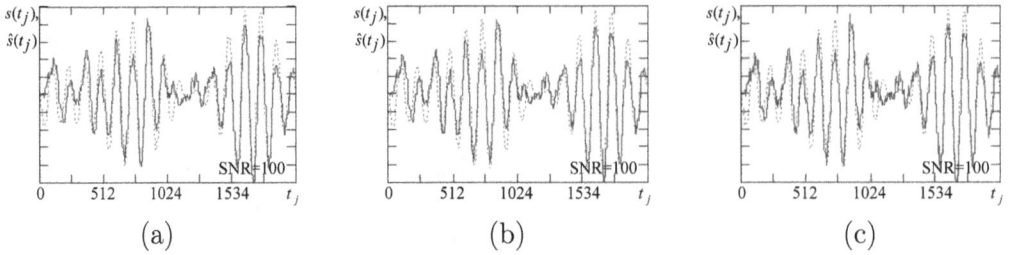

(a)   (b)   (c)

FIGURE 3.1.9   Signal $s(t)$ (dotted line) and its estimate $\hat{s}(t)$ (solid line) in the outputs of the corresponding filters (the case of Weibull distribution W of interference (noise) envelopes): (a) Huber filter; (b) truncated mean filter; (c) Hodges-Lehmann filter

filter; MF – median filter; HF – Huber filter; trMF – truncated mean filter; HLF – Hodges-Lehmann filter.

Useful signal $s(t)$ used in simulating is amplitude-modulated bandpass signal with oscillation period equal to $T_0=128$. Appearance of the signal $s(t)$ (dotted line) and its estimate $\hat{s}(t)$ (solid line) in the outputs of corresponding filters on SNR equal to 100 and envelopes of interference (noise) $n(t)$ with Weibull distribution W ($m_1^2/m_2=0.8$) are shown in Fig. 3.1.9a,b,c. Correlation coefficient $r(s(t),\hat{s}(t))$ between the signal $s(t)$ and its estimator $\hat{s}(t)$ within the pointed case for corresponding filters are equal to: homogeneous filter – 0.915; median filter – 0.87; Huber, truncated mean, and Hodges-Lehmann filter – 0.91.

Obtained results shown in Figs. 3.1.6a,b,c, 3.1.7a,b,c, 3.1.8a,b,c reveal the following qualitative features of filtering that are very similar for all three classes of interference (noise) envelope distribution. The relative filtering error δ of homogeneous filter faintly depends on interference (noise) envelope distribution (i.e., on relation m_1^2/m_2), excepting Cauchy one. On the contrary, for the filters realizing filtering in the space $\mathcal{L}(+,\vee,\wedge)$ with L-group properties, there exists such a dependence, so that the less m_1^2/m_2 is, the better filtering quality index δ is. Huber and truncated mean filters have rather close values of relative filtering error δ.

As follows from Figs. 3.1.6, 3.1.7, 3.1.8, Huber, Hodges-Lehmann, and truncated mean filters are more efficient than homogeneous filter in all three types of interference (noise) envelopes in the interval $m_1^2/m_2 \in [0.2, 0.7]$. Huber and truncated mean filters provide highly close filtering quality indices. In the worst case distributions ($m_1^2/m_2=0.9$) and rather great SNRs (from 100 and greater), Huber, Hodges-Lehmann, and truncated mean filters are slightly inferior to homogeneous filter in filtering quality. Median filter provides a gain with respect to homogeneous filter in the interval $m_1^2/m_2 \in [0.2, 0.7]$ within gamma and Weibull classes of envelope distributions. Within lognormal class of interference (noise) envelope distribution, median filter provides a gain in more narrow interval $m_1^2/m_2 \in [0.2, 0.65]$. In the worst case distributions ($m_1^2/m_2=0.9$), median filter is considerably inferior to homogeneous filter.

Figs. 3.1.10a,b,c, 3.1.11a,b,c, 3.1.12a,b,c depict dependences of gain (loss) $\Delta = 10\lg(\delta_{\text{hmF}}/\delta)$ on the ratio m_1^2/m_2 for aforementioned four filters forming the

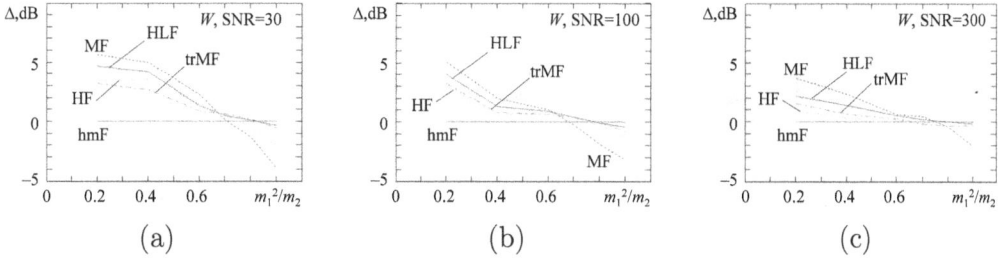

FIGURE 3.1.10   Dependences of gain (loss) $\Delta = 10\lg(\delta_{\mathrm{hmF}}/\delta)$ on the ratio m_1^2/m_2 in the case of Weibull distribution W of interference (noise) envelope and the following SNRs: (a) SNR=30; (b) SNR=100; (c) SNR=300

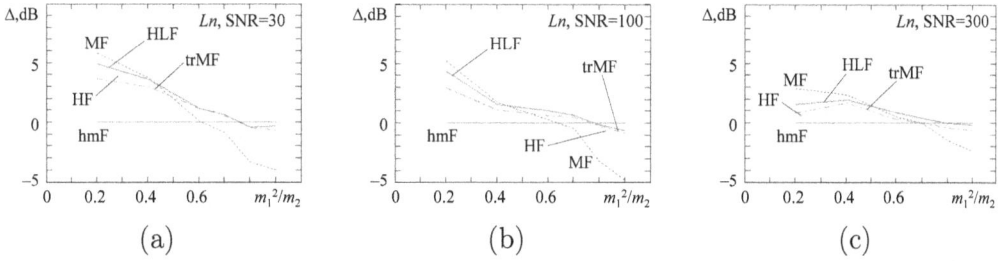

FIGURE 3.1.11   Dependences of gain (loss) $\Delta = 10\lg(\delta_{\mathrm{hmF}}/\delta)$ on the ratio m_1^2/m_2 in the case of lognormal distribution Ln of interference (noise) envelope and the following SNRs: (a) SNR=30; (b) SNR=100; (c) SNR=300

signal estimators according to the equations (3.1.3a,b,c,d) with respect to homogeneous filter hmF (3.1.2). Here δ, δ_{hmF} are relative filtering error of an arbitrary (from four filters mentioned above) and homogeneous hmF filters. The results are obtained by simulation of processing additive mixture $x(t)$ of signal and interference (noise). Interference (noise) envelopes have Weibull W, lognormal Ln, and gamma Γ distributions. SNRs are equal to SNR=30, 100, 300, respectively. For the purpose of plotting these dependences we use the simulation results shown in Figs. 3.1.6, 3.1.7, 3.1.8.

Dependences of gain (loss) $\Delta = 10\lg(\delta_{\mathrm{hmF}}/\delta)$ obtained when filtering the observed process $x(t)$ in one of four mentioned filters shown in Figs. 3.1.10, 3.1.11, 3.1.12 reveal the following qualitative features of filtering.

First, Huber, Hodges-Lehmann, and truncated mean filters are more efficient than homogeneous filter for all three types of interference (noise) envelope distributions used within simulation in the interval $m_1^2/m_2 \in [0.2, 0.7]$, so that Huber and truncated mean filters provide highly close filtering quality indices. In worst case distributions (m_1^2/m_2=0.9) and relatively large SNRs (from 100 and larger), Huber, Hodges-Lehmann, and truncated mean filters are inferior to homogeneous one in 1.5 dB. Second, median filter provides a gain in filtering with respect to homogeneous one in interval $m_1^2/m_2 \in [0.2, 0.7]$ when distributions of interference (noise) envelopes belong to gamma and Weibull classes. In lognormal class of interference

(a)   (b)   (c)

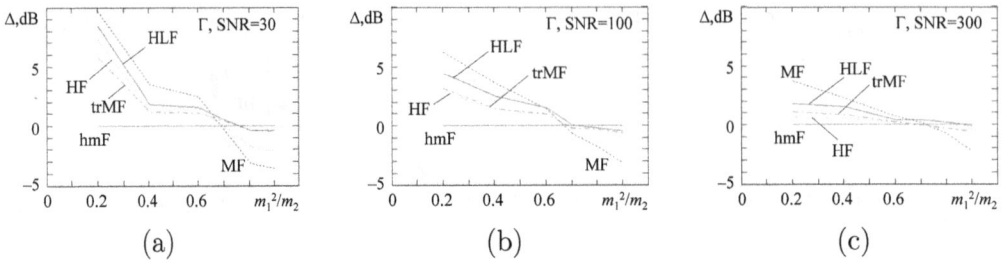

FIGURE 3.1.12   Dependences of gain (loss) $\Delta = 10\lg(\delta_{\mathrm{hmF}}/\delta)$ on the ratio m_1^2/m_2 in the case of gamma distribution Γ of interference (noise) envelope and the following SNRs: (a) – SNR=30; (b) – SNR=100; (c) – SNR=300

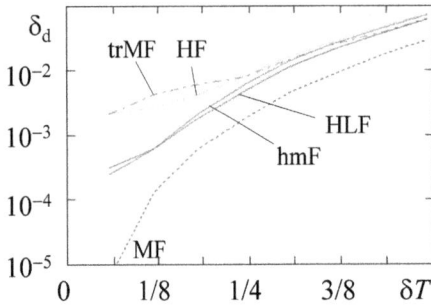

FIGURE 3.1.13   Dependences $\delta_{\mathrm{d}}(\delta T)$ of relative dynamic filtering error on relative size of temporal window δT

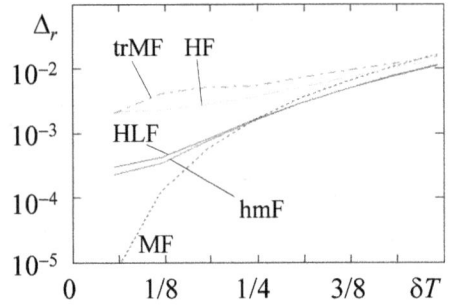

FIGURE 3.1.14   Dependences of difference $\Delta_r = 1 - r(s(t), \hat{s}(t))\big|_{n(t)=0}$ on relative size of temporal window δT

(noise) envelope distribution, median filter provides a gain in more narrow interval $m_1^2/m_2 \in [0.2, 0.65]$. In the worst case distributions (m_1^2/m_2=0.9), median filter is inferior to homogeneous one in up to 5 dB. Third, the less SNR is, the greater a gain $\Delta = 10\lg(\delta_{\mathrm{hmF}}/\delta)$ is, that is provided by filters realizing processing in the space $\mathcal{L}(+, \vee, \wedge)$ with L-group properties,.

One can get an idea on a contribution of relative dynamic filtering error δ_{d}

$$\delta_{\mathrm{d}} = \mathbf{M}\{(\hat{s}(t) - s(t))^2\}/2D_s\big|_{n(t)=0} \tag{3.1.5}$$

to relative filtering error $\delta = \mathbf{M}\{(\hat{s}(t) - s(t))^2\}/2D_s$ (3.1.4) by the dependences $\delta_{\mathrm{d}}(\delta T)$ of relative dynamic filtering error δ_{d} on relative size of temporal window $\delta T = (n-1)\Delta t/T_0$, ($T_0$ is a period of carrier of band-pass signal) in which the processing of the observed signal $x(t)$ is realized within the model (3.1.1). These dependences are obtained by statistical modeling in the absence of interference (noise) and shown in Fig. 3.1.13. Notations depicted in Fig. 3.1.13 correspond to analogous notations in other figures: hmF – homogeneous filter; MF – median filter; HF – Huber filter; trMF – truncated mean filter; HLF – Hodges-Lehmann filter.

Comparing the dependences $\delta_{\mathrm{d}}(\delta T)$, one can reveal the following features. On small sizes of temporal window δT ($\delta T \leq 1/4$), the best values of relative dynamic filtering error δ_d are provided by (in increasing order of error): median filter, then Hodges-Lehmann and homogeneous filter, then Huber and truncated mean filters. In the case of large sizes of temporal window δT ($\delta T \geq 3/8$), the best values of relative dynamic filtering error δ_d, as before, are provided by (in increasing order of error): Hodges-Lehmann and homogeneous filters, then median filter, then Huber and truncated mean filters. Taking into account the fact, that relative dynamic filtering error δ_d can insufficiently adequately demonstrate negative impact of temporal window size δT on a quality of signal estimate formed by nonlinear processing units, it is expedient considering the dependences of differences $\Delta_r = 1 - r(s(t), \hat{s}(t))\big|_{n(t)=0}$ on relative size of temporal window δT in the absence of interference (noise). Here $r(s(t), \hat{s}(t))\big|_{n(t)=0}$ is correlation coefficient between the signal $s(t)$ and its estimate $\hat{s}(t)$ in the output of the filters forming the estimators according to the equations (3.1.2), (3.1.3a,b,c,d). Such dependences for five considered filters were obtained by simulation and shown in Fig. 3.1.14.

Here, similarly, in the case of small sizes of temporal window δT ($\delta T \leq 1/4$), the best quality indices are provided by: median filter, then Hodges-Lehmann and homogeneous filters, then Huber and truncated mean filter. In relatively large sizes of temporal window δT ($\delta T \geq 3/8$), unlike the previous results depicted in Fig. 3.1.13, higher correlation coefficients $r(s(t), \hat{s}(t))\big|_{n(t)=0}$ are provided by Hodges-Lehmann and homogeneous filters, then median filter, then Huber and truncated mean filters. Truncated mean filter provides the worst values of correlation coefficient $r(s(t), \hat{s}(t))\big|_{n(t)=0}$. Some discrepancies in the impact of large sizes of temporal window δT ($\delta T \geq 3/8$) on relative dynamic filtering error δ_d and correlation coefficient for some filters can be explained by nonlinear filter affecting on high-frequency components of PSDs of the signals and are rather noticeable for median filter.

Summarizing the aforementioned, notice the following. Filters based on algorithms of signal processing in the space $\mathcal{L}(+, \vee, \wedge)$ with L-group properties, i.e., those that form the estimators (3.1.3a,b,c,d), provide a gain in filtering quality as compared to homogeneous filter in a wide class of distributions describing the behavior of interference (noise) of impulse type that possess the ratio m_1^2/m_2 within the interval $0 < m_1^2/m_2 \leq 0.7$, so that the less ratio m_1^2/m_2 is, the greater gain is. Apparently, these filters based on algorithms of signal processing in the space $\mathcal{L}(+, \vee, \wedge)$ with L-group properties possess the same advantages with respect to all linear filters based on algorithms of signal processing in linear signal space $\mathcal{LS}(+)$.

As follows from analyzing the results of filter operation, filters forming the signal estimators (3.1.2), (3.1.3a,b,c,d), being quasi-optimal, provide satisfactory filtering quality indices in the case of relatively large SNRs.

In the same time, it is quite interest considering such filters and signal filtering (extraction) algorithms operating in spaces with L-group properties, that would approximate in quality indices of processing to known optimal filters and filtering algorithms, such, for instance, as Wiener filter.

Discussing this problem is contained in the following section.

3.2   Adaptive Filtering Algorithms Based on L-group Operations

The subject of further consideration is exploring adaptive algorithms of signal filtering based on L-group operations. Adaptive filters and algorithms are usually used when: (a) conditions of receiving the useful signal (for instance, PSD of interference signal) are changed in time; (b) it is necessary to solve the following problems: system identification, channel equalization, signal delay alignment in multichannel systems, echo cancellation, linear prediction, signal enhancement, etc. [92, 94, 95, 100–102]. In particular, we consider adaptive filtering algorithms based on known algorithms of Wiener filter and Kalman filter, and also some varieties of least squares filters (LMS, RLS). On the other hand, we consider adaptive filtering algorithms based on L-group operations, that are analogues of mentioned above algorithms, and also compare their efficiency. Algorithms of signal processing based on L-group operations can be constructed, first, by the method of mapping of linear signal space into signal space with lattice properties, and second, on the basis of a measure of statistical interrelation (MSI).

The resulting process $y(t)$ in the output of a generalized adaptive filter (see Fig. 3.1.1) is determined by the relationship:

$$y(t) = \sum_{j=0}^{N-1} w_j(t)x(t - \Delta t \cdot j), \qquad (3.2.1)$$

where $x(t - \Delta t \cdot j)$ is a delayed copy of the received signal $x(t) = s(t) + n(t)$ taken from the output of j-th outlet of transversal filter; $w_j(t)$ is a weight coefficient in j-th processing channel that is, in general case, the function of time, and is chosen on the basis of one or another optimality criterion; Δt is a sample interval of transversal filter; t is time parameter; j is an index of a sample of the processed signal $\{x(t_j) = x(t - \Delta t \cdot j)\}$; $s(t)$ is useful signal; $n(t)$ is interference signal; N is a number of samples used in processing.

The equation (3.2.1) can be written in the form of scalar product of two vectors $\mathbf{w}(t)$ and $\mathbf{x}(t)$:

$$y(t) = \mathbf{w}^T(t)\mathbf{x}(t); \qquad (3.2.2)$$

$$\mathbf{w}_{\mathrm{opt}}(t) = \arg\min_{t \in T_d}[F(y(t), d(t))], \quad \mathbf{w}(t) = [w_0(t), \dots, w_{N-1}(t)]^T; \qquad (3.2.2a)$$

$$\mathbf{x}(t) = [x(t), x(t - \Delta t), \dots, x(t - \Delta t \cdot (N - 1))]^T, \qquad (3.2.2b)$$

where $d(t)$ is a desired signal; $F(y(t), d(t))$ is some functional of a desired signal $d(t)$ and resulting process $y(t)$ in the output of generalized adaptive filter; $\mathbf{w}(t)$ is a vector of weight coefficients; T_d is a domain of desired signal $d(t)$.

3.2.1   Signal Filtering Algorithms Based on Method of Mapping of Linear Space into Space with Lattice Properties

Remind, that signal space $\mathcal{L}(\vee, \wedge)$ with lattice properties can be realized by a mapping of linear space $\mathcal{LS}(+)$ into space $\mathcal{L}(\vee, \wedge)$ with lattice properties, in such a way that the results of interaction $x_\vee(t)$, $x_\wedge(t)$ between signal $s(t)$ and interference $n(t)$ are defined by the relationships [24, (7.7.1a,b)]:

$$x_\vee(t) = s(t) \vee n(t) = \{[s(t) + n(t)] + |s(t) - n(t)|\}/2; \tag{3.2.3a}$$

$$x_\wedge(t) = s(t) \wedge n(t) = \{[s(t) + n(t)] - |s(t) - n(t)|\}/2, \tag{3.2.3b}$$

that are implied from the known equations [29, § XIII.3;(14)], [29, § XIII.4;(22)].

Method of mapping of linear space into space with lattice properties that is determined by the relationships (3.2.3a,b) can be realized on the basis of known adaptive filters. Further we consider some possible variants of practical realization of this method using known adaptive filtering algorithms.

Before stating the main material of the section, we formulate and prove the following lemma that will be used later.

Lemma 3.2.1. *In L-group* $\mathcal{L}(+, \vee, \wedge)$ *for any* $a, b \in \mathcal{L}(+, \vee, \wedge)$ *the following identity holds:*

$$F(a, b) = (a + |b|) \wedge (b + |a|) \vee 0 + (a - |b|) \vee (b - |a|) \wedge 0 = a + b$$

Proof. Introduce the following notations:

$$G = (a + |b|) \wedge (b + |a|); \quad H = (a - |b|) \vee (b - |a|); \tag{3.2.4a}$$

$$\Delta_+ = |a| + |b|; \quad \Delta_- = ||a| - |b||. \tag{3.2.4b}$$

Then the expressions (3.2.4a) can be written in the form:

$$G = (|a| \cdot \operatorname{sgn}(a) + |b|) \wedge (|b| \cdot \operatorname{sgn}(b) + |a|); \tag{3.2.5a}$$

$$H = (|a| \cdot \operatorname{sgn}(a) - |b|) \vee (|b| \cdot \operatorname{sgn}(b) - |a|). \tag{3.2.5b}$$

Complete the following table for $F(a, b)$, G, H depending on the signs of variables a, b, $a + b$.

TABLE 3.2.1   Values $F(a, b)$, G, H depending on the signs of variables a, b, $a + b$

$a + b$	a	b	G	H	$G \vee 0$	$H \wedge 0$	$F(a, b)$
$+$	$+$	$+$	$\Delta_+ \wedge \Delta_+$	$-\Delta_- \vee \Delta_-$	$\Delta_+ \vee 0$	0	$a + b$
$+$	$\pm$	$\mp$	$\Delta_\pm \wedge \Delta_\mp$	$\pm\Delta_\mp \vee \mp\Delta_\pm$	$\Delta_- \vee 0$	0	$a + b$
$-$	$-$	$-$	$-\Delta_- \wedge \Delta_-$	$-\Delta_+ \vee -\Delta_+$	0	$-\Delta_+ \wedge 0$	$a + b$
$-$	$\pm$	$\mp$	$\pm\Delta_\pm \wedge \mp\Delta_\mp$	$-\Delta_\mp \vee -\Delta_\pm$	0	$-\Delta_- \wedge 0$	$a + b$

Based on relationships in the last column of the table and the signs of variables a, b, the values $F(a, b)$ are equal to $a + b$. $\qquad\square$

For generalized adaptive filter based on L-group operations with forming vector of weight coefficients $\mathbf{w}(t)$ (3.2.2a), a general algorithm of pairwise forming the estimators $\hat{x}_\vee(t)$, $\hat{x}_\wedge(t)$ and $\hat{x}'_\vee(t)$, $\hat{x}'_\wedge(t)$ of join (3.2.3a) and meet (3.2.3b) take the form:

$$\hat{x}_\vee(t) = A_0(t) + |A_1(t)|; \tag{3.2.6a}$$

$$\hat{x}_\wedge(t) = A_0(t) - |A_1(t)|; \tag{3.2.6b}$$

$$\hat{x}'_\vee(t) = A_1(t) + |A_0(t)|; \tag{3.2.7a}$$

$$\hat{x}'_\wedge(t) = A_1(t) - |A_0(t)|; \tag{3.2.7b}$$

$$A_0(t) = \sum_{j=0}^{(N-2)/2} w_{2j}(t) \cdot x(t - \Delta t \cdot 2j); \tag{3.2.8a}$$

$$A_1(t) = \sum_{j=1}^{N/2} w_{2j-1}(t) \cdot x(t - \Delta t \cdot (2j-1)); \tag{3.2.8b}$$

where $w_j(t)$ is a weight coefficient in j-th processing channel; N is an even number of samples; $A_0(t)$ and $A_1(t)$ are the results of weight processing the observed stochastic process $x(t)$ that are formed on the basis of signals in the outputs of even and odd channels of transversal filter, respectively.

Two pairs of the estimators (3.2.6a,b) and (3.2.7a,b) are necessary for the purpose of complete using information contained in the observed process $x(t)$ in the input of the filter.

Then the following theorem holds.

Theorem 3.2.1. *For the filter that realizes signal processing according to algorithm:*

$$z(t) = [\hat{x}_\vee(t) \wedge \hat{x}'_\vee(t)] \vee 0 + [\hat{x}_\wedge(t) \vee \hat{x}'_\wedge(t)] \wedge 0, \tag{3.2.9}$$

the latter is equivalent to algorithm (3.2.2), where vector of weight coefficients $\mathbf{w}(t)$ is determined by the relationship (3.2.2a):

$$z(t) \equiv y(t). \tag{3.2.10}$$

Here the functions from the right part of the identity (3.2.9) are determined by the relationships (3.2.6)...(3.2.8).

Proof. Taking into account the equations (3.2.6a,b), (3.2.7a,b), write the identity (3.2.9) in the form:

$$z(t) = (A_0(t) + |A_1(t)|) \wedge (A_1(t) + |A_0(t)|) \vee 0 +$$
$$+ (A_0(t) - |A_1(t)|) \vee (A_1(t) - |A_0(t)|) \wedge 0. \tag{3.2.11}$$

Then, basing on Lemma (3.2.1) and the expressions (3.2.8a,b), the relationship (3.2.11) can be written in the form of the following identity:

$$z(t) = (A_0(t) + |A_1(t)|) \wedge (A_1(t) + |A_0(t)|) \vee 0 +$$
$$+ (A_0(t) - |A_1(t)|) \vee (A_1(t) - |A_0(t)|) \wedge 0 =$$
$$= A_0(t) + A_1(t) = y(t).$$

$\square$

Distributivity of lattice $\mathcal{L}(\vee, \wedge)$ of L-group $\mathcal{L}(+, \vee, \wedge)$ implies the corollary from Theorem 3.2.1.

Corollary 3.2.1. *For the filter that realizes signal processing according to the algorithm:*

$$z'(t) = [\hat{x}_\vee(t) \vee 0] \wedge [\hat{x}'_\vee(t)) \vee 0] + [\hat{x}_\wedge(t) \wedge 0] \vee [\hat{x}'_\wedge(t)) \wedge 0], \qquad (3.2.12)$$

the latter is equivalent to algorithm (3.2.2), in which vector of weight coefficients $\mathbf{w}(t)$ *is determined by the relationship (3.2.2a):*

$$z'(t) \equiv y(t).$$

Theorem (3.2.1) contains important theoretical inference. First, optimal processing algorithm corresponding to optimal solution (3.2.2a) is not the sole one. Second, apparently, all signal processing algorithms known in Optimal Filtering Theory have their own equivalent formulations written in terms of L-groups. Third, efficiency of signal processing algorithms, based on L-group operations and determined by Theorem (3.2.1), is equivalent to efficiency of optimal signal processing algorithms of linear filters synthesized within linear signal space.

3.2.2 Adaptive Filtering Algorithms Based on Measure of Statistical Interrelation

The introduced three measures of statistical interrelation (MSI) (1.4.10...1.4.12) based on three generalized metrics (1.3.25...1.3.27), respectively, define three corresponding variants of constructing generalized adaptive filters in signal space with lattice properties, so that the resulting process in the output of filter (see Fig. 3.1.1) is determined by the following relationships, respectively:

$$y_p(t) = N - \sum_{j=0}^{N-1} [\mathrm{sgn}(w_j(t) \vee x(t_j)) - \mathrm{sgn}(w_j(t) \wedge x(t_j))]; \qquad (3.2.13)$$

$$y_s(t) = \sum_{j=0}^{N-1} [|x(t_j)| + |w_j(t)|] -$$

$$- \sum_{j=0}^{N-1} |x(t_j) - w_j(t)|[\mathrm{sgn}(w_j(t) \vee x(t_j)) - \mathrm{sgn}(w_j(t) \wedge x(t_j))]; \qquad (3.2.14)$$

$$y_{l_1}(t) = \sum_{j=0}^{N-1} [|x(t_j) + w_j(t)| - |x(t_j) - w_j(t)|], \qquad (3.2.15)$$

where $x(t_j) = x(t - \Delta t \cdot j)$ is a delayed copy of received signal $x(t) = s(t) + n(t)$ taken from the output of j-th outlet of transversal filter; $s(t)$ is useful signal, $s(t) \equiv d(t)$; $d(t)$ is desired signal; $n(t)$ is interference signal; $w_j(t)$ is a weight coefficient in j-th processing channel that is, in general case, a function of time and is chosen on the basis one or another optimality criterion; Δt is a sampling interval of transversal filter; t is time parameter; j is an index of samples of the processed signal $\{x(t_j) = x(t - \Delta t \cdot j)\}$; N is a number of samples used in processing.

3.3 Wiener Filter Based on *L*-group Operations

Wiener filter operating according to criterion of minimum variance of error signal $e(t) = d(t) - y(t)$ in the output of a system forms optimal solution for a vector of weight coefficients $\mathbf{w}(t)$ [95, (6.10)], [102, (18.24)], [100, (10.14)]:

$$\mathbf{w}(t) = [w_0(t), \ldots, w_{N-1}(t)]^T = \mathbf{R}_{xx}^{-1}(t)\mathbf{P}_{dx}(t), \qquad (3.3.1)$$

where $\mathbf{R}_{xx}(t)$ is correlation matrix of vector $\mathbf{x}(t) = [x(t), x(t - \Delta t), \ldots, x(t - \Delta t \cdot (N-1))]^T$ (3.2.2b) of the observed process $x(t)$, $x(t) = d(t) + n(t)$ equal to $\mathbf{R}_{xx}(t) = \mathbf{M}[\mathbf{x}(t)\mathbf{x}^T(t)]$; $\mathbf{P}_{dx}(t)$ is correlation vector equal to $\mathbf{P}_{dx}(t) = \mathbf{M}[d(t)\mathbf{X}(t)]$; $\mathbf{M}[*]$ is a symbol of mathematical expectation; $n(t)$ is an additive quasi-white Gaussian noise.

Alternative variant of statistical formulation of Wiener filter for vectors of the observed signal $\mathbf{x}(t)$ and desired signal $\mathbf{d}(t) = [d(t), d(t-\Delta t), \ldots, d(t-\Delta t \cdot (N-1))]^T$ can be written in the form (see, for instance, [95, (6.14),(6.19)]):

$$\mathbf{y} = \hat{\mathbf{d}} = \mathbf{X} \cdot \mathbf{w}; \qquad (3.3.2)$$

$$\mathbf{w} = (\mathbf{X}^T\mathbf{X})^{-1}\mathbf{X}^T\mathbf{d}; \qquad (3.3.2a)$$

$$\mathbf{X} = \|x_{j,k}\| = \left\|
\begin{array}{cccc}
x_0 & x_{-1} & \cdots & x_{1-M} \\
x_1 & x_0 & \cdots & x_{2-M} \\
x_2 & x_1 & \cdots & x_{3-M} \\
\vdots & \vdots & \ddots & \vdots \\
x_{N-1} & x_{N-2} & \cdots & x_{N-M}
\end{array}
\right\|, \qquad (3.3.2b)$$

where $\mathbf{X}$ is a $N \times M$ matrix of input data; $x_j = x(t_j) = x(t - \Delta t \cdot j)$; $\mathbf{w}$ is $M \times 1$ vector of weight coefficients; $j = 1 - M, 2 - M, \ldots, N - 1$.

Hereinafter we consider overdetermined system, i.e., $N > M$. For ergodic stochastic process in the input of the filter, the solution (3.3.2a) asymptotically converges to Wiener solution:

$$\lim_{N \to \infty} [\mathbf{w} = (\mathbf{X}^T\mathbf{X})^{-1}\mathbf{X}^T\mathbf{d}] = \mathbf{R}_{xx}^{-1}(t)\mathbf{P}_{dx}(t). \qquad (3.3.3)$$

3.3.1 Wiener Filter Based on Method of Mapping of Linear Space into Space with Lattice Properties

For Wiener filter based on *L*-group operations (or, simply, *L*-group Wiener filter) and forming vector of weight coefficients by direct inversion of correlation matrix

$\mathbf{R}_{xx}(t)$, general filtering algorithm with pairwise forming the estimators $\hat{x}_\vee(t), \hat{x}_\wedge(t)$ and $\hat{x}'_\vee(t), \hat{x}'_\wedge(t)$ of join (3.2.3a) and meet (3.2.3b), respectively, is determined by the relationships (3.2.9), (3.2.6)...(3.2.8). Weight coefficient $w_j(t)$ in j-th processing channel is based on optimal solution for vector of weight coefficients $\mathbf{w}(t)$ determined by the relationship (3.3.1), (3.3.2a).

Application of Theorem 3.2.1 with respect to Wiener filter based on L-group operations is illustrated by the following example.

Example 3.3.1. Fig. 3.3.1a and 3.3.1b depicts stochastic processes $x(t) = s(t) + n(t)$ $(s(t) = d(t))$; $y(t)$, respectively, in the input and output of Wiener filter realizing the algorithm (3.3.2). SNR q^2 (in power units) is equal to $q^2 = 10$. The signals $x(t)$, $y(t)$ are shown by solid line, and useful signal $s(t)$ – by dotted line. Interference signal $n(t)$ is quasi-white Gaussian noise. The ratio of upper bound frequency $f_{n,\max}$ of interference signal PSD to carrier frequency f_0 of useful signal is equal to $f_{n,\max}/f_0 = 32$. Number of samples N, used in signal processing according to algorithm determined by the relationships (3.3.2) and algorithms determined by the relationships (3.2.9), (3.2.6...3.2.8), (3.3.2), is equal to 64.

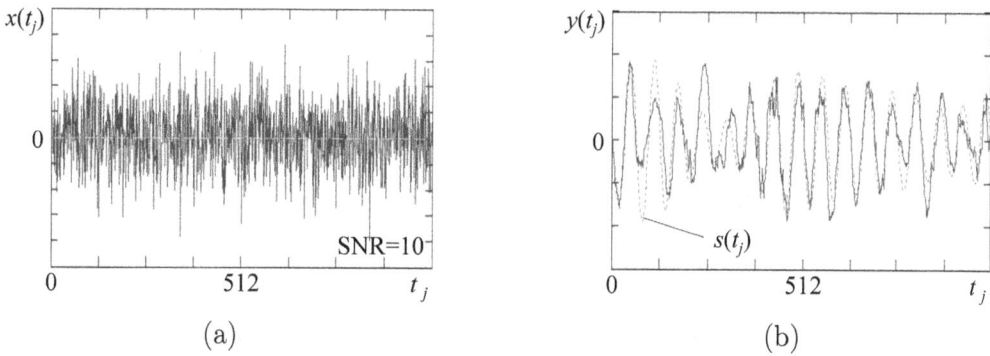

FIGURE 3.3.1   Stochastic processes (a) $x(t)$ in the input of Wiener filter; (b) $y(t)$ in the output of Wiener filter

Fig. 3.3.2a, b illustrates the estimates $\hat{x}_\vee(t), \hat{x}_\wedge(t)$ of join (3.2.6a) and meet (3.2.6b), respectively, formed by Wiener filter based on L-group operations (solid line) and useful signal (dotted line). Fig. 3.3.3a, b depicts the estimates $\hat{x}'_\vee(t), \hat{x}'_\wedge(t)$ of join (3.2.7a) and meet (3.2.7b) formed by Wiener filter based on L-group operations (solid line), respectively, and useful signal $s(t)$ (dotted line).

Fig. 3.3.4 illustrates the functions $g(t) = \hat{x}_\vee(t) \wedge \hat{x}'_\vee(t)$ (dashed line), $h(t) = \hat{x}_\wedge(t) \vee \hat{x}'_\wedge(t)$ (solid line) figuring in the equation (3.2.9) and formed on the basis of estimators (3.2.6a,b); (3.2.7a,b), and also useful signal $s(t)$ (dotted line). For convenience of visual perception, the functions $g(t)$ and $h(t)$ are shifted vertically at ± 1, respectively. Fig. 3.3.5 depict the resulting signal $z(t)$ in the output of processing unit determined by the equation (3.2.9), and also useful signal $s(t)$ (dotted line). The signal $z(t)$ in the output of Wiener filter, based on L-group operations and performing the algorithm (3.2.9), (3.2.6)...(3.2.8) (so that vector of weight coefficients is determined by the relationship (3.3.2a)), completely repeats the signal

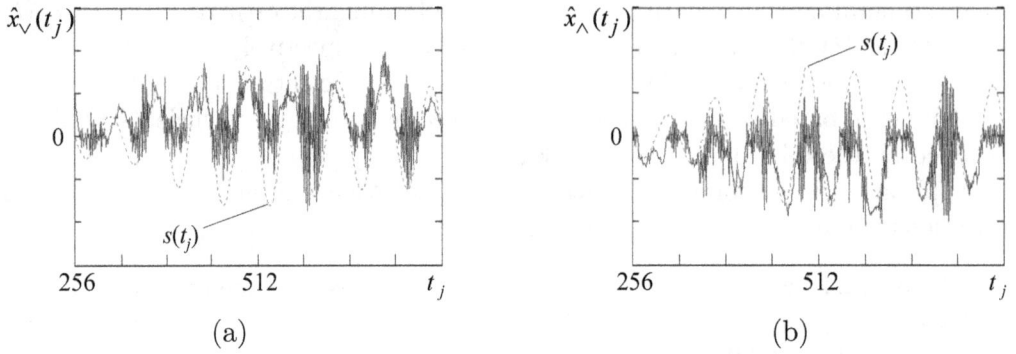

FIGURE 3.3.2   Estimates: (a) $\hat{x}_\vee(t)$ of join (3.2.6a); (b) $\hat{x}_\wedge(t)$ of meet (3.2.6b)

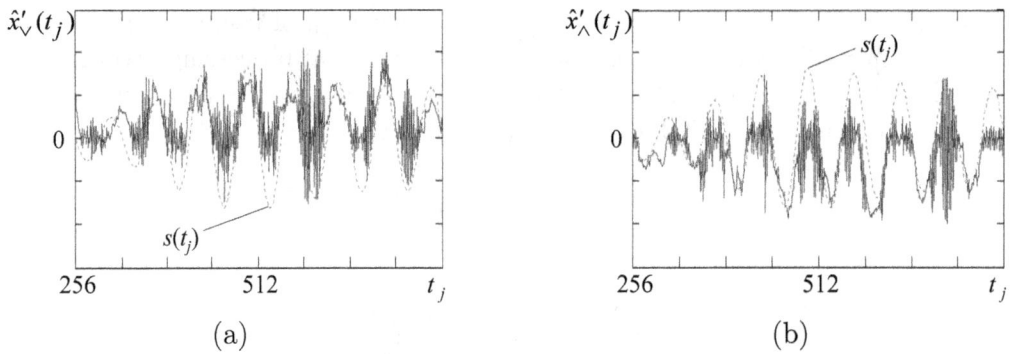

FIGURE 3.3.3   Estimates: (a) $\hat{x}'_\vee(t)$ of join (3.2.7a); (b) $\hat{x}'_\wedge(t)$ of meet (3.2.7b)

$y(t)$ in the output of classic Wiener filter realizing the algorithm (3.3.2) (see Fig. 3.3.1b), according to the identity (3.2.10) of Theorem 3.2.1: $z(t) \equiv y(t)$.

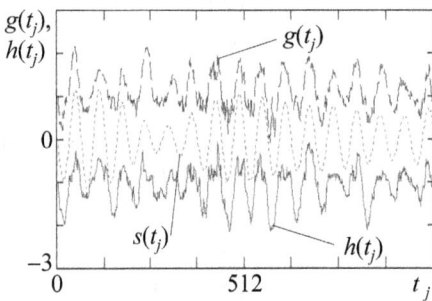

FIGURE 3.3.4   Functions $g(t) = \hat{x}_\vee(t) \wedge \hat{x}'_\vee(t)$, $h(t) = \hat{x}_\wedge(t) \vee \hat{x}'_\wedge(t)$

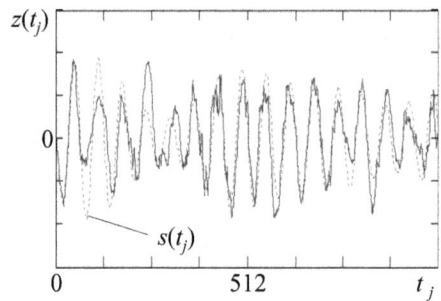

FIGURE 3.3.5   Resulting signal $z(t)$ in the output of L-group Wiener filter

Correlation coefficients between useful signal $s(t)$ and the signal $z(t)$ in the output of processing unit take the values $r[z(t), s(t)] = 0.85\ldots0.93$. $\triangledown$

3.3.2 Wiener Filter Based on Measure of Statistical Interrelation

In this subsection, we consider variants of constructing Wiener filter based on the notion of MSI that are similar in their essence to the filter (3.2.1), but realized in signal space with L-group properties.

3.3.2.1 Wiener Filter Based on Measure of Statistical Interrelation Concerned with Pseudometric

Variant of statistical formulation of Wiener filter based on MSI (1.4.10) concerned with pseudometric (1.3.25) is defined by the following relationships:

$$\mathbf{z} = \hat{\mathbf{d}} = \lambda_{\text{opt}} \mathbf{X} \cdot \mathbf{w}; \tag{3.3.4}$$

$$\mathbf{w} = \mathbf{N}_{xx}^{-1} \mathbf{P}_{dx}; \tag{3.3.4a}$$

$$\mathbf{N}_{xx} = \left\| N_{i,k}^{xx} \right\|; \quad \mathbf{P}_{dx} = \left\| P_k^{dx} \right\|; \tag{3.3.4b}$$

$$N_{i,k}^{xx} = N - \sum_{j=0}^{N-1} [\text{sgn}(x_{j,i} \vee x_{j,k}) - \text{sgn}(x_{j,i} \wedge x_{j,k})]; \tag{3.3.4c}$$

$$P_k^{dx} = N - \sum_{j=0}^{N-1} [\text{sgn}(x_{j,i} \vee d_j) - \text{sgn}(x_{j,i} \wedge d_j)]; \tag{3.3.4d}$$

$$\lambda_{\text{opt}} = \arg \min_{\lambda} [(\mathbf{d} - \lambda \mathbf{X}\mathbf{w})^T (\mathbf{d} - \lambda \mathbf{X}\mathbf{w})], \tag{3.3.4e}$$

where $\mathbf{z}$ is $N \times 1$ vector of signal in the output of filter; $\mathbf{X} = \|x_{j,k}\|$ is a $N \times M$ matrix of input data (3.3.2b); $\mathbf{w}$ is $M \times 1$ vector of weight coefficients; $\mathbf{N}_{xx}$ is $M \times M$ MSI matrix of input data $\mathbf{X}$ (3.3.2b); $\mathbf{P}_{dx}$ is $M \times 1$ MSI vector between input data $\mathbf{X}$ and vector of desired signal $\mathbf{d}$; λ_{opt} is weight coefficient minimizing a squared l_2-norm of a difference $\mathbf{d} - \lambda \mathbf{X}\mathbf{w}$, $\lambda_{\text{opt}} > 0$.

Another variant of statistical formulation of Wiener filter based on MSI that avoids performing operation of matrix multiplication (3.3.4a) is defined by the following relationships, respectively:

$$\mathbf{z}' = \hat{\mathbf{d}} = -\lambda_{\text{opt}} \mathbf{X} \cdot \mathbf{w}'; \tag{3.3.5}$$

$$\mathbf{w}' = [w_k']; \quad w_k' = F(\mathbf{N}_{xx}^{-1}, \mathbf{P}_{dx}, M); \tag{3.3.5a}$$

$$F(\mathbf{A}, \mathbf{b}, M) = c_k = M - \sum_{i=0}^{M-1} [\text{sgn}(A_{i,k} \vee b_i) - \text{sgn}(A_{i,k} \wedge b_i)]; \tag{3.3.5b}$$

$$\lambda_{\text{opt}}' = \arg \min_{\lambda'} [(\mathbf{d} + \lambda' \mathbf{X}\mathbf{w}')^T (\mathbf{d} + \lambda' \mathbf{X}\mathbf{w}')], \tag{3.3.5c}$$

where $F(\mathbf{A}, \mathbf{b}, M)$ is a function mapping a matrix $A = \|A_{i,k}\|$ and vector $b = [b_i]$ into a vector $c = [c_k]$; $i = 0, 1, \ldots, M-1$; $k = 0, 1, \ldots, M-1$; $\lambda_{\text{opt}}' > 0$.

Properties of MSI (1.4.10) concerned with pseudometric (1.3.25) are such that the signal $\lambda_{\text{opt}} \mathbf{X} \cdot \mathbf{w}'$ is in antiphase with respect to desired signal $\mathbf{d}$.

The third variant of statistical formulation of Wiener filter based on MSI, which, in the case of sufficiently large values of SNRs (SNR $\geqslant 15 \, \text{dB}$), avoids performing operation of matrix inversion in (3.3.5a), can be defined by the following relationships:

$$\mathbf{z}'' = \hat{\mathbf{d}} = \lambda_{\text{opt}} \mathbf{X} \cdot \mathbf{w}''; \tag{3.3.6}$$

$$\mathbf{w}'' = [w_k'']; \quad w_k'' = F(\mathbf{I}(M), \mathbf{P}_{dx}, M); \tag{3.3.6a}$$

$$\lambda_{\text{opt}}'' = \arg\min_{\lambda''}[(\mathbf{d} - \lambda'' \mathbf{X} \mathbf{w}'')^T (\mathbf{d} - \lambda'' \mathbf{X} \mathbf{w}'')], \tag{3.3.6b}$$

where the function $F(\mathbf{A}, \mathbf{b}, M)$ is determined by the relationship (3.3.5b); $\mathbf{I}(M)$ identity matrix with dimensionality $M \times M$: $I_{i,k} = 1$, if $i = k$, $I_{i,k} = 0$, if $i \neq k$.

If $M = 64$, $N = 1024$, L-group Wiener filter algorithms (3.3.4), (3.3.5), (3.3.6) are inferior to classic Wiener filter (3.3.2) in noise variance in the output of filter up to 1.5 dB on SNR=20 dB, and up to 3 dB on SNR=30 dB.

Notice that operation of matrix multiplication between matrix $\mathbf{X}$ and vector of weight coefficients in algorithms (3.3.4), (3.3.5), (3.3.6) can be substituted for the function $H(\mathbf{A}, \mathbf{b}, M)$:

$$H(\mathbf{A}, \mathbf{b}, M) = c_j = M - \sum_{k=0}^{M-1} [\text{sgn}(A_{j,k} \vee b_k) - \text{sgn}(A_{j,k} \wedge b_k)], \tag{3.3.7}$$

that maps a matrix $A = \|A_{i,k}\|$ and vector $b = [b_i]$ into a vector $c = [c_k]$, $j = 0, 1, \ldots, N - 1$. This replacement is, however, accompanied with additional losses.

3.3.2.2 Wiener Filter Based on Measure of Statistical Interrelation Concerned with Semimetric

Variant of statistical formulation of Wiener filter based on MSI (1.4.11) concerned with semimetric (1.3.26) is defined by the following relationships:

$$\mathbf{z} = \hat{\mathbf{d}} = \lambda_{\text{opt}} \mathbf{X} \cdot \mathbf{w}; \tag{3.3.8}$$

$$\mathbf{w} = \mathbf{N}_{xx}^{-1} \mathbf{P}_{dx}; \tag{3.3.8a}$$

$$\mathbf{N}_{xx} = \left\| N_{i,k}^{xx} \right\|; \quad \mathbf{P}_{dx} = \left\| P_k^{dx} \right\|; \tag{3.3.8b}$$

$$N_{i,k}^{xx} = \sum_{j=0}^{N-1} [|x_{j,i}| + |x_{j,k}| - |x_{j,i} - x_{j,k}|(\text{sgn}(x_{j,i} \vee x_{j,k}) - \text{sgn}(x_{j,i} \wedge x_{j,k}))]; \tag{3.3.8c}$$

$$P_k^{dx} = \sum_{j=0}^{N-1} [|x_{j,i}| + |d_j| - |x_{j,i} - d_j|(\text{sgn}(x_{j,i} \vee d_j) - \text{sgn}(x_{j,i} \wedge d_j))]; \tag{3.3.8d}$$

$$\lambda_{\text{opt}} = \arg\min_{\lambda}[(\mathbf{d} - \lambda \mathbf{X} \mathbf{w})^T (\mathbf{d} - \lambda \mathbf{X} \mathbf{w})], \tag{3.3.8e}$$

where all notations have the same meaning that those in the relationships (3.3.4a)...(3.3.4d).

Another variant of statistical formulation of Wiener filter based on MSI, that allows avoiding operation of matrix multiplication (3.3.8a), is defined by the following relationships:

$$\mathbf{z}' = \hat{\mathbf{d}} = -\lambda_{\text{opt}}\mathbf{X} \cdot \mathbf{w}'; \tag{3.3.9}$$

$$\mathbf{w}' = [w_k']; \ w_k' = F(\mathbf{N}_{xx}^{-1}, \mathbf{P}_{dx}, M); \tag{3.3.9a}$$

$$F(\mathbf{A}, \mathbf{b}, M) = c_k = \sum_{i=0}^{M-1} (|A_{i,k}| + |b_i|)$$

$$- \sum_{i=0}^{M-1} |A_{i,k} - b_i|[\text{sgn}(A_{i,k} \vee b_i) - \text{sgn}(A_{i,k} \wedge b_i)]; \tag{3.3.9b}$$

$$\lambda_{\text{opt}}' = \arg\min_{\lambda'}[(\mathbf{d} + \lambda'\mathbf{X}\mathbf{w}')^T(\mathbf{d} + \lambda'\mathbf{X}\mathbf{w}')], \tag{3.3.9c}$$

where $F(\mathbf{A}, \mathbf{b}, M)$ is a function mapping a matrix $A = \|A_{i,k}\|$ and vector $b = [b_i]$ into a vector $c = [c_k]$; $i = 0, 1, \ldots, M - 1$; $k = 0, 1, \ldots, M - 1$; $\lambda_{\text{opt}}' > 0$.

Properties of MSI (1.4.11) concerned with semimetric (1.3.26) are such that the signal $\lambda_{\text{opt}}\mathbf{X} \cdot \mathbf{w}'$ is in antiphase with respect to desired signal $\mathbf{d}$.

The third variant of statistical formulation of Wiener filter based on MSI that, in the case of sufficiently large SNRs (SNR $\geqslant$ 15 dB), allow avoiding operation of matrix inversion in (3.3.8a), can be written in the following form:

$$\mathbf{z}'' = \hat{\mathbf{d}} = \lambda_{\text{opt}}\mathbf{X} \cdot \mathbf{w}''; \tag{3.3.10}$$

$$\mathbf{w}'' = [w_k'']; \quad w_k'' = F(\mathbf{I}(M), \mathbf{P}_{dx}, M); \tag{3.3.10a}$$

$$\lambda_{\text{opt}}'' = \arg\min_{\lambda''}[(\mathbf{d} - \lambda''\mathbf{X}\mathbf{w}'')^T(\mathbf{d} - \lambda''\mathbf{X}\mathbf{w}'')], \tag{3.3.10b}$$

where the function $F(\mathbf{A}, \mathbf{b}, M)$ is determined by the relationship (3.3.9b); $\mathbf{I}(M)$ is identity matrix with dimensionality $M \times M$: $I_{i,k} = 1$, if $i = k$, $I_{i,k} = 0$, if $i \neq k$.

If $M = 64$, $N = 1024$, L-group Wiener filter algorithms,(3.3.8), (3.3.9), (3.3.10) are inferior to classic Wiener filter (3.3.2) in noise variance in the filter output up to 1 dB on SNR=20 dB, and up to 2 dB on SNR=30 dB.

3.3.2.3 Wiener Filter Based on Measure of Statistical Interrelation Concerned with l_1-metric

Variant of statistical formulation of Wiener filter based on MSI (1.4.12) concerned with l_1-metric (1.3.27) is defined by the following relationships:

$$\mathbf{z} = \hat{\mathbf{d}} = \lambda_{\text{opt}}\mathbf{X} \cdot \mathbf{w}; \tag{3.3.11}$$

$$\mathbf{w} = \mathbf{N}_{xx}^{-1}\mathbf{P}_{dx}; \tag{3.3.11a}$$

$$\mathbf{N}_{xx} = \|N_{i,k}^{xx}\|; \quad \mathbf{P}_{dx} = \|P_k^{dx}\|; \tag{3.3.11b}$$

$$N_{i,k}^{xx} = \sum_{j=0}^{N-1} [|x_{j,i} + x_{j,k}| - |x_{j,i} - x_{j,k}|]; \tag{3.3.11c}$$

$$P_k^{dx} = \sum_{j=0}^{N-1} [|x_{j,i} + d_j| - |x_{j,i} - d_j|]; \tag{3.3.11d}$$

$$\lambda_{\text{opt}} = \arg\min_{\lambda}[(\mathbf{d} - \lambda\mathbf{X}\mathbf{w})^T(\mathbf{d} - \lambda\mathbf{X}\mathbf{w})], \tag{3.3.11e}$$

where all notations have the same meaning that those in the relationships (3.3.4a)...(3.3.4d).

In consequence of specific properties of MSI matrix (3.3.11b) whose MSI is concerned with l_1-metric, variant of formulation of Wiener filter, which allows avoiding operation of matrix multiplication (3.3.11a), by an analogy with algorithm (3.3.9), operates with essential losses, thus, its realization is not expedient.

Variant of formulation of Wiener filter based on MSI, which, in the case of sufficiently large SNRs (SNR $\geqslant 15$ dB), allows avoiding operation of matrix inversion in (3.3.11a), can be written in the following form:

$$\mathbf{z}'' = \hat{\mathbf{d}} = \lambda_{\text{opt}} \mathbf{X} \cdot \mathbf{w}''; \tag{3.3.12}$$

$$\mathbf{w}'' = [w_k'']; \quad w_k'' = F(\mathbf{I}(M), \mathbf{P}_{dx}, M); \tag{3.3.12a}$$

$$F(\mathbf{A}, \mathbf{b}, M) = c_k = \sum_{i=0}^{M-1} |A_{i,k} + b_i| - \sum_{i=0}^{M-1} |A_{i,k} - b_i|; \tag{3.3.12b}$$

$$\lambda_{\text{opt}}'' = \arg\min_{\lambda''} [(\mathbf{d} - \lambda'' \mathbf{X} \mathbf{w}'')^T (\mathbf{d} - \lambda'' \mathbf{X} \mathbf{w}'')], \tag{3.3.12c}$$

where $\mathbf{I}(M)$ is identity matrix with dimensionality $M \times M$: $I_{i,k} = 1$, if $i = k$, $I_{i,k} = 0$, if $i \neq k$.

If $M = 64$, $N = 1024$, L-group algorithm (3.3.11) is inferior to Wiener filter (3.3.2) in noise variance in the filter output up to 1 dB on SNR=20 dB, and up to 2.5 dB on SNR=30 dB, whereas L-group algorithm (3.3.12) is inferior to Wiener filter (3.3.2) in noise variance in the filter output up to 1.6 dB on SNR=20 dB, and up to 3.6 dB on SNR=30 dB.

3.4   Least Mean Squares Filter Based on L-group Operations

Least mean squares (LMS) filter is based on steepest descent method [101] and estimating vector of weight coefficients without direct inversion of correlation matrix (see formula (3.3.1)). LMS filter forms an output signal $y(t)$ and optimal solution for vector of weight coefficients $\mathbf{w}(t)$ defined by the following relationships [95, (7.102)], [102, (18.30)], [100, (10.20)]:

$$y(t) = \mathbf{w}^T(t)\mathbf{X}(t); \tag{3.4.1}$$

$$\mathbf{w}(t_j) = [w_0(t_j), \ldots, w_{M-1}(t_j)]^T = \mathbf{w}(t_{j-1}) + \mu \cdot e(t_{j-1})\mathbf{X}(t_{j-1}); \tag{3.4.1a}$$

$$e(t_j) = d(t_j) - \mathbf{w}^T(t_j)\mathbf{X}(t_j); \tag{3.4.1b}$$

$$\mathbf{X}(t) = [X(t), X(t - \Delta t), \ldots, X(t - \Delta t \cdot (M - 1))]^T, \tag{3.4.1c}$$

where $\mathbf{X}(t)$ is $M \times 1$ vector composed of $\mathbf{x}(t)$; $\mathbf{x}(t) = [x(t), x(t - \Delta t), \ldots, x(t - \Delta t \cdot (N - 1))]^T$ is $N \times 1$ vector (3.2.2b) of the observed process $x(t)$, $x(t) = s(t) + n(t)$; $\mathbf{w}(t)$ is $M \times 1$ vector; μ is an adaptation step size; $e(t)$ is error signal; $s(t)$ is useful signal, $s(t) \equiv d(t)$; $d(t)$ is desired signal; $n(t)$ is interference signal; Δt is a sampling interval of transversal filter; M, N are numbers of samples used in processing.

3.4.1 Least Mean Squares Filter Based on the Method of Mapping of Linear Space into Space with Lattice Properties

For LMS filter based on L-group operations (or, simply, L-group LMS filter), general filtering algorithm with forming pairwise estimators $\hat{x}_\vee(t), \hat{x}_\wedge(t)$ and $\hat{x}'_\vee(t), \hat{x}'_\wedge(t)$ of join (3.2.3a) and meet (3.2.3b) is defined by the relationships (3.2.9), (3.2.6)...(3.2.8). Weight coefficient $w_j(t)$ in j-th processing channel is formed on the basis of optimal solution for vector of weight coefficients $\mathbf{w}(t)$ determined by the relationship (3.4.1a). Thus, complete L-group LMS algorithm based on the method of mapping of linear space into space with lattice properties is defined by the following relationships:

$$z(t) = [\hat{X}_\vee(t) \wedge \hat{X}'_\vee(t)] \vee 0 + [\hat{X}_\wedge(t) \vee \hat{X}'_\wedge(t)] \wedge 0; \qquad (3.4.2)$$

$$\hat{X}_\vee(t) = A_0(t) + |A_1(t)|, \quad \hat{X}_\wedge(t) = A_0(t) - |A_1(t)|; \qquad (3.4.2a)$$

$$\hat{X}'_\vee(t) = A_1(t) + |A_0(t)|, \quad \hat{X}'_\wedge(t) = A_1(t) - |A_0(t)|; \qquad (3.4.2b)$$

$$A_0(t) = \sum_{j=0}^{(M-2)/2} w_{2j}(t) \cdot X(t - \Delta t \cdot 2j); \qquad (3.4.2c)$$

$$A_1(t) = \sum_{j=1}^{M/2} w_{2j-1}(t) \cdot X(t - \Delta t \cdot (2j - 1)); \qquad (3.4.2d)$$

$$\mathbf{w}(t_j) = [w_0(t_j), \ldots, w_{M-1}(t_j)]^T = \mathbf{w}(t_{j-1}) + \mu \cdot e(t_{j-1})\mathbf{X}(t_{j-1}), \qquad (3.4.2e)$$

where error signal $e(t)$ is formed according to (3.4.1b).

Application of Theorem 3.2.1 with respect to LMS filter based on L-group operations is illustrated by the following example.

Example 3.4.1. Fig. 2.4.1a and 2.4.1b depict stochastic processes $x(t) = s(t) + n(t)$ $(s(t) = d(t))$; $y(t)$, respectively, in the input and output of LMS filter realizing algorithm (3.4.1). SNR q^2 (in power units) is equal to $q^2 = 10$. The signals $x(t), y(t)$ are shown by solid line, and useful signal $s(t)$ – by dotted line. Interference signal $n(t)$ is quasi-white Gaussian noise. The ratio of upper bound frequency $f_{n,\max}$ of interference signal PSD to carrier frequency f_0 of useful signal is equal to $f_{n,\max}/f_0 = 32$. Number of samples M used in signal processing according to classic LMS filtering algorithm determined by the relationships (3.4.1), and L-group LMS filtering algorithm determined by the relationships (3.4.2) is equal to 64.

Estimates of join and meet $\hat{X}_\vee(t), \hat{X}_\wedge(t)$ (3.4.2a) and also $\hat{X}'_\vee(t), \hat{X}'_\wedge(t)$ (3.4.2b), formed by recursive LMS filter based on the method of mapping of linear space into space with lattice properties, have the appearances that are similar to the corresponding estimates shown in Figs. 3.3.2, 3.3.3.

Fig. 2.4.2 demonstrates the functions $g(t) = \hat{X}_\vee(t) \wedge \hat{X}'_\vee(t)$ (dashed line), $h(t) = \hat{X}_\wedge(t) \vee \hat{X}'_\wedge(t)$ (solid line) figuring in the equation (3.4.2) and based on the estimators (3.4.2a) and (3.4.2b), and also useful signal $s(t)$ (dotted line). For convenience of visual perception, the functions $g(t)$ and $h(t)$ are shifted vertically at ± 1, respectively.

Fig. 2.4.3 depicts the resulting signal $z(t)$ in the output of processing unit determined by the equation (3.4.2) and also useful signal $s(t)$ (dotted line). The signal $z(t)$ in the output of LMS filter based on L-group operations and realized the algorithm (3.4.2) completely repeats the signal $y(t)$ in the output of classic LMS filter realizing the algorithm (3.4.1) (see Fig. 3.4.1b) according to identity (3.2.10) of Theorem 3.2.1.

Correlation coefficients between useful signal $s(t)$ and the signal $z(t)$ in the filter output take the values in the interval $r[z(t), s(t)] = 0.75 \ldots 0.85$. $\triangledown$

3.4.2   Least Mean Squares Filter Based on Measure of Statistical Interrelation

In this section, we consider variants of LMS filters based on MSIs that are similar to the filter (3.4.1), but realized in signal space with L-group properties.

3.4.2.1   Least Mean Squares Filter Based on Measure of Statistical Interrelation Concerned with Pseudometric

LMS filter based on MSI (1.4.10) concerned with pseudometric (1.3.25) forms the signal $z(t)$ in the filter output according to the following relationships defining corresponding L-group algorithm:

$$z(t) = N_p[\mathbf{w}(t), \mathbf{X}(t)]; \tag{3.4.3}$$

$$\mathbf{w}(t_j) = [w_0(t_j), \ldots, w_{N-1}(t_j)]^T = \mathbf{w}(t_{j-1}) + \mu \cdot e(t_{j-1})\mathbf{X}(t_{j-1}); \tag{3.4.3a}$$

$$e(t_j) = d(t_j) - k \cdot N_p[\mathbf{w}(t_j), \mathbf{X}(t_j)]; \tag{3.4.3b}$$

$$N_p(\mathbf{a}, \mathbf{b}) = M - \sum_{j=0}^{M-1} [\mathrm{sgn}(a_j \vee b_j) - \mathrm{sgn}(a_j \wedge b_j)], \tag{3.4.3c}$$

where $N_p(\mathbf{a}, \mathbf{b})$ is MSI (1.4.10) between vectors $\mathbf{a} = [a_j]$, $\mathbf{b} = [b_j]$ concerned with pseudometric (1.3.25); k is some constant, $k > 0$, in practice this constant is chosen from integer powers of 2: $k = 2^q$; the rest of variables have the same meaning that those from the relationships (3.4.1).

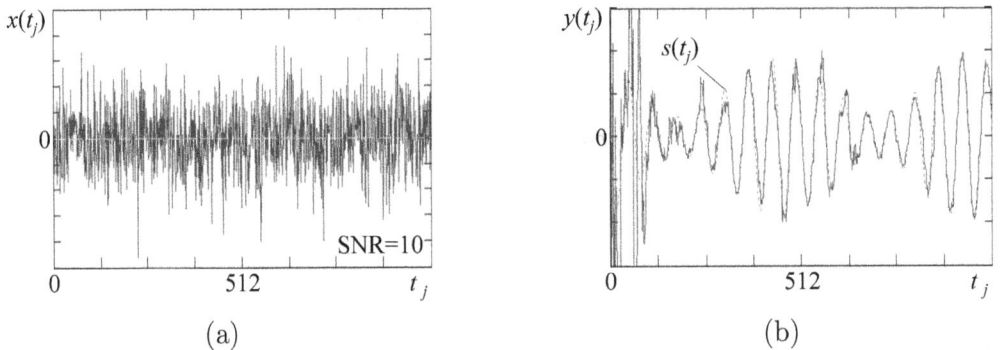

FIGURE 3.4.1   Stochastic processes: (a) $x(t)$ in the input of LMS filter; (b) $y(t)$ in the output of classic LMS filter

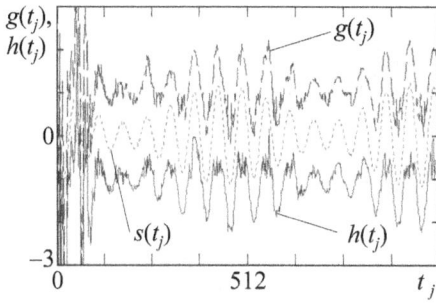

FIGURE 3.4.2 Functions $g(t) = \hat{X}_\vee(t) \wedge \hat{X}'_\vee(t)$, $h(t) = \hat{X}_\wedge(t) \vee \hat{X}'_\wedge(t)$

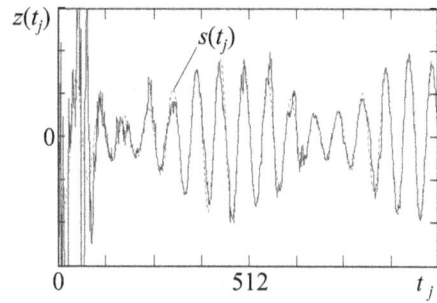

FIGURE 3.4.3 Resulting signal $z(t)$ in the output of L-group LMS filter

3.4.2.2 Least Mean Squares Filter Based on Measure of Statistical Interrelation Concerned with Semimetric

LMS filter based on MSI (1.4.11) concerned with semimetric (1.3.26) forms the signal $z(t)$ in the filter output according to the following relationships defining corresponding L-group algorithm:

$$z(t) = N_s[\mathbf{w}(t), \mathbf{X}(t)]; \tag{3.4.4}$$

$$\mathbf{w}(t_j) = [w_0(t_j), \ldots, w_{N-1}(t_j)]^T = \mathbf{w}(t_{j-1}) + \mu \cdot e(t_{j-1})\mathbf{X}(t_{j-1}); \tag{3.4.4a}$$

$$e(t_j) = d(t_j) - k \cdot N_s[\mathbf{w}(t_j), \mathbf{X}(t_j)]; \tag{3.4.4b}$$

$$N_s(\mathbf{a}, \mathbf{b}) = \sum_{j=0}^{M-1} (|a_j| + |b_j|)$$

$$- \sum_{j=0}^{M-1} |a_j - b_j|[\operatorname{sgn}(a_j \vee b_j) - \operatorname{sgn}(a_j \wedge b_j)], \tag{3.4.4c}$$

where $N_s(\mathbf{a}, \mathbf{b})$ is MSI (1.4.11) between vectors $\mathbf{a} = [a_j]$, $\mathbf{b} = [b_j]$ concerned with semimetric (1.3.26); k is some constant, $k > 0$, in practice this constant is chosen from integer powers of 2: $k = 2^q$; the rest of variables have the same meaning the ones from the relationships (3.4.1).

3.4.2.3 Least Mean Squares Filter Based on Measure of Statistical Interrelation Concerned with l_1-metric

LMS filter based on MSI (1.4.12) concerned with l_1-metric (1.3.27) forms the signal $z(t)$ in the filter output according to the following relationships defining corresponding L-group algorithm:

$$z(t) = N_{l_1}[\mathbf{w}(t), \mathbf{X}(t)]; \tag{3.4.5}$$

$$\mathbf{w}(t_j) = [w_0(t_j), \ldots, w_{N-1}(t_j)]^T = \mathbf{w}(t_{j-1}) + \mu \cdot e(t_{j-1})\mathbf{X}(t_{j-1}); \tag{3.4.5a}$$

$$e(t_j) = d(t_j) - k \cdot N_{l_1}[\mathbf{w}(t_j), \mathbf{X}(t_j)]; \qquad (3.4.5b)$$

$$N_{l_1}(\mathbf{a}, \mathbf{b}) = \sum_{j=0}^{M-1} [|a_j + b_j| - |a_j - b_j|]; \qquad (3.4.5c)$$

where $N_{l_1}(\mathbf{a}, \mathbf{b})$ is MSI (1.4.12) between vectors $\mathbf{a} = [a_j]$, $\mathbf{b} = [b_j]$ concerned with l_1-metric (1.3.27); k is some constant, $k > 0$, in practice this constant is chosen from integer powers of 2: $k = 2^q$; the rest of variables have the same meaning that those from the relationships (3.4.1a,b).

LMS filters based on MSI and realizing algorithms (3.4.3)...(3.4.5) can have efficiency that is rather close to efficiency of their classic analogue (3.4.1), so that their adaptation process is more robust.

3.5   Recursive Least Squares Filter Based on L-group Operations

Recursive least squares (RLS) filter is a variant of Wiener filter implementation (see formula (3.3.1)), in which estimation of a vector of weight coefficients is realized recursively without direct inversion of correlation matrix and optimal solution for vector of weight coefficients $\mathbf{w}(t)$ is formed according the following relationships [95, (7.78)...(7.81)], [102, (21.37)...(21.39)], [100, (10.32)]:

$$y(t) = \mathbf{w}^T(t)\mathbf{X}(t); \qquad (3.5.1)$$

$$\mathbf{w}(t_j) = [w_0(t_j), \dots, w_{M-1}(t_j)]^T = \mathbf{w}(t_{j-1}) + e(t_j)\mathbf{k}(t_j); \qquad (3.5.1a)$$

$$e(t_j) = d(t_j) - \mathbf{w}^T(t_{j-1})\mathbf{X}(t_j); \qquad (3.5.1b)$$

$$\mathbf{k}(t_j) = \frac{\lambda^{-1}\boldsymbol{\Phi}_{xx}(t_{j-1})\mathbf{X}(t_j)}{1 + \lambda^{-1}\mathbf{X}^T(t_j)\boldsymbol{\Phi}_{xx}(t_{j-1})\mathbf{X}(t_j)}; \qquad (3.5.1c)$$

$$\boldsymbol{\Phi}_{xx}(t_j) = \lambda^{-1}\boldsymbol{\Phi}_{xx}(t_{j-1}) - \lambda^{-1}\mathbf{k}(t_j)\mathbf{X}^T(t_j)\boldsymbol{\Phi}_{xx}(t_{j-1}); \qquad (3.5.1d)$$

$$\boldsymbol{\Phi}_{xx}(t_0) = \mathbf{R}_{xx}^{-1}(t_0) = c \cdot \mathbf{I}, \quad \mathbf{I} = \|\delta_{ik}\|; \qquad (3.5.1e)$$

$$\mathbf{w}(t_0) = [1, 1, \dots, 1]^T; \qquad (3.5.1f)$$

$$\mathbf{X}(t) = [X(t), X(t - \Delta t), \dots, X(t - \Delta t \cdot (M-1))]^T, \qquad (3.5.1g)$$

where $\mathbf{X}(t)$ is $M \times 1$ vector composed of a vector $\mathbf{x}(t)$; $\mathbf{x}(t) = [x(t), x(t - \Delta t), \dots, x(t - \Delta t \cdot (N-1))]^T$ is a vector (3.2.2b) of the observed process $x(t)$, $x(t) = d(t) + n(t)$; $e(t)$ is error signal; $d(t)$ is a desired signal; $n(t)$ is interference signal; M is a number of samples used in processing; $\boldsymbol{\Phi}_{xx}(t) = \mathbf{R}_{xx}^{-1}(t)$; $\mathbf{R}_{xx}(t)$ is correlation matrix of vector $\mathbf{X}(t)$ equal to $\mathbf{R}_{xx}(t) = \mathbf{M}[\mathbf{X}(t)\mathbf{X}^T(t)]$; $\mathbf{M}[*]$ is a symbol of mathematical expectation; λ is a forgetfulness coefficient; the relationship (3.5.1c) determines $M \times 1$ gain vector; the relationship (3.5.1d) determines recursive modification of inverse correlation matrix $\boldsymbol{\Phi}_{xx}(t) = \mathbf{R}_{xx}^{-1}(t)$; the relationships (3.5.1e), (3.5.1f) determine the initial conditions for correlation matrix and weight coefficient vector, respectively; $\mathbf{I}$ is identity matrix; δ_{ik} is Kronecker symbol:

$\delta_{ik} = 1$, if $i = k$; $\delta_{ik} = 0$, if $i \neq k$; c=const; time parameter t_j takes discrete values $j = 0, 1, \ldots, N - 1$; $t_j - t_{j-1} = \Delta t$, Δt is a sampling interval.

3.5.1 Recursive Least Squares Filter Based on the Method of Mapping of Linear Space into Space with Lattice Properties

For RLS filter based on the method of mapping of linear space into space with lattice properties, the general filtering algorithm forming the pairwise estimators $\hat{x}_\vee(t), \hat{x}_\wedge(t)$ and $\hat{x}'_\vee(t), \hat{x}'_\wedge(t)$ of join (3.2.2a) and meet (3.2.2b) is defined by the relationships (3.2.9), (3.2.6)...(3.2.8). Weight coefficient $w_j(t)$ in j-th processing channel is formed on the basis of optimal solution for vector of weight coefficients $\mathbf{w}(t)$ determined by the relationship (3.5.1a). Thus, complete L-group RLS algorithm based on the method of mapping of linear space into space with lattice properties is defined by the following relationships:

$$z(t) = [\hat{X}_\vee(t) \wedge \hat{X}'_\vee(t)] \vee 0 + [\hat{X}_\wedge(t) \vee \hat{X}'_\wedge(t)] \wedge 0; \qquad (3.5.2)$$

$$\hat{X}_\vee(t) = A_0(t) + |A_1(t)|, \quad \hat{X}_\wedge(t) = A_0(t) - |A_1(t)|; \qquad (3.5.2a)$$

$$\hat{X}'_\vee(t) = A_1(t) + |A_0(t)|, \quad \hat{X}'_\wedge(t) = A_1(t) - |A_0(t)|; \qquad (3.5.2b)$$

$$A_0(t) = \sum_{j=0}^{(M-2)/2} w_{2j}(t) \cdot X(t - \Delta t \cdot 2j); \qquad (3.5.2c)$$

$$A_1(t) = \sum_{j=1}^{M/2} w_{2j-1}(t) \cdot X(t - \Delta t \cdot (2j - 1)); \qquad (3.5.2d)$$

$$\mathbf{w}(t_j) = [w_0(t_j), \ldots, w_{M-1}(t_j)]^T = \mathbf{w}(t_{j-1}) + e(t_j)\mathbf{k}(t_j); \qquad (3.5.2e)$$

$$e(t_j) = d(t_j) - \mathbf{w}^T(t_{j-1})\mathbf{X}(t_j); \qquad (3.5.2f)$$

$$\mathbf{k}(t_j) = \frac{\lambda^{-1}\boldsymbol{\Phi}_{xx}(t_{j-1})\mathbf{X}(t_j)}{1 + \lambda^{-1}\mathbf{X}^T(t_j)\boldsymbol{\Phi}_{xx}(t_{j-1})\mathbf{X}(t_j)}; \qquad (3.5.2g)$$

$$\boldsymbol{\Phi}_{xx}(t_j) = \lambda^{-1}\boldsymbol{\Phi}_{xx}(t_{j-1}) - \lambda^{-1}\mathbf{k}(t_j)\mathbf{X}^T(t_j)\boldsymbol{\Phi}_{xx}(t_{j-1}); \qquad (3.5.2h)$$

$$\boldsymbol{\Phi}_{xx}(t_0) = \mathbf{R}_{xx}^{-1}(t_0) = c \cdot \mathbf{I}, \quad \mathbf{I} = \|\delta_{ik}\|; \qquad (3.5.2i)$$

$$\mathbf{w}(t_0) = [1, 1, \ldots, 1]^T; \qquad (3.5.2j)$$

$$\mathbf{X}(t) = [X(t), X(t - \Delta t), \ldots, X(t - \Delta t \cdot (M - 1))]^T, . \qquad (3.5.2k)$$

Application of Theorem 3.2.1 with respect to RLS filter based on L-group operations is illustrated by the following example.

Example 3.5.1. Fig. 3.5.1a and 3.5.1b depict stochastic processes $x(t) = s(t)+n(t)$ $(s(t) = d(t))$, $y(t)$ in the input and output of RLS filter realizing algorithm (3.5.1), respectively. SNR q^2 (in power units) is equal to $q^2 = 10$. The signals $x(t)$, $y(t)$ are shown by solid line, and useful signal $s(t)$ – by dotted line. Interference signal $n(t)$ is quasi-white Gaussian noise. The ratio of upper bound frequency $f_{n,\max}$ of interference signal PSD to a carrier frequency f_0 of useful signal is equal to $f_{n,\max}/f_0 = 32$. A number of samples M used when processing according to classic

(a)   (b)

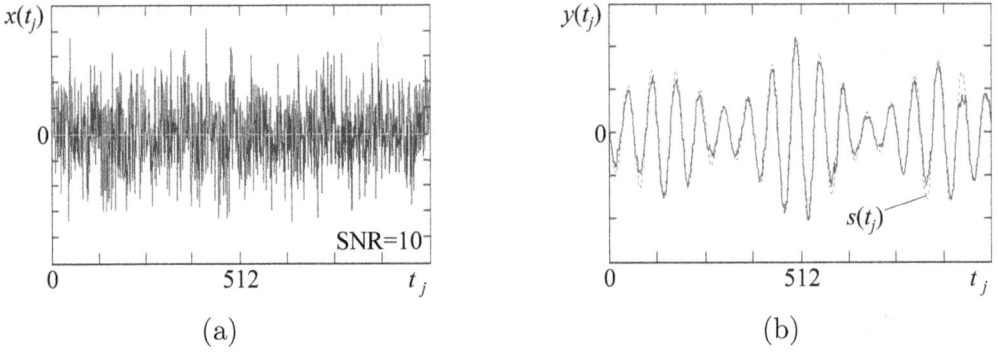

FIGURE 3.5.1   Stochastic processes: (a) $x(t)$ in the input of RLS filter; (b) $y(t)$ in the output of classic RLS filter

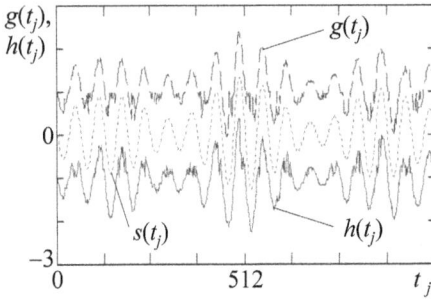

FIGURE 3.5.2   Functions $g(t) = \hat{X}_\vee(t) \wedge$ FIGURE 3.5.3   Resulting signal $z(t)$ in the
$\hat{X}'_\vee(t)$, $h(t) = \hat{X}_\wedge(t) \vee \hat{X}'_\wedge(t)$ output of L-group RLS filter

RLS algorithm defined by the relationships (3.5.1) and L-group RLS algorithm determined by the relationships (3.5.2) is equal to 64.

The estimators of join and meet $\hat{X}_\vee(t), \hat{X}_\wedge(t)$ (3.5.2a) and also $\hat{X}'_\vee(t), \hat{X}'_\wedge(t)$ (3.5.2b) formed by RLS filter based on the method of mapping of linear space into space with lattice properties have the form corresponding to the estimates shown in Figs. 3.5.2, 3.5.3.

Fig. 3.5.2 illustrates the functions $g(t) = \hat{X}_\vee(t) \wedge \hat{X}'_\vee(t)$ (dashed line) and $h(t) = \hat{X}_\wedge(t) \vee \hat{X}'_\wedge(t)$ (solid line) figuring in the equation (3.5.2) and formed on the basis of the estimators (3.5.2a,b), and also useful signal $s(t)$ (dotted line). For convenience of visual perception, the functions $g(t)$ and $h(t)$ are shifted vertically at ± 1, respectively.

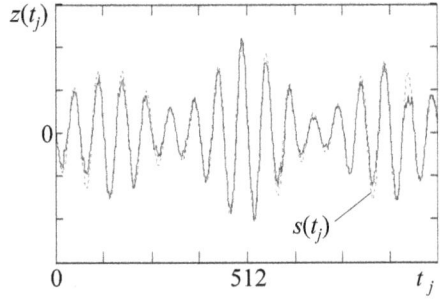

Fig. 3.5.3 depicts the resulting signal $z(t)$ in the output of L-group RLS filter defined by the equation (3.5.2) and also useful signal $s(t)$ (dotted line). The signal $z(t)$ in the output of RLS filter based on L-group operations and realizing the algorithm (3.5.2), according to the identity (3.2.10) of Theorem 3.2.1, completely repeats the signal $y(t)$ in the output of classic RLS filter realizing algorithm (3.5.1) (see Fig. 3.5.1b).

Correlation coefficients between useful signal $s(t)$ and the signal $z(t)$ in the filter output take the values $r[z(t), s(t)] = 0.85 \ldots 0.99$. $\triangledown$

3.5.2 Recursive Least Squares Filter Based on Measures of Statistical Interrelation

RLS filter based on measure of statistical interrelation (MSI) is a variant of Wiener filter (see the formula (3.3.1)) in which estimating vector of weight coefficients is realized recursively without direct inversion of correlation matrix, so that optimal solution for vector of weight coefficients $\mathbf{w}$ is defined by the following relationships:

$$\mathbf{w}(t_j) = [w_0(t_j), \ldots, w_{M-1}(t_j)]^T = \mathbf{w}(t_{j-1}) + e(t_j)\mathbf{k}(t_j); \tag{3.5.3a}$$

$$e(t_j) = d(t_j) - \mathbf{w}^T(t_{j-1})\mathbf{X}(t_j); \tag{3.5.3b}$$

$$\mathbf{k}(t_j) = \frac{\lambda^{-1}\boldsymbol{\Phi}_{xx}(t_{j-1})\mathbf{X}(t_j)}{1 + \lambda^{-1}\mathbf{X}^T(t_j)\boldsymbol{\Phi}_{xx}(t_{j-1})\mathbf{X}(t_j)}; \tag{3.5.3c}$$

$$\boldsymbol{\Phi}_{xx}(t_j) = \lambda^{-1}\boldsymbol{\Phi}_{xx}(t_{j-1}) - \lambda^{-1}\mathbf{N}[\mathbf{N}(\mathbf{k}(t_j), \mathbf{X}^T(t_j)), \boldsymbol{\Phi}_{xx}(t_{j-1})]; \tag{3.5.4a}$$

$$\mathbf{N}(\mathbf{A}, \mathbf{B}) = \left\|N^*_{i,k}\right\|; \ N^*_{i,k} = N^*[(\mathbf{A}^T)^{<i>}, \mathbf{B}^{<k>}]; \tag{3.5.4b}$$

$$\boldsymbol{\Phi}_{xx}(t_0) = c \cdot \mathbf{I}, \ \mathbf{I} = [\delta_{ik}]; \tag{3.5.5a}$$

$$\mathbf{w}(t_0) = [1, 1, \ldots, 1]^T, \tag{3.5.5b}$$

where $\mathbf{N}(\mathbf{A}, \mathbf{B}) = \left\|N^*_{i,k}\right\|$ is a matrix function of two matrices (or vectors) $\mathbf{A}, \mathbf{B}$ with dimensionalities $M \times L$ and $L \times M$, respectively; $(\mathbf{A}^T)^{<i>}$ is i-th column vector of transposed matrix $\mathbf{A}$; $\mathbf{B}^{<k>}$ is k-th column vector of matrix $\mathbf{B}$; $i = 0, \ldots, M-1$, $k = 0, \ldots, M-1$; $N^*_{i,k} = N^*[(\mathbf{A}^T)^{<i>}, \mathbf{B}^{<k>}, m]$ is an element of matrix function $\mathbf{N}(\mathbf{A}, \mathbf{B}) = \left\|N^*_{i,k}\right\|$ that is calculated on the basis of corresponding MSI defined by one of the following three relationships, respectively:

$$N_p(\mathbf{a}, \mathbf{b}) = L - \sum_{j=0}^{L-1} [\operatorname{sgn}(a_j \vee b_j) - \operatorname{sgn}(a_j \wedge b_j)]; \tag{3.5.6a}$$

$$N_s(\mathbf{a}, \mathbf{b}) = \sum_{j=0}^{L-1} (|a_j| + |b_j|) - \sum_{j=0}^{L-1} |a_j - b_j|[\operatorname{sgn}(a_j \vee b_j) - \operatorname{sgn}(a_j \wedge b_j)]; \tag{3.5.6b}$$

$$N_{l_1}(\mathbf{a}, \mathbf{b}) = \sum_{j=0}^{L-1} [|a_j + b_j| - |a_j - b_j|], \tag{3.5.6c}$$

where $N_p(\mathbf{a}, \mathbf{b})$, $N_s(\mathbf{a}, \mathbf{b})$, $N_{l_1}(\mathbf{a}, \mathbf{b})$ are MSIs concerned with pseudometric (1.3.25), semimetric (1.3.26), and l_1-metric (1.3.27), respectively; $\mathbf{a} = [a_j]$, $\mathbf{b} = [b_j]$ are vectors; $L = M$ or $L = 1$; the rest variables have the same sense that those from the relationships (3.5.1).

Thus, complete RLS filtering algorithm based on one of MSIs (3.5.6) is defined by the relationships (3.5.3)...(3.5.5), so that matrix function $\mathbf{N}(\mathbf{A}, \mathbf{B}) = \left\| N_{i,k}^* \right\|$ (3.5.4b) is calculated according to the formulas (3.5.6). We distinguish RLS filters based on MSIs (3.5.6) concerned with pseudometric (1.3.25), semimetric (1.3.26), and l_1-metric (1.3.27), respectively.

Results of simulating operation of RLS filters based on MSIs (3.5.6) allow claiming that such L-group filter can be not worse than classic RLS filter operating according to algorithm (3.5.1).

3.6   Kalman Filter Based on Method of Mapping of Linear Space into Space with Lattice Properties

Kalman filter algorithm is based on two equations: state equation and observation equation that are defined by two following relationships, respectively [99], [95, (7.1), (7.2)], [97, (13.12), (13.57)]:

$$\mathbf{s}(t_j) = \mathbf{\Phi}(t_j)\mathbf{s}(t_{j-1}) + \mathbf{e}(t_j); \qquad (3.6.1a)$$

$$\mathbf{x}(t_j) = \mathbf{H}(t_j)\mathbf{s}(t_j) + \mathbf{n}(t_j), \qquad (3.6.1b)$$

where $\mathbf{s}(t_j)$ is useful signal vector with dimensionality $L \times 1$: $\mathbf{s}(t_j) = [s_1(t_j), \ldots, s_L(t_j)]^T$; $\mathbf{\Phi}(t_j)$ is a state-transition matrix with dimensionality $L \times L$ describing signal transition from one state to another; $\mathbf{e}(t_j)$ is $L \times 1$ disturbance vector determining random deflections of a given state model, this vector determines $L \times L$ correlation matrix $\mathbf{R}_{ee}(t_j)$; $\mathbf{x}(t_j)$ is an observation vector with dimensionality $M \times 1$; $\mathbf{H}(t_j)$ is $M \times L$ matrix determining distortions in an information transmitting channel; $\mathbf{n}(t_j)$ is a vector of additive noise with dimensionality $M \times 1$ and $M \times M$ correlation matrix $\mathbf{R}_{nn}(t_j)$.

Then Kalman filtering algorithm supposes finding recursive estimator of useful signal [95, (7.29)...(7.32)], [97, (13.58)...(13.62)]:

$$\hat{\mathbf{s}}(t_j) = \mathbf{\Phi}(t_j)\hat{\mathbf{s}}(t_{j-1}) + \mathbf{K}(t_{j-1})[\mathbf{x}(t_{j-1}) - \mathbf{H}(t_{j-1})\hat{\mathbf{s}}(t_{j-1})]; \qquad (3.6.2)$$

$$\mathbf{K}(t_{j-1}) = \mathbf{\Phi}(t_{j-1})\mathbf{R}_{\hat{s}\hat{s}}(t_{j-1})\mathbf{H}^T(t_{j-1}) \times$$
$$\times [\mathbf{H}(t_{j-1})\mathbf{R}_{\hat{s}\hat{s}}(t_{j-1})\mathbf{H}^T(t_{j-1}) + \mathbf{R}_{nn}(t_{j-1})]^{-1}; \qquad (3.6.2a)$$

$$\mathbf{R}_{\hat{s}\hat{s}}(t_j) = [\mathbf{\Phi}(t_j) - \mathbf{K}(t_{j-1})\mathbf{H}(t_{j-1})]\mathbf{R}_{\hat{s}\hat{s}}(t_{j-1})[\mathbf{\Phi}(t_j) -$$
$$-\mathbf{K}(t_{j-1})\mathbf{H}(t_{j-1})]^T + \mathbf{R}_{ee}(t_j) + \mathbf{K}(t_{j-1})\mathbf{R}_{nn}(t_{j-1})\mathbf{K}^T(t_{j-1}); \qquad (3.6.2b)$$

$$\mathbf{R}_{\hat{s}\hat{s}}(t_0) = c \cdot \mathbf{I}, \quad \mathbf{I} = \|\delta_{ik}\|; \qquad (3.6.2c)$$

$$\hat{\mathbf{s}}(t_0) = [0, 0, \ldots, 0]^T, \qquad (3.6.2d)$$

where the relation (3.6.2a) determines amplification coefficient of the filter $\mathbf{K}(t_j)$ with dimensionality $L \times M$; the relationship (3.6.2b) determines correlation matrix of signal estimator $\mathbf{R}_{\hat{s}\hat{s}}(t_j)$ with dimensionality $L \times L$; the relationships (3.6.2d), (3.6.2c) determine the initial conditions for signal estimator and its correlation matrix, respectively; $\mathbf{I}$ is identity matrix; δ_{ik} is Kronecker symbol: $\delta_{ik} = 1$, if $i = k$; $\delta_{ik} = 0$, if $i \neq k$; c=const; time parameter t_j takes discrete values $j = 0, 1, \ldots, N-1$; $t_j - t_{j-1} = \Delta t$, Δt is sampling interval.

For Kalman filter based on L-group operations, filtering algorithm differ from generalized algorithm defined by the relationships (3.2.9), (3.2.6)...(3.2.8) for two reasons. First, useful signal $\mathbf{s}(t_j)$ is a vector with dimensionality $L \times 1$: $\mathbf{s}(t_j) = [s_1(t_j), \ldots, s_L(t_j)]^T$. Second, unlike generalized filter defined by vector of weight coefficients $\mathbf{w}(t_j)$ (3.2.2a) with dimensionality $N \times 1$, Kalman filter is defined by the relationship (3.6.2) that can be written in the form:

$$\hat{\mathbf{s}}(t_j) = \mathbf{W}_s(t_j)\hat{\mathbf{s}}(t_{j-1}) + \mathbf{W}_x(t_{j-1})\mathbf{x}(t_{j-1}); \tag{3.6.3}$$

$$\mathbf{W}_s(t_j) = \mathbf{\Phi}(t_j) - \mathbf{K}(t_{j-1})\mathbf{H}(t_{j-1}); \tag{3.6.3a}$$

$$\mathbf{W}_x(t_j) = \mathbf{K}(t_j), \tag{3.6.3b}$$

where $\mathbf{W}_s(t_j)$ is weight matrix with dimensionality $L \times L$; $\mathbf{W}_x(t_j)$ is weight matrix with dimensionality $L \times M$.

Taking into account the relationships (3.6.3), Kalman filter based on L-group operations is defined by vector (but not scalar) stochastic functions $\mathbf{A}_0(t_j) = [A_{0,1}(t_j), \ldots, A_{0,L}(t_j)]^T$, $\mathbf{A}_1(t_j) = [A_{1,1}(t_j), \ldots, A_{1,L}(t_j)]^T$:

$$\begin{cases} \mathbf{A}_0(t_j) = \mathbf{W}_s(t_j)\hat{\mathbf{s}}(t_{j-1}) + \mathbf{W}_x(t_{j-1})\mathbf{x}(t_{j-1}); & (a) \\ \mathbf{A}_1(t_j) = [0, \ldots, 0]^T; & (b) \\ j = 2 \cdot i, \quad i = 0, 1, 2, \ldots; & (c) \end{cases} \tag{3.6.4}$$

$$\begin{cases} \mathbf{A}_0(t_j) = [0, \ldots, 0]^T; & (a) \\ \mathbf{A}_1(t_j) = \mathbf{W}_s(t_j)\hat{\mathbf{s}}(t_{j-1}) + \mathbf{W}_x(t_{j-1})\mathbf{x}(t_{j-1}); & (b) \\ j = 2 \cdot i + 1, \quad i = 0, 1, 2, \ldots, & (c) \end{cases} \tag{3.6.5}$$

so that forming pairwise estimators $\hat{x}_\vee(t), \hat{x}_\wedge(t)$ and $\hat{x}'_\vee(t), \hat{x}'_\wedge(t)$ of join (3.2.3a) and meet (3.2.3b) is realized with a help of the following identities:

$$\hat{x}_{\vee,k}(t_j) = A_{0,k}(t_j) + |A_{1,k}(t_j)|; \tag{3.6.6a}$$

$$\hat{x}_{\wedge,k}(t_j) = A_{0,k}(t_j) - |A_{1,k}(t_j)|, \tag{3.6.6b}$$

$$\hat{x}'_{\vee,k}(t_j) = A_{1,k}(t_j) + |A_{0,k}(t_j)|; \tag{3.6.7a}$$

$$\hat{x}'_{\wedge,k}(t_j) = A_{1,k}(t_j) - |A_{0,k}(t_j)|, \tag{3.6.7b}$$

where $k = 1, \ldots, L$.

Then the following theorem holds.

Theorem 3.6.1. *For Kalman filter based on L-group operations and realizing the processing according to algorithm:*

$$z_k(t_j) = [\hat{x}_{\vee,k}(t_j) \wedge \hat{x}'_{\vee,k}(t_j)] \vee 0 + [\hat{x}_{\wedge,k}(t_j) \vee \hat{x}'_{\wedge,k}(t_j)] \wedge 0; \quad k = 1, \ldots, L, \tag{3.6.8}$$

that is equivalent to algorithm (3.6.2) in which weight matrices $\mathbf{W}_s(t_j)$ and $\mathbf{W}_x(t_j)$ are determined by the expressions (3.6.3a), (3.6.3b), respectively:

$$\mathbf{z}(t_j) = [z_k(t_j)]\,|_{k=1,\ldots,L} = [z_1(t_j), \ldots, z_L(t_j)]^T \equiv \hat{\mathbf{s}}(t_j). \qquad (3.6.9)$$

Here the functions situated in the right part of identity (3.6.8) are defined by the expressions (3.6.6a,b), (3.6.7a,b), (3.6.4), (3.6.5).

Proof of Theorem 3.6.1 is analogous to proof of Theorem 3.2.1.

Theorem 3.6.1 is illustrated by the following example.

Example 3.6.1. Let $x(t)$ be additive mixture of useful signal $s(t)$ and interference $n(t)$ that acts in the filter input:

$$x(t) = s(t) + n(t), \qquad (3.6.10)$$

so that useful signal $s(t)$ is amplitude-modulated bandpass signal determined by the expression:

$$s(t) = A(t)\cos(\omega_0 t + \varphi), \qquad (3.6.10a)$$

where $A(t)$ is amplitude of useful signal $s(t)$; $\omega_0 = 2\pi f_0$ is a cyclic frequency of carrier frequency f_0 of useful signal $s(t)$; φ is an unknown nonrandom initial phase of useful signal.

For useful signal vector $\mathbf{s}(t_j)$ (3.6.1a), taking into account (3.6.10a), the following relationship holds:

$$\mathbf{s}'(t_j) = \mathbf{\Phi}_0 \cdot \mathbf{s}(t_j), \qquad (3.6.11)$$

where $\mathbf{s}'(t_j) = [s'(t_j),\, s''(t_j)]^T$ is a vector of derivative of useful signal vector $\mathbf{s}(t_j)$; $\mathbf{s}(t_j) = [s(t_j),\, s'(t_j)]^T$; $\mathbf{\Phi}_0 = \begin{bmatrix} 0 & 1 \\ -\omega_0^2 & 0 \end{bmatrix}$.

Then, despite nonlinearity of the function (3.6.10a), on linear approximation when Δt is sufficiently small, one can consider that useful signal vector $\mathbf{s}(t_j)$ and its derivation $\mathbf{s}'(t_j)$ are coupled by the following relationship:

$$\mathbf{s}(t_j) = \mathbf{s}(t_{j-1}) + \mathbf{s}'(t_{j-1})\Delta t = \mathbf{s}(t_{j-1}) + \Delta t \mathbf{\Phi}_0 \mathbf{s}(t_{j-1}). \qquad (3.6.12)$$

Then, taking into account (3.6.12), state-transition matrix $\mathbf{\Phi}(t_j)$ from the state equation (3.6.1a) takes the following form:

$$\mathbf{\Phi}(t_j) = \mathbf{I} + \Delta t \mathbf{\Phi}_0 = \begin{bmatrix} 1 & 0 \\ 0 & 1 \end{bmatrix} + \begin{bmatrix} 0 & 1 \\ -\omega_0^2 & 0 \end{bmatrix} \cdot \Delta t. \qquad (3.6.13)$$

Taking into account that $\mathbf{H}(t_j) = [1\quad 0]$, observation equation (3.6.1b) is simplified:

$$x(t_j) = s(t_j) + n(t_j). \qquad (3.6.14)$$

Fig. 3.6.1a and 3.6.1b illustrates stochastic processes $x(t) = s(t) + n(t)$; $y(t)$, respectively, in the input and output of Kalman filter realizing the algorithm (3.6.2). SNR q^2 (in power units) is equal to $q^2 = 10$. The signals $x(t)$, $y(t)$ are shown by

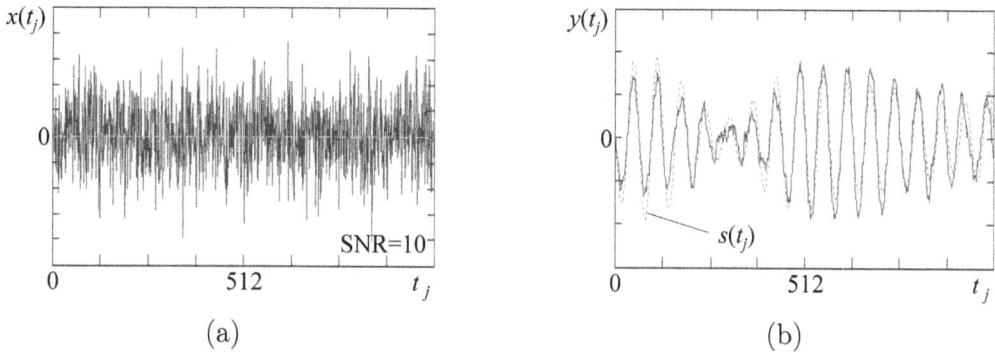

FIGURE 3.6.1   Stochastic processes: (a) $x(t)$ in the input of Kalman filter; (b) $y(t)$ in the output of Kalman filter

solid line and useful signal $s(t)$ – by dotted line. Interference signal $n(t)$ is quasi-white Gaussian noise. The ratio of upper bound frequency $f_{n,\max}$ of interference signal PSD to carrier frequency f_0 of useful signal is equal to $f_{n,\max}/f_0 = 32$.

Fig. 3.6.2a, 3.6.2b depict the estimates $\hat{x}_\vee(t)$, $\hat{x}_\wedge(t)$ of join (3.6.6a) and meet (3.6.6b) (solid line) formed by Kalman filter based on L-group operations, correspondingly, and useful signal $s(t)$ (dotted line). Fig. 3.6.3a, 3.6.3b depicts the estimates $\hat{x}'_\vee(t)$, $\hat{x}'_\wedge(t)$ of join (3.6.7a) and meet (3.6.7b) (solid line) formed by Kalman filter based on L-group operations, correspondingly, and useful signal $s(t)$ (dotted line).

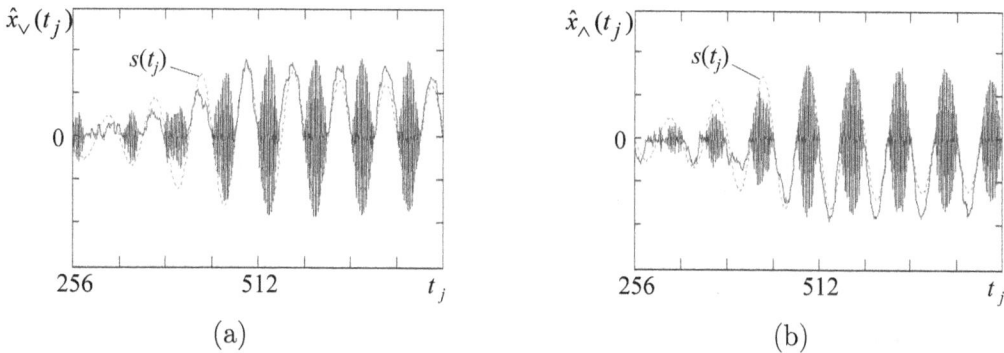

FIGURE 3.6.2   Estimates: (a) $\hat{x}_\vee(t)$ of join (3.6.6a); (b) $\hat{x}_\wedge(t)$ of meet (3.6.6b)

Fig. 3.6.4 illustrates the functions $g(t) = \hat{x}_\vee(t) \wedge \hat{x}'_\vee(t)$ (dashed line), $h(t) = \hat{x}_\wedge(t) \vee \hat{x}'_\wedge(t)$ (solid line) figuring in the equation (3.6.8) and formed on the basis of estimators (3.6.6a,b), (3.6.7a,b), and also useful signal $s(t)$ (dotted line). For convenience of visual perception, the functions $g(t)$ and $h(t)$ are shown to be shifted vertically at ± 1 with respect to zero, respectively.

Fig. 3.6.5 depicts the resulting signal $z(t)$ (solid line) in the filter output defined by the equation (3.6.8) and also useful signal $s(t)$ (dotted line). Notice, despite the fact that general solution for Kalman filter based on L-group operations is described by vector stochastic process (3.6.3), the solution for this concrete example is

(a)   (b)

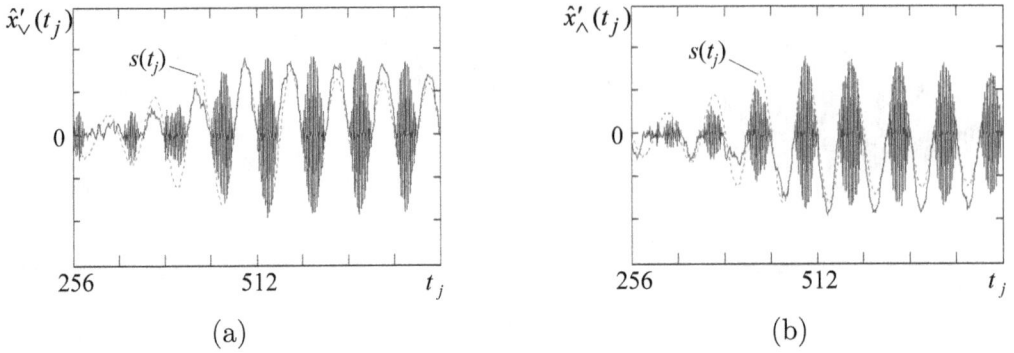

FIGURE 3.6.3   Estimates: (a) $\hat{x}'_\vee(t)$ of join (3.6.7a); (b) $\hat{x}'_\wedge(t)$ of meet (3.6.7b)

described by scalar stochastic process. The signal $z(t)$ in the output of Kalman filter, based on L-group operations and realizing the algorithm (3.6.8), (3.6.4...3.6.7), (3.6.3a,b), completely repeats the signal $y(t)$ in the output of classic Kalman filter realizing the algorithm (3.6.2) (see Fig. 3.6.1b), according to identity (3.6.9) of Theorem 3.6.1.

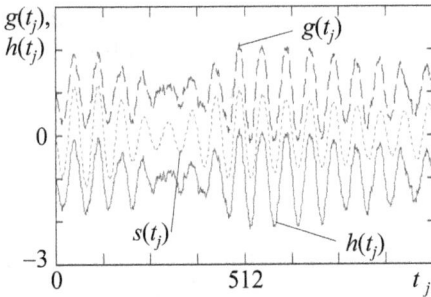

FIGURE 3.6.4   Functions $g(t) = \hat{x}_\vee(t) \wedge \hat{x}'_\vee(t)$, $h(t) = \hat{x}_\wedge(t) \vee \hat{x}'_\wedge(t)$

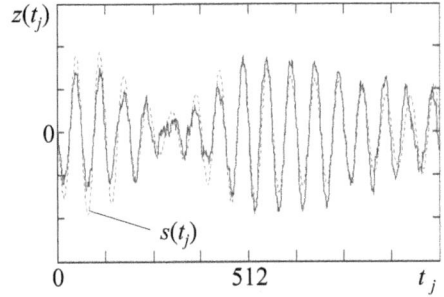

FIGURE 3.6.5   Resulting signal $z(t)$ in L-group Kalman filter output

Correlation coefficients between the useful signal $s(t)$ and the signal $z(t)$ in the filter output take the values $r[z(t), s(t)] = 0.85\ldots0.95$. $\triangledown$

3.7 L-group Composite Filter

Consider additive mixture $x(t)$ of useful signal $s(t)$ and interference $n(t)$ acting in the filter input in the form (3.6.10), so that the signal $s(t)$ is amplitude-modulated harmonic oscillation of the form (3.6.10a).

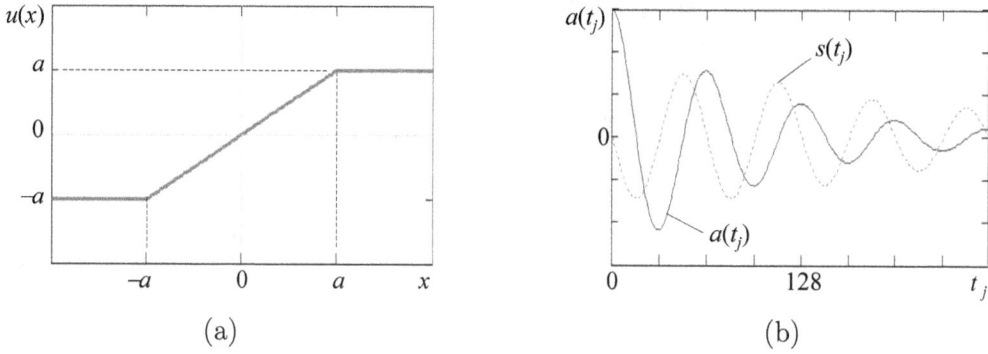

FIGURE 3.7.1   (a) Amplitude characteristic of limiter; (b) impulse response of the filter

Composite filter is constructed according to limiter – band-pass filter scheme and forms the following quasi-optimal estimator $\hat{s}(t)$ of the signal $s(t)$:

$$\hat{s}(t_j) = y(t_j) = \sum_{i=0}^{j} a(t_{j-i})u(t_j); \tag{3.7.1}$$

$$u(t_j) = u[x(t_j)] = [(x(t_j) \wedge a) \vee 0] + [(x(t_j) \vee (-a) \wedge 0]; \tag{3.7.1a}$$

$$a(t_j) = \frac{1}{\beta \cdot T_0} \exp\left(-\frac{t_j}{\alpha \cdot T_0}\right) \cos(\omega_0 t_j + \varphi_a) \cdot 1(t_j), \tag{3.7.1b}$$

where $u(t_j) = u[x(t_j)]$ is amplitude characteristic of a limiter shown in Fig. 3.7.1a; a=const, $a > A_{\max}(t)$; $A_{\max}(t)$ is a maximal value of amplitude $A(t)$ of the signal (3.6.10a); $a(t_j)$ is impulse response of the filter depicted in Fig. 3.7.1b; time parameter index j accepts discrete values $j = 0, 1, \ldots, N - 1$; $t_j - t_{j-1} = \Delta t$, Δt is a sample interval; α, β=const; $\vee$, $\wedge$ are join and meet operations of L-group, respectively; $1(t)$ is Heaviside unit step function; $\omega_0 = 2\pi f_0$ is a cyclic frequency of useful signal $s(t)$; T_0 is a period of oscillation of useful signal carrier f_0: $T_0 = 1/f_0$.

For composite filter based on the method of mapping of linear space into space with lattice properties, filtering algorithm forming pairwise estimators $\hat{x}_\vee(t), \hat{x}_\wedge(t)$ and $\hat{x}'_\vee(t), \hat{x}'_\wedge(t)$ of join (3.2.3a) and meet (3.2.3b) is defined by the relationships that are similar to (3.2.9), (3.2.6)...(3.2.7), respectively:

$$z(t_j) = [\hat{x}_\vee(t_j) \wedge \hat{x}'_\vee(t_j)] \vee 0 + [\hat{x}_\wedge(t_j) \vee \hat{x}'_\wedge(t_j)] \wedge 0; \tag{3.7.2}$$

$$\hat{x}_\vee(t_j) = A_0(t_j) + |A_1(t_j)|; \tag{3.7.3a}$$

$$\hat{x}_\wedge(t_j) = A_0(t_j) - |A_1(t_j)|; \tag{3.7.3b}$$

$$\hat{x}'_\vee(t_j) = A_1(t_j) + |A_0(t_j)|; \tag{3.7.4a}$$

$$\hat{x}'_\wedge(t_j) = A_1(t_j) - |A_0(t_j)|. \tag{3.7.4b}$$

The functions figuring in the formulas (3.7.3)...(3.7.4) are formed on the basis of the following relationships:

$$A_0(t_j) = \sum_{i=0}^{j} [1 - \text{mod}_2 i] a(t_{j-i}) u(t_j); \tag{3.7.5a}$$

$$A_1(t_j) = \sum_{i=0}^{j} \text{mod}_2 i \cdot a(t_{j-i}) u(t_j), \tag{3.7.5b}$$

where $u(t_j)$ is amplitude characteristic of a limiter; $a(t_j)$ is impulse response of the filter defined by the formulas (3.7.1a), (3.7.1b), respectively; $\text{mod}_2(2i+1) = 1$; $\text{mod}_2(2i) = 0$.

Application of Theorem 3.2.1 with respect to composite filter based on the method of mapping of linear space into space with lattice properties is illustrated by the following example.

Example 3.7.1. Fig. 3.7.2a depicts stochastic process $x(t) = s(t) + n(t)$ in the inputs of composite filter realizing the algorithm (3.7.2)...(3.7.5) and Kalman filter (3.6.2). The input process $x(t)$ is the same for these both filters and is used for their comparison. Fig. 3.7.2b depicts the signal in the output of Kalman filter realizing the algorithm (3.6.2). SNR q^2 (in power units) is equal to $q^2 = 10$. The signals $x(t)$, $y(t)$ are shown by solid line, and useful signal $s(t)$ – by dotted line. Interference signal $n(t)$ is a quasi-white Gaussian noise. The ratio of upper bound frequency $f_{n,\text{max}}$ of interference signal to a carrier frequency f_0 of useful signal is equal to $f_{n,\text{max}}/f_0 = 32$.

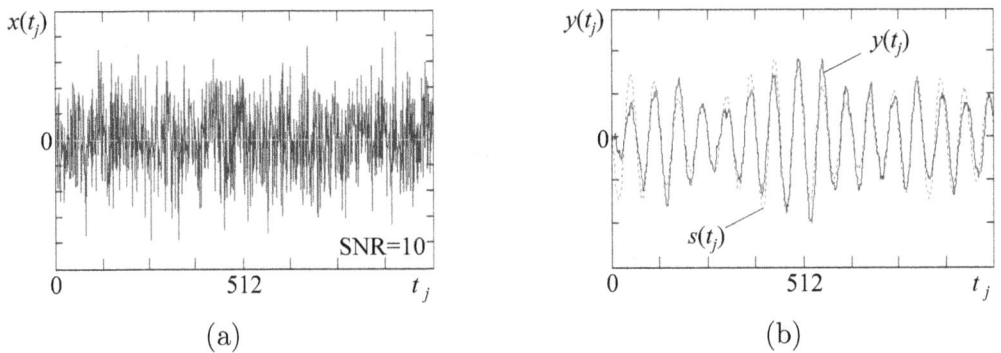

FIGURE 3.7.2 Stochastic processes: (a) $x(t)$ in the input of composite filter; (b) $y(t)$ in the output of Kalman filter

Functions $g(t) = \hat{x}_\vee(t) \wedge \hat{x}'_\vee(t)$ (dashed line) and $h(t) = \hat{x}_\wedge(t) \vee \hat{x}'_\wedge(t)$ (solid line) (3.7.2) formed by composite filter based on the method of mapping of linear space into space with lattice properties are shown in Fig. 3.3.3.

Fig. 3.7.4 illustrates the resulting signal $z(t)$ in the filter output (solid line) defined by the equation (3.7.2) and also useful signal $s(t)$ (dotted line). The signal $z(t)$ in the output of composite filter, realizing the algorithm (3.7.2)...(3.7.5), completely repeats the signal in the output of composite filter realizing the algorithm (3.7.1), (3.7.1a,b), according to identity (3.2.9) of Theorem 3.2.1. Signal processing quality of L-group composite filter slightly differs from processing quality of Kalman filter operating under the same conditions, inasmuch as difference in relative filtering error does not exceed 1.5 dB.

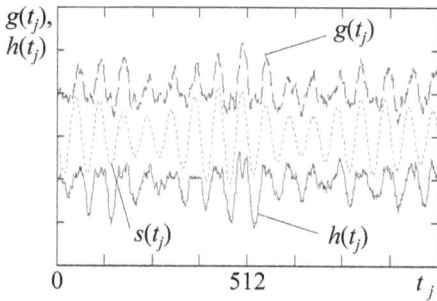

FIGURE 3.7.3   Functions $g(t) = \hat{x}_\vee(t) \wedge$ $\hat{x}'_\vee(t)$ (dashed line), $h(t) = \hat{x}_\wedge(t) \vee \hat{x}'_\wedge(t)$ (solid line)

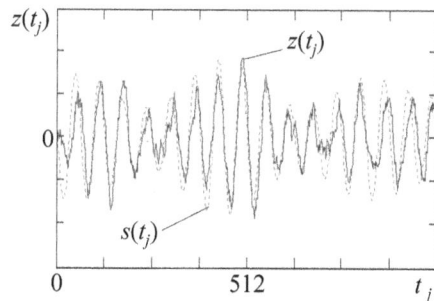

FIGURE 3.7.4   Resulting signal $z(t)$ in the filter output defined by equation (3.7.2)

Correlation coefficients between useful signal $s(t)$ and the signal $z(t)$ in the filter output take the values $r[s(t), z(t)]=0.79\ldots 0.87$. $\triangledown$

3.8   Robust Filtering Algorithms Based on Measures of Statistical Interrelation

The feature of filters considered in the Sections 3.2...3.6 is determined by the fact that depending on variable reception conditions, for instance, changes of power of received useful signal and/or interference (noise), according to their algorithm, these filters have to adapt themselves to external environment by adjusting weight coefficients. Residual noise variance in the output of a filter can essentially vary within a wide interval that is not always acceptable. Filtering algorithms based on L-group operations can provide a constancy of noise variance in the output of a filter when changing a power of additive mixture of signal and interference (noise) in the input of filter. Such algorithms are discussed within this section.

The introduced three MSIs (1.4.10...1.4.12) concerned with three corresponding metrics (1.3.25...1.3.27) determine three variants of filter constructing in space with lattice properties, so that the resulting process $y(t)$ in the output of a filter is defined by the following relationships, respectively:

$$y_p(t_j) = \sum_{i=0}^{j}[1 - (\operatorname{sgn}(h(t_{j-1}) \vee p \cdot x(t_i)) - \operatorname{sgn}(h(t_{j-i}) \wedge p \cdot x(t_i)))]; \qquad (3.8.1)$$

$$y_s(t_j) = \sum_{i=0}^{j} [|h(t_{j-1})| + p|x(t_i)|] -$$

$$- \sum_{i=0}^{j} |h(t_{j-1}) - p \cdot x(t_i)|[\text{sgn}(h(t_{j-1}) \vee x(t_i)) - \text{sgn}(h(t_{j-1}) \wedge x(t_i))]; \quad (3.8.2)$$

$$y_{l_1}(t_j) = \sum_{i=0}^{j} [|a(t_{j-1}) + p \cdot x(t_i)| - |a(t_{j-1}) - p \cdot x(t_i)|], \qquad (3.8.3)$$

where $x(t_j) = x(t - \Delta t \cdot j)$ is a delayed copy of received signal $x(t) = s(t) + n(t)$ taken from the output of j-th outlet of transversal filter; time parameter index j takes discrete values $j = 0, 1, \ldots, N - 1$; $t_j - t_{j-1} = \Delta t$, Δt is a sampling interval (time delay interval of transversal filter), so that $\{x(t_j) = x(t - \Delta t \cdot j)\}$; $s(t)$ is amplitude-modulated bandpass signal (see Fig. 3.7.1b); $n(t)$ is interference signal; N is a number of samples used in processing; p is some constant, $p > 0$; $h(t_i)$ is impulse response of filters (3.8.1), (3.8.2), and $a(t_i)$ is impulse response of the filter (3.8.3) shown in Fig. 3.7.1b, both defined by the following relationships, respectively:

$$h(t_j) = \frac{1}{\beta \cdot T_0} \exp\left(-\frac{t_j}{\alpha \cdot T_0}\right) \cos(\omega_0 t_j + \varphi_h)[1(t_j) - 1(t_j - \beta \cdot T_0)]; \quad (3.8.4a)$$

$$a(t_j) = \frac{1}{\beta \cdot T_0} \exp\left(-\frac{t_j}{\alpha \cdot T_0}\right) \cos(\omega_0 t_j + \varphi_a) \cdot 1(t_j), \qquad (3.8.4b)$$

α, β=const; $\vee$, $\wedge$ are operations of join and meet of L-group, respectively; $\omega_0 = 2\pi f_0$ is a cyclic frequency of useful signal $s(t)$; T_0 is a period of oscillation of useful signal carrier f_0: $T_0 = 1/f_0$; $1(t)$ is Heaviside unit step function; $\beta \cdot T_0 << t_{N-1}$.

The filters (3.8.1), (3.8.2) possess impulse responses with the following property $h(t_j) = 0$, $t_j \in]\beta \cdot T_0, t_{N-1}]$, otherwise in the presence of sole useful signal $s(t)$ with finite duration ($s(t)$=0 if $t \notin T_s = [t_0, t_0 + T]$) in the inputs of these filters (in the absence of interference (noise) $n(t)$=0, in the outputs of the filters (3.8.1), (3.8.2) there appears a signal ($y(t) \neq 0, y_s(t) \neq 0$) on $t >> t_0 + T$. In more details, this effect is discussed in Chapter 8.

Fig. 3.8.1 illustrates the dependences of relative filtering error

$$\delta_p = D[y_p(t) - s(t)]/(2D[s(t)]),$$

$$\delta_{l_1} = D[y_{l_1}(t) - s(t)]/(2D[s(t)])$$

on a constant p that figures in filtering algorithms (3.8.1), (3.8.3), respectively. The plots are shown in logarithmic scale in both axes for the case when: $s(t)$ is amplitude-modulated signal (see Fig. 3.7.1b); $n(t)$ is interference signal in the form of quasi-white Gaussian noise; $T_0/\Delta t = 2f_{max}/f_0 = 64$ (f_{max} is upper bound frequency of PSD of quasi-white noise $n(t)$); SNRs q^2 (in power units) are equal to $q^2 = 10, 100$; variance $D[x(t)]$ of the process $x(t)$ in the filter input is constant: $D[x(t)]$=const. Algorithm (3.8.1), as follows from its definition and Fig. 3.8.1, is invariant with respect to a variance of the process $x(t)$ in the filter input (for any

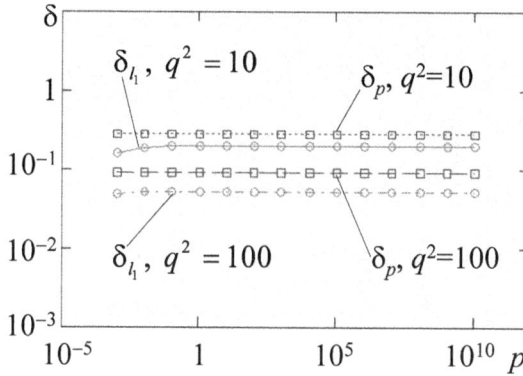

FIGURE 3.8.1   Dependences of relative filtering error δ, δ_{l_1} for the filters (3.8.1), (3.8.3), respectively

values of parameter p and $D[x(t)]=$const) providing a constancy of relative filtering error. Algorithm (3.8.3), as follows from Fig. 3.8.1, is invariant with respect to a variance of process $x(t)$ in the filter input providing a constancy of relative filtering error if the following condition holds:

$$\|p \cdot x(t_j)\|_2^2 / \|a(t_j)\|_2^2 > 25,$$

i.e., when $p > 0.1$, where $\|a(t_j)\|_2^2 = \sum_j a^2(t_j)$ is squared l_2-norm of a function; $a(t_j)$ is impulse response of the filter (3.8.3) determined by the relationship (3.8.4b). A payment for this parametric invariance is a lesser efficiency as compared with optimal filters. Thus, for instance, in the presence of normal noise, the filter defined by the relationship (3.8.1) is inferior to Kalman filter in filtering error $D[y(t) - s(t)]$ up to 4.0 and 3.0 dB on SNR $q^2 = 10$, 100, respectively; and the filter defined by the relationship (3.8.3) is inferior to Kalman filter in a filtering error up to 3 and 2 dB on SNR $q^2 = 10$, 100, respectively. As for the filter defined by the relationship (3.8.2), it does not possess invariance property loosing to Kalman filter up to 1.25 dB on SNR $q^2 = 10$, 100 in the presence of normal noise.

4

Signal Detection, Classification, and Resolution in Spaces with L–group Properties

4.1 Nonparametric Detection Algorithms in Signal Space with L–group Properties

4.1.1 Formulating Signal Detection Problem

Consider the model of interaction between a signal $s(t)$ and interference (noise) $n(t)$ in signal space $\mathcal{L}(+, \vee, \wedge)$ with L-group properties that is described by operation of addition «+» of L-group:

$$x(t) = \theta s(t) + n(t), \ \theta \in \{0, 1\}, \ t \in T_{\text{obs}}; \tag{4.1.1}$$

$$s(t) = 0, \ t \notin T_s : \ T_s = [t_0, t_0 + T] \subset T_{\text{obs}}, \tag{4.1.1a}$$

where T_{obs} is observation interval of the signal $s(t)$; T_s is a domain of the signal $s(t)$; t_0 is time of signal arrival; T is a duration of the signal $s(t)$; θ is unknown nonrandom parameter taking only two values from the set $\{0, 1\}$, $\theta \in \{0, 1\}$: $\theta = 0$ (signal is absent) and $\theta = 1$ (signal is present).

Problem of signal detection in the presence of interference (noise) lies in making a decision, using some criterion, concerning signal presence or absence in the observed process $x(t)$, or, in other words, a decision $\hat{\theta}$ concerning a value of unknown parameter θ in the observation interval T_{obs} [93, 103–110]. Within the last variant of signal detection problem statement, signal detection is considered to be equivalent to estimation of unknown nonrandom parameter θ.

Depending on a presence and content of prior information on statistical characteristics of the observed stochastic processes, Signal Processing Theory distinguishes two types of prior uncertainties: parametric and nonparametric.

In the case of parametric prior uncertainty, known probability density functions (PDFs) are described by some parameters that can be changed in random manner in time. To provide robust functioning (with respect to electromagnetic environment conditions), signal processing algorithms must realize preliminary estimation of unknown interference parameters with following exploiting these estimates to normalize the observed signals. Such detectors are called adaptive ones. Adaptation appears to be more complicated, if several parameters or type of distribution are unknown.

In the case of nonparametric prior uncertainty, expressions for PDFs of interference and a mixture between signal and interference are unknown, so that there

DOI: 10.1201/9781003275855-4

are no concrete data on these PDFs. In this situation, detection algorithms can rely on nonparametric statistical hypothesis testing [111–113].

Thus, depending on available prior information concerning these or those characteristics of useful and interference signals, both parametric and nonparametric algorithms can be used to solve the detection problem. Parametric algorithms cease to be optimal, if an assumption on a concrete kind of interference (noise) distribution, within which this algorithm is synthesized, does not hold. Under such conditions of nonparametric prior uncertainty, when a class of actual interference (noise) is wider than a given class of distributions, nonparametric algorithms of detection are of a practice interest.

As a rule, the more prior information on useful signal is used in its detection algorithm, the better detection quality indices are provided when processing the signals. Signal processing quality achieved during detection is defined by detection quality indices in the form of conditional probabilities of false alarm and correct detection. To carry out a comparative analysis for a pair of detection algorithms, asymptotic relative efficiency is used.

The subject of our discussion in the following subsections is quality indices of signal detection algorithms operating in signal spaces with L-group properties within the case of additive interaction between useful signal and interference (noise). As it is shown below, all known nonparametric detection algorithms can be formulated in terms of L-groups.

When processing discrete signals, the model of interaction between useful signal $s(t)$ and interference (noise) $n(t)$ has a form that is similar to (4.1.1):

$$x(t_j) = \theta s(t_j) + n(t_j), \ \theta \in \{0, 1\}; \tag{4.1.2}$$

$$t_j \in T_{\text{obs}}, \ s(t_j) = 0, \ t_j \notin T_s : \ T_s \subset T_{\text{obs}},$$

where $t_j = t - j\Delta t$, $j = 0, 1, \ldots, N - 1$, $N \in \mathbf{N}$, $\mathbf{N}$ is the set of natural numbers; T_{obs} is observation interval of the signal $s(t)$; $T_s = [t_0, t_0 + T]$ is a domain of the signal $s(t)$; $\{n(t_j)\}$ are independent samples of interference (noise) from a class of distributions with symmetric PDF $p_n(z) = p_n(-z)$; $\{s(t_j)\}$ are the samples of useful signal $s(t)$; $\{x(t_j)\}$ are the samples of the observed stochastic process $x(t)$; t_j is discrete time parameter; Δt is a sampling interval providing independence of interference samples $\{n(t_j)\}$; j is an index of time parameter t_j of the samples $\{n(t_j)\}$, $\{x(t_j)\}$.

Instantaneous values (temporal samples) of the signal $\{s(t_j)\}$ and interference $\{n(t_j)\}$ are the elements of sample space $\mathcal{L}(+, \vee, \wedge)$ with L-group properties: $s(t_j), n(t_j) \in \mathcal{L}(+, \vee, \wedge)$.

In this case, the problem of signal detection of deterministic signal $s(t)$ in the presence of additive interference (noise) $n(t)$ is formulated as the problem of testing statistical hypothesis H_0 concerning the fact that independent sample $\mathbf{x} = [x(t_0), \ldots, x(t_j), \ldots, x(t_{N-1})]^T$ from realization of the observed stochastic process $x(t)$ belongs to interference (noise) distribution with PDF

$$p(\mathbf{x}/H_0) = \prod_{j=0}^{N-1} p(x_j/H_0) \tag{4.1.3}$$

against the alternative H_1 concerning the fact that this sample $\mathbf{x}$ belongs to additive mixture of the signal and interference (noise) with PDF

$$p(\mathbf{x}/H_1) = \prod_{j=0}^{N-1} p(x_j - s_j/H_1), \qquad (4.1.4)$$

where $x_j = x(t_j)$, $s_j = s(t_j)$.

Detection algorithm is *nonparametric*, if conditional probability of false alarm preserves its constant value in the case of hypothesis H_0, i.e. for all samples belonging to nonparametric class with PDF (4.1.3).

Nonparametric algorithms are synthesized on the heuristic basis using special statistics that, in the case of homogeneous and independent sample used for testing the hypothesis H_0, possess nonparametric property with respect to this hypothesis. The most essential constraint is the assumption on homogeneity and independence of the sample, therefore, the discussed algorithms of signal detection possess nonparametric property only in the case of stationarity of interference and independence of the sample.

Quantitative characteristic of a decision rule (detection algorithm) α_A with respect to another decision rule (detection algorithm) α_B is *asymptotic relative efficiency* (ARE) $e(\alpha_A, \alpha_B)$ in the case if signal-to-noise ratio (SNR) q^2 tends to zero $q^2 \to 0$ and a sample size N tends to infinity $N \to \infty$. ARE is a limit of a ratio between two sample sizes $N_B = N_B(F, D)$ and $N_A = N_A(F, D)$ that are used for performing the algorithms α_B, α_A, respectively, to provide given conditional probabilities of false alarm F and correct detection D if $N_B \to \infty$, $N_A \to \infty$:

$$e(\alpha_A, \alpha_B) = \lim_{N_A, N_B \to \infty} \frac{N_B(F, D)}{N_A(F, D)}. \qquad (4.1.5)$$

Algorithm α_A is more efficient than algorithm α_B, if $e(\alpha_A, \alpha_B) > 1$.

In some cases, ARE does not depend on conditional probabilities of false alarm F and correct detection D and is a universal mean for comparative analyzing two detection algorithms on SNR $q^2 \to 0$. This circumstance provides an advantage of this asymptotic approach as compared to the direct evaluation of detection quality indices D, F.

Usually, nonparametric detection algorithms are compared with linear correlation algorithm

$$y_{LC}(t_j) = \sum_{i=0}^{j} x(t_i)s(t_i) \underset{d_0}{\overset{d_1}{\underset{<}{>}}} l_0, \qquad (4.1.6)$$

which, as is known, calculating a scalar product between the samples $\{x(t_j)\}$ and $\{s(t_j)\}$ of the observed process $x(t)$ and the signal $s(t)$, is the optimal (under Neyman-Pearson criterion) detection algorithm of deterministic signal in the presence of additive Gaussian interference (noise) and uses independent and homogeneous sample of observations.

In its quality indices of detection, algorithm (4.1.6) is equivalent to one which calculates operation of convolution between the observed process $x(t)$ and impulse response $h(t)$ of a linear filter (that is called *matched filter*):

$$y_{MF}(t_j) = \sum_{i=0}^{j} x(t_i) h(t_{j-i}) \overset{d_1}{\underset{d_0}{\gtrless}} l_0, \qquad (4.1.7)$$

where $h(t_j) = s(t_0 + T - t_j)$ is impulse response of the filter matched with the signal $s(t)$; $t_0 + T$ is time of signal ending $s(t)$ (see relationship (4.1.1a)); l_0 is a threshold of detection; d_0, d_1 are decisions on signal absence or presence, respectively.

4.1.2　Sign Detection Algorithm

Within the formulated nonparametric detection problem, sign algorithm of detection of deterministic signal $s(t)$ in the presence of additive interference (noise) $n(t)$ makes a decision d_1 (or d_0) on signal presence (absence) in the result of calculating *scalar product* between sign functions of a vector $\mathbf{x} = [x_i]$ of the observed process $x(t)$ and a vector $\mathbf{s} = [s_i]$ of the signal $s(t)$, comparing it with a threshold l_0, and is defined by the following relationship:

$$y_{SC}(t_j) = \sum_{i=0}^{j} \operatorname{sgn}(x(t_i)) \operatorname{sgn}(s(t_i)) \overset{d_1}{\underset{d_0}{\gtrless}} l_0. \qquad (4.1.8)$$

Here we refer this algorithm to the class of detection algorithms in signal spaces with L-group properties (or, simply, the class of L-group algorithms), bearing in mind that a normal form of notation of the function $\operatorname{sgn}(t)$ can be written using join and meet operations of L-group (1.2.5):

$$\operatorname{sgn}(t) = -1 \vee [1 \wedge k \cdot t], \quad k >> 1.$$

Algorithm (4.1.8) is referred to a class of correlation ones. Its substantial analogue is sign algorithm based on matched filtering of useful signal which is determined by the following relationship:

$$y_{SC/MF}(t_j) = \sum_{i=0}^{j} \operatorname{sgn}(x(t_i)) \operatorname{sgn}(h(t_{j-i})) \overset{d_1}{\underset{d_0}{\gtrless}} l_0, \qquad (4.1.9)$$

where $h(t_j) = s(t_0 + T - t_j)$ is impulse response of a filter matched with the signal $s(t)$; $t_0 + T$ is time of a signal ending $s(t)$ (see the relationship (4.1.1a)).

When determining the efficiency of nonparametric and others detection algorithms, an important role belongs to those notions that define some relationships between different norms of a signal in normed spaces. In practice, one can consider the relationships only between l_1-, l_2-, l_3-, and l_4-norms $\|s\|_1$, $\|s\|_2$, $\|s\|_3$, $\|s\|_4$ of the signal $s(t)$, respectively. Since here we consider only a bandpass signals, introduce the main relationships between l_1-, l_2-, l_3-, and l_4-norms $\|s\|_1$, $\|s\|_2$, $\|s\|_3$, $\|s\|_4$ of a bandpass signal $s(t)$ with a rectangular envelope and amplitude A. In this case l_1-, l_2-, l_3-, and l_4 norms $\|s\|_1$, $\|s\|_2$, $\|s\|_3$, $\|s\|_4$ of a bandpass signal $s(t)$ are defined by the following relationships, respectively:

$$\|s\|_1 = \sum_{k=0}^{K-1} |s_k| = A \cdot K \cdot k_1; \tag{4.1.10a}$$

$$\|s\|_2^2 = \sum_{k=0}^{K-1} |s_k|^2 = A^2 \cdot K \cdot k_2; \tag{4.1.10b}$$

$$\|s\|_3^3 = \sum_{k=0}^{K-1} |s_k|^3 = A^3 \cdot K \cdot k_3; \tag{4.1.10c}$$

$$\|s\|_4^4 = \sum_{k=0}^{K-1} |s_k|^4 = A^4 \cdot K \cdot k_4, \tag{4.1.10d}$$

where k_1, k_2, k_3, k_4 are coefficients determined by the relationships:

$$k_1 = \tfrac{1}{\pi} \int_0^{\pi} \sin(t)dt = \tfrac{2}{\pi}; \tag{4.1.11a}$$

$$k_2 = \tfrac{1}{\pi} \int_0^{\pi} \sin^2(t)dt = \tfrac{1}{2}; \tag{4.1.11b}$$

$$k_3 = \tfrac{1}{\pi} \int_0^{\pi} \sin^3(t)dt = \tfrac{4}{3\pi}; \tag{4.1.11c}$$

$$k_4 = \tfrac{1}{\pi} \int_0^{\pi} \sin^4(t)dt = \tfrac{3}{8}, \tag{4.1.11d}$$

To describe characteristics of a bandpass signal $s(t)$, introduce a norm-factor F_{nm} as a quantity determining a ratio between l_n- and l_m-norms $\|s\|_n$, $\|s\|_m$ of a signal $s(t)$, $n, m \in \{1, 2, 3\}$, $n < m$:

$$F_{nm} = \frac{\|s\|_n}{\|s\|_m} = k_{nm} K^{\left(\frac{1}{n} - \frac{1}{m}\right)} = \frac{k_n^{1/n}}{k_m^{1/m}} K^{\left(\frac{1}{n} - \frac{1}{m}\right)}, \tag{4.1.12}$$

where $k_{nm} = k_n^{1/n}/k_m^{1/m}$ is a coefficient of a ratio (4.1.11); K is a sample size.

Thus, for instance, coefficients of ratios k_{12}, k_{23}, k_{24}, and also norm-factors F_{12}, F_{23}, F_{24} for the pairs composed of l_1-, l_2-, l_3-, l_4-norms are, respectively, equal to:

$$k_{12} = \frac{2\sqrt{2}}{\pi}; \; k_{23} = \left(\frac{3\pi}{4}\right)^{\frac{1}{3}} \frac{1}{\sqrt{2}}; \; k_{24} = \left(\frac{8}{3}\right)^{\frac{1}{4}} \frac{1}{\sqrt{2}}; \tag{4.1.13a}$$

$$F_{12} = \frac{2\sqrt{2}}{\pi} K^{\frac{1}{2}}; \ F_{23} = \left(\frac{3\pi}{4}\right)^{\frac{1}{3}} \frac{1}{\sqrt{2}} K^{\frac{1}{6}}; \ F_{24} = \left(\frac{2}{3}\right)^{\frac{1}{4}} K^{\frac{1}{4}}. \tag{4.1.13b}$$

Is known, that ARE $e_{SC,LC}$ of sign detection algorithm (4.1.8) with respect to linear correlation one (4.1.6) for radiofrequency (RF) bandpass signals is determined by the relationship [112, (2.80)], [114, (16.18)]:

$$e_{SC,LC} = 4D_n p^2(0) k_{12}^2, \tag{4.1.14}$$

where $p(0)$ is a value of PDF $p(x)$ of interference (noise) $n(t)$ in the point $x = 0$; D_n is a variance of interference (noise) $n(t)$; k_{12} is a coefficient of a ratio between l_1-, l_2-norms of a signal $s(t)$ (see the formula (4.1.13a)).

AREs $e_{SC,LC}$ of sign detection algorithm (4.1.8) with respect to linear correlation one (4.1.6), calculated by the formula (4.1.14) for the distributions listed in the table 2.6.2, are shown in the table 4.1.1. Thus, for instance, in normal or Laplace distribution of interference (noise), AREs $e_{SC,LC}$ are equal to 0.516 and 1.621, respectively.

4.1.3   Rank Detection Algorithm

Insufficient efficiency of sign detection algorithm for some interference (noise) distributions is explained by the fact that it uses rather small part of information contained in initial realization, inasmuch as this algorithm takes into account just their signs ignoring the values of the samples $\{x(t_j)\}$. Rank detection algorithms are deprived of this disadvantage, since they take into account relative values of the samples $\{x(t_j)\}$ from the set

$$\mathbf{x} = [x(t_0), \dots, x(t_j), \dots, x(t_{N-1})]^T = [x_0, \dots, x_j, \dots, x_{N-1}]^T.$$

Rank algorithms use *rank statistics*. Rank R_k of an element x_k of the sample $\mathbf{x}$ is an order argument of this element in a variational series, hence, each element can be associated with order statistic $x^{(R_k)}$ of variational series. Rank R_k of the element x_k from the sample $\mathbf{x}$ with a size N can be written over Heaviside unit step function:

$$R_k = \sum_{i=0}^{N-1} 1(x_k - x_i), \ k = 0, 1, \dots, N - 1. \tag{4.1.15}$$

Knowing that normal form of notation of the function $1(t)$ can be represented by join and meet operations of L-group (1.2.4):

$$1(t) = 0 \vee [1 \wedge k \cdot t], \ k >> 1, \tag{4.1.16}$$

the formulas (4.1.15) and (4.1.16) imply that the ranks are statistics of differences of sample values written in terms of L-group.

Within the formulated problem of nonparametric detection, rank algorithm of detection of deterministic signal $s(t)$ in the presence of additive interference (noise) $n(t)$ makes a decision d_1 (or d_0) concerning the signal presence (absence) in the result of calculating *scalar product* between some function g of a rank vector $\mathbf{R} = [R_i]$ (4.1.15) of the observed process $x(t)$ and a vector $\mathbf{s} = [s_i]$ of the signal $s(t)$, comparing it with a threshold l_0, and is defined by the following relationship:

$$y_R(t_j) = \sum_{i=0}^{j} g(R_i)s(t_i) \underset{d_0}{\overset{d_1}{\underset{<}{>}}} l_0. \tag{4.1.17}$$

Particular kinds of rank algorithm are: median (if $g(t) = \mathrm{sgn}(t - 0.5N)$); Wilcoxon (if $g(t) = t$); van der Waerden (if $g(t) = \Phi^{-1}(t/N)$, $\Phi(t) = \frac{1}{\sqrt{2\pi}} \int\limits_{-\infty}^{t} \exp\left[-\frac{x^2}{2}\right]$).

Algorithm (4.1.17) is referred to a class of correlation algorithms. Its substantial analogue is rank algorithm based on matched filtering of useful signal that is defined by the formula:

$$y_{R/MF}(t_j) = \sum_{i=0}^{j} g(R_i)h(t_{j-i}) \underset{d_0}{\overset{d_1}{\underset{<}{>}}} l_0, \tag{4.1.18}$$

where $h(t_j) = s(t_0 + T - t_j)$ is impulse response of a filter matched with the signal $s(t)$; $t_0 + T$ is time of signal ending (see the formula (4.1.1a)).

ARE of rank Wilcoxon algorithm ($g(t) = t$) (4.1.17) with respect to linear correlation one (4.1.6) for RF bandpass signals with zero mean is defined by the relationship [114, (16.47a)]:

$$e_{RW,LC} = 12D_n \left(\int\limits_{-\infty}^{\infty} p^2(x)dx \right)^2, \tag{4.1.19}$$

where $p(x)$ is PDF of interference (noise) $n(t)$; D_n is a variance of interference (noise) $n(t)$.

AREs $e_{RW,LC}$ of rank Wilcoxon detection algorithm (4.1.17) with respect to linear correlation one (4.1.6), obtained according to formula (4.1.19) for distributions listed in the table 2.6.2, are shown in the table 4.1.1. Thus, for instance, for normal or Laplace distributions of interference (noise), AREs $e_{RW,LC}$ are equal to $3/\pi \approx 0.955$ and 1.5, respectively.

4.1.4   Sign-rank Detection Algorithm

Sign-rank algorithm of detection of bandpass deterministic signal $s(t)$ in the presence of additive interference (noise) $n(t)$ makes a decision d_1 (or d_0) concerning the signal presence (or absence) in the result of calculating *scalar product* between rank vector $\mathbf{R} = [R_k]^T$ of the observed process $x(t)$ and sign function of a vector $\mathbf{s} = [s_i]$ of the signal $s(t)$, its comparing with a threshold l_0), and is defined by the following relationship:

$$y_{SR}(t_j) = \sum_{i=0}^{j} R_i \, \mathrm{sgn}(s(t_i)) \underset{d_0}{\overset{d_1}{\underset{<}{>}}} l_0, \tag{4.1.20}$$

where $h(t_j) = s(t_0 + T - t_j)$ is impulse response of a filter matched with the signal $s(t)$; $t_0 + T$ is time of signal ending (see formula (4.1.1a)).

Algorithm (4.1.20) is referred to a class of correlation algorithms. Its substantial analogue is sign-rank algorithm based on matched filtering of useful signal:

$$y_{SR/MF}(t_j) = \sum_{i=0}^{j} R_i \, \mathrm{sgn}(h(t_{j-i})) \overset{d_1}{\underset{d_0}{\gtrless}} l_0, \qquad (4.1.21)$$

where $h(t_j) = s(t_0 + T - t_j)$ is impulse response of a filter matched with the signal $s(t)$; $t_0 + T$ is time of signal ending (see the formula (4.1.1a)).

ARE of sign-rank algorithm (4.1.20) with respect to linear correlation one (4.1.6) for RF bandpass signals with zero mean is defined by relationship [114, (16.47)]:

$$e_{SR,LC} = 12 D_n k_{12}^2 \left(\int_{-\infty}^{\infty} p^2(x) dx \right)^2, \qquad (4.1.22)$$

where $p(x)$ is PDF of interference (noise) $n(t)$; D_n is a variance of interference (noise) $n(t)$; k_{12} is coefficient of the ratio between l_1-, l_2-norms of the signal $s(t)$ (see the formula (4.1.13a)).

AREs $e_{RW,LC}$ of sign-rank detection algorithm (4.1.20) with respect to linear correlation algorithm (4.1.6), calculated according to the formula (4.1.22) for distributions listed in the table 2.6.2, are shown in the table 4.1.1. Thus, for instance, for normal or Laplace distributions of interference (noise), AREs $e_{RW,LC}$ are equal to 0.774 and 1.216, respectively.

4.1.5   Short Summary Concerning Sign, Rank, and Sign-rank Detection Algorithms

Sign (4.1.8), (4.1.9); rank (4.1.17), (4.1.18); and sign-rank (4.1.20), (4.1.21) algorithms of detection of deterministic RF bandpass signals can be represented in terms of L-group operations that allow refer these nonparametric detection algorithms to a class of L-group algorithms.

One can get an idea on appearance of the signals in the outputs of matched filters performing the operation of convolution (4.1.7), (4.1.9), (4.1.18), (4.1.21) by Fig. 4.1.1a,b,c,d, respectively. The signals correspond to the following conditions: in the filter input, there is an additive interference in the form of quasi-white Gaussian noise with the ratio of interference bandwidth ΔF_n to carrier frequency f_0 of useful signal $s(t)$ equal to $\Delta F_n / f_0 = 16$; useful signal $s(t)$ (shown by dotted line) is 8-element binary phase-coded signal; SNR $2E/N_0$ is equal to $2E/N_0 = 50$.

AREs of sign, rank, and sign-rank detection algorithms with respect to linear correlation algorithm (4.1.6) for different interference distributions from the table 2.6.1 are shown in the table 4.1.1.

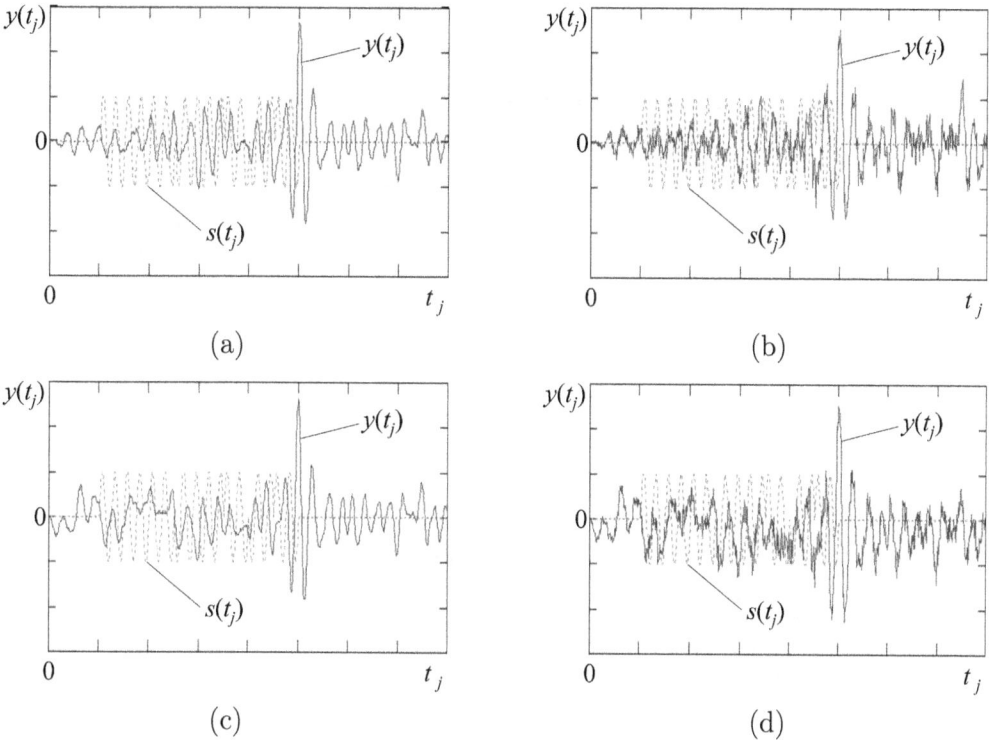

FIGURE 4.1.1   Signals in the outputs of the filters: (a) linear (4.1.7); (b) sign (4.1.9); (c) rank (4.1.18); (d) sign-rank (4.1.21)

TABLE 4.1.1   AREs of sign, rank, and sign-rank detection algorithms

Distribution notation	Distribution	Algorithm		
		Sign	Rank	Sign-rank
$N(0, b)$	normal	0.516	0.955	0.774
$T(0, 0.01, 3)$	Tukey	0.550	1.009	0.818
$T(0, 0.05, 3)$	Tukey	0.675	1.196	0.969
$L(0, b)$	logistic	0.667	1.097	0.889
$DE(0, b)$	Laplace	1.621	1.50	1.216
$St(0, 5)$	Student	0.779	1.241	1.006
$St(0, 3)$	Student	1.311	1.896	1.536
$C(0, b)$	Cauchy	∞	∞	∞
$U(0, b)$	uniform	0.270	1.000	0.810
$Tr(0, b)$	triangular	0.540	0.889	0.720

Conclusions

1. Discussed nonparametric algorithms of signal detection (4.1.9), (4.1.18), (4.1.21) can be represented in terms of algebraic operations of L–group.

2. Discussed nonparametric algorithms of signal detection provide a constant level of conditional probability of false alarm. This circumstance allow referring these detectors to a class of *constant false alarm rate processors*.

3. The efficiency of nonparametric algorithms of signal detection in the presence of Gaussian interference (noise) is lower than efficiency of classic optimal detection algorithm (4.1.7) based on matched filter. At the same time, nonparametric algorithms provide more efficient signal detection in the presence of impulse type interference (noise) with "heavy tails" of a distribution than classic optimal detection algorithm based on matched filter (4.1.7) in linear signal space. Nonparametric detection algorithms are inferior to classic optimal detection algorithm (4.1.7) in efficiency in the presence of harmonic type interference (noise) with relatively "light tails" of a distribution.

4.2 Notion of Generalized Matched Filter

The central notion for considering the problems of signal detection, classification, and resolution is the notion of matched filter [103–109].

Remind, that *matched filter* (or a filter matched with a given deterministic signal) is a *linear filter* maximizing signal-to-noise ratio (SNR) in the filter output in the presence of additive mixture of a given deterministic signal and white Gaussian noise in the filter input. Thus, classic matched filter is defined over solving an optimization problem, so that impulse response $h(t)$ of a filter matched with the signal $s(t)$ takes the form: $h(t_j) = s(t_0 + T - t_j)$, where $t_0 + T$ is time of signal ending $s(t)$ (see (4.1.1a)). Here we introduce the similar notion for signal spaces with arbitrary properties, not using any optimality criteria, but basing only on the notion of *generalized measure of statistical interrelation* (MSI), introduced in Subsection 1.4.3, and also the notion of *algorithm*.

Definition 4.2.1. *Generalized matched filter* is a filter whose algorithm assumes calculating a generalized MSI $N(\mathbf{x}, \mathbf{h})$ between the observed discrete stochastic process $\mathbf{x} = [x(t_j)]$ and impulse response of a filter $\mathbf{h} = [h(t_j)]$:

$$z(t_j) = N(\mathbf{x}, \mathbf{h}) = F[\rho(\mathbf{x}, \mathbf{h})].$$

Remind that generalized MSI $N(\mathbf{x}, \mathbf{h})$ (see subsection 1.4.3) is defined as some monotone decreasing function F of generalized metric $\rho(\mathbf{x}, \mathbf{h})$ related to that signal space in which signal processing is realized, so that impulse response $h(t)$ of a filter matched with the signal $s(t)$ is equal to: $h(t_j) = s(t_0 + T - t_j); j = 0, 1, \ldots, N - 1.$

The introduced three MSIs (1.4.10...1.4.12) based on three generalized metrics (1.3.25...1.3.27) define three corresponding variants of constructing generalized matched filters in signal space with L-group properties, so that the resulting processes in the filter outputs are defined by the following relationships, respectively:

$$z_p(t_j) = \sum_{i=0}^{j} [1 - (\mathrm{sgn}(h(t_{j-i}) \vee x(t_i)) - \mathrm{sgn}(h(t_{j-i}) \wedge x(t_i)))] \cdot |\mathrm{sgn}(h(t_{j-i}))|;$$

(4.2.1)

$$z_s(t_j) = \sum_{i=0}^{j} [|h(t_{j-i})| + |x(t_i)|]$$

$$- \sum_{i=0}^{j} |h(t_{j-i}) - x(t_i)| [\mathrm{sgn}(h(t_{j-i}) \vee x(t_i)) - \mathrm{sgn}(h(t_{j-i}) \wedge x(t_i))]; \quad (4.2.2)$$

$$z_{l_1}(t_j) = \sum_{i=0}^{j} [|h(t_{j-i}) + x(t_i)| - |h(t_{j-i}) - x(t_i)|], \quad (4.2.3)$$

where $x(t_j) = x(t - \Delta t \cdot j)$ is a delayed copy of the received signal $x(t) = s(t) + n(t)$ taken from the output of j-th outlet of transversal filter; index j of time parameter t_j takes discrete values $j = 0, 1, \ldots, N-1$; Δt is a sampling interval (interval between the adjacent outlets of transversal filter) $\Delta t = t_j - t_{j-1}$; $n(t)$ is interference (noise) signal; N is a number of samples used in processing; $h(t_j)$ is impulse response of the filters (4.2.1)...(4.2.3) matched with useful signal $s(t)$.

For generalized MSI (1.4.9) that defines classic matched filter (4.1.7) and also generalized matched filters (4.2.1)...(4.2.3), one can point corresponding equivalent short forms of MSI representation.

TABLE 4.2.1   Short forms of representation of generalized matched filters

Short forms of representation of generalized MSIs N($\mathbf{x}, \mathbf{h}$)	Generalized metrics $\rho(\mathbf{x}, \mathbf{y})$
$N_{l_2}(\mathbf{x}, \mathbf{h}) = \sum_{i=1}^{j} x(t_i) \cdot h(t_{j-i})$	$\left(\sum_{i=1}^{n} (x_i - y_i)^2 \right)^{1/2}$
$N_p(\mathbf{x}, \mathbf{h}) = \sum_{i=1}^{j} \mathrm{sgn}(x(t_i)) \cdot \mathrm{sgn}(h(t_{j-i}))$	$\sum_{i=1}^{n} [\mathrm{sgn}(x_i \vee y_i) - \mathrm{sgn}(x_i \wedge y_i)]$
$N_{l_1}(\mathbf{x}, \mathbf{h}) = \sum_{i=1}^{j} HF[x(t_i), h(t_{j-i})]$	$\sum_{i=1}^{n} |x_i - y_i|$
$N_s(\mathbf{x}, \mathbf{h}) = \sum_{i=1}^{j} T[x(t_i), h(t_{j-i})]$	$\sum_{i=1}^{n} |x_i - y_i| [\mathrm{sgn}(x_i \vee y_i) - \mathrm{sgn}(x_i \wedge y_i)]$

Here: $HF(x, a)$ is notation for Huber function determined by the formula (1.2.8); $T(x, a) = x\,\mathrm{sgn}(a) + a\,\mathrm{sgn}(x)$.

According to Definition 4.2.1, *generalized algorithm of detection* of deterministic signal in signal space with L-group properties can be written in the following form:

$$z(t_j) = \mathbf{N}(\mathbf{x}, \mathbf{h}) = F[\rho(\mathbf{x}, \mathbf{h})] \underset{d_0}{\overset{d_1}{\underset{<}{\gtrless}}} l_0, \qquad (4.2.4)$$

where l_0 is a threshold of detection.

Within the following sections we obtain quality indices of deterministic signal detection determining the efficiency of detection algorithms based on matched filtering algorithms (4.2.1)...(4.2.3), and also compare them with detection quality indices of known classic algorithm based on matched filtering (4.1.7). At the same time, we assume that signal processing is realized within general model of detection determined by the relationship (4.1.2).

Remind, that matched filtering (4.1.7) in linear space with scalar product requires performing $N_\times = N^2$ operations of multiplication and $N_+ = N^2$ operations of addition, where N is a number of processed samples.

Generalized matched filtering (4.2.1) in pseudometric space require performing $N_+ = 2N^2$ operations of addition, $N_{\vee,\wedge} = 6N^2$ operations of join/ meet, or equivalently, $N_+ = 2N^2$ operations of addition, $N_{\vee,\wedge} = 2N^2$ operations of join/meet, and $N_{sgn} = 2N^2$ operations of sign function.

Generalized matched filtering (4.2.2) in semimetric space requires performing $N_+ = 4N^2$ operations of addition, $N_{\vee,\wedge} = 9N^2$ operations of join/ meet, $N_{s\_inv} = 3N^2$ operations of sign inverting, and up to $N_{shift} = N^2$ operations of binary digit shift, or equivalently, $N_+ = 4N^2$ operations of addition, $N_{|*|} = 3N^2$ operations of modulus, $N_{\vee,\wedge} = 2N^2$ operations of join/ meet, $N_{sgn} = 2N^2$ operations of sign function, and up to $N_{shift} = N^2$ operations of binary digit shift.

Generalized matched filtering (4.2.3) in l_1-metric space requires performing $N_+ = 3N^2$ operations of addition, $N_{\vee,\wedge} = 3N^2$ operations of join/ meet, $N_{s\_inv} = N^2$ operations of sign inverting, or equivalently, $N_+ = 3N^2$ operations of addition and $N_{|*|} = 2N^2$ operations of modulus.

Generalized data concerning a required number of algebraic operations for performing generalized matched filtering in sample spaces of different types: linear space with scalar product (4.1.7), pseudometric space (4.2.1), semimetric space (4.2.2), and l_1-metric space (4.2.3) are contained in the table 4.2.2.

Notice, that a computational rate of performing generalized matched filtering in sample spaces with L-group properties (4.2.1), (4.2.2), (4.2.3), unlike their analogue in linear space (4.1.7), does not depend on a processed data width. Numerical relationships from the table 4.2.2 determining the required number of algebraic operations allow concluding that it is preferable (from a standpoint of providing a higher data processing rate on a given computational performance of a system) to perform generalized matched filtering in spaces with L-group properties, so that under some conditions, an achieved quality of signal processing in these spaces can be better than in linear spaces with scalar product.

TABLE 4.2.2   Required number of algebraic operations for performing generalized matched filtering in sample spaces of different types

Operation	Linear space	Pseudometric space	Semimetric space	l_1-metric space
Multiplication	N^2			
Addition	N^2	$2N^2$	$4N^2$	$3N^2$
Join/meet		$6N^2$	$9N^2$	$3N^2$
Sign inverting			$3N^2$	N^2
Binary digit shift			up to N^2	

Conclusions

1. The introduced notion of generalized matched filter allow from the unified standpoint formulating signal processing algorithms in a space with arbitrary generalized metric and algebraic properties.

2. Algorithm of generalized matched filter is defined by generalized measure of statistical interrelation that is determined by generalized metric of signal space in which signal processing is realized.

3. Algorithm of generalized matched filter can be represented by two forms: complete and short ones (see Table 4.2.1).

4.3   Detection Algorithm Based on Measure of Statistical Interrelation Concerned with Pseudometric

Algorithm of deterministic signal detection based on MSI (1.4.10) concerned with pseudometric (1.3.25) is defined by the following relationship:

$$z_p(t_j) = \sum_{i=0}^{j} [1 - (\mathrm{sgn}(h(t_{j-i}) \vee x(t_i)) - \mathrm{sgn}(h(t_{j-i}) \wedge x(t_i)))] \times$$

$$\times |\mathrm{sgn}(h(t_{j-i}))| \begin{array}{c} d_1 \\ > \\ < \\ d_0 \end{array} l_0. \quad (4.3.1)$$

Taking into account known interrelation between a signal and impulse response of matched filter $h(t_j) = s(t_0 + T - t_j)$, filter type detection algorithm (4.3.1) is equivalent to correlation type detection algorithm that is determined by the expression:

$$z_{p,C}(t_j) = \sum_{i=0}^{j} [1 - (\text{sgn}(s(t_i) \vee x(t_i)) - \text{sgn}(s(t_i) \wedge x(t_i)))] \times$$

$$\times |\text{sgn}(s(t_i))| \underset{d_0}{\overset{d_1}{\underset{<}{\overset{>}{\lessgtr}}}} l_0, \quad (4.3.2)$$

so that the following identity holds:

$$z_p(t_j) = z_{p,C}(t_j), \ t_j = t_0 + T. \quad (4.3.3)$$

Remind, that the multiplier $|\text{sgn}(h(t_{j-i}))|$ in (4.3.1) (or $|\text{sgn}(s(t_i))|$ in (4.3.2)) is necessary to provide physical realizability of the filter.

Taking into account the identity

$$1 - (\text{sgn}(a \vee b) - \text{sgn}(a \wedge b)) = \text{sgn}(a)\,\text{sgn}(b), \quad (4.3.4)$$

one can easily to conclude that detection algorithms (4.3.1) and (4.1.9), and also (4.3.2) and (4.1.8) are equivalent:

$$z_p(t_j) \equiv y_{SC/MF}(t_j); \quad (4.3.5a)$$

$$z_{p,C}(t_j) \equiv y_{SC}(t_j), \quad (4.3.5b)$$

Find conditional PDFs $p(z(t_j)/H_0)|_{t_j=t_0+T}$, $p(z(t_j)/H_1)|_{t_j=t_0+T}$ in the output of generalized matched filter (4.3.1) at the time instant $t_j = t_0 + T$ taking into account joint fulfillment of the identities (4.3.3) and (4.3.5b). Indexes of hypotheses H_0, H_1 correspond to values of unknown nonrandom parameter θ from the model (4.1.2), $\theta \in \{0, 1\}$: $\theta = 0$ (signal is absent) and $\theta = 1$ (signal is present).

For convenience of analysis within the initial relationship (4.1.8), we introduce auxiliary variables corresponding to a shift of coordinate system origin into a point t_0:

$$z = \sum_{k=0}^{K-1} y_k = \sum_{k=0}^{K-1} \text{sgn}(x_k)\,\text{sgn}(s_k); \quad (4.3.6a)$$

$$y_k = \text{sgn}(x_k)\,\text{sgn}(s_k); \quad (4.3.6b)$$

$$t_k = t_j - t_0; \ k = 0, 1, \ldots, K - 1; \ t_{K-1} = T; \ K - 1 = T/\Delta t, \quad (4.3.6c)$$

where t_0 and $t_0 + T$ are time of signal arrival and ending, respectively (see formula (4.1.1a); T is a duration of the signal $s(t)$; Δt is a sampling interval providing independence of interference (noise) samples $\{n(t_j)\}$ within the model (4.1.2).

Notice that random variable y_k (4.3.6b) possesses generalized Bernoulli distribution:

$$p_y(x/H_i) = p_k \delta(x - a_k) + q_k \delta(x + a_k); \quad (4.3.7)$$

$$p_k = F_n(\theta|s_k|, 0); \ q_k = F_n(-\theta|s_k|, 0); \ q_k = 1 - p_k; \quad (4.3.7a)$$

$$a_k = |\text{sgn}(s_k)| = 1, \quad (4.3.7b)$$

where $p_y(x/H_i)$ is conditional PDF of random variable y_k (4.3.6b) for a hypothesis H_i, $i \in \{0,1\}$; $F_n(x,c)$ is CDF of interference (noise) $n(t)$ with zero mean $c = 0$; c is a center of symmetry of CDF $F_n(x,c)$; $\theta = i \in \{0,1\}$ is a parameter of the model (4.1.2).

Then, random variable z (4.3.6a), that is the sum of K independent and identically distributed random variables with Bernoulli distribution, possesses generalized binomial distribution $p_z(x/H_i)$ and characteristic function (CF) $\Phi_z(u)$:

$$\Phi_z(u) = \prod_{k=0}^{K-1} \Phi_y(u) = \prod_{k=0}^{K-1} (p_k \exp(ja_k u) + q_k \exp(-ja_k u)), \qquad (4.3.8)$$

where $\Phi_y(u)$ is CF of random variable y_k (4.3.6b) with PDF (4.3.7); $j = \sqrt{-1}$.

Using CF $\Phi_z(u)$, one can determine expectation m_z and variance D_z of random variable z (4.3.6a) as the first two cumulants κ_1, κ_2, respectively (see, for instance [36, (18.3.27)]):

$$m_z = \kappa_1 = (j)^{-1}[\ln \Phi_z(u)]'|_{u=0} = \sum_{k=0}^{K-1} a_k(p_k - q_k) = 2\sum_{k=0}^{K-1} (p_k - 0.5); \qquad (4.3.9)$$

$$D_z = \kappa_2 = (j)^{-2}[\ln \Phi_z(u)]''|_{u=0} = 4\sum_{k=0}^{K-1} a_k^2 p_k q_k = 4\sum_{k=0}^{K-1} p_k q_k. \qquad (4.3.10)$$

When instantaneous value of the signal $s_k = s(t_k)$ is sufficiently small, basing on the formula (4.3.7a) find approximate values of probabilities p_k, q_k:

$$p_k \approx 0.5 + |s_k|p_n(0); \quad q_k \approx 0.5 - |s_k|p_n(0), \qquad (4.3.11)$$

where $p_n(x)$ is PDF of interference (noise) $n(t)$.

Then, substituting (4.3.11) into (4.3.9) and (4.3.10), we obtain, that the expectation m_z and variance D_z of random variable z (4.3.6a) are, correspondingly, equal to:

$$m_z = 2p_n(0) \sum_{k=0}^{K-1} |s_k| = 2p_n(0) \|s\|_1; \qquad (4.3.12)$$

$$D_z = 4\sum_{k=0}^{K-1} [0.5 + |s_k|p_n(0)][0.5 - |s_k|p_n(0)] = K - 4p_n^2(0) \|s\|_2^2, \qquad (4.3.13)$$

where $\|s\|_1$, $\|s\|_2$ are l_1- and l_2-norms of the signal $s(t)$, respectively.

Here we mean that condition of weak signal presence always holds, since this condition provides non-negativity of probability q_k (4.3.11):

$$|s_k|p_n(0) < 0.5. \qquad (4.3.14)$$

Considering as a useful signal $s(t)$ some RF bandpass signal with rectangular envelope and amplitude A and supposing that interference $n(t)$ has a variance D_n, the condition (4.3.14) is written in the form of the following implication:

$$|s_k|p_n(0) \leqslant Ap_n(0) < 0.5 \Rightarrow q^2 < K/(8D_n p_n^2(0)), \qquad (4.3.15)$$

where q^2 is SNR equal to:

$$q^2 = \frac{A^2 K}{2D_n} = \frac{2E}{N_0}, \qquad (4.3.16)$$

E, N_0 are energy of the signal $s(t)$ and power spectral density (PSD) of interference (noise) $n(t)$, respectively.

Norm-factor F_{12} (4.1.12) and (4.1.13b) allow writing SNR (4.3.16) over l_1-norm $\|s\|_1$ of the signal $s(t)$:

$$q^2 = \frac{\|s\|_2^2}{D_n} = \frac{\|s\|_1^2}{F_{12}^2 D_n}. \qquad (4.3.17)$$

Then, taking into account (4.3.17), the expressions (4.3.12) and (4.3.13) for expectation m_z and variance D_z of random variable z (4.3.6a) in the output of detector can be represented in the following form:

$$m_z(q) = 2p_n(0)\|s\|_1 = 2p_n(0)\sqrt{D_n}F_{12} \cdot q = 2p_n(0)\sqrt{D_n}k_{12}\sqrt{K} \cdot q; \qquad (4.3.18)$$

$$D_z(q) = K - 4p_n^2(0)\|s\|_2^2 = K - 4p_n^2(0)D_n q^2. \qquad (4.3.19)$$

It is known, that if $\sum_{k=0}^{K-1} p_k q_k > 25$ [115], that always holds when processing sufficiently lengthy signals, random variable z (4.3.6a) with generalized binomial distribution $p_z(x/H_i)$ can be approximated by Gaussian random variable with expectation m_z and variance D_z that are determined by the formulas (4.3.18) and (4.3.19), respectively:

$$\sum_{k=0}^{K-1} p_k q_k > 25 \Rightarrow p_z(x/H_i) \sim N(m_z, D_z). \qquad (4.3.20)$$

Determine a sample size K that provides *normalization condition* (4.3.20). Taking into account the relationships (4.3.11), compile the triple inequality:

$$\sum_{k=0}^{K-1} p_k q_k \geqslant \sum_{k=0}^{K-1} [0.25 - |s_k|^2 p_n^2(0)] \geqslant K[0.25 - A^2 p_n^2(0)] > 25.$$

Taking into account the formula (4.3.16), the composed inequality can be written in the form of normalization condition for generalized binomial distribution $p_z(x/H_i)$ (4.3.20), when a sample size must be greater than some quantity $K_{\text{n.c.}}$:

$$K > K_{\text{n.c.}} = 100 + 8D_n p_n^2(0)q^2. \qquad (4.3.20a)$$

Here we notice that normalization condition (4.3.20a) impose larger constraints on a sample size K, than the relation (4.3.15) providing fulfillment of the condition (4.3.14).

Therefore, quality indices of detection in the form of conditional probabilities of false alarm F and correct detection D are determined by the following relationships, respectively:

$$F = \int_{l_0}^{\infty} p_z(x/H_0)dx = 1 - F_z[l_0, m_z(0), D_z(0)]; \qquad (4.3.21a)$$

$$D(q) = \int\limits_{l_0}^{\infty} p_z(x/H_1)dx = 1 - F_z[l_0, m_z(q), D_z(q)]; \qquad (4.3.21b)$$

$$l_0 = F_z^{-1}[1 - F, m_z(0), D_z(0)], \qquad (4.3.21c)$$

where $p_z(x/H_i)$ is conditional PDF of random variable z (4.3.6a) for hypothesis H_i, $i \in \{0,1\}$ that is determined according to assumption on normality (4.3.20) by the formula:

$$p_z(x/H_i) = \frac{1}{\sqrt{2\pi D_z(q)}} \exp\left(-\frac{(x - m_z(q))^2}{2D_z(q)}\right), \qquad (4.3.22)$$

so that $q = 0$ if H_0, $q > 0$ if H_1; l_0 is a threshold of detection; $q = \sqrt{2E/N_0}$ is a detection parameter; $m_z(q)$, $D_z(q)$ are expectation and variance of distribution $p_z(x/H_i)$ determined by relationships (4.3.18), (4.3.19), respectively; $F_z[l_0, m_z(q), D_z(q)]$ is CDF of random variable z (4.3.6a) which corresponds to the conditional PDF $p_z(x/H_i)$:

$$F_z[y, m_z(q), D_z(q)] = \int\limits_{-\infty}^{y} p_z(x/H_i)dx,$$

$F_z^{-1}[y, m_z(q), D_z(q)]$ is an inverse function with respect to CDF $F_z[y, m_z(q), D_z(q)]$; $m_z(0) = 0$, $D_z(0) = K$.

In the absence of the signal $s(t) = 0$ when $p_k = q_k = 0.5$, CF $\Phi_z(u)$ of random variable z (4.3.6a) is determined by the expression:

$$\Phi_z(u)|_{s(t)=0} = \frac{1}{2^K}[\exp(-ju) + \exp(ju)]^K. \qquad (4.3.23)$$

The last formula (4.3.23) implies that distribution $p_z(x/H_0)$ of statistic z (4.3.6a) in the absence of the signal $s(t) = 0$ does not depend on interference (noise) distribution and described by generalized binomial distribution with parameters $(K, 0.5)$. Thus, algorithm (4.3.1) of detection of deterministic signal $s(t)$ is a nonparametric one.

Detection curves determining quality indices of detection (4.3.21a,b) of L-group algorithm (4.3.1) in the presence of Gaussian interference (noise) $n(t)$ within the model (4.1.2) are shown in Fig. 4.3.1 by dashed line. For convenience of comparative analysis, the dependences $D(q)$ describing classic algorithm of deterministic signal detection (4.1.7) are shown in Fig. 4.3.1 by solid lines.

Remind, that quality indices of detection F_{cls}, D_{cls} describing classic optimal algorithm (4.1.7) are defined by the relationships:

$$F_{cls} = 1 - \Phi(l_0, 0, 1); \qquad (4.3.24a)$$

$$D_{cls}(q) = 1 - \Phi(l_0, q, 1); \qquad (4.3.24b)$$

$$l_0 = \Phi^{-1}(1 - F_{cls}, 0, 1), \qquad (4.3.24c)$$

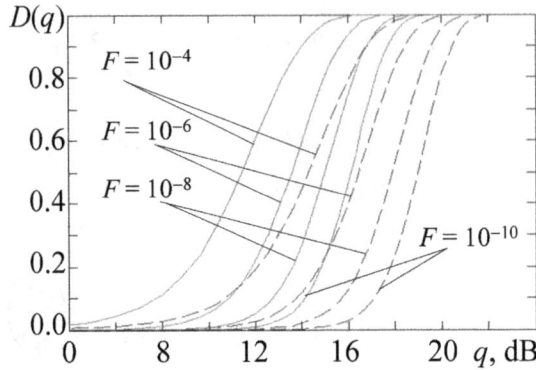

FIGURE 4.3.1   Detection curves determining quality indices of detection of *L*-group algorithm (4.3.1) (dashed line), and classic optimal algorithm (4.1.7) (solid line)

where $\Phi(y, m_x, D_x) = \frac{1}{\sqrt{2\pi D_x}} \int\limits_{-\infty}^{y} \exp\left(-\frac{(x-m_x)^2}{2D_x}\right); \ \Phi^{-1}(y, m_x, D_x)$ is an inverse function with respect to CDF $\Phi(y, m_x, D_x)$.

As follows from the dependences (4.3.21b), detection algorithm (4.3.1) based on MSI concerned with pseudometric is inferior to classic linear detection algorithm (4.1.7) in up to 3 dB in the presence of normal interference (noise). The mentioned loss, nevertheless, is compensated by invariance property of conditional probability of false alarm F with respect to parametric and nonparametric prior uncertainty conditions. Here we notice that quality indices of detection (4.3.21a,b), determined in the SNR interval $q^2 = 0 \ldots 23.5$ dB ($q = 0 \ldots 15$), do not practically depend on a sample size K when the condition (4.3.20a) holds in the presence of normal interference (noise):

$$K > K_{\text{n.c.}} = 100 + 8 D_n p_n^2(0) q^2 \approx 387.$$

Basing on the relationships (4.3.18), (4.3.19), and (4.3.21a,b), one can determine quality indices of signal detection in the presence of non-Gaussian interference with arbitrary distribution and a finite variance.

Find ARE $e_{MSI(pm),MF}$ of detection algorithm (4.3.1) based on MSI (1.4.10) concerned with pseudometric (1.3.25) in regard to classic linear detection algorithm (4.1.7).

On the basis of quality indices of detection F_{cls}, D_{cls} characterizing classic algorithm (4.1.7) and defined by the relationships (4.3.25a,b):

$$F = 1 - \Phi(l_0, 0, 1); \ D = 1 - \Phi(l_0 - q, 0, 1),$$

it is easily to determine the difference between two quantiles x_F, x_D of standard normal distribution:

$$\Delta_{MF} = x_F - x_D|_{MF} = l_0 - (l_0 - q) = q. \tag{4.3.25}$$

Similarly, for detection algorithm (4.3.1) based on MSI concerned with pseudometric, using quality indices of detection F, D that are determined by the relationships

(4.3.21a,b):

$$F = 1 - \Phi\left(\frac{l_0}{\sqrt{D_z(q)}}, 0, 1\right); \; D = 1 - \Phi\left(\frac{l_0 - m_z(q)}{\sqrt{D_z(q)}}, 0, 1\right)$$

one can find a limit of difference of standard normal distribution quantiles x_F, x_D:

$$\lim_{K \to \infty} \Delta_{MSI(pm)} = \lim_{K \to \infty} \left(x_F - x_D|_{MSI(pm)}\right)$$

$$= \lim_{K \to \infty} \left(\frac{l_0}{\sqrt{D_z(0)}} - \frac{l_0 - m_z(q)}{\sqrt{D_z(q)}}\right) = 2p_n(0)\sqrt{D_n}k_{12}q. \quad (4.3.26)$$

Then, taking into account (4.3.25), ARE $e_{MSI(pm),MF}$ of detection algorithm (4.3.1) based on MSI concerned with pseudometric in regard to classic linear detection algorithm (4.1.7) is equal to squared ratio of limit differences of quantiles x_F, x_D of the corresponding distributions:

$$e_{MSI(pm),MF} = \left(\frac{\lim_{K \to \infty} \Delta_{MSI(pm)}}{\Delta_{MF}}\right)^2 = 4p_n^2(0)D_nk_{12}^2. \quad (4.3.27)$$

Formula (4.3.27) implies that ARE $e_{MSI(pm),MF}$ of detection algorithm (4.3.1) based on MSI concerned with pseudometric in respect of classic linear detection algorithm (4.1.7) is equivalent to ARE $e_{SC,LC}$ of sign algorithm of detection (4.1.8) obtained for RF bandpass signals. Formula (4.3.27) implies that detection algorithm (4.3.1) based on MSI concerned with pseudometric is inferior to classic linear detection algorithm (4.1.7) (when a sample size is infinitely large and interference (noise) has a normal distribution) in

$$10|\lg(4p_n^2(0)D_nk_{12}^2)| = 10|\lg(4D_nk_{12}^2/(2\pi D_n))| = 10\left|\lg\left[\frac{2}{\pi}\left(\frac{2\sqrt{2}}{\pi}\right)^2\right]\right| \approx 2.873\,\text{dB}.$$

The results of evaluating ARE $e_{MSI(pm),MF}$ of detection algorithm (4.3.2) based on MSI concerned with pseudometric in regard to classic linear detection algorithm (4.1.7) for different interference distributions from the table 2.6.1 are shown in the table 4.4.1.

Conclusions

1. Detection algorithm (4.3.1) based on MSI concerned with pseudometric is equivalent to sign algorithm of detection (4.1.9), is referred to a class of nonparametric algorithms, is defined in terms of L–group operations, and can be realized without performing operation of multiplication.

2. The obtained relationships (4.3.21a,b) allow evaluating quality indices of signal detection for algorithm (4.3.1) operating in the presence of interference (noise) with arbitrary distribution.

3. The obtained relationship (4.3.27) allow evaluating AREs of signal detection algorithm (4.3.1) operating in the presence of interference with arbitrary distribution in regard to classic linear detection algorithm (4.1.7) that operates in linear signal space. ARE (4.3.27) of L–group detection algorithm (4.3.1) is equivalent to ARE (4.1.14) of sign algorithm of detection (4.1.9).

4.4   Detection Algorithm Based on Measure of Statistical Interrelation Concerned with l_1-metric

Algorithm of deterministic signal detection based on measure of statistical interrelation (1.4.12) concerned with l_1-metric (1.3.27) is defined by the following relationship:

$$z_{l_1}(t_j) = \sum_{i=0}^{j} \tfrac{1}{2}[|x(t_i) + h(t_{j-i})| - |x(t_i) - h(t_{j-i})|] \underset{d_0}{\overset{d_1}{\underset{<}{>}}} l_0. \qquad (4.4.1)$$

Taking into account the known relation between a signal and impulse response of matched filter $h(t_j) = s(t_0 + T - t_j)$, filter type detection algorithm (4.4.1) is equivalent to correlation type detection algorithm which is defined by the relationship:

$$z_{l_1,C}(t_j) = \sum_{i=0}^{j} g[x(t_i), s(t_i)] = \sum_{i=0}^{j} \tfrac{1}{2}[|x(t_i) + s(t_i)| - |x(t_i) - s(t_i)|] \underset{d_0}{\overset{d_1}{\underset{<}{>}}} l_0, \qquad (4.4.2)$$

so that the identity holds:

$$z_{l_1}(t_j) = z_{l_1,C}(t_j), \ t_j = t_0 + T. \qquad (4.4.3)$$

To simplify the following argumentation, we notice that a function $g(x,a)$ figuring in detection algorithm (4.4.2) is equivalent to Huber function $HF(x,a)$ (2.3.3) and (2.3.4):

$$g(x,a) = \tfrac{1}{2}[|x + a| - |x - a|] \equiv HF(x,a); \qquad (4.4.4a)$$

$$HF(x,a) = |a| \wedge (x \cdot \mathrm{sgn}(a) \vee -|a|) = -|a| \vee (x \cdot \mathrm{sgn}(a) \wedge |a|), \qquad (4.4.4b)$$

where x is a main variable; a is a limiting variable.

Find conditional PDFs $p(z(t_j)/H_0)|_{t_j=t_0+T}$, $p(z(t_j)/H_1)|_{t_j=t_0+T}$ in the output of generalized matched filter (4.4.1) in the instant $t_0 + T$ taking into account joint fulfillment of the identities (4.4.3) and (4.4.4a). Indexes of hypotheses H_0, H_1 correspond to the values of unknown nonrandom parameter θ from the model (4.1.2), $\theta \in \{0,1\}$: $\theta = 0$ (signal is absent) and $\theta = 1$ (signal is present).

For convenience of analysis within the initial relationship (4.4.2), introduce auxiliary variables corresponding to a shift of coordinate system origin into a point t_0:

$$z = \sum_{k=0}^{K-1} y_k = \sum_{i=0}^{K-1} \tfrac{1}{2}[|x_k + s_k| - |x_k - s_k|]; \qquad (4.4.5a)$$

$$y_k = \tfrac{1}{2}[|x_k + s_k| - |x_k - s_k|] = |s_k| \wedge (x_k \operatorname{sgn}(s_k) \vee -|s_k|); \qquad (4.4.5b)$$

$$t_k = t_j - t_0; \; k = 0, 1, \dots, K-1; \; t_{K-1} = T; \; K-1 = T/\Delta t, \qquad (4.4.5c)$$

where t_0 and $t_0 + T$ are time of signal arrival and ending, respectively (see formula (4.1.1a); T is a duration of the signal; Δt is a sampling interval providing independence of the samples $\{n(t_j)\}$ of interference (noise) $n(t)$ within the model (4.1.2).

Notice, that random variable y_k (4.4.5b) possesses mixture continuous-discrete distribution $p_y(x/H_i)$:

$$p_y(x/H_i) = p_n(x - \theta \cdot s_k)[1(x + |s_k|) - 1(x - |s_k|)]$$
$$+ p_{1k}\delta(x - |s_k|) + p_{2k}\delta(x + |s_k|); \quad (4.4.6a)$$

$$p_{1k} = 1 - F_n(|s_k|, \theta s_k); \; p_{2k} = F_n(-|s_k|, \theta s_k), \qquad (4.4.6b)$$

where $p_y(x/H_i)$ is conditional PDF of random variable y_k (4.4.5b) for hypothesis H_i, $i \in \{0,1\}$; $F_n(x,c)$ is CDF of interference (noise) $n(t)$; c is a center of symmetry of distribution $F_n(x,c)$; $\theta = i \in \{0,1\}$ is a parameter of the model (4.1.2).

In the case of sufficiently weak signal (when SNR $q^2 \leqslant 100$), distribution (4.4.6) can be approximated by the following mixture continuous-discrete distribution:

$$p_y(x/H_i) \approx p_{0k} \cdot p_U(x) + (p_{1k} + p_{2k}) \cdot p_B(x/H_i); \qquad (4.4.7)$$
$$p_U(x) = [1(x + |s_k|) - 1(x - |s_k|)]/(2|s_k|); \qquad (4.4.7a)$$
$$p_B(x/H_i) = p_k\delta(x - |s_k|) + q_k\delta(x + |s_k|); \qquad (4.4.7b)$$
$$p_k = \frac{p_{1k}}{p_{1k} + p_{2k}}, \; q_k = \frac{p_{2k}}{p_{1k} + p_{2k}}, \; q_k = 1 - p_k; \qquad (4.4.7c)$$
$$p_{0k} = 1 - (p_{1k} + p_{2k}), \qquad (4.4.7d)$$

where $p_U(x)$ is PDF of uniform distribution; $p_B(x/H_i)$ is conditional PDF of generalized Bernoulli distribution with parameter p_k; the probabilities p_{1k}, p_{2k} are determined by the relationships (4.4.6a) and depend on the parameter θ.

Formula (4.4.7) implies that in the absence of the signal $s(t) = 0$, distribution $p_z(x/H_0)$ (4.4.5a) does not depend on interference (noise) distribution. Thus, algorithm (4.4.1) of detection of deterministic signal $s(t)$ is nonparametric.

For random variable y_k (4.4.5b) with mixture continuous-discrete distribution $p_y(x/H_i)$ (4.4.7), one can find expectation m_y and variance D_y [116–121]:

$$m_y = p_{0k} \cdot m_U + (p_{1k} + p_{2k}) \cdot m_B; \qquad (4.4.8)$$

$$D_y = p_{0k} \cdot [D_U + m_U^2] + (p_{1k} + p_{2k}) \cdot [D_B + m_B^2] - m_y^2; \qquad (4.4.9)$$

$$m_U = 0, \ D_U = 4|s_k|^2/12; \qquad (4.4.10a)$$

$$m_B = |s_k|(p_k - q_k), \ D_B = 4|s_k|^2 p_k q_k, \qquad (4.4.10b)$$

where m_U, D_U are expectation and variance of uniform distribution $p_U(x)$ (4.4.7a), respectively; m_B, D_B are expectation and variance of generalized Bernoulli distribution $p_B(x/H_i)$ (4.4.7b), respectively; p_{0k}, p_{1k}, p_{2k} are probabilities that depend on parameter θ and are determined by the relationships (4.4.7d) and (4.4.6a), respectively; p_k, q_k are probabilities that depend on parameter θ and are determined by the relationships (4.4.7c).

Taking into account the relationships (4.4.7c) and (4.4.10a,b), formulas for expectation m_y (4.4.8) and variance D_y (4.4.9) take the form:

$$m_y = (p_{1k} - p_{2k}) \cdot |s_k|; \qquad (4.4.11)$$

$$D_y = p_{0k} \cdot |s_k|^2/3 + (p_{1k} + p_{2k}) \cdot |s_k|^2 - (p_{1k} - p_{2k})^2 \cdot |s_k|^2. \qquad (4.4.12)$$

In the absence of useful signal $s(t)$ ($\theta = 0$) in the additive mixture $x(t)$ (4.1.2), probabilities p_{1k}, p_{2k} are equal:

$$p_{1k}|_{\theta=0} = p_{2k}|_{\theta=0} = 1 - F_n(|s_k|, 0) = F_n(-|s_k|, 0), \qquad (4.4.13)$$

therefore, in this case, expectation m_y (4.4.11) and variance D_y (4.4.12) are, respectively, equal to:

$$m_y|_{\theta=0} = 0; \qquad (4.4.14)$$

$$D_y|_{\theta=0} = p_{0k} \cdot |s_k|^2/3 + (p_{1k} + p_{2k}) \cdot |s_k|^2. \qquad (4.4.15)$$

In the presence of useful signal $s(t)$ ($\theta = 1$) in the additive mixture $x(t)$ (4.1.2), probabilities p_{1k}, p_{2k} are, respectively, equal to:

$$p_{1k} = 1 - F_n(|s_k|, |s_k|) = 0.5; \ p_{2k} = F_n(-|s_k|, |s_k|) < 0.5. \qquad (4.4.16)$$

Considering the value $|s_k|$, figuring in the formulas of probabilities p_{1k}, $p_{2k}|_{\theta=0}$ (4.4.13) and p_{1k}, $p_{2k}|_{\theta=1}$ (4.4.16), to be sufficiently small, the probabilities p_{0k}, p_{1k}, p_{2k} are, respectively, equal to:

$$p_{0k}|_{\theta=0} = 2|s_k|p_n(0); \qquad (4.4.17a)$$

$$p_{1k}|_{\theta=0} = p_{2k}|_{\theta=0} = 0.5 - |s_k|p_n(0); \qquad (4.4.17b)$$

$$p_{0k}|_{\theta=1} = 2|s_k|p_n(0); \qquad (4.4.18a)$$

$$p_{1k}|_{\theta=1} = 0.5; \qquad (4.4.18b)$$

$$p_{2k}|_{\theta=1} = 0.5 - 2|s_k|p_n(0), \qquad (4.4.18c)$$

where $p_n(0)$ is a value of PDF $p_n(x)$ of interference $n(t)$ in the point $x = 0$.

Here we mean that the condition of weak signal presence always holds:

$$|s_k|p_n(0) < 0.25. \qquad (4.4.19)$$

Considering as a useful signal $s(t)$ some RF bandpass signal with rectangular envelope and amplitude A and assuming that interference (noise) $n(t)$ possesses a variance D_n, we rewrite the condition (4.4.19) in the form of the following implication:

$$|s_k|p_n(0) \leqslant Ap_n(0) < 0.25 \Rightarrow q^2 < K/(16D_n p_n^2(0)), \qquad (4.4.20)$$

where q^2 is SNR that is determined by (4.3.16).

Substituting the values of probabilities p_{0k}, p_{1k}, p_{2k} from the formulas (4.4.17) and (4.4.18) into the formulas (4.4.11) and (4.4.12), we obtain the resulting expressions for expectation m_y and variance D_y of random variable y_k (4.4.5b) with mixture continuous-discrete distribution $p_y(x/H_i)$ (4.4.7) in the absence ($\theta = 0$) and the presence ($\theta = 1$) of useful signal $s(t)$ in additive mixture $x(t)$ (4.1.2):

$$m_y|_{\theta=0} = 0; \qquad (4.4.21a)$$

$$D_y|_{\theta=0} = |s_k|^2 - \tfrac{4}{3}p_n(0) \cdot |s_k|^3; \qquad (4.4.21b)$$

$$m_y|_{\theta=1} = 2p_n(0) \cdot |s_k|^2; \qquad (4.4.22a)$$

$$D_y|_{\theta=0} = |s_k|^2 - \tfrac{4}{3}p_n(0) \cdot |s_k|^3 - 4p_n^2(0) \cdot |s_k|^4. \qquad (4.4.22b)$$

When processing sufficiently lengthy signals, random variable z (4.4.5a) in the form of sum of K independent and identically distributed random variables y_k with mixture continuous-discrete distribution $p_y(x/H_i)$ can be approximated by Gaussian random variable

$$p_z(x/H_i) \sim N(m_z, D_z) \qquad (4.4.23)$$

with expectation m_z and variance D_z that are determined by the following relationships, respectively:

$$m_z|_{\theta=0} = \sum_{k=0}^{K-1} m_y|_{\theta=0} = 0; \qquad (4.4.24a)$$

$$D_z|_{\theta=0} = \sum_{k=0}^{K-1} D_y|_{\theta=0} = \sum_{k=0}^{K-1} |s_k|^2 - \tfrac{4}{3}p_n(0) \sum_{k=0}^{K-1} |s_k|^3 =$$
$$= \|s\|_2^2 - \tfrac{4}{3}p_n(0) \|s\|_3^3; \quad (4.4.24b)$$

$$m_z|_{\theta=1} = \sum_{k=0}^{K-1} m_y|_{\theta=1} = 2p_n(0) \cdot \sum_{k=0}^{K-1} |s_k|^2 = 2p_n(0) \cdot \|s\|_2^2; \qquad (4.4.25a)$$

$$D_z|_{\theta=1} = \sum_{k=0}^{K-1} D_y|_{\theta=1} = \sum_{k=0}^{K-1} |s_k|^2 - \tfrac{4}{3}p_n(0) \sum_{k=0}^{K-1} |s_k|^3 - 4p_n^2(0) \sum_{k=0}^{K-1} |s_k|^4 =$$
$$= \|s\|_2^2 - \tfrac{4}{3}p_n(0) \|s\|_3^3 - 4p_n^2(0) \|s\|_4^4. \quad (4.4.25b)$$

Taking into account the equivalence of norms in normed space [59] and relations between norms defined by norm-factor F_{nm} (4.1.12), we rewrite the relationships (4.4.24) and (4.4.25) using SNR q^2 (4.3.16) related with l_2-norm $\|s\|_2$:

$$q^2 = A^2 K/(2D_n) = \|s\|_2^2/D_n = 2E/N_0. \tag{4.4.26}$$

Thus, taking into account the formulas (4.1.13a,b), the resulting relationships for expectation m_z and variance D_z from (4.4.24) and (4.4.25) take the form:

$$m_z(q = 0) = 0; \tag{4.4.27a}$$

$$D_z(q = 0) = D_n q^2 - \tfrac{4}{3} p_n(0) \frac{D_n^{3/2} q^3}{K^{1/2} k_{23}^3}; \tag{4.4.27b}$$

$$m_z(q) = 2p_n(0) \cdot D_n q^2; \tag{4.4.28a}$$

$$D_z(q) = D_n q^2 - \tfrac{4}{3} p_n(0) \frac{D_n^{3/2} q^3}{K^{1/2} k_{23}^3} - 4p_n^2(0) \frac{D_n^2 q^4}{K \cdot k_{24}^4}. \tag{4.4.28b}$$

Taking into account asymptotic normality of distribution $p_z(x/H_i)$ (4.4.23) of random variable z, quality indices of detection in the form of conditional probabilities of false alarm F and correct detection D are determined by the following relationships, respectively:

$$F = \int_{l_0}^{\infty} p_z(x/H_0)dx = 1 - F_z[l_0, m_z(0), D_z(0)]; \tag{4.4.29a}$$

$$D(q) = \int_{l_0}^{\infty} p_z(x/H_1)dx = 1 - F_z[l_0, m_z(q), D_z(q)]; \tag{4.4.29b}$$

$$l_0 = F_z^{-1}[1 - F, m_z(0), D_z(0)], \tag{4.4.29c}$$

where $p_z(x/H_i)$ is conditional PDF of random variable z (4.4.5a) for hypothesis H_i, $i \in \{0,1\}$ determined, according to assumption on normality (4.4.23), by the formula:

$$p_z(x/H_i) = \frac{1}{\sqrt{2\pi D_z(q)}} \exp\left(-\frac{(x - m_z(q))^2}{2D_z(q)}\right), \tag{4.4.30}$$

so that $q = 0$ if H_0, $q > 0$ if H_1; l_0 is a threshold of detection; $q = \sqrt{2E/N_0}$ is a detection parameter; $m_z(q)$, $D_z(q)$ are expectation and variance of distribution $p_z(x/H_i)$ determined by the relationships (4.4.27), (4.4.28), respectively; $F_z[y, m_z(q), D_z(q)]$ is CDF of random variable z that corresponds to conditional PDF $p_z(x/H_i)$ (4.4.30):

$$F_z[y, m_z(q), D_z(q)] = \int_{-\infty}^{y} p_z(x/H_i)dx,$$

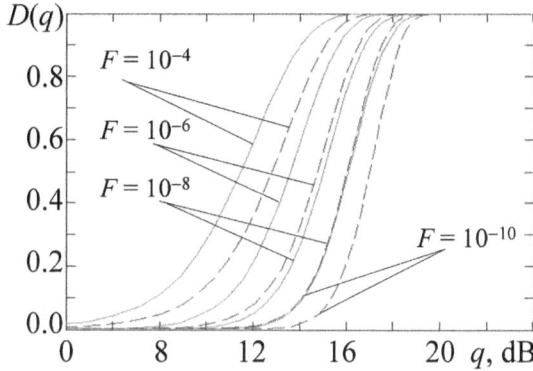

FIGURE 4.4.1 Detection curves determining quality indices of detection of L-group algorithm (4.4.1) (dashed line) and classic algorithm (4.1.7) (solid line)

$F_z^{-1}[y, m_z(q), D_z(q)]$ is an inverse function with respect to CDF $F_z[y, m_z(q), D_z(q)]$.

Detection curves determining quality indices of detection (4.4.29a,b) of L-group algorithm (4.4.1) operating in the presence of Gaussian interference $n(t)$ within the model (4.1.2) are shown in Fig. 4.4.1 by dashed line. For convenience of comparative analysis, dependence $D(q, F)$ related to classic detection algorithm (4.1.7) is shown in Fig. 4.4.1 by solid line.

Dependences $D(q, F)$ depicted in Fig. 4.4.1 are related to a sample size K which is close to minimal: $K \sim K_{\min} = 16 D_n p_n^2(0) q^2$ (see the inequality (4.4.20)). As follows from the dependences $D(q, F)$ (4.4.29b), detection algorithm (4.4.1) based on MSI concerned with l_1-metric is inferior to classic linear detection algorithm (4.1.7) in up to 1 dB when a sample size K is close to minimal $K \sim K_{\min}$: $K_{\min} = 8q^2/\pi \approx 430$.

Notice that using relationships (4.4.27), (4.4.28), and (4.4.29a,b), one can determine quality indices of signal detection in the presence of non-Gaussian interference with arbitrary distribution and finite variance.

For detection algorithm (4.4.1) based on MSI concerned with l_1-metric, using quality indices of detection F, D that are determined by the relationships (4.4.29a,b):

$$F = 1 - \Phi \left(\frac{l_0}{\sqrt{D_z(q)}}, 0, 1 \right) ; \quad D = 1 - \Phi \left(\frac{l_0 - m_z(q)}{\sqrt{D_z(q)}}, 0, 1 \right),$$

one can find a limit of difference of quantiles x_F, x_D of standard normal distribution:

$$\lim_{K \to \infty} \Delta_{MSI(m)} = \lim_{K \to \infty} \left(x_F - x_D|_{MSI(m)} \right)$$

$$= \lim_{K \to \infty} \left(\frac{l_0}{\sqrt{D_z(0)}} - \frac{l_0 - m_z(q)}{\sqrt{D_z(q)}} \right) = 2p_n(0)\sqrt{D_n}q. \quad (4.4.31)$$

Then, taking into account (4.3.25), ARE $e_{MSI(m),MF}$ of detection algorithm (4.4.1) based on MSI concerned with l_1-metric in relation to classic linear detection algorithm (4.1.7) is equal to squared ratio of limit differences of quantiles x_F, x_D of the corresponding distributions:

$$e_{MSI(m),MF} = \left(\frac{\lim\limits_{K \to \infty} \Delta_{MSI(m)}}{\Delta_{MF}} \right)^2 = 4p_n^2(0)D_n. \qquad (4.4.32)$$

Formula (4.3.27) implies that ARE $e_{MSI(m),MF}$ of detection algorithm (4.4.1) based on MSI concerned with l_1-metric in relation to classic linear detection algorithm (4.1.7), obtained for RF bandpass signal, is slightly higher than ARE $e_{SC,LC}$ (4.1.14) of sign algorithm of detection (4.1.8) in relation to classic correlation algorithm (4.1.7). Formula (4.4.32) implies that in the case of infinitely large sample and normal distribution of interference, detection algorithm (4.4.1) based on MSI concerned with l_1-metric is inferior to classic linear detection algorithm (4.1.7) in

$$10|\lg(4p_n^2(0)D_n)| = 10|\lg(4D_n/(2\pi D_n))| = 10|\lg\left[\tfrac{2}{\pi}\right]| \approx 1.96 \text{ dB}.$$

Thus, loss in signal energy of algorithm (4.4.1) with respect to classic algorithm (4.1.7) in the presence of Gaussian interference is $0.9\ldots 1.96$ dB depending on a sample size K.

AREs $e_{MSI(m),MF}$ of detection algorithm (4.4.1) based on MSI concerned with l_1-metric in relation to classic linear detection algorithm (4.1.7) for different interference distributions from the Table 2.6.1 are shown in the Table 4.4.1.

TABLE 4.4.1   AREs of detection algorithms (4.3.1) and (4.4.1) based on MSIs concerned with pseudometric and l_1-metric, respectively

Notation of distribution	Distribution	Algorithm	
		MSI$_p$ (4.3.1)	MSI$_{l_1}$ (4.4.1)
$N(0,b)$	Normal	0.516	0.637
$T(0,0.01,3)$	Tukey	0.550	0.678
$T(0,0.05,3)$	Tukey	0.675	0.833
$L(0,b)$	Logistic	0.667	0.823
$DE(0,b)$	Laplace	1.621	2.00
$St(0,5)$	Student	0.779	0.961
$St(0,3)$	Student	1.311	1.617
$C(0,b)$	Cauchy	∞	∞
$U(0,b)$	Uniform	0.270	0.333
$Tr(0,b)$	Triangular	0.540	0.667

The Table 4.4.1 gives initial information on AREs of detection algorithms (4.3.1) and (4.4.1) for pointed types of distributions. Now let us try to get a clear idea on ARE (4.4.32) of algorithm (4.4.1) for the distribution families of interferences (noises) with Weibull, gamma, and lognormal PDFs of their envelopes determined by the relationships (2.7.5...2.7.7), respectively. For this purpose we exploit known formula that establishes the relation between PDF $p_n(x)$ of interference (noise)

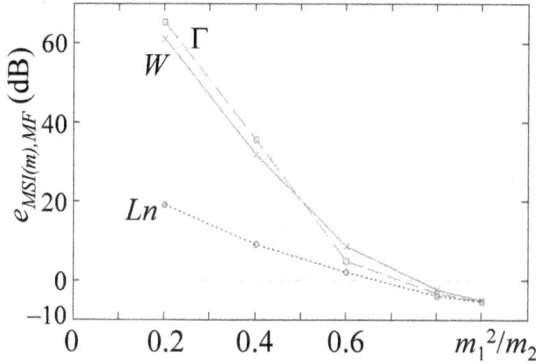

FIGURE 4.4.2   AREs (4.4.34) of detection algorithm (4.4.1) with respect to classic algorithm (4.1.7) for Weibull, gamma, and lognormal distribution families of envelopes

instantaneous values and PDF of its envelope $p_E(z)$ [52, (3.3.82)]:

$$p_n(x) = \frac{1}{\pi} \int\limits_{|x|}^{\infty} \frac{p_E(z)}{\sqrt{z^2 - x^2}} \mathrm{d}z. \tag{4.4.33}$$

We use this formula for numerical calculating the values $p_n(0)$ figuring in the formula of ARE (4.4.32), assuming that value of a interference (noise) variance D_n from the same formula is equal to: $D_n = 0.5m_2$, where m_2 is the second moment of interference (noise) envelope: $m_2 = \int\limits_0^{\infty} z^2 p_E(z)\mathrm{d}z$.

Fig. 4.4.2 illustrates the dependences of AREs (4.4.32) (in dB) characterizing detection algorithm (4.4.1)

$$e_{MSI(m),MF} = 10\lg[4p_n^2(0)D_n] \text{ (dB)}, \tag{4.4.34}$$

based on MSI concerned with l_1-metric, of the ratio m_1^2/m_2 of squared expectation to the second moment for Weibull, gamma, and lognormal distribution families $\{p_E(z; m_1^2/m_2)\}$ of interference (noise) envelopes determined by the relationships (2.7.5...2.7.7). Dependences for Weibull, gamma, and lognormal distribution families are shown by solid, dashed, and dotted lines, respectively.

As can be seen from the figure, in the presence of harmonic type interference (noise) when m_1^2/m_2 takes its values in the interval $m_1^2/m_2 \in [0.68\ldots0.77, 0.9]$, AREs are below zero and L-group detection algorithm (4.4.1) is inferior to matched filter algorithm (4.1.7) in up to 5 dB. On the contrary, in the presence of impulse type interference (noise) when m_1^2/m_2 takes its values in the interval $m_1^2/m_2 \in [0.2, 0.68\ldots0.77]$, algorithm (4.4.1) provides a gain in ARE with respect to classic detection algorithm (4.1.7), so that the less m_1^2/m_2 is, the greater an obtained gain is. The same types of dependences of AREs are inherent to detection algorithm (4.3.1) based on MSI concerned with pseudometric, but these dependences pass a little bit below (in about 0.9 dB) with respect to the corresponding dependences shown in Fig. 4.4.2.

Conclusions

1. Detection algorithm (4.4.1) based on MSI concerned with l_1-metric relates to a class of nonparametric algorithms, is defined in terms of L–group operations, and can be realized without performing operation of multiplication.

2. The obtained relationships (4.4.29a,b) allow evaluating quality indices of signal detection in the presence of interference (noise) with arbitrary distribution when using algorithm (4.4.1).

3. The obtained relationship (4.4.32) allow evaluating ARE of signal detection algorithm (4.4.1) in the presence of interference (noise) with arbitrary distribution with respect to classic algorithm of matched filter (4.1.7) operating in linear signal space.

4.5   Detection Algorithm Based on Measure of Statistical Interrelation Concerned with Semimetric

Algorithm of deterministic signal detection based on measure of statistical interrelation concerned with semimetric (1.3.26) is defined by the following relationship:

$$z_s(t_j) = \sum_{i=0}^{j} [|x(t_i)| + |h(t_{j-i})|]$$

$$- \sum_{i=0}^{j} (\mathrm{sgn}(h(t_{j-i}) \vee x(t_i)) - \mathrm{sgn}(h(t_{j-i}) \wedge x(t_i)))|x(t_i) - h(t_{j-i})|] \begin{array}{c} d_1 \\ > \\ < \\ d_0 \end{array} l_0. \quad (4.5.1)$$

Taking into account the known relation between the signal and impulse response of matched filter $h(t_j) = s(t_0 + T - t_j)$, filter type detection algorithm (4.5.1) is equivalent to correlation type detection algorithm that is determined by the relationship:

$$z_{s,C}(t_j) = \sum_{i=0}^{j} g[x(t_i), s(t_i)] = \sum_{i=0}^{j} [|x(t_i) + s(t_i)|$$

$$- \sum_{i=0}^{j} (\mathrm{sgn}(s(t_i) \vee x(t_i)) - \mathrm{sgn}(s(t_i) \wedge x(t_i)))|x(t_i) - s(t_i)|] \begin{array}{c} d_1 \\ > \\ < \\ d_0 \end{array} l_0, \quad (4.5.2)$$

so that the identity holds:

$$z_s(t_j) = z_{s,C}(t_j), \; t_j = t_0 + T. \quad (4.5.3)$$

To simplify the next discussion, we notice (see table 4.2.1) that the function $g(x, a)$ figuring in detection algorithm (4.5.2) is equivalent to the function:

$$g(x, a) = a \cdot \text{sgn}(x) + x \cdot \text{sgn}(a). \tag{4.5.4}$$

Find conditional PDFs $p(z(t_j)/H_0)|_{t_j=t_0+T}$, $p(z(t_j)/H_1)|_{t_j=t_0+T}$ in the output of generalized matched filter (4.5.1) in time instant $t_j = t_0 + T$, taking into account joint fulfillment of the identities (4.5.3) and (4.5.4). Indexes of hypotheses H_0, H_1 correspond to the values of unknown nonrandom parameter θ from the model (4.1.2), $\theta \in \{0, 1\}$: $\theta = 0$ (signal is absent) and $\theta = 1$ (signal is present).

For convenience of analysis within the initial relationship (4.5.2), we introduce the auxiliary variables corresponding a shift of coordinate system origin into a point t_0:

$$z = \sum_{k=0}^{K-1} y_k = \sum_{k=0}^{K-1} [s_k \cdot \text{sgn}(x_k) + x_k \cdot \text{sgn}(s_k)]; \tag{4.5.5a}$$

$$y_k = s_k \cdot \text{sgn}(x_k) + x_k \cdot \text{sgn}(s_k); \tag{4.5.5b}$$

$$t_k = t_j - t_0; \ k = 0, 1, \ldots, K-1; \ t_{K-1} = T; \ K - 1 = T/\Delta t, \tag{4.5.5c}$$

where t_0 and $t_0 + T$ are time of signal arrival and ending, respectively (see formula (4.1.1a)); T is a duration of signal $s(t)$; Δt is a sampling interval providing independence of samples $\{n(t_j)\}$ of interference (noise) $n(t)$ within the model (4.1.2).

Basing on the features of the function (4.5.4), we notice that depending on hypothesis H_i, $i \in \{0, 1\}$ and the sign of signal sample s_k, random variable y_k (4.5.5b) takes the following values with probabilities p_k, q_k, respectively:

$$y_k = \begin{cases} (x_k + s_k) \cdot \text{sgn}(s_k), \ p_k; \\ (x_k - s_k) \cdot \text{sgn}(s_k), \ q_k; \end{cases} \tag{4.5.6}$$

$$p_k = F_n(\theta|s_k|, 0); \ q_k = F_n(-\theta|s_k|, 0) \tag{4.5.6a}$$

$$p_k + q_k = 1; \tag{4.5.6b}$$

where $F_n(x, c)$ is CDF of interference (noise) $n(t)$; c is a center of distribution $F_n(x, c)$; $\theta \in \{0, 1\}$ is a parameter of the model (4.1.2) that is equal: $\theta = 1$ (signal is present); $\theta = 0$ (signal is absent).

Random variable y_k (4.5.5b), taking the values according to the relationships (4.5.6), (4.5.6a), possesses generalized Bernoulli distribution $p_y(x/H_i)$:

$$p_y(x/H_i) = p_k \delta(x - a_k) + q_k \delta(x - b_k); \tag{4.5.7}$$

$$a_k = (1 + \theta)|s_k| + n_k \cdot \text{sgn}(s_k); \tag{4.5.7a}$$

$$b_k = (\theta - 1)|s_k| + n_k \cdot \text{sgn}(s_k); \tag{4.5.7b}$$

where $p_y(x/H_i)$ is conditional PDF of random variable y_k (4.5.5b) for hypothesis H_i, $i \in \{0, 1\}$; probabilities p_k, q_k are determined by the relationships (4.5.6a); $\delta(x)$ is Dirac delta function.

Then random variable z (4.5.5a) that is sum of K independent random variables with Bernoulli distribution possesses a generalized binomial distribution $p_z(x/H_i)$ and characteristic function (CF) $\Phi_z(u)$:

$$\Phi_z(u) = \prod_{k=0}^{K-1} \Phi_y(u) = \prod_{k=0}^{K-1} (p_k \exp(ja_k u) + q_k \exp(jb_k u)), \qquad (4.5.8)$$

where $\Phi_y(u)$ is CF of random variable y_k (4.5.5b) with PDF (4.5.7); $j = \sqrt{-1}$.

Using CF $\Phi_z(u)$, one can find expectation m_z and variance D_z of random variable z (4.3.7a) in the form of two first cumulants κ_1, κ_2, respectively (see, for instance, [36, (18.3.27)]):

$$m_z = \kappa_1 = (j)^{-1} [\ln \Phi_z(u)]'|_{u=0}$$

$$= \sum_{k=0}^{K-1} (a_k p_k + b_k q_k) = \sum_{k=0}^{K-1} (p_k(a_k - b_k) + b_k); \qquad (4.5.9)$$

$$D_z = \kappa_2 = (j)^{-2} [\ln \Phi_z(u)]''|_{u=0} = \sum_{k=0}^{K-1} (a_k + b_k)^2 p_k q_k. \qquad (4.5.10)$$

If instantaneous value of the signal $s_k = s(t_k)$ is sufficiently small, from (4.5.6a) we obtain approximate values of probabilities p_k, q_k:

$$p_k \approx 0.5 + \theta |s_k| p_n(0); \quad q_k \approx 0.5 - \theta |s_k| p_n(0), \qquad (4.5.11)$$

Substituting the relationships (4.5.11) and (4.5.7a,b) into the equations (4.5.9) and (4.5.10), we obtain the relationships determining expectation m_z and variance D_z of random variable z (4.5.5a), respectively:

$$m_z|_{\theta=0} = 0; \qquad (4.5.12a)$$

$$D_z|_{\theta=0} = \sum_{k=0}^{K-1} 4n_k^2 \left(0.25 - |s_k|^2 p_n^2(0)\right)$$

$$\approx \|n\|_2^2 \left(1 - 4p_n^2(0)\frac{\|s\|_2^2}{K}\right) = D_n K \left(1 - 4p_n^2(0)\frac{D_n q^2}{K}\right); \qquad (4.5.12b)$$

$$m_z|_{\theta=1} = \sum_{k=0}^{K-1} [2|s_k|(0.5 + |s_k| p_n(0))] = \|s\|_1 + 2p_n(0) \cdot \|s\|_2^2$$

$$= F_{12} \|s\|_2 + 2p_n(0) \cdot \|s\|_2^2 = k_{12}\sqrt{D_n K q^2} + 2p_n(0) D_n K q^2; \qquad (4.5.13a)$$

$$D_z|_{\theta=1} = \sum_{k=0}^{K-1} 4x_k^2 \left(0.25 - |s_k|^2 p_n^2(0)\right) \approx \|x\|_2^2 \left(1 - 4p_n^2(0)\frac{\|s\|_2^2}{K}\right)$$

$$= \left(\|s\|_2^2 + \|n\|_2^2\right)\left(1 - 4p_n^2(0)\frac{\|s\|_2^2}{K}\right)$$

$$= D_n K \left(1 + \frac{q^2}{K}\right)\left(1 - 4p_n^2(0)\frac{D_n q^2}{K}\right); \qquad (4.5.13b)$$

where $\|s\|_1$, $\|s\|_2$ are l_1-, l_2-norms of the signal $s(t)$, respectively; $p_n(x)$, D_n are PDF and variance of interference (noise) $n(t)$, respectively; k_{12}, F_{12} are coefficients determined by formulas (4.1.13a,b), respectively; q^2 is SNR determined by the formula (4.3.17).

If $\sum\limits_{k=0}^{K-1} p_k q_k > 25$ [115], that always holds when processing sufficiently lengthy signals, random variable z (4.5.5a) with generalized binomial distribution $p_z(x/H_i)$ can be approximated by Gaussian random variable with expectation m_z and variance D_z that are determined by the formulas (4.5.12) and (4.5.13), respectively:

$$\sum_{k=0}^{K-1} p_k q_k > 25 \Rightarrow p_z(x/H_i) \sim N(m_z, D_z). \qquad (4.5.14)$$

Probabilities p_k, q_k (4.5.11) are determined by the relationships that are identical to (4.3.11), therefore the *normalization condition* of generalized binomial distribution $p_z(x/H_i)$ (4.5.14) is similar to the relationship (4.3.20a):

$$K > K_{\text{n.c.}} = 100 + 8D_n p_n^2(0) q^2. \qquad (4.5.14a)$$

Taking into account asymptotic normality of distribution $p_z(x/H_i)$ (4.5.14) of random variable z, quality indices of detection in the form of conditional probabilities of false alarm F and correct detection D are determined by the following relationships, respectively:

$$F = \int\limits_{l_0}^{\infty} p_z(x/H_0)dx = 1 - F_z[l_0, m_z(0), D_z(0)]; \qquad (4.5.15a)$$

$$D(q) = \int\limits_{l_0}^{\infty} p_z(x/H_1)dx = 1 - F_z[l_0, m_z(q), D_z(q)]; \qquad (4.5.15b)$$

$$l_0 = F_z^{-1}[1 - F, m_z(0), D_z(0)], \qquad (4.5.15c)$$

where $p_z(x/H_i)$ is conditional PDF of random variable (4.5.5a) for hypothesis H_i, $i \in \{0,1\}$ determined according to the assumption on normality (4.5.14) by the formula:

$$p_z(x/H_i) = \frac{1}{\sqrt{2\pi D_z(q)}} \exp\left(-\frac{(x - m_z(q))^2}{2D_z(q)}\right), \qquad (4.5.16)$$

so that $q = 0$ if H_0, $q > 0$ if H_1; l_0 is a threshold of detection; $q = \sqrt{2E/N_0}$ is parameter of detection; $m_z(q)$, $D_z(q)$ are expectation and variance of distribution $p_z(x/H_i)$ determined by the relationships (4.5.12), (4.5.13), respectively; $F_z[y, m_z(q), D_z(q)]$ is CDF of random variable z that corresponds to conditional PDF $p_z(x/H_i)$:

$$F_z[y, m_z(q), D_z(q)] = \int\limits_{-\infty}^{y} p_z(x/H_i)dx,$$

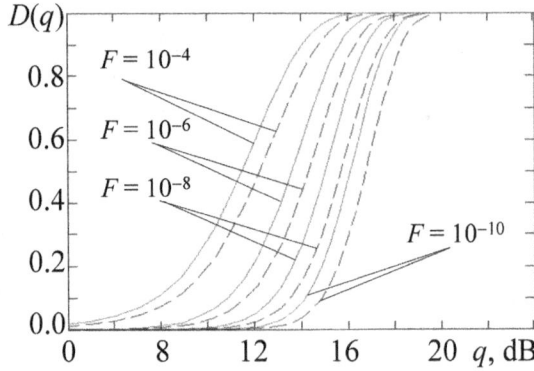

FIGURE 4.5.1   Detection curves determining quality indices of detection of L-group algorithm (4.5.1) (dashed line) and classic algorithm (4.1.7) (solid line)

where $F_z^{-1}[y, m_z(q), D_z(q)]$ is inverse function with respect to CDF $F_z[y, m_z(q), D_z(q)]$.

Detection curves, determining quality indices of signal detection (4.5.15a,b) of L-group algorithm (4.5.1) operating in the presence of Gaussian interference $n(t)$ within the model (4.1.2), are shown in Fig. 4.5.1 by dashed line. For convenience of comparative analysis, dependences $D(q, F)$ describing classic algorithm of deterministic signal detection (4.1.7) are shown in Fig. 4.5.1 by solid line.

The dependences $D(q, F)$ depicted in Fig. 4.5.1 are shown for the case of sample size $K >> K_{n.c.}$, where $K_{n.c.}$ is determined by the inequality (4.5.14a). As follows from the dependences $D(q, F)$ (4.5.15b), detection algorithm (4.5.1) based on MSI concerned with semimetric is inferior to classic linear detection algorithm in up to 1 dB.

Notice that using the relationships (4.5.12), (4.5.13), and (4.5.15a,b), one can determine quality indices of signal detection in the presence of non-Gaussian interference with arbitrary distribution and finite variance.

For detection algorithm (4.5.1) based on MSI concerned with semimetric, using quality indices of detection F, D that are determined by the relationships (4.5.15a,b):

$$F = 1 - \Phi\left(\frac{l_0}{\sqrt{D_z(q)}}, 0, 1\right); \quad D = 1 - \Phi\left(\frac{l_0 - m_z(q)}{\sqrt{D_z(q)}}, 0, 1\right),$$

one can find a limit of differences of quantiles x_F, x_D of standard normal distribution:

$$\lim_{K \to \infty} \Delta_{MSI(sm)} = \lim_{K \to \infty} \left(x_F - x_D|_{MSI(sm)}\right)$$

$$= \lim_{K \to \infty} \left(\frac{l_0}{\sqrt{D_z(0)}} - \frac{l_0 - m_z(q)}{\sqrt{D_z(q)}}\right) = k_{12}q. \quad (4.5.17)$$

Then, taking into account (4.3.25), ARE $e_{MSI(sm),MF}$ of detection algorithm (4.5.1) based on MSI concerned with semimetric in relation to classic detection algorithm

(4.1.7) is equal to squared ratio of limit differences of quantiles x_F, x_D of the corresponding distributions:

$$e_{MSI(sm),MF} = \left(\frac{\lim\limits_{K\to\infty} \Delta_{MSI(sm)}}{\Delta_{MF}} \right)^2 = k_{12}^2 = \frac{8}{\pi^2}. \qquad (4.5.18)$$

Formula (4.5.18) implies that in the case of infinitely large sample and normal distribution of interference (noise), detection algorithm (4.5.1) based on MSI concerned with semimetric is inferior to classic linear detection algorithm (4.1.7) in

$$|10\lg(k_{12}^2)| = |10\lg\left[\tfrac{8}{\pi^2}\right]| \approx 0.912 \text{ dB}.$$

Thus, in arbitrary interference (noise) distribution, the loss in signal energy for algorithm (4.5.1) with respect to classic linear detection algorithm (4.1.7) does not exceed 1 dB.

Conclusions

1. Detection algorithm (4.5.1) based on MSI concerned with semimetric is formulated in terms of L–group operations and can be realized without performing operation of multiplication.

2. The obtained relationships (4.5.15a,b) allow evaluating quality indices of signal detection in the presence of interference (noise) with arbitrary distribution when using algorithm (4.5.1).

3. The obtained relationship (4.5.18) allow evaluating AREs of L–group detection algorithm (4.5.1) operating in the presence of interference (noise) with arbitrary distribution with respect to classic detection algorithm based on matched filter (4.1.7) operating in linear signal space.

4.6 Classification Algorithms in Signal Spaces with L-group Properties

Consider the model of interaction between a signal $s_i(t)$ from a set $S = \{s_i(t)\}$, $i = 1, \ldots, m$ and interference (noise) $n(t)$ in signal space with L-group properties $\mathcal{L}(+, \vee, \wedge)$ which is described by operation of addition «+» of L-group:

$$x(t) = s_i(t) + n(t), \ t \in T_s, \qquad (4.6.1)$$

where $T_s = [t_0, t_0 + T]$ is domain of the signal $s_i(t)$; t_0 is known time of arrival of the signal $s_i(t)$; T is a duration of the signal $s_i(t)$; $m \in \mathbf{N}$, $\mathbf{N}$ is set of natural numbers.

The problem of *classification* of the signals in the presence of interference (noise) lies in making a decision (using some criterion) that allows distinguishing which one from a set of the signals $S = \{s_i(t)\}$, $i = 1, \ldots, m$ is contained in the observed process $x(t)$ [38, 104, 114, 122–128]. Assumptions on the fact that one or another signal is considered to be transmitted are formalized in the form of statistical hypotheses $\{H_i\}$, $i = 1, \ldots, m$. The problem of classification of signals from the set $S = \{s_i(t)\}$, $i = 1, \ldots, m$ in the presence of interference (noise) $n(t)$ is multi-alternative variant of statistical hypothesis testing. In this case, a solution d_k is accepting hypothesis H_k and rejecting the rest of hypotheses H_i, $i \neq k$.

Hereinafter we consider symmetric information transmitting system exploiting deterministic orthogonal signals, so that the following relationships holds:

$$p_{\mathrm{apr}}(s_i) = 1/m; \; \|s_i(t)\|_p = \|s_j(t)\|_p; \; E_i = E_j = E;$$

$$p(s_i/s_j) = \mathbf{P}_{ij} = p(s_j/s_i) = \mathbf{P}_{ji} = \mathbf{P}_{e,1},$$

where $p_{\mathrm{apr}}(s_i)$ is prior probability of receiving the signal $s_i(t)$; E_i is energy of the signal $s_i(t)$; $\|s_i(t)\|_p = \left(\sum\limits_{j=0}^{N-1} |s_i(t_j)|^p\right)^{1/p}$ is p-norm of the signal $s_i(t)$; $p(s_j/s_i)$ is conditional probability of wrong decision on receiving the signal $s_j(t)$ under condition that the signal $s_i(t)$ is received; m is a number of orthogonal signals used in information transmitting system.

The subject of further consideration is quality indices of signal classification algorithms operating in signal spaces with L-group properties in the case of additive interaction between useful and interference (noise) signals.

Signal processing is realized according to generalized algorithm of classification of deterministic orthogonal signals in the signal space with L-group properties which is defined by the following relationships in the assumption on receiving the signal $s_k(t)$:

$$\underset{i \in I; \; s_i(t) \in S}{\arg\max} \; z_i(t = t_0 + T)\big|_{H_k} \overset{d_k}{=} \hat{k}; \tag{4.6.2a}$$

$$z_i(t = t_0 + T) = \mathrm{N}(\mathbf{x}, \mathbf{h}_i) = F[\rho(\mathbf{x}, \mathbf{h}_i)], \tag{4.6.2b}$$

where H_k is hypothesis on receiving the signal $s_k(t)$ within the observed stochastic process $x(t)$; d_k is decision on accepting the hypothesis H_k with choosing the number of processing channel $\hat{k}$, in which maximal value of the signal $z_i(t)$ is observed in the output of matched filter in time instant $t = t_0 + T$; i is a number of processing channel; $\mathrm{N}(\mathbf{x}, \mathbf{h}_i)$ is generalized MSI between the observed stochastic process $\mathbf{x} = [x(t_j)]$ and impulse response $\mathbf{h}_i = [h_i(t_j)]$ of the filter matched with the signal $s_i(t)$; F is some monotone decreasing function; $\rho(\mathbf{x}, \mathbf{h}_i)$ is generalized metric of a signal space in which processing is realized; $j = 0, 1, \ldots, K - 1$.

Find an expression for symbol error probability describing demodulation algorithm based on the relationships (4.6.2a,b) in the assumption on asymptotic normality of random variable $z_i(t)$ (4.6.2b) in the output of generalized matched filter. This supposition is confirmed by the results obtained in the previous sections for corresponding filters.

We assume that, within the model (4.6.1) of interaction between the signals from the set $S = \{s_i(t)\}$, $i = 1, \ldots, m$ and interference (noise) $n(t)$, reception of the signal $s_k(t)$ is realized in the input of a filter:

$$x(t) = s_k(t) + n(t), \; t \in T_s, \tag{4.6.3}$$

so that the signals from the set $S = \{s_i(t)\}$, $i = 1, \ldots, m$ are deterministic orthogonal with equal energies $E_i = E$ and durations $T_i = T$, and interference $n(t)$ is a noise with unknown distribution and PSD N_0 with its upper bound frequency Δf.

Obtain an expression for symbol error probability when classifying orthogonal signals according to the approach stated, for instance, in [129].

Denote joint conditional PDF of random variables $z_i(t)$, $i = 1, \ldots, m$, in the outputs of partial processing channels of demodulator by $p(z_1, \ldots, z_m/s_k)$, so that $t = t_0 + T$ and there is the signal $s_k(t)$ in the observed process (4.6.3). This demodulator realizes algorithm based on the relationships (4.6.2a,b). According to the mentioned algorithm, conditional probability of correct classification is equal to:

$$\mathbf{P}_{kk} = \int\limits_{-\infty}^{\infty} dz_k \int\limits_{-\infty}^{z_k} \cdots \int\limits_{-\infty}^{z_k} p(z_1, \ldots, z_m/s_k) dz_1 \ldots dz_m. \tag{4.6.4}$$

Instantaneous values of stochastic processes $z_i(t)$ (4.6.2b) in the outputs of processing channels $i = 1, \ldots, m$, containing information on MSI in time instant t equal to $t = t_0 + T$, are considered to be independent, jointly Gaussian, with expectations m_{z_i} and variances D_{z_i} that will be determined later depending on concrete type of generalized matched filter (4.6.2b).

Hence, we assume that joint conditional PDF of random variables $z_i(t)$ (4.6.2b), $i = 1, \ldots, m$ in the outputs of partial processing channels is equal to:

$$p(z_1, \ldots, z_m/s_k) = (2\pi D_{z_i})^{-(m-1)/2} \exp\left[-\frac{1}{2D_{z_i}} \sum_{i \neq k} (z_i - m_{z_i})^2\right]$$

$$\times (2\pi D_{z_k})^{-1/2} \exp\left[-\frac{1}{2D_{z_k}} (z_k - m_{z_k})^2\right]. \tag{4.6.5}$$

Substituting (4.6.5) into (4.6.4) along with successive changing a variable:

$$x = (z_k - m_{z_i})/\sqrt{D_{z_i}}; \; z_k = \sqrt{D_{z_i}} x + m_{z_i}, \tag{4.6.6}$$

we obtain:

$$\mathbf{P}_{kk} = (2\pi D_{z_k})^{-1/2} \int\limits_{-\infty}^{\infty} \exp\left[-\frac{1}{2D_{z_k}} (z_k - m_{z_k})^2\right] dz_k (2\pi D_{z_i})^{-(m-1)/2}$$

$$\times \underbrace{\int\limits_{-\infty}^{z_k} \cdots \int\limits_{-\infty}^{z_k} \exp\left[-\frac{1}{2D_{z_i}} \sum_{i \neq k} (z_i - m_{z_i})^2\right] dz_1 \ldots dz_m}_{m-1}$$

$$= (2\pi D_{z_k})^{-1/2} \int\limits_{-\infty}^{\infty} \exp\left[-\frac{1}{2D_{z_k}} (z_k - m_{z_k})^2\right] \Phi^{m-1}(x) dz_k$$

$$= \frac{1}{\sqrt{2\pi}} \sqrt{\frac{D_{z_i}}{D_{z_k}}} \int\limits_{-\infty}^{\infty} \exp\left[-\frac{1}{2}\left(\frac{z_k}{\sqrt{D_{z_k}}} - \frac{m_{z_k}}{\sqrt{D_{z_k}}} \right)^2 \right] \Phi^{m-1}(x)dx$$

$$= \frac{1}{\sqrt{2\pi}} \sqrt{\frac{D_{z_i}}{D_{z_k}}} \int\limits_{-\infty}^{\infty} \exp\left[-\frac{1}{2}\left(\frac{\sqrt{D_{z_i}}}{\sqrt{D_{z_k}}} x - \frac{(m_{z_k} - m_{z_i})}{\sqrt{D_{z_k}}} \right)^2 \right] \Phi^{m-1}(x)dx, \quad (4.6.7)$$

where $\Phi(x) = \frac{1}{\sqrt{2\pi}} \int\limits_{-\infty}^{x} \exp\left[-u^2/2 \right] du$ is probability integral.

According to the last variant of formula for probability of correct classification $\mathbf{P}_{kk}$ (4.6.7), the expression for symbol error probability $\mathbf{P}_s = 1 - \mathbf{P}_{kk}$ of classification of orthogonal signals in signal space with L-group properties has the following appearance:

$$\mathbf{P}_s = \sum_{i=1}^{m} (1 - \mathbf{P}_{ii}) p_{\mathrm{apr}}(s_i) = 1 - \mathbf{P}_{kk} =$$

$$= 1 - \frac{1}{\sqrt{2\pi}} \sqrt{\frac{D_{z_i}}{D_{z_k}}} \int\limits_{-\infty}^{\infty} \exp\left[-\frac{1}{2}\left(\frac{\sqrt{D_{z_i}}}{\sqrt{D_{z_k}}} x - \frac{(m_{z_k} - m_{z_i})}{\sqrt{D_{z_k}}} \right)^2 \right] \Phi^{m-1}(x)dx, \quad (4.6.8)$$

where m_{z_i}, m_{z_k} and D_{z_i}, D_{z_k} are expectations and variances in i-th and k-th processing channels under condition of receiving the signal $s_k(t)$.

In the case if generalized algorithm of signal classification (4.6.2) assumes exploiting classic matched filter (4.1.7) that calculates generalized MSI $N_{l_2}(\mathbf{x}, \mathbf{h}_i)$ in the form of scalar product between two vectors $\mathbf{x}, \mathbf{h}_i$ in each processing channel (see Table 4.2.1):

$$\arg\max_{i \in I;\, s_i(t) \in S} y_{MF,i}(t = t_0 + T)\big|_{H_k} \overset{d_k}{=} \hat{k}; \quad (4.6.9a)$$

$$y_{MF}(t_j) = N_{l_2}(\mathbf{x}, \mathbf{h}_i) = \sum_{i=0}^{j} x(t_i) h(t_{j-i}), \quad (4.6.9b)$$

then as follows from the relationship (4.6.8), when classifying the deterministic orthogonal signals in the presence of additive white Gaussian noise, symbol error probability is defined by the relationship:

$$\mathbf{P}_s(q) = 1 - \frac{1}{\sqrt{2\pi}} \int\limits_{-\infty}^{\infty} \exp\left[-\frac{1}{2}\left(x - \sqrt{2q^2} \right)^2 \right] \Phi^{m-1}(x)dx, \quad (4.6.10)$$

so that expectations and variances from the formula (4.6.8) are, respectively, equal to: $m_{z_i} = 0$, $m_{z_k} = 2q^2 = 2E/N_0$; $D_{z_i} = D_{z_k} = 2q^2 = 2E/N_0$; $q^2 = E/N_0$ is SNR.

The result (4.6.10) for symbol error probability $\mathbf{P}_s(q)$ is well known and its upper bound can be approximated by the expression [38, (4.107)]:

$$\mathbf{P}_s(q) \leqslant (m - 1)[1 - \Phi(q)].$$

Taking into account the known relationship $q^2 = q_b^2 \log_2 m$ that associates energy per symbol with energy per bit, the expression for bit error probability $\mathbf{P}_b(q)$ describing classic algorithm of signal classification (4.6.9) takes the form:

$$\mathbf{P}_b(q) = 1 - \frac{1}{\sqrt{2\pi}} \int\limits_{-\infty}^{\infty} \exp\left[-\frac{1}{2}\left(x - \sqrt{2q_b^2 \log_2 m}\right)^2\right] \Phi^{m-1}(x)dx, \qquad (4.6.11)$$

where $q_b^2 = E_b/N_0 = E/(N_0 \log_2 m) = q^2/\log_2 m$; E_b is energy per bit of the signal $s_i(t)$; E is signal energy (energy per symbol) of the signal $s_i(t)$; N_0 is PSD of interference (noise); m is a number of orthogonal signals used in information transmitting system.

Further, using the relationship (4.6.8) and also the results obtained within the Sections 4.3...4.5, we find expressions determining symbol and bit error probabilities describing algorithms of orthogonal signal classification based on generalized MSIs (4.6.2). To obtain symbol error probability (4.6.8), taking into account the known interrelation between signal detection and classification, we use the results obtained within Sections 4.3...4.5, that allows essentially reducing the stated material not going into details.

4.6.1 Classification Algorithm Based on Measure of Statistical Interrelation Concerned with Pseudometric

Algorithm of classification of deterministic orthogonal signals based on measure of statistical interrelation (1.4.10) concerned with pseudometric (1.3.25) is defined according to the general expression (4.6.2) by the following relationships:

$$\underset{i \in I;\, s_i(t) \in S}{\arg\max}\; z_{p,i}(t = t_0 + T)|_{H_k} \overset{d_k}{=} \hat{k}; \qquad (4.6.12a)$$

$$z_{p,i}(t = t_0 + T) = \mathrm{N}_p(\mathbf{x}, \mathbf{h}_i) = F[\rho_p(\mathbf{x}, \mathbf{h}_i)]; \qquad (4.6.12b)$$

$$z_{p,i}(t_j) = \sum_{l=0}^{j} [1 - (\mathrm{sgn}(h_i(t_{j-l}) \vee x(t_l)) - \mathrm{sgn}(h_i(t_{j-l}) \wedge x(t_l)))], \qquad (4.6.12c)$$

where variables from these relationships are similar to those that figure in formulas (4.6.2a,b).

Taking into account the known interrelation between the signal and impulse response of matched filter $h_i(t_j) = s_i(t_0 + T - t_j)$, filter type classification algorithm (4.6.12c) is equivalent to correlation type classification algorithm that is defined by the expression:

$$z_{p,i}^C(t_j) = \sum_{l=0}^{j} [1 - (\mathrm{sgn}(s_i(t_l) \vee x(t_l)) - \mathrm{sgn}(s_i(t_l) \wedge x(t_l)))], \qquad (4.6.13)$$

so that the identity holds:

$$z_{p,i}(t_j) = z_{p,i}^C(t_j),\; t_j = t_0 + T. \qquad (4.6.14)$$

By the analogy with (4.3.6), taking into account the relation (4.3.4), within the initial relationship (4.6.13) we introduce auxiliary variables corresponding to a shift of coordinate system origin into a point t_0:

$$z_i = \sum_{l=0}^{K-1} y_{i,l} = \sum_{l=0}^{K-1} \operatorname{sgn}(x_l) \operatorname{sgn}(s_{i,l}); \qquad (4.6.15a)$$

$$y_{i,l} = \operatorname{sgn}(x_l) \operatorname{sgn}(s_{i,l}); \qquad (4.6.15b)$$

$$t_l = t_j - t_0; \; l = 0, 1, \ldots, K-1; \; t_{K-1} = T; \; K-1 = T/\Delta t; \; i = 1, 2, \ldots, m, \quad (4.6.15c)$$

where t_0 and $t_0 + T$ are time of signal arrival end ending, respectively (see formula (4.6.1); T is a duration of signal $s_i(t)$; Δt is a sampling interval providing independence of interference (noise) samples $\{n(t_j)\}$ within the model (4.6.1).

Taking into account the relationships (4.3.18), (4.3.19), in the case of receiving the signal $s_k(t)$ (4.6.3), the expressions for expectations m_{z_i}, m_{z_k} and variances D_{z_i}, D_{z_k} of random variables z_i, z_k (4.6.15a) in the outputs of i-th and k-th processing channels can be written in the following form, respectively:

$$m_{z_i}(q) = 0; \qquad (4.6.16a)$$

$$m_{z_k}(q) = 2p_n(0) \|s_k\|_1 = 2p_n(0)\sqrt{D_n} F_{12} \cdot \sqrt{2}q$$
$$= 2\sqrt{2}p_n(0)\sqrt{D_n} k_{12}\sqrt{K} \cdot q; \quad (4.6.16b)$$

$$D_{z_i}(q) = D_{z_k}(q) = K - 4p_n^2(0) \|s_k\|_2^2 = K - 8p_n^2(0)D_n q^2, \qquad (4.6.17)$$

where $q^2 = E/N_0$ is SNR (it differs from SNR $q^2 = 2E/N_0$ used in the relationships (4.3.18), (4.3.19)); $p_n(x)$ is PDF of interference (noise) $n(t)$; D_n is a variance of interference (noise) $n(t)$; $\|s_k\|_1$ is l_1-norm of the signal $s_k(t)$; F_{12} is a norm-factor (4.1.12), (4.1.13b); k_{12} is coefficient (4.1.13a); K is a sample size used when forming the sum (4.6.15a).

Substituting expectations m_{z_i}, m_{z_k} and variances D_{z_i}, D_{z_k} into the formula (4.6.8), we obtain the resulting expression for symbol error probability $\mathbf{P}_s(q)$ describing L-group algorithm of orthogonal signals classification (4.6.12):

$$\mathbf{P}_s(q) = 1 - \frac{1}{\sqrt{2\pi}}$$
$$\times \int_{-\infty}^{\infty} \exp\left[-\frac{1}{2}\left(x - \frac{2\sqrt{2}p_n(0)\sqrt{D_n} k_{12}\sqrt{K} \cdot q}{\sqrt{K - 8p_n^2(0)D_n q^2}}\right)^2\right] \Phi^{m-1}(x)dx. \quad (4.6.18)$$

Then, taking into account known relationship $q^2 = q_b^2 \log_2 m$, the expression for bit error probability $\mathbf{P}_b(q_b)$ describing orthogonal signal classification algorithm

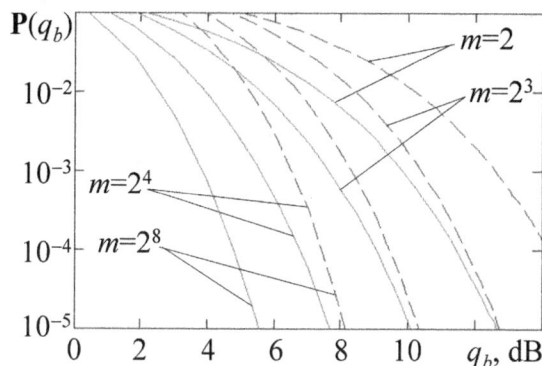

FIGURE 4.6.1   Dependences of bit error probability $\mathbf{P}_b(q_b)$ for L-group algorithm (4.6.12) (dashed line) and classic algorithm (4.6.9) (solid line)

(4.6.12) takes the form:

$$\mathbf{P}_b(q_b) = 1 - \frac{1}{\sqrt{2\pi}}$$

$$\times \int\limits_{-\infty}^{\infty} \exp\left[-\frac{1}{2}\left(x - \frac{2\sqrt{2}p_n(0)\sqrt{D_n}k_{12}\sqrt{Kq_b^2\log_2 m}}{\sqrt{K - 8p_n^2(0)D_nq_b^2\log_2 m}}\right)^2\right]\Phi^{m-1}(x)dx. \quad (4.6.19)$$

Fig. 4.6.1 depicts the dependences of bit error probability $\mathbf{P}_b(q_b)$ on SNR per bit $q_b^2 = E_b/N_0$, $E_b = E/\log_2 m$, $q^2 = q_b^2\log_2 m$, obtained by known relationship (4.6.11) (solid line), and also calculated by the formula (4.6.19) (dashed line) for orthogonal signals from the set $S = \{s_i(t)\}$, $i = 1, \ldots, m$, $m = 2, 4, 16, 256$. Dependences shown in Fig. 4.6.1 are obtained in assumption on normality of interference (noise) distribution. Noise-immunity of demodulator classifying orthogonal signals in signal space with L-group properties is characterized by loss that does not exceed 3 dB in necessary SNR with respect to classic result (4.6.11) that corresponds to results obtained in Section 4.3.

4.6.2   Classification Algorithm Based on Measure of Statistical Interrelation Concerned with l_1-metric

According to general expression (4.6.2), algorithm of classification of deterministic orthogonal signals based on MSI (1.4.12) concerned with l_1-metric (1.3.27) is defined by the following relationships:

$$\underset{i\in I; s_i(t)\in S}{\arg\max}\; z_{l_1,i}(t = t_0 + T)|_{H_k} \overset{d_k}{=} \hat{k}; \quad (4.6.20a)$$

$$z_{l_1,i}(t = t_0 + T) = N_{l_1}(\mathbf{x}, \mathbf{h}_i) = F[\rho_{l_1}(\mathbf{x}, \mathbf{h}_i)]; \quad (4.6.20b)$$

$$z_{l_1,i}(t_j) = \sum_{l=0}^{j} [|x(t_l) + h(t_{j-l})| - |x(t_l) - h(t_{j-l})|], \quad (4.6.20c)$$

where variables figuring in these relationships are similar to the same variables that figure in formulas (4.6.2a,b).

Taking into account the known relation between a signal and impulse response of matched filter $h_i(t_j) = s_i(t_0 + T - t_j)$, filter type classification algorithm (4.6.20c) is equivalent to correlation type classification algorithm that is determined by the expression:

$$z^C_{l_1,i}(t_j) = \sum_{l=0}^{j} [|x(t_l) + s_i(t_l)| - |x(t_l) - s_i(t_l)|], \tag{4.6.21}$$

so that the following identity holds:

$$z_{l_1,i}(t_j) = z^C_{l_1,i}(t_j), \ t_j = t_0 + T. \tag{4.6.22}$$

By the analogy with (4.4.5), for convenience of analysis, within the initial relationship (4.6.21), we introduce auxiliary variables corresponding to a shift of coordinate system origin into a point t_0:

$$z_i = \sum_{l=0}^{K-1} y_{i,l} = \sum_{l=0}^{K-1} [|x_l + s_{i,l}| - |x_l - s_{i,l}|]; \tag{4.6.23a}$$

$$y_{i,l} = |x_l + s_{i,l}| - |x_l - s_{i,l}|; \tag{4.6.23b}$$

$$t_l = t_j - t_0; \ l = 0, 1, \ldots, K-1; \ t_{K-1} = T; \ K-1 = T/\Delta t; \ i = 1, 2, \ldots, m, \tag{4.6.23c}$$

where t_0 and $t_0 + T$ are time of signal arrival and ending, respectively (see the formula (4.6.1); T is a duration of the signal $s_i(t)$; Δt is a sampling interval providing independence of interference samples $\{n(t_j)\}$ within the model (4.6.1).

According to relationships (4.4.25) and (4.4.28), in the case of receiving the signal $s_k(t)$ (4.6.3), the expressions for expectations m_{z_i}, m_{z_k} and variances D_{z_i}, D_{z_k} of random variables z_i, z_k (4.6.23a) in the outputs of i-th and k-th processing channels can be written in the following form, respectively:

$$m_{z_i}(q) = 0; \tag{4.6.24a}$$

$$m_{z_k}(q) = 2p_n(0) \cdot \|s_k\|^2_2 = 4p_n(0) \cdot D_n q^2; \tag{4.6.24b}$$

$$D_{z_i}(q) = D_{z_k}(q) = \|s_k\|^2_2 - \tfrac{4}{3}p_n(0)\|s_k\|^3_3 - 4p^2_n(0)\|s_k\|^4_4$$

$$= 2D_n q^2 - \tfrac{8\sqrt{2}}{3}p_n(0)\frac{D_n^{3/2}q^3}{K^{1/2}k^3_{23}} - 16p^2_n(0)\frac{D_n^2 q^4}{K \cdot k^4_{24}}, \tag{4.6.25}$$

where $q^2 = E/N_0$ is SNR; $p_n(x)$ is PDF of interference (noise) $n(t)$; D_n is a variance of interference (noise) $n(t)$; $\|s_k\|_2$, $\|s_k\|_3$, $\|s_k\|_4$ are l_2-, l_3-, and l_4-norms of the signal $s_k(t)$, respectively; k_{23}, k_{24} are coefficients (4.1.13a); K is a sample size that used when forming the sum (4.6.23a).

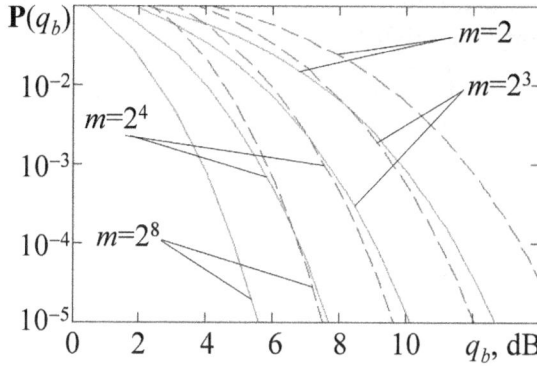

FIGURE 4.6.2 Dependences of bit error probability $\mathbf{P}_b(q_b)$ for L-group algorithm (4.6.20) (dashed line) and classic algorithm (4.6.9) (solid line)

Substituting the values of expectations m_{z_i}, m_{z_k} and variances D_{z_i}, D_{z_k} into the formula (4.6.8), we obtain the resulting expression for symbol error probability $\mathbf{P}_s(q)$ describing L-group algorithm of orthogonal signal classification (4.6.20):

$$\mathbf{P}_s(q) = 1 - \frac{1}{\sqrt{2\pi}} \int\limits_{-\infty}^{\infty} \exp\left[-\frac{1}{2}(x - A(q))^2\right] \Phi^{m-1}(x) dx; \qquad (4.6.26)$$

$$A(q) = \frac{4p_n(0) \cdot D_n q^2}{\sqrt{2D_n q^2 - \frac{8\sqrt{2}}{3} p_n(0) \frac{D_n^{3/2} q^3}{K^{1/2} k_{23}^3} - 16 p_n^2(0) \frac{D_n^2 q^4}{K \cdot k_{24}^4}}}. \qquad (4.6.26a)$$

Then, taking into account the relationship $q^2 = q_b^2 \log_2 m$, the expression for bit error probability $\mathbf{P}_b(q_b)$ describing L-group algorithm of orthogonal signal classification (4.6.20) takes the form:

$$\mathbf{P}_b(q_b) = 1 - \frac{1}{\sqrt{2\pi}} \int\limits_{-\infty}^{\infty} \exp\left[-\frac{1}{2}(x - B(q_b))^2\right] \Phi^{m-1}(x) dx; \qquad (4.6.27)$$

$$B(q_b) = \frac{4p_n(0) \cdot D_n q_b^2 \log_2 m}{\sqrt{2D_n q_b^2 \log_2 m - \frac{8\sqrt{2}}{3} p_n(0) \frac{D_n^{3/2} (q_b^2 \log_2 m)^{1.5}}{K^{1/2} k_{23}^3} - 16 p_n^2(0) \frac{D_n^2 (q_b^2 \log_2 m)^2}{K \cdot k_{24}^4}}}. \qquad (4.6.27a)$$

Fig. 4.6.2 illustrates the dependences of bit error probability $\mathbf{P}_b(q_b)$ on SNR per bit $q_b^2 = E_b/N_0$, $E_b = E/\log_2 m$, $q^2 = q_b^2 \log_2 m$, calculated by known relationship (4.6.11) (solid line), and also obtained according to the formula (4.6.27) (dashed line) for orthogonal signals from the set $S = \{s_i(t)\}$, $i = 1, \ldots, m$, $m = 2, 4, 16, 256$. Dependences shown in Fig. 4.6.2 are obtained in the assumption on interference (noise) Gaussianity. Noise immunity of L-group demodulator, classifying orthogonal signals in signal space with L-group properties according to algorithm (4.6.20), is characterized by loss that does not exceed 2 dB in SNR with respect to classic result (4.6.11), that corresponds to the results obtained within Section 4.4.

4.6.3   Classification Algorithm Based on Measure of Statistical Interrelation Concerned with Semimetric

Classification algorithm, intended for distinguishing deterministic orthogonal signals and based on measure of statistical interrelation (1.4.11) concerned with semimetric (1.3.26), is defined according to general expression (4.6.2) by the following relationships:

$$\arg\max_{i\in I;\, s_i(t)\in S} z_{s,i}(t = t_0 + T)|_{H_k} \stackrel{d_k}{=} \hat{k}; \tag{4.6.28a}$$

$$z_{s,i}(t = t_0 + T) = \mathrm{N}_s(\mathbf{x}, \mathbf{h}_i) = F[\rho_s(\mathbf{x}, \mathbf{h}_i)]; \tag{4.6.28b}$$

$$z_{s,i}(t_j) = \sum_{l=0}^{j} [|x(t_l)| + |h_i(t_{j-l})|] - \sum_{l=0}^{j} [|x(t_l) + h_i(t_{j-l})| - |x(t_l) - h_i(t_{j-l})|]$$
$$\times (\mathrm{sgn}(h_i(t_{j-l}) \vee x(t_l)) - \mathrm{sgn}(h_i(t_{j-l}) \wedge x(t_l))), \tag{4.6.28c}$$

where variables situated in these relationships are similar to the same variables figuring in the formulas (4.6.2a,b).

Taking into account the known relation between a signal and impulse response of matched filter $h_i(t_j) = s_i(t_0 + T - t_j)$, filter type classification algorithm (4.6.28c) is equivalent to correlation type classification algorithm that is defined by the expression:

$$z_{s,i}^C(t_j) = \sum_{l=0}^{j} [|x(t_l)| + |s_i(t_l)|] - \sum_{l=0}^{j} [|x(t_l) + s_i(t_l)| - |x(t_l) - s_i(t_l)|]$$
$$\times (\mathrm{sgn}(s_i(t_l) \vee x(t_l)) - \mathrm{sgn}(s_i(t_l) \wedge x(t_l))), \tag{4.6.29}$$

so that the identity holds:

$$z_{s,i}(t_j) = z_{s,i}^C(t_j), \quad t_j = t_0 + T. \tag{4.6.30}$$

By the analogy with (4.5.5), taking into account the relation (4.5.4), within the initial relationship (4.6.29), we introduce auxiliary variables corresponding to a shift of coordinate system origin into a point t_0:

$$z_i = \sum_{k=0}^{K-1} y_{i,l} = \sum_{l=0}^{K-1} [s_{i,l} \cdot \mathrm{sgn}(x_l) + x_l \cdot \mathrm{sgn}(s_{i,l})]; \tag{4.6.31a}$$

$$y_{i,l} = s_{i,l} \cdot \mathrm{sgn}(x_l) + x_l \cdot \mathrm{sgn}(s_{i,l}); \tag{4.6.31b}$$

$$t_l = t_j - t_0; \; l = 0, 1, \ldots, K-1; \; t_{K-1} = T; \; K-1 = T/\Delta t; \; i = 1, 2, \ldots, m, \tag{4.6.31c}$$

where t_0 and $t_0 + T$ are time of signal arrival and ending, respectively (see relationship (4.6.1); T is a duration of the signal $s_i(t)$; Δt is a sampling interval providing independence of interference (noise) samples $\{n(t_j)\}$ within the model (4.6.1).

Taking into account the relationships (4.5.13), in the case of receiving the signal $s_k(t)$ (4.6.3), the expressions for expectations m_{z_i}, m_{z_k} and variances D_{z_i}, D_{z_k} of random variables z_i, z_k (4.6.31a) in the outputs of i-th and k-th processing channels can be written in the following form, respectively:

$$m_{z_i}(q) = 0; \qquad (4.6.32a)$$

$$m_{z_k}(q) = F_{12}\,\|s_k\|_2 + 2p_n(0)\cdot\|s_k\|_2^2 = k_{12}\sqrt{2D_nKq^2} + 4p_n(0)D_nKq^2; \quad (4.6.32b)$$

$$D_{z_i}(q) = D_{z_k}(q) = \left(\|s_k\|_2^2 + \|n\|_2^2\right)\left(1 - 4p_n^2(0)\frac{\|s_k\|_2^2}{K}\right)$$
$$= D_nK\left(1 + \frac{2q^2}{K}\right)\left(1 - 8p_n^2(0)\frac{D_nq^2}{K}\right). \quad (4.6.33)$$

where $q^2 = E/N_0$ is SNR; $p_n(x)$ is PDF of interference (noise) $n(t)$; D_n is a variance of interference (noise) $n(t)$; $\|s_k\|_2$ is l_2-norm of the signal $s_k(t)$; k_{12} is coefficient (4.1.13a); K is a sample size that is used when forming the sum (4.6.31a).

Substituting the values of expectations m_{z_i}, m_{z_k} (4.6.32) and variances D_{z_i}, D_{z_k} (4.6.33) into the formula (4.6.8), we obtain the resulting expression for symbol error probability $\mathbf{P}_s(q)$ describing L-group algorithm of orthogonal signal classification (4.6.28):

$$\mathbf{P}_s(q) = 1 - \frac{1}{\sqrt{2\pi}}\int_{-\infty}^{\infty}\exp\left[-\frac{1}{2}(x - A(q))^2\right]\Phi^{m-1}(x)dx; \qquad (4.6.34)$$

$$A(q) = \frac{k_{12}\sqrt{2D_nKq^2} + 4p_n(0)D_nKq^2}{D_nK\left(1 + \frac{2q^2}{K}\right)\left(1 - 8p_n^2(0)\frac{D_nq^2}{K}\right)}. \qquad (4.6.34a)$$

Then, taking into account the relation $q^2 = q_b^2\log_2 m$, the expression for bit error probability $\mathbf{P}_b(q_b)$ when realizing L-group algorithm of orthogonal signal classification (4.6.28) takes the form:

$$\mathbf{P}_b(q_b) = 1 - \frac{1}{\sqrt{2\pi}}\int_{-\infty}^{\infty}\exp\left[-\frac{1}{2}(x - A(q_b))^2\right]\Phi^{m-1}(x)dx; \qquad (4.6.35)$$

$$A(q_b) = \frac{k_{12}\sqrt{2D_nKq_b^2\log_2 m} + 4p_n(0)D_nKq_b^2\log_2 m}{D_nK\left(1 + \frac{2q_b^2\log_2 m}{K}\right)\left(1 - 8p_n^2(0)\frac{D_nq_b^2\log_2 m}{K}\right)}. \qquad (4.6.35a)$$

Fig. 4.6.3 depicts the dependences of bit error probability $\mathbf{P}_b(q_b)$ on SNR per bit $q_b^2 = E_b/N_0$, $E_b = E/\log_2 m$, $q^2 = q_b^2\log_2 m$, calculated by known relationship (4.6.11) (solid line), and also obtained according to the formula (4.6.35) (dashed line) for orthogonal signals from the set $S = \{s_i(t)\}$, $i = 1,\ldots,m$, $m = 2, 4, 16, 256$. Dependences shown in Fig. 4.6.3 are obtained under the assumption on interference (noise) Gaussianity. Noise immunity of demodulator classifying orthogonal signals in signal space with L-group properties is characterized by loss that does not exceed 1 dB in SNR with respect to classic result (4.6.11), that well corresponds to the results obtained within Section 4.5.

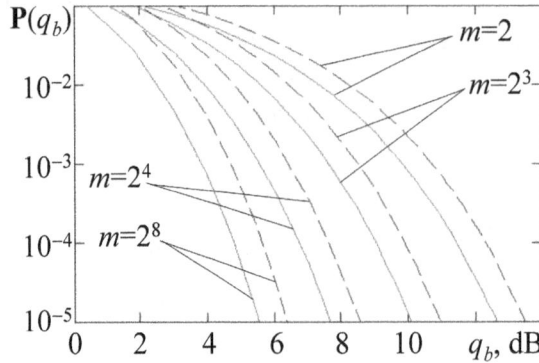

FIGURE 4.6.3   Dependences of bit error probability $\mathbf{P}_b(q_b)$ for L-group algorithm (4.6.28) (dashed line) and classic algorithm (4.6.9) (solid line)

4.6.4   Asymptotic Relative Efficiency of L-group Classification Algorithms

Notice that the symbol error probabilities $\mathbf{P}_s(q)$ for both classic algorithm of signal classification (4.6.9) and algorithms of signal classification in space with L-group properties (4.6.12), (4.6.20), (4.6.28) are described similarly by the function $F(u, m)$:

$$\mathbf{P}_s(q) = 1 - F(f(q), m); \tag{4.6.36}$$

$$F(q, m) = \frac{1}{\sqrt{2\pi}} \int\limits_{-\infty}^{\infty} \exp\left[-\frac{1}{2}(x - f(q))^2\right] \Phi^{m-1}(x) dx. \tag{4.6.36a}$$

that possesses the properties of CDF:

$$\lim_{u \to -\infty} F(u, m) = 0; \quad \lim_{u \to \infty} F(u, m) = 1; \tag{4.6.36b}$$

$$F(u'', m) \geqslant F(u', m), \text{ if } u'' \geqslant u'. \tag{4.6.36c}$$

Besides, the equality holds:

$$F(0, m) = 1/m.$$

For classic algorithm of signal classification (4.6.9) and algorithms of signal classification in spaces with L-group properties (4.6.12), (4.6.20), (4.6.28), taking into account the relationships (4.6.10) and (4.6.18), (4.6.26), (4.6.34), the functions $f(q)$ figuring in compound function $F(q, m)$ (4.6.36a) are, respectively, equal to:

$$f_{MF}(q) = \sqrt{2q^2}; \tag{4.6.37}$$

$$f_P(q) = \frac{2\sqrt{2}p_n(0)\sqrt{D_n}k_{12}\sqrt{K} \cdot q}{\sqrt{K - 8p_n^2(0)D_n q^2}}; \tag{4.6.38a}$$

$$f_{l_1}(q) = \cfrac{4p_n(0) \cdot D_n q^2}{\sqrt{2D_n q^2 - \frac{8\sqrt{2}}{3}p_n(0)\frac{D_n^{3/2}q^3}{K^{1/2}k_{23}^3} - 16p_n^2(0)\frac{D_n^2 q^4}{K \cdot k_{24}^4}}}; \qquad (4.6.38b)$$

$$f_s(q) = \frac{k_{12}\sqrt{2D_n K q^2} + 4p_n(0)D_n K q^2}{D_n K \left(1 + \frac{2q^2}{K}\right)\left(1 - 8p_n^2(0)\frac{D_n q^2}{K}\right)}, \qquad (4.6.38c)$$

where $q^2 = E/N_0$ is SNR; $p_n(x)$ is PDF of interference (noise) $n(t)$; D_n is a variance of interference (noise) $n(t)$; k_{12}, k_{23}, k_{24} are coefficients (4.1.13a); K is a sample size that is used when forming the sums (4.6.15a), (4.6.23a), (4.6.31a).

Then, taking into account the relationship (4.6.36) that is common for symbol error probabilities $\mathbf{P}_s(q)$ (4.6.10) and (4.6.18), (4.6.26), (4.6.34), asymptotic relative efficiency (ARE) $e_{MSI,MF}$ of algorithm of orthogonal signal classification in space with L-group properties with respect to classic algorithm of signal classification (4.6.9) is defined by the expression:

$$e_{MSI,MF} = \lim_{K \to \infty}\left(\frac{f_{MSI}(q)}{f_{MF}(q)}\right)^2, \qquad (4.6.39)$$

where $f_{MSI}(q)$ is one of the functions $f_p(q)$, $f_{l_1}(q)$, $f_s(q)$ determined by the relationships (4.6.38a,b,c), respectively.

Substituting the functions $f_p(q)$ (4.6.38a), $f_{l_1}(q)$ (4.6.38b), $f_s(q)$ (4.6.38c) into the expression (4.6.39) and calculating the limit, we obtain AREs of algorithms of deterministic orthogonal signal classification in space with L-group properties (4.6.12), (4.6.20), (4.6.28):

$$e_{MSI(pm),MF} = 4p_n^2(0)D_n k_{12}^2; \qquad (4.6.40a)$$

$$e_{MSI(m),MF} = 4p_n^2(0)D_n; \qquad (4.6.40b)$$

$$e_{MSI(sm),MF} = k_{12}^2 = 8/\pi^2, \qquad (4.6.40c)$$

Comparing the obtained relationships with the expressions determining AREs of algorithms of signal detection (4.3.27), (4.4.32), (4.5.18) in space with L-group properties, as can be seen, they completely correspond to each other. Remind that AREs numerical values of detection algorithms based on MSIs with respect to classic linear detection algorithm (4.1.7) for different interference (noise) distributions from the Table 2.6.1 are shown in the Table 4.4.1. Formula (4.6.40c) implies that useful signal energy loss of L-group algorithm (4.6.28) with respect to classic linear algorithm (4.6.9) does not exceed 1 dB in the presence of interference (noise) with arbitrary distribution.

Conclusions

1. Signal classification algorithms (4.6.12), (4.6.20), (4.6.28), based on MSI concerned with pseudometric, l_1-metric, and semimetric, respectively, are formulated in terms of L–group operations, and do not require performing operation of multiplication.

2. The obtained relationships (4.6.19), (4.6.27), (4.6.35) allow evaluating bit error probability describing the discussed *L*–group algorithms of orthogonal signals classification (4.6.12), (4.6.20), (4.6.28), respectively, operating in the presence of interference (noise) with arbitrary distribution.

3. The obtained relationships (4.6.40a,b,c) allow evaluating AREs of *L*–group classification algorithms (4.6.12), (4.6.20), (4.6.28) with respect to classic algorithm of signal classification (4.6.9) operating in the presence of interference (noise) with arbitrary distribution.

4.7 Algorithms of Signal Resolution in Space with *L*-group Properties

4.7.1 Ambiguity Function and Mismatching Function of Signals

Ambiguity function [130] introduced by P. M. Woodward as a generalized characteristic of resolution accurately describes the properties of all the signals applied in practice. Ambiguity function is a bivariate function $\Psi(\tau, F)$ of time delay τ and Doppler frequency shift F that is defined by the following relationships [71,72,131]:

$$\Psi(\tau, F) = |\psi(\tau, F)| = \left| \frac{1}{2E} \int\limits_{-\infty}^{\infty} \dot{S}(t)\overline{\dot{S}(t-\tau)} \exp(-j2\pi Ft)\,\mathrm{d}t \right|; \qquad (4.7.1)$$

$$\dot{S}(t) = S(t)\exp(j\varphi(t)); \qquad (4.7.1a)$$

$$s(t) = S(t)\exp(j\Phi(t)); \qquad (4.7.1b)$$

$$\Phi(t) = \omega_0 t + \varphi(t); \qquad (4.7.1c)$$

$$s_R(t) = \mathrm{Re}[s(t)]; \quad s_I(t) = \mathrm{Im}[s(t)]; \qquad (4.7.1d)$$

$$s_I(t) = s_R(t) * \frac{1}{\pi t} = \mathcal{H}[s_R(t)];$$
$$s_R(t) = -s_I(t) * \frac{1}{\pi t} = \mathcal{H}^{-1}[s_I(t)]; \qquad (4.7.1e)$$

$$E = \frac{1}{2} \int\limits_{-\infty}^{\infty} |\dot{S}(t)|^2 \,\mathrm{d}t, \qquad (4.7.1f)$$

where $\dot{S}(t)$, $S(t)$ are complex envelope and envelope of analytic signal $s(t)$, $S(t) = \sqrt{s_R^2(t) + s_I^2(t)}$, respectively; $\Phi(t)$ is instantaneous phase of the analytic signal $s(t)$, $\Phi(t) = \mathrm{arctg}[s_I(t)/s_R(t)]$; $\omega_0 = 2\pi f_0$, f_0 is a central frequency of signal spectrum; $s_R(t)$, $s_I(t)$ are real signal and the signal coupled with it by Hilbert transform $\mathcal{H}$ (4.7.1e), respectively; E is an energy of real signal $s_R(t)$; $*$ is operation of convolution.

Taking into account known interrelation between signal resolution in parameters $\lambda = [\tau, F]^T$ and their estimation, we notice that in the case of preliminary averaging ambiguity function over uniformly distributed initial phase of the signal $s(t)$, elements of Fisher information matrix $\boldsymbol{\Phi} = \|\Phi_{ik}\|$, $i, k \in \{1, 2\}$ (see, for instance, [97, (3.21)], [95, (4.114)]) are calculated using the relationship [132, (14.10)], [133, (4.59),(4.58)]:

$$\Phi_{ik} = -q^2 \frac{\partial^2}{\partial \lambda_i \partial \lambda_k} \Psi(\lambda), \tag{4.7.2}$$

where $\lambda = [\lambda_1, \lambda_2]^T = [\tau, F]^T$ is a vector of estimated (resolution) parameters; $\Psi(\lambda) = \Psi(\tau, F)$ is ambiguity function determined by the relationship (4.7.1); $q^2 = 2E/N_0$ is SNR, E, N_0 are energy of real signal $s_R(t)$ and power spectral density (PSD) of noise, respectively.

Thus, from the standpoint of *Signal Theory*, ambiguity function is a convenient mean for analyzing characteristics of signal resolution and signal parameter estimation. However, utilizing ambiguity function as an analytical tool for theoretical investigations of signal resolution problem in non-Euclidean spaces in general and in signal spaces with lattice properties in particular is impossible due to specificity of the properties of the signals and the algorithms of their processing in these signal spaces. Along with a "signal" approach to exploring signal resolution based on ambiguity function (4.7.1), there exist so-called "filter" approach that relies on the notion of *mismatching function* and represents the standpoint of *Signal Processing Theory*.

Parameter λ' of received signal $s'(t, \lambda')$ is usually mismatched with respect to parameter λ of the expected useful signal $s(t, \lambda)$ matched with some filter used for signal processing. The effects of mismatching take place during signal detection and exert influence on both signal resolution and estimation of signal parameters. Mismatching can be evaluated over the signal $w(\lambda', \lambda)$ in the output of signal processing unit matched with the expected signal $s(t, \lambda)$. When a received signal $s'(t, \lambda')$ with a mismatched value of parameter λ' is present in the input of the processing unit in the absence of interference (noise), the output response $w(\lambda', \lambda)$ is called a mismatching function [132, 133]:

$$w(\lambda', \lambda) = w(\tau, F) = \int\limits_{-\infty}^{\infty} s(t)h(\tau - t)\exp(-j2\pi F t)\,\mathrm{d}t; \tag{4.7.3}$$

$$\tau = t_0' - t_0; \quad F = f' - f, \tag{4.7.3a}$$

where $h(t)$ is impulse response of a filter matched with the expected signal $s(t, \lambda)$; $\lambda' = [t_0', f']^T$, $\lambda = [t_0, f]^T$ are vector parameters: t_0', t_0; f', f are times of arrival and central frequencies of spectra of the received $s'(t, \lambda')$ and expected $s(t, \lambda)$ signals, respectively.

The function determined in such a way is called a *time-frequency mismatching function* $w(\lambda', \lambda)$ of a processing unit, if vector parameter of the expected signal includes two scalar parameters: time of signal arrival t_0' and frequency f' transformed owing to Doppler effect [132]. Vector parameter λ' of received signal is expressed over two scalar parameters $t_0' = t_0 + \tau$ and $f' = f + F$, where τ and F are mismatchings in time delay and frequency shift, respectively.

Along with mismatching function, in works [132, (9.1)], [133, (4.52)] *normalized mismatching function* $\rho(\lambda', \lambda)$ is introduced:

$$\rho(\lambda', \lambda) = \frac{w(\lambda', \lambda)}{\sqrt{w(\lambda', \lambda')w(\lambda, \lambda)}}$$

$$= \rho(\tau, F) = \frac{1}{E} \int\limits_{-\infty}^{\infty} s(t)h(\tau - t)\exp(-j2\pi Ft)\,\mathrm{d}t. \quad (4.7.4)$$

In the case, when all unknown parameters are estimated (i.e., time of arrival t_0' and a central frequency f' of spectrum of the received signal), the elements of Fisher information matrix $\mathbf{\Phi} = \|\Phi_{ik}\|$, $i, k \in \{1, 2\}$ are calculated using the relationships [133, (4.52), (4.53)]:

$$\Phi_{ik} = -q^2 \frac{\partial^2}{\partial \lambda_i' \partial \lambda_k'} \rho(\lambda'), \quad (4.7.5)$$

where $\rho(\lambda') = \rho(t_0', f')$ is time-frequency mismatching function of the received signal $s'(t, \lambda')$; $\lambda' = [t_0', f']^T$ is vector parameter of the received signal $s'(t, \lambda')$.

Then, a variance $D\{\hat{\lambda}/\lambda\}$ of maximum likelihood estimator $\hat{\lambda}$ of scalar parameter λ in the absence of non-informational parameters is defined by the relationship [132, (14.12)], [133, (4.60)]:

$$D\{\hat{\lambda}/\lambda\} = -1/[q^2\rho''(0)], \quad q^2 >> 1, \quad (4.7.6)$$

and in the case if signal initial phase is uniformly distributed in the interval $[-\pi, \pi]$, a variance of estimator $\hat{\lambda}$ is determined by the relationship [133, (4.61)]:

$$D\{\hat{\lambda}/\lambda\} = -1/[q^2\Psi''(0)], \quad q^2 >> 1. \quad (4.7.7)$$

Thus, increasing accuracy of maximum likelihood estimator of scalar parameter λ is provided by increasing the signal energy, i.e., by the quantity q^2, and on the other hand, by exploiting coherent signals and also those signals whose ambiguity function and mismatching function possess as sharp peak as possible (have greater (in absolute value) negative second derivative) in a point $\lambda = 0$.

4.7.2   Signal Resolution. General Considerations

Consider the model of additive interaction between the signals $s_m(t, \lambda)$ from a set $S = \{s_m(t, \lambda)\}$, $m = 1, \dots, M$ and interference (noise) $n(t)$ in signal space with L-group properties:

$$x(t) = \left(\sum_{m=1}^{M} \theta_m s_m(t; \lambda)\right) + n(t), \quad t \in T_{\text{obs}}, \quad (4.7.8)$$

where T_{obs} is observation interval of the signals $s_m(t, \lambda)$; θ_m is parameter taking the values from the set $\{0, 1\}$: $\theta_m \in \{0, 1\}$; λ is unknown nonrandom scalar parameter of the signal $s_m(t, \lambda)$ taking the values on a set Λ: $\lambda \in \Lambda$; $M \in \mathbf{N}$, $\mathbf{N}$ is set of natural numbers.

The problem of signal *resolution-classification* in the presence of interference (noise) lies in making a decision (using some criterion) that allows distinguishing which one of signal combinations from a set $S = \{s_m(t, \lambda)\}$, $m = 1, \ldots, M$ is contained in the observed process $x(t)$. If the signals from a set $S = \{s_m(t, \lambda)\}$, $m = 1, \ldots, M$ are such that the formulated problem has the solutions on any of 2^M signal combinations, then one can claim that the signals $\{s_m(t, \lambda)\}$ are resolvable in a parameter λ.

It is obvious that the formulated problem of signal resolution-classification is equivalent to the problem of signal classification from a set $S = \{s_k(t, \lambda)\}$, $k = 1, 2, \ldots, 2^M$.

Within the process of resolution-classification, one can consider *resolution-detection* of a useful signal $s_m(t, \lambda)$ in the presence of interference (noise) $n(t)$ and other interfering signals $\{s_j(t, \lambda)\}$, $j \neq m$, $j = 1, \ldots, M$. Thus, in the case of resolution-detection, the problem of signal processing is the establishing the presence of the m-th signal in the observed process $x(t)$ (4.7.8), i.e., its separate detection.

If the type of signal combination (4.7.8) from a set $S = \{s_m(t, \lambda)\}$ is established and information extraction lies in estimation of their parameters taken separately, then the problem of signal processing turns to *resolution-estimation*.

4.7.3 Signal Resolution in Linear Space

By the following definition we introduce characteristic of resolution for deterministic (coherent) signals in linear space.

Definition 4.7.1. *Time-frequency mismatching function* $w_{l_2}(t_i, f_k)$ *of deterministic signal* $s(t)$ *in linear space with scalar product (with* l_2*-metric) is a bivariate function of discrete time and frequency that taking into account the expression* (4.7.3) *is defined by the following relationship:*

$$w_{l_2}(t_i, f_k) = \sum_{n=0}^{i} s(t_n)\overline{h(t_i - t_n)} \cdot \exp\left(-j2\pi \frac{f_k - f_0}{f_{\max}} \frac{t_n}{\Delta t}\right); \tag{4.7.9}$$

$$t_n = n \cdot \Delta t; \quad f_k = k\Delta f; \tag{4.7.9a}$$

$$i = 0, 1, \ldots, N - 1; \quad k = 0, 1, \ldots, K; \tag{4.7.9b}$$

$$f_{\max} = K\Delta f, \tag{4.7.9c}$$

where $\{s(t_n)\}$ is a set of samples of analytic signal $s(t)$ (4.7.1b), $t_n \in T_{\text{obs}}$, $s(t) = 0$, $t \notin T_s$: $T_s = [t_0, t_0 + T] \subset T_{\text{obs}}$; T_{obs} is an observation interval of the signal $s(t)$; T_s is domain of the signal $s(t)$; t_0 is time of arrival of the signal $s(t)$; T is a duration of the signal $s(t)$; $h(t)$ is impulse response of the filter matched with the signal $s(t)$; $\overline{h(t)}$ is a function conjugate to the function $h(t)$; Δt, Δf are sampling intervals of the signal $s(t)$ in time and frequency domains, respectively; f_0 is a central frequency of signal spectrum; $K = 2K'$, $N, K' \in \mathbb{N}$; $j = \sqrt{-1}$.

In coordinate system (i, k) of indexes of time and frequency samples, the expression (4.7.9) for time-frequency mismatching function takes the form:

$$w_{l_2}(i, k) = \sum_{n=0}^{i} s(n)\overline{h(i - n)} \cdot \exp\left(-j2\pi \tfrac{k-k_0}{K} n\right), \qquad (4.7.10)$$

where k_0 is an index of frequency sample corresponding to a central frequency f_0 of signal spectrum: $k_0 = f_0/\Delta f$.

Notice, that in the instant of signal ending (on $t_1 = t_0 + T$), time-frequency mismatching function $w_{l_2}(i, k)$ (4.7.10) possesses delta-function in the neighborhood of central frequency f_0 of signal spectrum:

$$w_{l_2}(i_1 = t_1/\Delta t, k) = \sum_{n=0}^{i_1} |s(n)|^2 \cdot \exp\left(-j2\pi \tfrac{k-k_0}{K} n\right)$$

$$= 2E \cdot \delta_{k,k_0} + S_k(1 - \delta_{k,k_0}), \qquad (4.7.11)$$

where E is an energy of real signal $\mathrm{Re}[s(t)]$; $\delta_{k,m}$ is Kronecker symbol: $\delta_{k,m} = 1$, $k = m$; $\delta_{k,m} = 0$, $k \neq m$, so that for a part of spectrum S_k that is outside of the frequency f_0, the relationship holds $S_k << 2E$.

A section of bivariate time-frequency mismatching function $w_{l_2}(i, k)$ (4.7.10) by a plane, that is perpendicular to the plane (i, k) and passes through a point $k = k_0$, represents an autocorrelation function (ACF) $R(i)$ of the signal $s(t)$:

$$w_{l_2}(i, k = k_0) = R(i) = \sum_{n=0}^{i} s(n)\overline{h(i - n)}, \qquad (4.7.12)$$

so that in the interval $t_i \in [t_1 - \tfrac{T_0}{4}, t_1 + \tfrac{T_0}{4}]$, the real part $\mathrm{Re}[R(t_i)]$ of ACF $R(t_i)$ of bandpass signal $s(t)$ is equal to:

$$\mathrm{Re}[R(t_i)] = 2E \cos[2\pi f_0(t - t_1)]; \qquad (4.7.13)$$

$$t_i \in [t_1 - \tfrac{T_0}{4}, t_1 + \tfrac{T_0}{4}],$$

where $t_1 = t_0 + T$.

Fig. 4.7.1a illustrates real part of the section $w_{l_2}(i, k = k_0)$ of bivariate time-frequency mismatching function (TFMF) $w_{l_2}(i, k)$ (4.7.10) of binary phase-coded signal $s(t)$ based on 7-element Barker code that is determined by the relationship (4.7.12). Fig. 4.7.1b illustrates the real part of the section $w_{l_2}(i_1 = t_1/\Delta t, k)$ (4.7.11) of bivariate TFMF $w_{l_2}(i, k)$ (4.7.10) by the plane, that is perpendicular to the plane (i, k) and passes trough the point $i_1 = t_1/\Delta t = (t_0+T)/\Delta t$. The numbers of time N and frequency $K+1$ samples of TFMF $w_{l_2}(i, k)$ for the cases depicted in Fig. 4.7.1a,b are, respectively, equal to: $N = 512$, $K + 1 = 257$ (see formula (4.7.9b)).

Any two copies of a signal whose mutual mismatching in time $\Delta\tau$ and frequency ΔF exceed a measure of filter resolution in time delay Δ_τ and frequency shift Δ_f, respectively:

$$|\Delta\tau| > \Delta_\tau; \quad |\Delta F| > \Delta_f, \qquad (4.7.14)$$

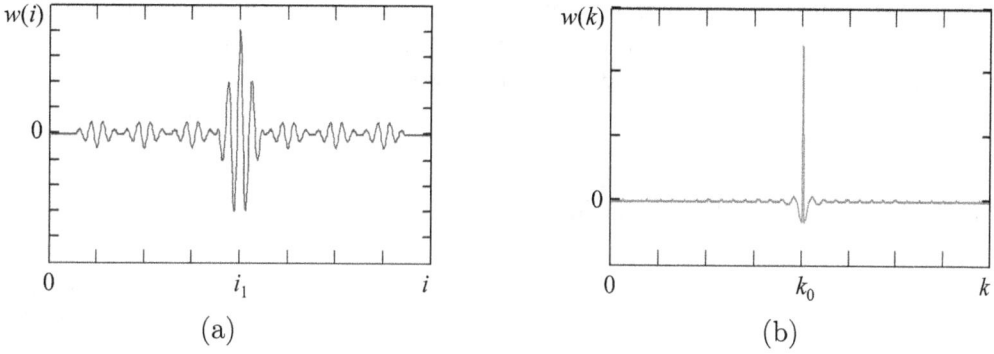

FIGURE 4.7.1 Real parts of the sections of TFMF in l_2-metric space: (a) $w_{l_2}(i, k = k_0)$; (b) $w_{l_2}(i_1 = t_1/\Delta t, k)$

are considered to be resolvable, so that the quantities Δ_τ, Δ_f are doubled minimal roots of the equations $w(\Delta\tau, 0) = 0.5$, $w(0, \Delta F) = 0.5$, respectively:

$$\Delta_\tau = 2 \inf_{\Delta\tau \in A_{\Delta\tau}} \{\arg[w(\Delta\tau, 0) = 0.5]\}; \qquad (4.7.15a)$$

$$\Delta_f = 2 \inf_{\Delta F \in A_{\Delta F}} \{\arg[w(0, \Delta F) = 0.5]\}, \qquad (4.7.15b)$$

where $w(\tau, F)$ is TFMF (4.7.3).

The relationships (4.7.13) and (4.7.11) imply that potential resolution of a filter matched with coherent RF bandpass signal in time delay Δ_τ and frequency shift Δ_f are defined by the relationships:

$$\Delta_\tau = T_0/3; \ \Delta_f = \Delta f, \qquad (4.7.16)$$

where $T_0 = 1/f_0$ is a period of oscillation of central frequency f_0 of the signal $s(t)$; $\Delta f = f_{\max}/K$ is a sampling interval in frequency (4.7.9c) of the signal $s(t)$.

Fig. 4.7.2a depicts the section $W'_{l_2}(i, k)$ of modulus $W_{l_2}(i, k) = |w_{l_2}(i, k)|$ of TFMF $w_{l_2}(i, k)$ (4.7.10) of binary phase-coded signal with 7-element Barker code at the level $0.1 \max[W_{l_2}(i, k)]$. This section $W'_{l_2}(i, k)$ is defined by the following relationship:

$$W'_{l_2}(i, k) = W_{l_2}(i, k) \vee A - A; \qquad (4.7.17)$$

$$A = p \max[W_{l_2}(i, k)]; \qquad (4.7.17a)$$

$$W_{l_2}(i, k) = |w_{l_2}(i, k)|, \qquad (4.7.17b)$$

where $\vee$ is join operation of L-group: $a \vee b = \max[a, b]$; p is a relative level of a section; A is absolute level of a section.

The section $W'_{l_2}(i, k)$ of modulus $W_{l_2}(i, k) = |w_{l_2}(i, k)|$ of TFMF $w_{l_2}(i, k)$ (4.7.10) shown in Fig. 4.7.2a corresponds to the case when the numbers of time N and frequency $K + 1$ samples are, respectively, equal to: $N = 512$, $K + 1 = 61$ (see the formula (4.7.9b)).

The section $W'_{l_2}(i,k)$ of modulus $W_{l_2}(i,k) = |w_{l_2}(i,k)|$ of TFMF $w_{l_2}(i,k)$ (4.7.10) defines potential filter resolution in time delay Δ'_τ and frequency shift Δ'_f for incoherent signals, so that the following relationships hold:

$$\Delta'_\tau \geq \Delta_\tau; \ \Delta'_f = \Delta_f, \tag{4.7.18}$$

where Δ_τ, Δ_f are potential resolutions in time delay and frequency shift for coherent signals.

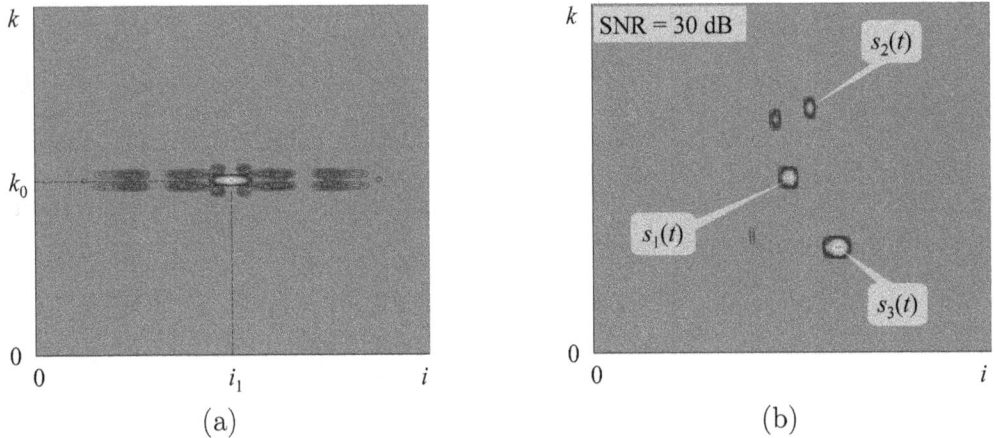

FIGURE 4.7.2   Sections $W'_{l_2}(i,k)$ of modulus $|w_{l_2}(i,k)|$ of TFMF in l_2-metric space: (a) $w_{l_2}(i,k)$ (4.7.10): a sole binary phase-coded signal; (b) $w_{l_2}(i,k)$ (4.7.20): three binary phase-coded signals

To provide resolution of m deterministic signals in the presence of interference (noise) (see the formula (4.7.8)), signal processing is based on calculating modified TFMF (4.7.9) according to the following relationship:

$$w_{l_2}(t_i, f_k) = \sum_{n=0}^{i} x(t_n) \left(\sum_{m=1}^{M} \overline{h_m(t_i - t_n)} \cdot \exp\left(-j2\pi \frac{f_k - f_{0m}}{f_{\max}} \frac{t_n}{\Delta t}\right) \right), \tag{4.7.19}$$

where $x(t)$ is observed stochastic process that is additive mixture of the resolvable signals $\{s_m(t)\}$ and interference (noise) $n(t)$ (see the formula (4.7.8)); $h_m(t)$ is impulse response of m-th signal $s_m(t)$; f_{0m} is a central frequency of spectrum of m-th signal $s_m(t)$; rest of variables from the formula (4.7.19) have the same meaning that those from the formula (4.7.9).

Within coordinate system (i,k) of indexes of time and frequency samples, the expression (4.7.19) for TFMF takes the form:

$$w_{l_2}(i,k) = \sum_{n=0}^{i} x(n) \left(\sum_{m=1}^{M} \overline{h_m(i - n)} \cdot \exp\left(-j2\pi \frac{k - k_{0m}}{K} n\right) \right), \tag{4.7.20}$$

where k_{0m} is an index of frequency sample corresponding to a central frequency f_{0m} of spectrum of m-th signal $s_m(t)$: $k_{0m} = f_{0m}/\Delta f$.

Fig. 4.7.2b illustrates the section $W'_{l_2}(i,k)$ of modulus $W_{l_2}(i,k) = |w_{l_2}(i,k)|$ of TFMF $w_{l_2}(i,k)$ (4.7.20) at the level $0.5\max[w_{l_2}(i,k)]$ when providing resolution of three binary phase-coded signals ($M = 3$) with 7-element Barker code in the presence of interference (noise) $n(t)$, where the function $W'_{l_2}(i,k)$ is calculated according to the relationships (4.7.17), so that $p = 0.5$. The ratio $\mathrm{SNR}_m = E_m/N_0$ of m-signal energy E_m to noise PSD N_0 is equal to $\mathrm{SNR}_m = 30$ dB, and numbers of time N and frequency $K + 1$ samples are, respectively, equal to $N = 512$, $K + 1 = 61$ (see formula (4.7.9b)).

4.7.4 Signal Resolution Algorithm Based on Measure of Statistical Interrelation Concerned with Pseudometric

Using a short form of representation of measure of statistical interrelation (MSI) in the corresponding signal space (see Table 4.2.1), one can define the notion of time-frequency mismatching function (TFMF) $w_p(t_i, f_k)$ that is a characteristic of deterministic signal resolution in space with pseudometric (generalized metric).

Definition 4.7.2. *Time-frequency mismatching function* of deterministic signal in signal space with pseudometric (generalized metric) is a function of discrete time and frequency that is defined by the following relationships:

$$w_p(t_i, f_k) = \sum_{n=0}^{i} [\mathrm{sgn}(\mathrm{Re}[s(t_n)])\,\mathrm{sgn}(\mathrm{Re}[\overline{h(t_i - t_n)}])$$
$$- \mathrm{sgn}(\mathrm{Im}[s(t_n)])\,\mathrm{sgn}(\mathrm{Im}[\overline{h(t_i - t_n)}]) + j \cdot \mathrm{sgn}(\mathrm{Re}[s(t_n)])\,\mathrm{sgn}(\mathrm{Im}[\overline{h(t_i - t_n)}])$$
$$+ j \cdot \mathrm{sgn}(\mathrm{Im}[s(t_n)])\,\mathrm{sgn}(\mathrm{Re}[\overline{h(t_i - t_n)}])] \cdot \exp\left(-j2\pi \frac{f_k - f_0}{f_{\max}} \frac{t_n}{\Delta t}\right); \quad (4.7.21)$$

$$t_n = n \cdot \Delta t; \; f_k = k\Delta f; \tag{4.7.21a}$$
$$i = 0, 1, \ldots, N - 1; \; k = 0, 1, \ldots, K; \tag{4.7.21b}$$
$$f_{\max} = K\Delta f, \tag{4.7.21c}$$

where $\{s(t_n)\}$ is a set of samples of analytic signal $s(t)$ (4.7.1b), $t_n \in T_{\mathrm{obs}}$, $s(t) = 0$, $t \notin T_s$: $T_s = [t_0, t_0 + T] \subset T_{\mathrm{obs}}$; T_{obs} is observation interval of the signal $s(t)$; T_s is domain of the signal $s(t)$; t_0 is time of arrival of the signal $s(t)$; T is a duration of the signal $s(t)$; $h(t)$ is impulse response of a filter matched with the signal $s(t)$; $\overline{h(t)}$ is a function conjugate to $h(t)$; Δt, Δf are sampling intervals of the signal $s(t)$ in time and frequency domains, respectively; f_0 is a central frequency of signal spectrum; $K = 2K'$, $N, K' \in \mathbb{N}$; $j = \sqrt{-1}$.

More complicated form of the expression (4.7.21) for TFMF $w_p(t_i, f_k)$ in signal space with pseudometric as against more simple TFMF $w_{l_2}(t_i, f_k)$ in signal space with scalar product is explained by the fact that sign function $\mathrm{sgn}(*)$ does not assume using complex numbers. Meanwhile, despite apparent complexity of representation, TFMF $w_p(t_i, f_k)$ is calculated on the basis of sign functions and modification and/ or preservation of operand signs, and allows avoiding operation of complex multiplication. Notice that, in signal space with pseudometric, TFMF $w_p(t_i, f_k)$ can be defined based on complete form of MSI representation. In this case TFMF $w_p(t_i, f_k)$ is calculated on the basis of L-group operations of join, meet, and addition.

Within the coordinate system (i, k) of indexes of time and frequency samples, the expression (4.7.8) for TFMF takes the form:

$$w_p(i, k) = \sum_{n=0}^{i} [\mathrm{sgn}(\mathrm{Re}[s(n)])\, \mathrm{sgn}(\mathrm{Re}[\overline{h(i-n)}])$$

$$- \mathrm{sgn}(\mathrm{Im}[s(n)])\, \mathrm{sgn}(\mathrm{Im}[\overline{h(i-n)}]) + j \cdot \mathrm{sgn}(\mathrm{Re}[s(n)])\, \mathrm{sgn}(\mathrm{Im}[\overline{h(i-n)}])$$

$$+ j \cdot \mathrm{sgn}(\mathrm{Im}[s(n)])\, \mathrm{sgn}(\mathrm{Re}[\overline{h(i-n)}])] \cdot \exp\left(-j2\pi \tfrac{k-k_0}{K} n\right), \quad (4.7.22)$$

where k_0 is an index of frequency sample corresponding to a central frequency f_0 of signal spectrum: $k_0 = f_0/\Delta f$.

At the instant of signal ending (at $t_1 = t_0 + T$), TFMF $w_p(i, k)$ (4.7.22) possesses delta-function in the neighborhood of central frequency f_0 of signal spectrum:

$$w_p(i_1 = t_1/\Delta t, k) = \sum_{n=0}^{i_1} 2\,\mathrm{sgn}(\mathrm{Re}[s(n)]) \cdot \mathrm{sgn}(\mathrm{Re}[\overline{h(i-n)}]) \cdot \exp\left(-j2\pi \tfrac{k-k_0}{K} n\right)$$

$$= S_{k_0} \cdot \delta_{k,k_0} + S_k(1 - \delta_{k,k_0}), \quad (4.7.23)$$

where $S_{k_0} = 2T/\Delta t$ is a doubled measure of real signal $\mathrm{Re}[s(t)]$ in signal space with pseudometric l_p; $\delta_{k,m}$ is Kronecker symbol: $\delta_{k,m} = 1$, $k = m$; $\delta_{k,m} = 0$, $k \neq m$, so that for the part of spectrum S_k that is outside the frequency f_0, the relationship holds $S_k << S_{k_0}$.

The section of bivariate TFMF $w_p(i, k)$ (4.7.22) by the plane that is perpendicular to the plane (i, k) and passes through the point $k = k_0$ is MSI $N_p(i)$ of the signal $s(t)$:

$$w_p(i, k = k_0) = N_p(i) = \sum_{n=0}^{i} [\mathrm{sgn}(\mathrm{Re}[s(n)])\, \mathrm{sgn}(\mathrm{Re}[\overline{h(i-n)}])$$

$$- \mathrm{sgn}(\mathrm{Im}[s(n)])\, \mathrm{sgn}(\mathrm{Im}[\overline{h(i-n)}]) + j \cdot \mathrm{sgn}(\mathrm{Re}[s(n)])\, \mathrm{sgn}(\mathrm{Im}[\overline{h(i-n)}])$$

$$+ j \cdot \mathrm{sgn}(\mathrm{Im}[s(n)])\, \mathrm{sgn}(\mathrm{Re}[\overline{h(i-n)}])], \quad (4.7.24)$$

so that, in the interval $t_i \in [t_1 - \tfrac{T_0}{4}, t_1 + \tfrac{T_0}{4}]$, the real part $\mathrm{Re}[N_p(t_i)]$ of MSI $N_p(t_i)$ of RF bandpass signal $s(t)$ is equal to:

$$\mathrm{Re}[N_p(t_i)] = S_{k_0} \cos[2\pi f_0(t - t_1)]; \quad (4.7.25)$$

$$t_i \in [t_1 - \tfrac{T_0}{4}, t_1 + \tfrac{T_0}{4}],$$

where $t_1 = t_0 + T$.

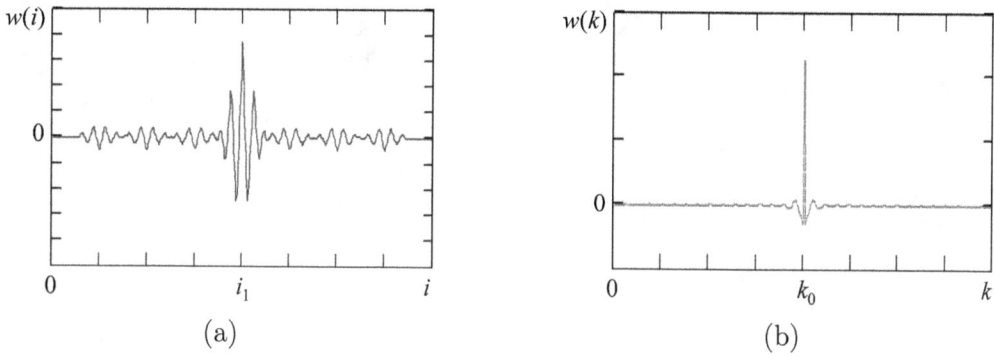

FIGURE 4.7.3   Real parts of the sections of TFMF in pseudometric space: (a) $w_p(i, k = k_0)$; (b) $w_p(i_1 = t_1/\Delta t, k)$

Fig. 4.7.3a illustrates the real part of the section $w_p(i, k = k_0)$ of bivariate TFMF $w_p(i, k)$ (4.7.22) of binary phase-coded signal $s(t)$ with 7-element Barker code that is defined by the relationship (4.7.24).

Fig. 4.7.3b depicts real part of the section $w_p(i_1 = t_1/\Delta t, k)$ (4.7.23) of bivariate TFMF $w_p(i, k)$ (4.7.22) by a plane that is perpendicular to the plane (i, k) and passes through the point $i_1 = t_1/\Delta t = (t_0 + T)/\Delta t$. The numbers of time N and frequency $K + 1$ samples of TFMF $w_p(i, k)$ for the cases shown in Fig. 4.7.3a,b are, respectively, equal to: $N = 512$, $K + 1 = 257$ (see the formula (4.7.21b)).

The relationships (4.7.25) and (4.7.23) imply that potential resolutions of a filter, matched with coherent bandpass signal, in time delay Δ_τ and frequency shift Δ_f are defined by the relationships:

$$\Delta_\tau = T_0/3; \ \Delta_f = \Delta f, \qquad (4.7.26)$$

where $T_0 = 1/f_0$ is a period of oscillation of central frequency f_0 of the signal $s(t)$; $\Delta f = f_{\max}/K$ is a sampling interval of the signal $s(t)$ in frequency domain (4.7.21c).

The relationships (4.7.26) imply that signal resolution in signal space with pseudometric is equivalent to potential resolution in signal space with scalar product (4.7.16).

Fig. 4.7.4a illustrates the section $W_p'(i, k)$ of modulus $W_p(i, k) = |w_p(i, k)|$ of TFMF $w_p(i, k)$ (4.7.22) of binary phase-coded signal with 7-element Barker code at the level $0.1 \max[W_p(i, k)]$. This section $W_p'(i, k)$ is defined by the following relationship:

$$W_p'(i, k) = W_p(i, k) \vee A - A; \qquad (4.7.27)$$
$$A = p \max[W_p(i, k)]; \qquad (4.7.27a)$$
$$W_p(i, k) = |w_p(i, k)|, \qquad (4.7.27b)$$

where $\vee$ is join operation of L-group: $a \vee b = \max[a, b]$; p is relative level of a section; A is absolute level of a section.

The section $W_p'(i, k)$ of modulus $W_p(i, k) = |w_p(i, k)|$ of TFMF $w_p(i, k)$ (4.7.22) shown in Fig. 4.7.4a corresponds to the case when the numbers of time N and

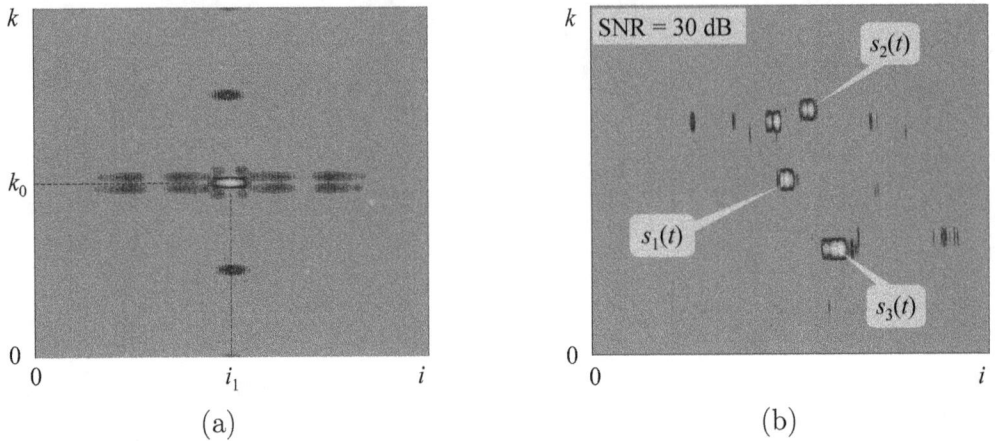

FIGURE 4.7.4 Sections $W'_p(i,k)$ of modulus $|w_p(i,k)|$ of TFMF in pseudometric space: (a) $w_p(i,k)$ (4.7.22): a sole binary phase-coded signal; (b) $w_p(i,k)$ (4.7.30): three binary phase-coded signals

frequency $K+1$ samples are, respectively, equal to: $N = 512$, $K+1 = 61$ (see formula (4.7.21b)). As well as any nonlinear transform, TFMF $w_p(i,k)$ (4.7.22) possesses harmonics on multiple frequencies.

The section $W'_p(i,k)$ of modulus $W_p(i,k) = |w_p(i,k)|$ of TFMF $w_p(i,k)$ (4.7.22) defines potential filter resolutions in time delay Δ'_τ and frequency shift Δ'_f for incoherent signals, so that the following relationships hold:

$$\Delta'_\tau \geq \Delta_\tau; \; \Delta'_f = \Delta_f, \tag{4.7.28}$$

where Δ_τ, Δ_f are potential resolutions in time delay and in frequency shift for coherent signals.

To provide resolution of m deterministic signals in the presence of interference (noise) (see formula (4.7.8)), signal processing is based on calculating the modified TFMF (4.7.21) according to the following relationship:

$$
\begin{aligned}
w_p&(t_i, f_k) \\
&= \sum_{n=0}^{i} \left[\text{sgn}(\text{Re}[x(t_n)]) \, \text{sgn}\left(\text{Re}\left[\sum_{m=1}^{M} \overline{h_m(t_i - t_n)} \cdot \exp\left(-j2\pi \tfrac{f_k - f_{0m}}{f_{\max}} \tfrac{t_n}{\Delta t}\right) \right] \right) \right. \\
&\quad - \text{sgn}(\text{Im}[x(t_n)]) \, \text{sgn}\left(\text{Im}\left[\sum_{m=1}^{M} \overline{h_m(t_i - t_n)} \cdot \exp\left(-j2\pi \tfrac{f_k - f_{0m}}{f_{\max}} \tfrac{t_n}{\Delta t}\right) \right] \right) \\
&\quad + j \cdot \text{sgn}(\text{Re}[x(t_n)]) \, \text{sgn}\left(\text{Im}\left[\sum_{m=1}^{M} \overline{h_m(t_i - t_n)} \cdot \exp\left(-j2\pi \tfrac{f_k - f_{0m}}{f_{\max}} \tfrac{t_n}{\Delta t}\right) \right] \right) \\
&\quad \left. + j \cdot \text{sgn}(\text{Im}[x(t_n)]) \, \text{sgn}\left(\text{Re}\left[\sum_{m=1}^{M} \overline{h_m(t_i - t_n)} \cdot \exp\left(-j2\pi \tfrac{f_k - f_{0m}}{f_{\max}} \tfrac{t_n}{\Delta t}\right) \right] \right) \right],
\end{aligned}
\tag{4.7.29}
$$

where $x(t)$ is the observed stochastic process that is additive mixture of the resolvable signals $\{s_m(t)\}$ and interference (noise) $n(t)$ (see the formula (4.7.8)); $h_m(t)$ is impulse response of m-th signal $s_m(t)$; f_{0m} is a central frequency of spectrum of m-th signal $s_m(t)$; rest of variables from the formula (4.7.29) have the same meaning that those from the formula (4.7.21).

Within coordinate system (i, k) of indexes of time and frequency samples, the expression (4.7.29) for TFMF takes the form:

$$
w_p(i, k) = \sum_{n=0}^{i} \left[\operatorname{sgn}(\operatorname{Re}[x(n)]) \operatorname{sgn} \left(\operatorname{Re} \left[\sum_{m=1}^{M} \overline{h_m(i - n)} \cdot \exp\left(-j2\pi \tfrac{k-k_{0m}}{K} n\right) \right] \right) \right.
$$
$$
- \operatorname{sgn}(\operatorname{Im}[x(n)]) \operatorname{sgn} \left(\operatorname{Im} \left[\sum_{m=1}^{M} \overline{h_m(i - n)} \cdot \exp\left(-j2\pi \tfrac{k-k_{0m}}{K} n\right) \right] \right)
$$
$$
+ j \cdot \operatorname{sgn}(\operatorname{Re}[x(n)]) \operatorname{sgn} \left(\operatorname{Im} \left[\sum_{m=1}^{M} \overline{h_m(i - n)} \cdot \exp\left(-j2\pi \tfrac{k-k_{0m}}{K} n\right) \right] \right)
$$
$$
\left. + j \cdot \operatorname{sgn}(\operatorname{Im}[x(n)]) \operatorname{sgn} \left(\operatorname{Re} \left[\sum_{m=1}^{M} \overline{h_m(i - n)} \cdot \exp\left(-j2\pi \tfrac{k-k_{0m}}{K} n\right) \right] \right) \right], \quad (4.7.30)
$$

where k_{0m} is an index of frequency sample corresponding to central frequency f_{0m} of spectrum of m-th signal $s_m(t)$: $k_{0m} = f_{0m}/\Delta f$.

Fig. 4.7.4b depicts the section $W_p'(i, k)$ of modulus $W_p(i, k) = |w_p(i, k)|$ of TFMF $w_p(i, k)$ (4.7.30) at the level $0.5 \max[w_p(i, k)]$ when providing resolution of three binary phase-coded signals ($M = 3$) with 7-elements Barker code in the presence of interference (noise) $n(t)$, where the function $W_p'(i, k)$ is calculated according to the relationships (4.7.27), so that $p = 0.5$. The ratio $\mathrm{SNR}_m = E_m/N_0$ of m-signal energy E_m to noise PSD N_0 is equal to $\mathrm{SNR}_m = 30$ dB, and the numbers of time N and frequency $K + 1$ samples are, respectively, equal to $N = 512$, $K + 1 = 61$ (see the formula (4.7.21b)).

4.7.5 Signal Resolution Algorithm Based on Measure of Statistical Interrelation Concerned with l_1-metric

Using a short form of MSI representation in the corresponding signal space (see Table 4.2.1), one can determine TFMF $w_{l_1}(t_i, f_k)$ that characterizes deterministic signal resolution in signal space with l_1-metric.

Definition 4.7.3. *Time-frequency mismatching function* $w_{l_1}(t_i, f_k)$ *of deterministic signal* $s(t)$ *in signal space with* l_1-*metric is a function of discrete time and frequency that is defined by the following relationships:*

$$
w_{l_1}(t_i, f_k) = \sum_{n=0}^{i} \exp\left(-j2\pi \tfrac{f_k - f_0}{f_{\max}} \tfrac{t_n}{\Delta t}\right) \cdot [HF(\operatorname{Re}[s(t_n)], \operatorname{Re}[\overline{h(t_i - t_n)}])
$$
$$
- HF(\operatorname{Im}[s(t_n)], \operatorname{Im}[\overline{h(t_i - t_n)}]) + j \cdot HF(\operatorname{Re}[s(t_n)], \operatorname{Im}[\overline{h(t_i - t_n)}])
$$
$$
+ j \cdot HF(\operatorname{Im}[s(t_n)], \operatorname{Re}[\overline{h(t_i - t_n)}])]; \quad (4.7.31)
$$

$$t_n = n \cdot \Delta t; \; f_k = k \Delta f; \tag{4.7.31a}$$

$$i = 0, 1, \ldots, N-1; \; k = 0, 1, \ldots, K; \tag{4.7.31b}$$

$$f_{\max} = K \Delta f, \tag{4.7.31c}$$

where $HF(x, a)$ is Huber function determined by the relationship (4.4.4b); $\{s(t_n)\}$ is a set of samples of analytic signal $s(t)$ (4.7.1b), $t_n \in T_{obs}$, $s(t) = 0$, $t \notin T_s$: $T_s = [t_0, t_0 + T] \subset T_{obs}$; T_{obs} is observation interval of the signal $s(t)$; T_s is a domain of the signal $s(t)$; t_0 is time of arrival of the signal $s(t)$; T is a duration of the signal $s(t)$; $h(t)$ is impulse response of a filter matched with the signal $s(t)$; $\overline{h(t)}$ is a function conjugate to $h(t)$; Δt, Δf are sampling intervals of the signal $s(t)$ in time and frequency domains, respectively; f_0 is a central frequency of signal spectrum; $K = 2K'$, $N, K' \in \mathbb{N}$; $j = \sqrt{-1}$.

More complicated form of the expression (4.7.31) for TFMF $w_{l_1}(t_i, f_k)$ in signal space with l_1-metric as compared with more simple TFMF $w_{l_2}(t_i, f_k)$ in signal space with scalar product is explained by the fact that Huber function $HF(x, a)$ does not assume using complex numbers. Meanwhile, despite apparent complexity of representation, TFMF $w_{l_1}(t_i, f_k)$ is calculated on the basis of function of modulus and allows avoiding operation of complex multiplication. Notice, that TFMF $w_{l_1}(t_i, f_k)$ in signal space with l_1-metric can be defined on the basis of complete form of MSI representation. In this case, TFMF $w_{l_1}(t_i, f_k)$ is calculated by L-group operations of join, meet, and addition.

Within coordinate system (i, k) of indexes of time and frequency samples, the expression (4.7.31) for TFMF takes the form:

$$w_{l_1}(i, k) = \sum_{n=0}^{i} \exp\left(-j2\pi \tfrac{k-k_0}{K} n\right) \times [HF(\mathrm{Re}[s(n)], \mathrm{Re}[\overline{h(i-n)}])$$
$$- HF(\mathrm{Im}[s(n)], \mathrm{Im}[\overline{h(i-n)}]) + j \cdot HF(\mathrm{Re}[s(n)], \mathrm{Im}[\overline{h(i-n)}])$$
$$+ j \cdot HF(\mathrm{Im}[s(n)], \mathrm{Re}[\overline{h(i-n)}])], \quad (4.7.32)$$

where k_0 is an index of frequency sample corresponding to a central frequency f_0 of signal spectrum: $k_0 = f_0 / \Delta f$.

At the instant of the signal ending (at $t_1 = t_0 + T$), TFMF $w_{l_1}(t_i, f_k)$ (4.7.32) possesses delta-function in the neighborhood of central frequency f_0 of signal spectrum:

$$w_{l_1}(i_1 = t_1/\Delta t, k) = \sum_{n=0}^{i_1} \exp\left(-j2\pi \tfrac{k-k_0}{K} n\right) \cdot [HF(\mathrm{Re}[s(n)], \mathrm{Re}[\overline{h(i-n)}])$$
$$- HF(\mathrm{Im}[s(n)], \mathrm{Im}[\overline{h(i-n)}]) + j \cdot HF(\mathrm{Re}[s(n)], \mathrm{Im}[\overline{h(i-n)}])$$
$$+ j \cdot HF(\mathrm{Im}[s(n)], \mathrm{Re}[\overline{h(i-n)}])] = S_{k_0} \cdot \delta_{k,k_0} + S_k(1 - \delta_{k,k_0}), \quad (4.7.33)$$

where $S_{k_0} = \sum_{n=0}^{i_1} 2HF(\mathrm{Re}[s(n)], \mathrm{Re}[\overline{h(i-n)}])$ is a doubled measure of real signal $\mathrm{Re}[s(t)]$ in signal space with l_1-metric; $\delta_{k,m}$ is Kronecker symbol: $\delta_{k,m} = 1$, $k = m$;

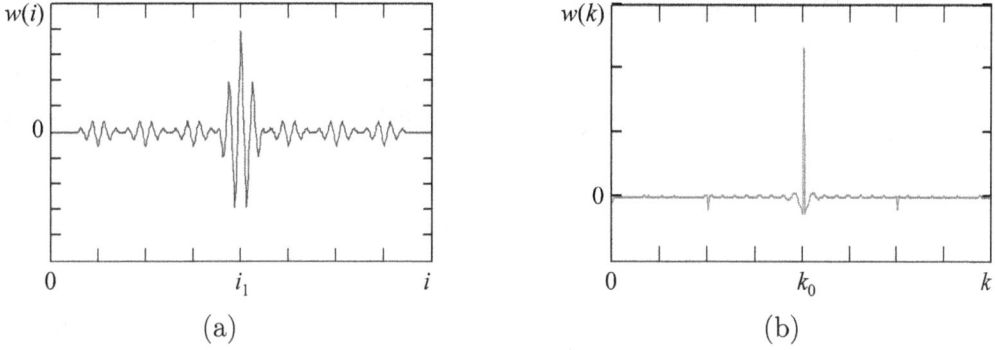

FIGURE 4.7.5   Real parts of the sections of TFMF in l_1-metric space: (a) $w_{l_1}(i, k = k_0)$; (b) $w_{l_1}(i_1 = t_1/\Delta t, k)$

$\delta_{k,m} = 0$, $k \neq m$, so that for a part of spectrum S_k that is outside the frequency f_0, the relationship holds $S_k << S_{k_0}$.

The section of TFMF $w_{l_1}(i, k)$ (4.7.32) by a plane that is perpendicular to the plane (i, k) and passes through the point $k = k_0$ is MSI $N_{l_1}(i)$ of the signal $s(t)$:

$$w_{l_1}(i, k = k_0) = N_{l_1}(i) = \sum_{n=0}^{i} [HF(\text{Re}[s(n)], \text{Re}[\overline{h(i-n)}])$$

$$- HF(\text{Im}[s(n)], \text{Im}[\overline{h(i-n)}]) + j \cdot HF(\text{Re}[s(n)], \text{Im}[\overline{h(i-n)}])$$

$$+ j \cdot HF(\text{Im}[s(n)], \text{Re}[\overline{h(i-n)}])], \quad (4.7.34)$$

so that in the interval $t_i \in [t_1 - \frac{T_0}{4}, t_1 + \frac{T_0}{4}]$, real part $\text{Re}[N_{l_1}(t_i)]$ of MSI $N_{l_1}(t_i)$ of RF bandpass signal $s(t)$ is equal to:

$$\text{Re}[N_{l_1}(t_i)] = S_{k_0} \cos[2\pi f_0(t - t_1)]; \quad (4.7.35)$$

$$t_i \in [t_1 - \frac{T_0}{4}, t_1 + \frac{T_0}{4}],$$

where $t_1 = t_0 + T$.

Fig. 4.7.5a illustrates real part of the section $w_{l_1}(i, k = k_0)$ of bivariate TFMF $w_{l_1}(i, k)$ (4.7.32) of binary phase-coded signal $s(t)$ with 7-elements Barker code that is defined by the relationship (4.7.34).

Fig. 4.7.5b depicts real part of the section $w_{l_1}(i_1 = t_1/\Delta t, k)$ (4.7.33) of bivariate TFMF $w_{l_1}(i, k)$ (4.7.32) by a plane that is perpendicular to the plane (i, k) and passes through the point $i_1 = t_1/\Delta t = (t_0 + T)/\Delta t$. The numbers of time N and frequency $K + 1$ samples of TFMF $w_{l_1}(i, k)$ for the cases depicted in Fig. 4.7.5a,b are, respectively, equal to: $N = 512$, $K + 1 = 257$ (see the formula (4.7.31b)).

The formulas (4.7.35) and (4.7.33) imply that potential resolutions of a filter, matched with coherent bandpass signal, in time delay Δ_τ and in frequency shift Δ_f are determined by the relationships:

$$\Delta_\tau = T_0/3; \quad \Delta_f = \Delta f, \quad (4.7.36)$$

where $T_0 = 1/f_0$ is an oscillation period of central frequency f_0 of the signal spectrum; $\Delta f = f_{\max}/K$ is a sampling interval of the signal $s(t)$ in frequency (4.7.31c).

The relationships (4.7.36) imply that signal resolution in signal space with l_1-metric is equivalent to potential signal resolution in signal space with scalar product (4.7.16).

Fig. 4.7.6a illustrates the section $W'_{l_1}(i,k)$ of modulus $W_{l_1}(i,k) = |w_{l_1}(i,k)|$ of TFMF $w_{l_1}(i,k)$ (4.7.32) of phase-coded signal with 7-element Barker code at the level $0.1 \max[W_{l_1}(i,k)]$ that is defined by the following relationship:

$$W'_{l_1}(i,k) = W_{l_1}(i,k) \vee A - A; \qquad (4.7.37)$$

$$A = p \max[W_{l_1}(i,k)]; \qquad (4.7.37a)$$

$$W_{l_1}(i,k) = |w_{l_1}(i,k)|, \qquad (4.7.37b)$$

where $\vee$ is join operation of L-group: $a \vee b = \max[a,b]$; p is relative level of a section; A is absolute level of a section.

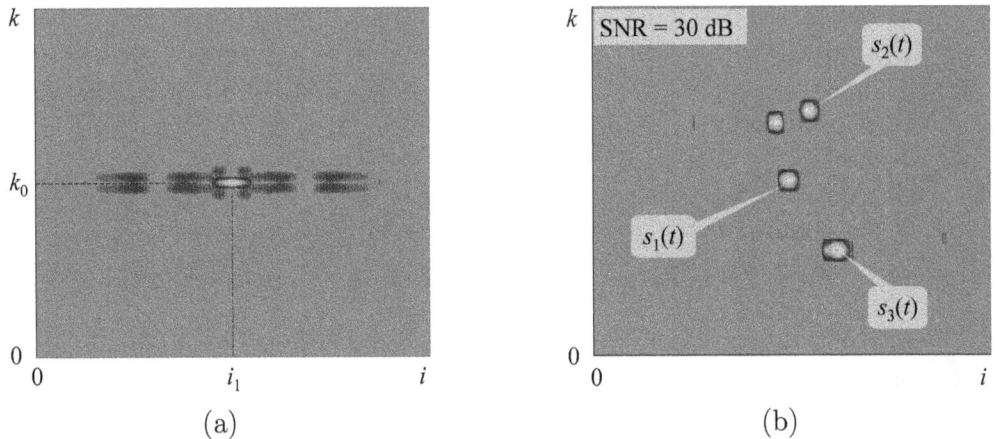

FIGURE 4.7.6   Sections $W'_{l_1}(i,k)$ of modulus $|w_{l_1}(i,k)|$ of TFMF in l_1-metric space: (a) $w_{l_1}(i,k)$ (4.7.32): a sole binary phase-coded signal; (b) $w_{l_1}(i,k)$ (4.7.40): three binary phase-coded signals

The section $W'_{l_1}(i,k)$ of modulus $W_{l_1}(i,k) = |w_{l_1}(i,k)|$ of TFMF $w_{l_1}(i,k)$ (4.7.32) shown in Fig. 4.7.6a corresponds to the case when the numbers of time N and frequency $K+1$ samples are, respectively, equal to: $N = 512$, $K+1 = 61$ (see the formula (4.7.31b)). As well as any nonlinear transformation, TFMF $w_{l_1}(i,k)$ (4.7.32) possesses harmonics on multiple frequencies.

The section $W'_{l_1}(i,k)$ of modulus $W_{l_1}(i,k) = |w_{l_1}(i,k)|$ of TFMF $w_{l_1}(i,k)$ (4.7.32) defines potential filter resolutions in time delay Δ'_τ and in frequency shift Δ'_f for incoherent signals, so that the following relationships hold:

$$\Delta'_\tau \geq \Delta_\tau; \ \Delta'_f = \Delta_f, \qquad (4.7.38)$$

where Δ_τ, Δ_f are potential resolutions in time delay and in frequency shift for coherent signals.

To provide resolution of m deterministic signal in the presence of interference (noise) (see the formula (4.7.8)), signal processing is based on calculating the modified TFMF (4.7.31) according to the following relationship:

$$w_{l_1}(t_i, f_k)$$

$$= \sum_{n=0}^{i} [HF\left(\operatorname{Re}[x(t_n)], \operatorname{Re}\left[\sum_{m=1}^{M} \overline{h_m(t_i - t_n)} \cdot \exp\left(-j2\pi \frac{f_k - f_{0m}}{f_{\max}} \frac{t_n}{\Delta t}\right)\right]\right)$$

$$- HF\left(\operatorname{Im}[x(t_n)], \operatorname{Im}\left[\sum_{m=1}^{M} \overline{h_m(t_i - t_n)} \cdot \exp\left(-j2\pi \frac{f_k - f_{0m}}{f_{\max}} \frac{t_n}{\Delta t}\right)\right]\right)$$

$$+ j \cdot HF\left(\operatorname{Re}[x(t_n)], \operatorname{Im}\left[\sum_{m=1}^{M} \overline{h_m(t_i - t_n)} \cdot \exp\left(-j2\pi \frac{f_k - f_{0m}}{f_{\max}} \frac{t_n}{\Delta t}\right)\right]\right)$$

$$+ j \cdot HF\left(\operatorname{Im}[x(t_n)], \operatorname{Re}\left[\sum_{m=1}^{M} \overline{h_m(t_i - t_n)} \cdot \exp\left(-j2\pi \frac{f_k - f_{0m}}{f_{\max}} \frac{t_n}{\Delta t}\right)\right]\right)], \quad (4.7.39)$$

where $x(t)$ is the observed stochastic process that is additive mixture of the resolvable signals $\{s_m(t)\}$ and interference (noise) $n(t)$ (see the formula (4.7.8)); $h_m(t)$ is impulse response of m-th signal $s_m(t)$; f_{0m} is a central frequency of spectrum of m-th signal $s_m(t)$; rest of variables from the formula (4.7.39) have the same meaning like those from the formula (4.7.31).

Within coordinate system (i, k) of indexes of time and frequency samples, the expression (4.7.39) for TFMF takes the form:

$$w_{l_1}(i, k)$$

$$= \sum_{n=0}^{i} [\operatorname{sgn}(\operatorname{Re}[x(n)]) \operatorname{sgn}\left(\operatorname{Re}\left[\sum_{m=1}^{M} \overline{h_m(i - n)} \cdot \exp\left(-j2\pi \frac{k - k_{0m}}{K} n\right)\right]\right)$$

$$- \operatorname{sgn}(\operatorname{Im}[x(n)]) \operatorname{sgn}\left(\operatorname{Im}\left[\sum_{m=1}^{M} \overline{h_m(i - n)} \cdot \exp\left(-j2\pi \frac{k - k_{0m}}{K} n\right)\right]\right)$$

$$+ j \cdot \operatorname{sgn}(\operatorname{Re}[x(n)]) \operatorname{sgn}\left(\operatorname{Im}\left[\sum_{m=1}^{M} \overline{h_m(i - n)} \cdot \exp\left(-j2\pi \frac{k - k_{0m}}{K} n\right)\right]\right)$$

$$+ j \cdot \operatorname{sgn}(\operatorname{Im}[x(n)]) \operatorname{sgn}\left(\operatorname{Re}\left[\sum_{m=1}^{M} \overline{h_m(i - n)} \cdot \exp\left(-j2\pi \frac{k - k_{0m}}{K} n\right)\right]\right)], \quad (4.7.40)$$

where k_{0m} is an index of frequency sample corresponding to a central frequency f_{0m} of spectrum of m-th signal $s_m(t)$: $k_{0m} = f_{0m}/\Delta f$.

Fig. 4.7.6b depicts the section $W'_{l_1}(i, k)$ of modulus $W_{l_1}(i, k) = |w_{l_1}(i, k)|$ of TFMF $w_{l_1}(i, k)$ (4.7.40) at the level $0.5 \max[w_{l_1}(i, k)]$ when providing resolution of three binary phase-coded signals ($M = 3$) with 7-element Barker code in the presence of interference (noise) $n(t)$, where the function $W'_{l_1}(i, k)$ is calculated according to the relationships (4.7.37), so that $p = 0.5$. The ratio $\mathrm{SNR}_m = E_m/N_0$ of m-signal energy E_m to noise PSD N_0 is equal to $\text{SNR}_m = 30$ dB, and the numbers of time N and frequency $K + 1$ samples are, respectively, equal to $N = 512$, $K + 1 = 61$ (see the formula (4.7.31b)).

4.7.6   Signal Resolution Algorithm Based on Measure of Statistical Interrelation Concerned with Semimetric

Using a short form of MSI representation in the corresponding signal space (see Table 4.2.1), one can define the notion of TFMF $w_s(t_i, f_k)$ that characterize deterministic signal resolution in signal space with semimetric.

Definition 4.7.4. *Time-frequency mismatching function* $w_s(t_i, f_k)$ *of deterministic signal* $s(t)$ *in signal space with semimetric (generalized metric) is a function of discrete time and frequency that is defined by the following relationship:*

$$w_s(t_i, f_k) = \sum_{n=0}^{i} \exp\left(-j2\pi \frac{f_k - f_0}{f_{\max}} \frac{t_n}{\Delta t}\right) \times [\text{Re}[s(t_n)] \, \text{sgn}(\text{Re}[\overline{h(t_i - t_n)}])$$

$$- \text{Im}[s(t_n)] \, \text{sgn}(\text{Im}[\overline{h(t_i - t_n)}]) + j \cdot \text{Re}[s(t_n)] \, \text{sgn}(\text{Im}[\overline{h(t_i - t_n)}])$$

$$+ j \cdot \text{Im}[s(t_n)] \, \text{sgn}(\text{Re}[\overline{h(t_i - t_n)}])$$

$$+ \text{Re}[\overline{h(t_i - t_n)}] \, \text{sgn}(\text{Re}[s(t_n)]) - \text{Im}[\overline{h(t_i - t_n)}] \, \text{sgn}(\text{Im}[s(t_n)])$$

$$+ j \cdot \text{Re}[\overline{h(t_i - t_n)}] \, \text{sgn}(\text{Im}[s(t_n)]) + j \cdot \text{Im}[\overline{h(t_i - t_n)}] \, \text{sgn}(\text{Re}[s(t_n)])]; \quad (4.7.41)$$

$$t_n = n \cdot \Delta t; \; f_k = k\Delta f; \tag{4.7.41a}$$

$$i = 0, 1, \ldots, N - 1; \; k = 0, 1, \ldots, K; \tag{4.7.41b}$$

$$f_{\max} = K\Delta f, \tag{4.7.41c}$$

where $\{s(t_n)\}$ is a set of samples of analytic signal $s(t)$ (4.7.1b), $t_n \in T_{\text{obs}}$, $s(t) = 0$, $t \notin T_s$: $T_s = [t_0, t_0 + T] \subset T_{\text{obs}}$; T_{obs} is an observation interval of the signal $s(t)$; T_s is a domain of the signal $s(t)$; t_0 is time of arrival of the signal $s(t)$; T is a duration of the signal $s(t)$; $h(t)$ is impulse response of a filter matched with the signal $s(t)$; $\overline{h(t)}$ is a function conjugate to $h(t)$; Δt, Δf are sampling intervals of the signal $s(t)$ in time and frequency domains, respectively; f_0 is a central frequency of signal spectrum; $K = 2K'$, $N, K' \in \mathbb{N}$; $j = \sqrt{-1}$.

More complicated form of the expression (4.7.41) for TFMF $w_s(t_i, f_k)$ in signal space with semimetric as compared with more simple TFMF $w_{l_2}(t_i, f_k)$ in signal space with scalar product is explained by the fact that sign function $\text{sgn}(*)$ does not assume using complex numbers. Meanwhile, despite apparent complexity of representation, TFMF $w_s(t_i, f_k)$ is calculated on the basis of sign functions, modification and/ or preservation of operand signs, function of modulus, and allows avoiding operation of complex multiplication. Notice, that in signal space with semimetric, TFMF $w_s(t_i, f_k)$ can be defined on the basis of complete form of MSI representation. In this case, TFMF $w_s(t_i, f_k)$ is calculated by L-group operations of join, meet, and addition.

Within coordinate system (i, k) of indexes of time and frequency samples, the expression (4.7.41) for TFMF takes the form:

$$w_s(i,k) = \sum_{n=0}^{i} \exp\left(-j2\pi\tfrac{k-k_0}{K}n\right) \cdot [\mathrm{Re}[s(n)]\,\mathrm{sgn}(\mathrm{Re}[\overline{h(i-n)}])$$

$$- \mathrm{Im}[s(n)]\,\mathrm{sgn}(\mathrm{Im}[\overline{h(i-n)}]) + j \cdot \mathrm{Re}[s(n)]\,\mathrm{sgn}(\mathrm{Im}[\overline{h(i-n)}])$$

$$+ j \cdot \mathrm{Im}[s(n)]\,\mathrm{sgn}(\mathrm{Re}[\overline{h(i-n)}])$$

$$+ \mathrm{Re}[\overline{h(i-n)}]\,\mathrm{sgn}(\mathrm{Re}[s(n)]) - \mathrm{Im}[\overline{h(i-n)}]\,\mathrm{sgn}(\mathrm{Im}[s(n)])$$

$$+ j \cdot \mathrm{Re}[\overline{h(i-n)}]\,\mathrm{sgn}(\mathrm{Im}[s(n)]) + j \cdot \mathrm{Im}[\overline{h(i-n)}]\,\mathrm{sgn}(\mathrm{Re}[s(n)])], \quad (4.7.42)$$

where k_0 is an index of frequency sample corresponding to a central frequency f_0 of signal spectrum: $k_0 = f_0/\Delta f$.

At the instant of signal ending (at $t_1 = t_0+T$), TFMF $w_s(i,k)$ (4.7.42) possesses delta-function in the neighborhood of central frequency f_0 of signal spectrum $s(t)$:

$$w_s(i_1 = t_1/\Delta t, k) = S_{k_0} \cdot \delta_{k,k_0} + S_k(1 - \delta_{k,k_0}), \quad (4.7.43)$$

where $S_{k_0} = \sum_{n=0}^{i_1} 2\{\mathrm{Re}[s(n)] \cdot \mathrm{sgn}(\mathrm{Re}[\overline{h(i-n)}]) + \mathrm{Re}[\overline{h(i-n)}] \cdot \mathrm{sgn}(\mathrm{Re}[s(n)]))$ is a doubled measure of real signal $\mathrm{Re}[s(t)]$ in the signal space with semimetric l_s; $\delta_{k,m}$ is Kronecker symbol: $\delta_{k,m} = 1$, $k = m$; $\delta_{k,m} = 0$, $k \neq m$, so that for a part of spectrum S_k that is outside of the frequency f_0, the relationship holds $S_k << S_{k_0}$.

The section of bivariate TFMF $w_s(i,k)$ (4.7.42) by a plane that is perpendicular to the plane (i,k) and passes through the point $k = k_0$ is MSI $N_s(i)$ of the signal $s(t)$:

$$w_s(i, k = k_0) = N_s(i) = \sum_{n=0}^{i} [\mathrm{Re}[s(n)]\,\mathrm{sgn}(\mathrm{Re}[\overline{h(i-n)}])$$

$$- \mathrm{Im}[s(n)]\,\mathrm{sgn}(\mathrm{Im}[\overline{h(i-n)}]) + j \cdot \mathrm{Re}[s(n)]\,\mathrm{sgn}(\mathrm{Im}[\overline{h(i-n)}])$$

$$+ j \cdot \mathrm{Im}[s(n)]\,\mathrm{sgn}(\mathrm{Re}[\overline{h(i-n)}])$$

$$+ \mathrm{Re}[\overline{h(i-n)}]\,\mathrm{sgn}(\mathrm{Re}[s(n)]) - \mathrm{Im}[\overline{h(i-n)}]\,\mathrm{sgn}(\mathrm{Im}[s(n)])$$

$$+ j \cdot \mathrm{Re}[\overline{h(i-n)}]\,\mathrm{sgn}(\mathrm{Im}[s(n)]) + j \cdot \mathrm{Im}[\overline{h(i-n)}]\,\mathrm{sgn}(\mathrm{Re}[s(n)])], \quad (4.7.44)$$

so that, in the interval $t_i \in [t_1 - \tfrac{T_0}{4}, t_1 + \tfrac{T_0}{4}]$, real part $\mathrm{Re}[N_s(t_i)]$ of MSI $N_s(t_i)$ of bandpass signal $s(t)$ is equal to:

$$\mathrm{Re}[N_s(t_i)] = S_{k_0} \cos[2\pi f_0(t - t_1)]; \quad (4.7.45)$$

$$t_i \in [t_1 - \tfrac{T_0}{4}, t_1 + \tfrac{T_0}{4}],$$

where $t_1 = t_0 + T$.

Fig. 4.7.7a illustrates real part of the section $w_s(i, k = k_0)$ of bivariate TFMF $w_s(i,k)$ (4.7.42) of binary phase-coded signal $s(t)$ with 7-element Barker code. This section $w_s(i, k = k_0)$ is defined by the relationship (4.7.44).

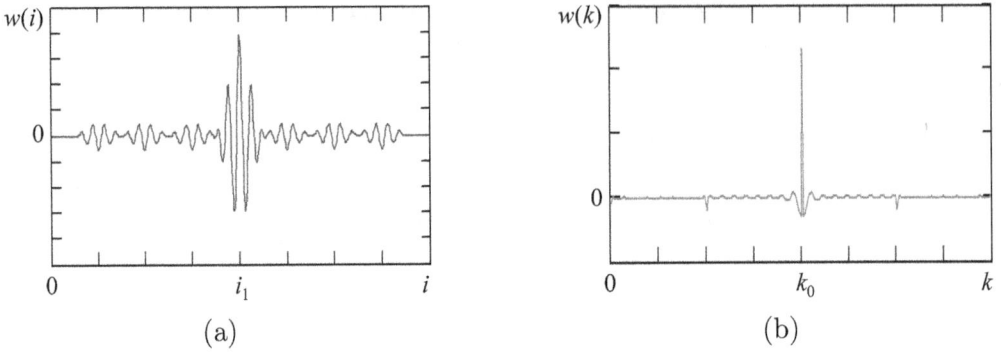

(a)   (b)

FIGURE 4.7.7   Real parts of the sections of TFMF in semimetric space: (a) $w_s(i, k = k_0)$;
(b) $w_s(i_1 = t_1/\Delta t, k)$

Fig. 4.7.7b depicts real part of the section $w_s(i_1 = t_1/\Delta t, k)$ (4.7.43) of bivariate TFMF $w_s(i, k)$ (4.7.42) by a plane that is perpendicular to the plane (i, k) and passes through the point $i_1 = t_1/\Delta t = (t_0 + T)/\Delta t$. The numbers of time N and frequency $K + 1$ samples of TFMF $w_s(i, k)$ for the cases shown in Fig. 4.7.7a,b are, respectively, equal to: $N = 512$, $K + 1 = 257$ (see the formula (4.7.41b)).

Relationships (4.7.45) and (4.7.43) imply that potential resolutions of a filter, matched with coherent bandpass signal, in time delay Δ_τ and in frequency shift Δ_f are defined by the relationships:

$$\Delta_\tau = T_0/3; \; \Delta_f = \Delta f, \tag{4.7.46}$$

where $T_0 = 1/f_0$ is an oscillation period of central frequency f_0 of signal spectrum; $\Delta f = f_{max}/K$ is a sampling interval of the signal $s(t)$ in frequency domain (4.7.41c).

The relationships (4.7.46) imply that potential signal resolution in signal space with semimetric is equivalent to potential signal resolution in linear signal space with scalar product (4.7.16).

Fig. 4.7.8a illustrates the section $W'_s(i, k)$ of modulus $W_s(i, k) = |w_s(i, k)|$ of TFMF $w_s(i, k)$ (4.7.42) of binary phase-coded signal with 7-element Barker code at the level $0.1 \max[W_s(i, k)]$ that is determined by the following relationship:

$$W'_s(i, k) = W_s(i, k) \vee A - A; \tag{4.7.47}$$
$$A = p \max[W_s(i, k)]; \tag{4.7.47a}$$
$$W_s(i, k) = |w_s(i, k)|, \tag{4.7.47b}$$

where $\vee$ is join operation of L-group: $a \vee b = \max[a, b]$; p is relative level of a section; A is absolute level of the section.

The section $W'_s(i, k)$ of modulus $W_s(i, k) = |w_s(i, k)|$ of TFMF $w_s(i, k)$ (4.7.42) shown in Fig. 4.7.8a corresponds to the case when the numbers of time N and frequency $K + 1$ samples are, respectively, equal to: $N = 512$, $K + 1 = 61$ (see the formula (4.7.41b)). As well as any nonlinear transformation, TFMF $w_s(i, k)$ (4.7.42) possesses harmonics on multiple frequencies.

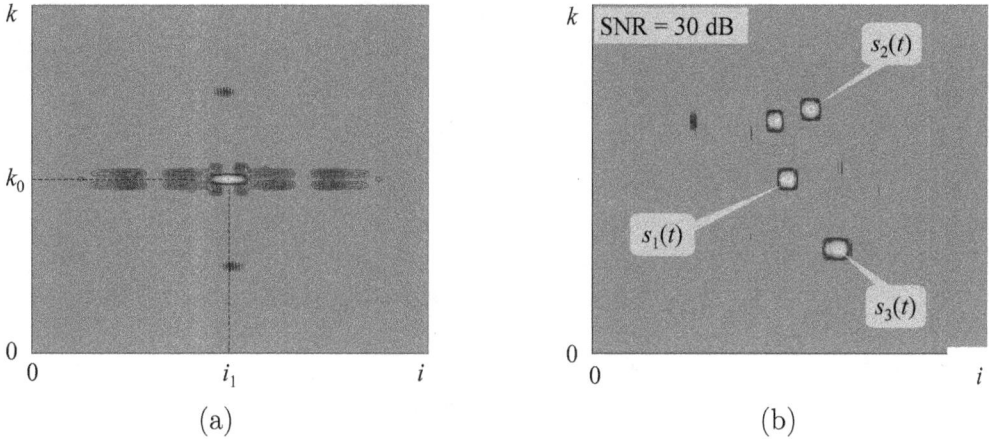

FIGURE 4.7.8   Sections $W'_s(i,k)$ of modulus $|w_s(i,k)|$ of TFMF in semimetric space: (a) — $w_s(i,k)$ (4.7.42): a sole binary phase-coded signal; (b) — $w_s(i,k)$ (4.7.50): three binary phase-coded signals

The section $W'_s(i,k)$ of modulus $W_s(i,k) = |w_s(i,k)|$ of TFMF $w_s(i,k)$ (4.7.42) defines potential filter resolutions in time delay Δ'_τ and frequency shift Δ'_f for incoherent signals, so that the following inequalities hold:

$$\Delta'_\tau \geq \Delta_\tau; \ \Delta'_f = \Delta_f, \tag{4.7.48}$$

where Δ_τ, Δ_f are potential resolutions in time delay and frequency shift for coherent signals.

To provide resolution of m deterministic signals in the presence of interference (noise) (see the formula (4.7.8)), signal processing is based on calculating modified TFMF (4.7.41) according to the following relationships:

$$
\begin{aligned}
&w_s(t_i, f_k) \\
&= \sum_{n=0}^{i} [\mathrm{Re}[x(t_n)]\,\mathrm{sgn}\left(\mathrm{Re}\left[\sum_{m=1}^{M}\overline{h_m(t_i - t_n)}\cdot\exp\left(-j2\pi\frac{f_k - f_{0m}}{f_{max}}\frac{t_n}{\Delta t}\right)\right]\right) \\
&\quad - \mathrm{Im}[x(t_n)]\,\mathrm{sgn}\left(\mathrm{Im}\left[\sum_{m=1}^{M}\overline{h_m(t_i - t_n)}\cdot\exp\left(-j2\pi\frac{f_k - f_{0m}}{f_{max}}\frac{t_n}{\Delta t}\right)\right]\right) \\
&\quad + j\cdot\mathrm{Re}[x(t_n)]\,\mathrm{sgn}\left(\mathrm{Im}\left[\sum_{m=1}^{M}\overline{h_m(t_i - t_n)}\cdot\exp\left(-j2\pi\frac{f_k - f_{0m}}{f_{max}}\frac{t_n}{\Delta t}\right)\right]\right) \\
&\quad + j\cdot\mathrm{Im}[x(t_n)]\,\mathrm{sgn}\left(\mathrm{Re}\left[\sum_{m=1}^{M}\overline{h_m(t_i - t_n)}\cdot\exp\left(-j2\pi\frac{f_k - f_{0m}}{f_{max}}\frac{t_n}{\Delta t}\right)\right]\right) \\
&\quad + \mathrm{Re}\left[\sum_{m=1}^{M}\overline{h_m(t_i - t_n)}\cdot\exp\left(-j2\pi\frac{f_k - f_{0m}}{f_{max}}\frac{t_n}{\Delta t}\right)\right]\,\mathrm{sgn}(\mathrm{Re}[x(t_n)]) \\
&\quad - \mathrm{Im}\left[\sum_{m=1}^{M}\overline{h_m(t_i - t_n)}\cdot\exp\left(-j2\pi\frac{f_k - f_{0m}}{f_{max}}\frac{t_n}{\Delta t}\right)\right]\,\mathrm{sgn}(\mathrm{Im}[x(t_n)])
\end{aligned}
$$

$$+ j \cdot \text{Re} \left[\sum_{m=1}^{M} \overline{h_m(t_i - t_n)} \cdot \exp\left(-j2\pi \frac{f_k - f_{0m}}{f_{\max}} \frac{t_n}{\Delta t}\right) \right] \text{sgn}(\text{Im}[x(t_n)])$$

$$+ j \cdot \text{Im} \left[\sum_{m=1}^{M} \overline{h_m(t_i - t_n)} \cdot \exp\left(-j2\pi \frac{f_k - f_{0m}}{f_{\max}} \frac{t_n}{\Delta t}\right) \right] \text{sgn}(\text{Re}[x(t_n)])], \quad (4.7.49)$$

where $x(t)$ is the observed stochastic process that is additive mixture of the resolvable signals $\{s_m(t)\}$ and interference (noise) $n(t)$ (see the formula (4.7.8)); $h_m(t)$ is impulse response of m-th signal $s_m(t)$; f_{0m} is a central frequency of spectrum of m-th signal $s_m(t)$; rest of variables from the formula (4.7.49) have the same meaning that those from the formula (4.7.41).

Within coordinate system (i, k) of indexes of time and frequency samples, the expression (4.7.49) for TFMF takes the form:

$$w_s(i, k) =$$

$$= \sum_{n=0}^{i} [\text{Re}[x(n)] \, \text{sgn} \left(\text{Re} \left[\sum_{m=1}^{M} \overline{h_m(i - n)} \cdot \exp\left(-j2\pi \frac{k - k_{0m}}{K} n\right) \right] \right)$$

$$- \text{Im}[x(n)] \, \text{sgn} \left(\text{Im} \left[\sum_{m=1}^{M} \overline{h_m(i - n)} \cdot \exp\left(-j2\pi \frac{k - k_{0m}}{K} n\right) \right] \right)$$

$$+ j \cdot \text{Re}[x(n)] \, \text{sgn} \left(\text{Im} \left[\sum_{m=1}^{M} \overline{h_m(i - n)} \cdot \exp\left(-j2\pi \frac{k - k_{0m}}{K} n\right) \right] \right)$$

$$+ j \cdot \text{Im}[x(n)] \, \text{sgn} \left(\text{Re} \left[\sum_{m=1}^{M} \overline{h_m(i - n)} \cdot \exp\left(-j2\pi \frac{k - k_{0m}}{K} n\right) \right] \right)$$

$$+ \text{Re} \left[\sum_{m=1}^{M} \overline{h_m(i - n)} \cdot \exp\left(-j2\pi \frac{k - k_{0m}}{K} n\right) \right] \text{sgn}(\text{Re}[x(n)])$$

$$- \text{Im} \left[\sum_{m=1}^{M} \overline{h_m(i - n)} \cdot \exp\left(-j2\pi \frac{k - k_{0m}}{K} n\right) \right] \text{sgn}(\text{Im}[x(n)])$$

$$+ j \cdot \text{Re} \left[\sum_{m=1}^{M} \overline{h_m(i - n)} \cdot \exp\left(-j2\pi \frac{k - k_{0m}}{K} n\right) \right] \text{sgn}(\text{Im}[x(n)])$$

$$+ j \cdot \text{Im} \left[\sum_{m=1}^{M} \overline{h_m(i - n)} \cdot \exp\left(-j2\pi \frac{k - k_{0m}}{K} n\right) \right] \text{sgn}(\text{Re}[x(n)])], \quad (4.7.50)$$

where k_{0m} is an index of frequency sample corresponding to a central frequency f_{0m} of spectrum of m-th signal $s_m(t)$: $k_{0m} = f_{0m}/\Delta f$.

Fig. 4.7.8b depicts the section $W_s'(i, k)$ of modulus $W_s(i, k) = |w_s(i, k)|$ of TFMF $w_s(i, k)$ (4.7.50) at the level $0.5 \max[w_s(i, k)]$ when providing resolution of three binary phase-coded signals ($M = 3$) with 7-element Barker code in the presence of interference (noise) $n(t)$, where the function $W_s'(i, k)$ is calculated according to the relationships (4.7.47), so that $p = 0.5$. The ratio $\text{SNR}_m = E_m/N_0$ of m-signal energy E_m to noise PSD N_0 is equal to $\mathrm{SNR}_m = 30$ dB, and numbers of time N and frequency $K + 1$ samples are, respectively, equal to $N = 512$, $K + 1 = 61$ (see the formula (4.7.41b)).

Conclusions

1. "Signal" and "filter" approaches to exploring the questions related to signal resolution are based on the notions of *ambiguity function* (4.7.1) and *mismatching function* (4.7.3), respectively. These "signal" and "filter" approaches to signal resolution represent the standpoints of *Signal Theory* and *Signal Processing Theory*, respectively.

2. By the relationships (4.7.9), (4.7.21), (4.7.31), (4.7.41) we introduce the notions of *time-frequency mismatching functions* of deterministic signal in linear signal space, and also in signal spaces with pseudometric, l_1-metric, and semimetric, respectively, that are measures of resolution of coherent signals in the corresponding spaces.

3. The obtained relationships (4.7.16), (4.7.26), (4.7.36), (4.7.46) defining potential resolution of generalized matched filters listed in the table 4.2.1 are equivalent.

5

Spectral Estimation and Spectral Analysis on L-groups

Spectral analysis is one of signal processing directions based on special methods and approaches that allow estimating components of investigated signal in frequency domain. Mathematical basis which associates the temporal signal with its representation in frequency domain includes transforms based on exploiting generalized Fourier series (1.6.4). Generalized Fourier series is constructed on one or another basis of orthogonal (as a rule, harmonic) functions. Spectral analysis is realized with respect to both the stochastic signals carrying some information and known deterministic signals that are observed in the presence of interference and noise. In this connection, signal spectral analysis is based on using certain statistical methods and algorithms whose choice is based on prior information concerning the explored signals. Correspondence degree between the properties of observed signals and available prior information on them determines the efficiency of used statistical methods and algorithms of spectral analysis.

Known methods of spectral analysis and estimation are well discussed in numerous works [134–137], [96, 138–140], [141–144], [145–147]. In this chapter we consider alternative algorithms of spectral estimation based on L-group operations that, being analogues of known algorithms, possess the same frequency resolution and could be realized by lesser computational resources. All methods of spectral estimation, based on L-group operations and involved in this chapter, are stated in comparison with their known linear analogues as applied to discrete time sequences (discrete signals). Here we notice, that methods of spectral estimation based on L-group operations do not allow obtaining an estimator of power spectral density (PSD), unlike PSD, a square under the function of L-group estimator does not characterize complete power of researched process. Inverse Fourier transform of L-group spectral estimator does not coincide with autocorrelation function (ACF) estimator of investigated stochastic process (signal). L-group estimator is considered to be spectral estimator in the sense that it describes relative intensities of frequency spectrum components, but L-group spectral estimator is not a PSD estimator. Advantage of L-group spectral estimators is that they provide heights of spectral component peaks that are proportional to powers of corresponding harmonics from analyzed process. Some L-group spectral estimators (for instance, correlogram and periodogram) are hyperspectral density estimators in the sense of the relationships (1.6.16)...(1.6.19).

As discussed in Section 1.6, nonlinear mappings (1.6.14a,b,c) provide, first, quite satisfactory (in adequacy) discrete signal representation in frequency domain, and

DOI: 10.1201/9781003275855-5

second, simplicity of calculations of corresponding measures of statistical interrelation (MSI) that do not require performing operation of multiplication. The last circumstance distinguishes the mappings (1.6.14a,b,c) from known linear transforms that allow transferring from temporal to frequency form of signal representation, thus, in some cases, one can consider algorithms of discrete signal hyperspectral representation based on L-group operations to be more preferable, though this inference requires special research.

5.1   Correlogram Method of Spectral Estimation Based on L–group Operations

Known literature [96, 134–136, 138–143] discusses indirect method of PSD estimation that supposes using infinite sequence of signal samples to obtain ACF estimator whose Fourier transform gives a desired PSD. Within correlogram methods of spectral estimation, PSD estimators are formed on the basis of correlation estimators.

PSD $P_x(f)$ of discrete stochastic signal $x(i)$ is discrete-time Fourier transform of its ACF $r_x(m)$ (see, for instance, [140, (5.20)]):

$$P_x(f) = \Delta t \cdot \sum_{m=-\infty}^{\infty} r_x(m) \exp(-j2\pi f \cdot m\Delta t), \qquad (5.1.1)$$

where $j = \sqrt{-1}$; Δt is a sample interval of initial continuous stochastic process $x(t)$; f is a discrete frequency parameter.

Correlogram method of spectral estimation is based on substituting a finite set of values of the estimator $\hat{r}_x(m)$ of ACF $r_x(m)$ (correlogram) into the expression (5.1.1). One of possible PSD estimators $\hat{P}_x(f)$ based on unbiased ACF estimator $\hat{r}_x(m)$ is defined by the following relationships (see, for instance, [140, (5.21),(5.9)]):

$$\hat{P}_x(f) = \Delta t \cdot \sum_{m=-L-1}^{L-1} \hat{r}_x(m) \exp(-j2\pi f \cdot m\Delta t); \qquad (5.1.2)$$

$$\hat{r}_x(m) = \frac{1}{N-m} \sum_{i=0}^{N-m-1} x(i+m)x^*(i); \qquad (5.1.2a)$$

$$\hat{r}_x(m) = \hat{r}_x^*(-m), \qquad (5.1.2b)$$

where N is a number of samples $\{x(i)\}$ of complex-valued stochastic process $x(t)$; $x^* = a - jb$, if $x = a + jb$; maximal index L of discrete time shift m figuring in the estimator (5.1.2) is chosen by the relationship $L \ll N$.

The estimator $\hat{r}_x(m)$, as well as ACF $r_x(m)$, is a conjugate symmetric function, thus, the property (5.1.2b) allow using non-negative indexes only, i.e. ACF $r_x(m)$ and its estimator $\hat{r}_x(m)$ are completely defined by N samples. An idea of using maximal index L of time shift m: $L \ll N$ was suggested by Blackman R.B. and Tukey J.W. [134] to avoid large values of a variance of the estimator (5.1.2a) in the case of large time shifts $L \to N$, inasmuch as these values of a variance give less robust PSD estimator. Taking into account known decomposition of complex exponent: $\exp(-j\varphi) = \cos(\varphi) - j\sin(\varphi)$ and the property (5.1.2b), the formula (5.1.2) can be rewritten in the form:

$$\hat{P}_x(f) = \frac{2T}{N}$$
$$\times \sum_{m=0}^{L-1} (\mathrm{Re}[\hat{r}_x(m)]\cos(2\pi f \cdot mT/N) + \mathrm{Im}[\hat{r}_x(m)]\sin(2\pi f \cdot mT/N)), \quad (5.1.3)$$

where $T/N = \Delta t$; T is a duration of complex-valued stochastic process $x(t)$; N is a number of samples $\{x(i)\}$ of stochastic process $x(t)$ taken in the interval $[0, T]$.

As follows from the obtained relationship (5.1.3), the correlogram estimator $\hat{P}_x(f)$ assumes calculating a sum of scalar products between real and imaginary parts of ACF estimator $\hat{r}_x(m)$ (5.1.2a), and on the other hand, harmonic functions $\cos(2\pi f \cdot mT/N)$, $\sin(2\pi f \cdot mT/N)$, respectively.

Taking into account known relation between measure of statistical interrelation (MSI) and scalar product of a pair of signals, considered in Chapter 1, by the analogy with the estimator (5.1.3), one can obtain correlogram estimators of hyperspectral density (HSD) $\hat{H}_p(f)$, $\hat{H}_s(f)$, $\hat{H}_{l_1}(f)$ based on MSIs estimators $\hat{N}_p(x; m)$ (1.4.13), $\hat{N}_s(x; m)$ (1.4.15), $\hat{N}_{l_1}(x; m)$ (1.4.17) that are defined by the relationships (1.6.17), (1.6.18), (1.6.19), respectively:

$$\hat{H}_p(f) = N_p(\mathrm{Re}[\hat{N}_p(x; m)], \cos F(f, m)) + N_p(\mathrm{Im}[\hat{N}_p(x; m)], \sin F(f, m)); \quad (5.1.4)$$

$$N_p[a(m), b(m)] = L - \sum_{m=0}^{L-1} [\mathrm{sgn}(a(m) \vee b(m)) - \mathrm{sgn}(a(m) \wedge b(m))]; \quad (5.1.4a)$$

$$\hat{H}_s(f) = N_s(\mathrm{Re}[\hat{N}_s(x; m)], \cos F(f, m)) + N_s(\mathrm{Im}[\hat{N}_s(x; m)], \sin F(f, m)); \quad (5.1.5)$$

$$N_s[a(m), b(m)] = \sum_{m=0}^{L-1} |a(m)| + \sum_{m=0}^{L-1} |b(m)|$$

$$- \sum_{m=0}^{L-1} |a(m) - b(m)| \cdot [\mathrm{sgn}(a(m) \vee b(m)) - \mathrm{sgn}(a(m) \wedge b(m))]; \quad (5.1.5a)$$

$$\hat{H}_{l_1}(f) = N_{l_1}(\mathrm{Re}[\hat{N}_{l_1}(x; m)], \cos F(f, m)) + N_{l_1}(\mathrm{Im}[\hat{N}_{l_1}(x; m)], \sin F(f, m)); \quad (5.1.6)$$

$$N_{l_1}[a(m), b(m)] = \sum_{m=0}^{L-1} |a(m) + b(m)| - \sum_{m=0}^{L-1} |a(m) - b(m)|; \quad (5.1.6a)$$

$$\cos F(f, m) = \cos(2\pi f \cdot mT/N); \quad \sin F(f, m) = \sin(2\pi f \cdot mT/N), \quad (5.1.7)$$

where $a(m)$, $b(m)$ are arbitrary functions.

Maximal index L of time shift m figuring in the estimators (5.1.4), (5.1.5), (5.1.6) is chosen basing on fulfillment of relation $L \ll N$. Notice, that as alternative estimators (5.1.4), (5.1.5), (5.1.6), one can use the estimators of hyperspectral densities $\hat{H}'_p(f)$, $\hat{H}'_s(f)$, $\hat{H}'_{l_1}(f)$ in the form of meet operation of the functions (5.1.4a), (5.1.5a), (5.1.6a), respectively:

$$\hat{H}'_p(f) = \mathrm{N}_p(\mathrm{Re}[\hat{\mathrm{N}}_p(x;m)], \cos\mathrm{F}(f,m)) \wedge \mathrm{N}_p(\mathrm{Im}[\hat{\mathrm{N}}_p(x;m)], \sin\mathrm{F}(f,m)); \quad (5.1.8a)$$

$$\hat{H}'_s(f) = \mathrm{N}_s(\mathrm{Re}[\hat{\mathrm{N}}_s(x;m)], \cos\mathrm{F}(f,m)) \wedge \mathrm{N}_s(\mathrm{Im}[\hat{\mathrm{N}}_s(x;m)], \sin\mathrm{F}(f,m)). \quad (5.1.8b)$$

$$\hat{H}'_{l_1}(f) = \mathrm{N}_{l_1}(\mathrm{Re}[\hat{\mathrm{N}}_{l_1}(x;m)], \cos\mathrm{F}(f,m)) \wedge \mathrm{N}_{l_1}(\mathrm{Im}[\hat{\mathrm{N}}_{l_1}(x;m)], \sin\mathrm{F}(f,m)). \quad (5.1.8c)$$

Here we notice that the term «correlogram» estimators with respect to the estimators of hyperspectral density (5.1.4), (5.1.5), (5.1.6) is used rather conditionally and is not quite correct. Besides, notice also, that using ACF estimator $\hat{r}_x(m)$ (5.1.2a) instead of the functions $\hat{\mathrm{N}}_p(x;m)$ (1.4.13), $\hat{\mathrm{N}}_s(x;m)$ (1.4.15), $\hat{\mathrm{N}}_{l_1}(x;m)$ (1.4.17) in the relationships (5.1.4), (5.1.5), (5.1.6), one can really obtain correlogram estimators in L-groups.

Having gotten the algorithms (5.1.4), (5.1.5), (5.1.6) of correlogram estimators of hyperspectral density $\hat{H}_p(f)$, $\hat{H}_s(f)$, $\hat{H}_{l_1}(f)$ based on MSIs (1.4.10), (1.4.11), (1.4.12), respectively, it is necessary, first, to find out how close are they with respect to classic correlogram estimator (5.1.3), and second, could they be used for estimating spectral composition of a mixture of harmonic signals and noise? Algorithms (5.1.4) and (5.1.5) are illustrated by the following example.

Example 5.1.1. Let $x(i)$ be additive mixture of M complex discrete harmonic signals observed in the presence of complex quasi-white Gaussian noise $n(i)$ with zero mean:

$$x(i) = \sum_{l=1}^{M} s_l(i) + n(i) = \sum_{l=1}^{M} A_l \exp[j2\pi f_l \cdot i + \varphi_l] + n(i), \quad (5.1.9)$$

where $i = 0, 1, \ldots, N-1$; $A_l = \mathrm{const}$, $f_l = \mathrm{const}$, φ_l is an unknown nonrandom initial phase taking its values in the interval $\varphi_l \in [0, 2\pi]$; $n(i) = n_c(i) + j \cdot n_s(i)$; $\mathbf{M}\{n_c^2(i)\} = \mathbf{M}\{n_s^2(i)\} = D_n/2$, $\mathbf{M}\{*\}$ is a symbol of mathematical expectation; D_n is a variance of quasi-white Gaussian noise $n(i)$ with zero mean.

Within the considered example, amplitudes of harmonics are chosen to be equal $A_l = A$, so that the relationships hold: $A^2/2 = D_n$; $F_n > 2\max_l\{f_l\}$, where D_n is a variance of noise $n(i)$; F_n is an upper bound frequency of PSD of quasi-white Gaussian noise $n(i)$; $M = 8$. The number of samples N of stochastic process $x(i)$ used when forming estimators of ACF $\hat{r}_x(m)$ (5.1.2a), MSIs $\hat{\mathrm{N}}_p(x;m)$ (1.4.13), and $\hat{\mathrm{N}}_s(x;m)$ (1.4.15) is equal to $N = 1024$. Maximal index L of MSI time shift m figuring in the estimators (5.1.4), (5.1.5) is equal to $L=128$.

Fig. 5.1.1 illustrates correlogram estimate of PSD $\hat{P}_x(f)$ of the mixture $x(i)$ (5.1.9) obtained by the relationship (5.1.3) for initial situation described above. True positions of the frequencies of harmonic signals from the mixture (5.1.9) are shown by the symbols "+".

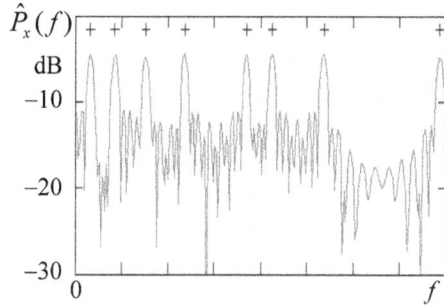

FIGURE 5.1.1   Correlogram estimate of PSD $\hat{P}_x(f)$ (5.1.3) of the mixture $x(i)$ (5.1.9)

Fig. 5.1.2a and 5.1.2b depict correlogram estimates of hyperspectral densities $\hat{H}_p(f)$ (5.1.4) and $\hat{H}_s(f)$ (5.1.5) of the signal $x(i)$ (5.1.9), correspondingly, for the situation described above. True positions of the frequencies of harmonic signals (5.1.9) are denoted by the symbols "+". Correlation coefficients between hyperspectral density estimates $\hat{H}_p(f)$ (5.1.4), $\hat{H}_s(f)$ (5.1.5), and correlogram estimate of PSD $\hat{P}_x(f)$ (5.1.3) are equal to: $r[\hat{H}_p(f), \hat{P}_x(f)]=0.765$ and $r[\hat{H}_s(f), \hat{P}_x(f)]=0.915$, respectively.

$\triangledown$

5.2   Periodogram Method of Spectral Estimation Based on L–group Operations

Known literature [96, 134–136, 138–143] considers a direct method of PSD estimation that assumes calculating squared modulus of Fourier transform for infinite sequence of signal samples using statistical averaging. PSD estimators based on such a direct method are called periodograms. The resulting function obtained without

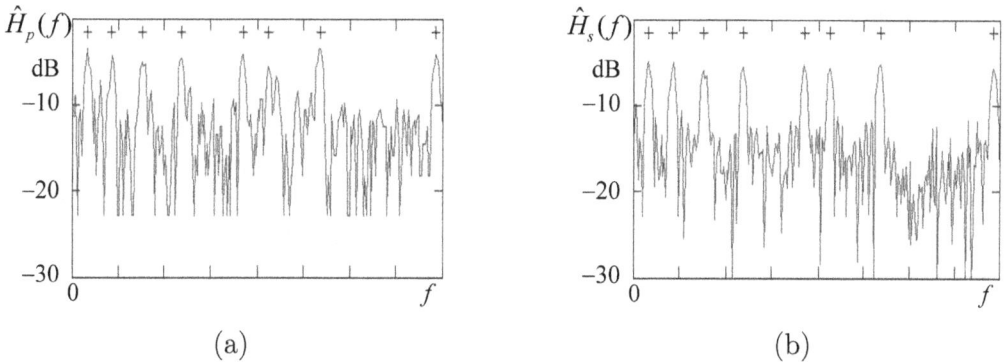

(a)   (b)

FIGURE 5.1.2   Correlogram estimates of hyperspectral densities: (a) $\hat{H}_p(f)$ (5.1.4); (b) $\hat{H}_s(f)$ (5.1.5)

such averaging and called sample spectrum is considered to be unsatisfactory due to statistical inconsistency of obtained estimator. In this section we briefly consider known methods of averaging that provide statistically consistent spectral estimators using a finite number of samples of initial signal.

PSD estimator $\hat{P}_x(f)$ of discrete stochastic signal $x(i)$ obtained by the direct method is discrete-time Fourier transform of the sample sequence $\{x(i)\}$ of the observed continuous signal $x(t)$ (see, for instance, [140, (5.31)]):

$$\hat{P}_x(f) = \frac{\Delta t}{N} \cdot \left| \sum_{i=0}^{N-1} x(i) \exp\left(-j2\pi f \cdot \frac{i\Delta t}{2N} \right) \right|^2, \tag{5.2.1}$$

where $j = \sqrt{-1}$; Δt is a sample interval of initial continuous stochastic process $x(t)$; f is discrete frequency parameter.

This expression represents the initial (non-modified) form of periodogram PSD estimator that is inconsistent, since operation of expectation is ignored. Therefore, to improve statistical properties of periodogram estimators, operations that replace ensemble averaging are used. Thus, for instance, Bartlett method [144]) supposes averaging in a set of periodograms obtained in segments of initial sample sequence. Within Welch method [145], Bartlett approach is used with respect to overlapping segments, so that data window is introduced to reduce estimator bias owing to a leakage effect. We consider both methods as applied to L-group algorithms.

5.2.1 Periodogram Method of Spectral Estimation Based on Measure of Statistical Interrelation

Taking into account known decomposition of complex exponent: $\exp(-j\varphi) = \cos(\varphi) - j\sin(\varphi)$, the formula (5.2.1) can be rewritten in the form:

$$\hat{P}_x(f) = \frac{\Delta t}{N} \cdot \left| \sum_{i=0}^{N-1} \left(\mathrm{Re}[x(i)] \cos\left(2\pi f \cdot \frac{i\Delta t}{2N} \right) + \mathrm{Im}[x(i)] \sin\left(2\pi f \cdot \frac{i\Delta t}{2N} \right) \right. \right.$$

$$\left. \left. +j\,\mathrm{Im}[x(i)] \cos\left(2\pi f \cdot \frac{i\Delta t}{2N} \right) - j\,\mathrm{Re}[x(i)] \sin\left(2\pi f \cdot \frac{i\Delta t}{2N} \right) \right) \right|^2, \tag{5.2.2}$$

where Δt is a sample interval of initial continuous stochastic process $x(t)$; N is a number of samples $\{x(i)\}$ of stochastic process $x(t)$ taken in the interval $[0, T]$.

As follows from the obtained expression (5.2.2), the estimator $\hat{P}_x(f)$ supposes calculating a sum of scalar products between real and imaginary parts of the samples $\{x(i)\}$ and, on the other hand, harmonic functions $\cos(2\pi f \cdot i\Delta t/(2N))$, $\sin(2\pi f \cdot i\Delta t/(2N))$, respectively.

Taking into account known relation between MSI and scalar product of a pair of signals considered in Chapter 1, by the analogy with the estimator (5.2.2), one can obtain periodogram estimators of hyperspectral densities (HSD) $\hat{H}_p(f)$ and $\hat{H}_s(f)$ based on MSIs (1.4.10) and (1.4.11), respectively, that are defined by the following relationships:

$$\hat{H}_p(f) = |N_p(\text{Re}[x(i)], \cos F(f, i)) + N_p(\text{Im}[x(i)], \sin F(f, i)) \tag{5.2.3}$$
$$+ j \cdot N_p(\text{Im}[x(i)], \cos F(f, i)) - j \cdot N_p(\text{Re}[x(i)], \sin F(f, i))|^2;$$

$$N_p(u(i), v(i)) = N - \sum_{i=0}^{N-1} [\text{sgn}(u(i) \vee v(i)) - \text{sgn}(u(i) \wedge v(i))]; \tag{5.2.3a}$$

$$\hat{H}_s(f) = |N_s(\text{Re}[x(i)], \cos F(f, i)) + N_s(\text{Im}[x(i)], \sin F(f, i)) \tag{5.2.4}$$
$$+ j \cdot N_s(\text{Im}[x(i)], \cos F(f, i)) - j \cdot N_s(\text{Re}[x(i)], \sin F(f, i))|^2;$$

$$N_s(u(i), v(i)) = \sum_{i=0}^{N-1} |u(i)| + \sum_{i=0}^{N-1} |v(i)| \tag{5.2.4a}$$

$$- \sum_{i=0}^{N-1} |u(i) - v(i)| \cdot [\text{sgn}(u(i) \vee v(i)) - \text{sgn}(u(i) \wedge v(i))];$$

$$\cos F(f, i) = \cos\left(\frac{\pi f \cdot (2i+1)\Delta t}{2N}\right); \quad \sin F(f, i) = \sin\left(\frac{\pi f \cdot (2i+1)\Delta t}{2N}\right), \tag{5.2.5}$$

where $N_p(u(i), v(i))$ and $N_s(u(i), v(i))$ are sample MSIs (1.4.10) and (1.4.11), respectively.

Notice, that according to (5.2.5), here as orthogonal harmonic functions we use the functions $\cos F(f, i)$, $\sin F(f, i)$ that are used in discrete cosine transform and discrete sine transform. They differ from the functions (5.1.7) used to obtain correlogram estimators of HSDs (5.1.4), (5.1.5), (5.1.6). Both variants are acceptable for using in *L*-group algorithms.

Having obtained the algorithms (5.2.3) and (5.2.4) of periodogram estimators of HSDs $\hat{H}_p(f)$ and $\hat{H}_s(f)$ based on MSIs (1.4.10) and (1.4.11), respectively, it is necessary to find out, first, how close are they with respect to classic periodogram estimator (5.2.1), and second, could they be used for estimating spectral structure of a mixture of harmonic signals and noise? Algorithms (5.2.3) and (5.2.4) are illustrated by the following example.

Example 5.2.1. Let $x(i)$ be additive mixture of M complex discrete harmonic signals observed in the presence of complex quasi-white Gaussian noise $n(i)$ with zero mean (5.1.9).

Within the considered example, amplitudes of harmonics are chosen to be equal $A_l = A$, so that the following relationships hold: $A^2/2 = D_n$, $F_n > 2\max_l\{f_l\}$, where D_n is a variance of noise $n(i)$; F_n is upper bound frequency of PSD of quasi-white Gaussian noise $n(i)$; $M = 15$. The number of samples N of stochastic process $x(i)$ used when forming the estimators (5.2.1), (5.2.3), (5.2.4) is equal to $N=1024$.

Fig. 5.2.1 depicts periodogram estimate $\hat{P}_x(f)$ of PSD of the mixture $x(i)$ (5.1.9) obtained as a result of calculating the relationship (5.2.1) for initial situation described above. True positions of the frequencies of harmonic signals from the mixture (5.1.9) are denoted by the symbols "+".

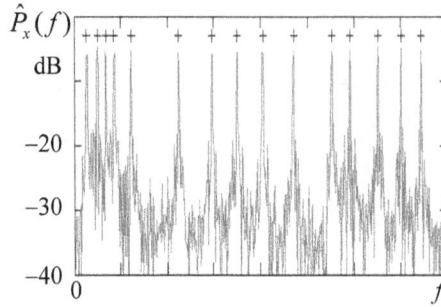

FIGURE 5.2.1   Periodogram estimate $\hat{P}_x(f)$ of PSD (5.2.1) of the mixture $x(i)$ (5.1.9)

Fig. 5.2.2a and 5.2.2b illustrate periodogram HSD estimates $\hat{H}_p(f)$ (5.2.3) and $\hat{H}_s(f)$ (5.2.4) of the signal $x(i)$ (5.1.9), respectively, for initial situation described above. True positions of the frequencies of harmonic signals (5.1.9) are denoted by the signs "+". Correlation coefficients between estimates of HSDs $\hat{H}_p(f)$ (5.2.3), $\hat{H}_s(f)$ (5.2.4), and periodogram PSD estimate $\hat{P}_x(f)$ (5.2.1) are equal to: $r[\hat{H}_p(f), \hat{P}_x(f)] = 0.973$ and $r[\hat{H}_s(f), \hat{P}_x(f)] = 0.995$, respectively.

$\triangledown$

5.2.2   Periodogram Method of Spectral Estimation Based on Measure of Statistical Interrelation: Bartlett Approach

According to Bartlett method, smoothing of sample spectrum is based on creating pseudo-ensemble of periodograms by partitioning a data sequence of N samples into N_I non-overlapping intervals with L samples in each, so that $N_I \cdot L \leqslant N$. Then, the p-th interval of samples $x^{(p)}(i)$ is determined by the relationship:

$$x^{(p)}(i) = x(i) \cdot I^{(p)}(i); \tag{5.2.6}$$

$$I^{(p)}(i) = 1(i - pL) - 1(i - (pL + L)), \tag{5.2.6a}$$

where $p = 0, \ldots, N_I-1$; $i = 0, \ldots, N-1$; $I^{(p)}(i)$ is interval function; $1(i)$ is Heaviside unit step function.

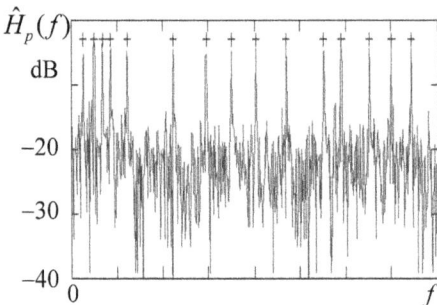

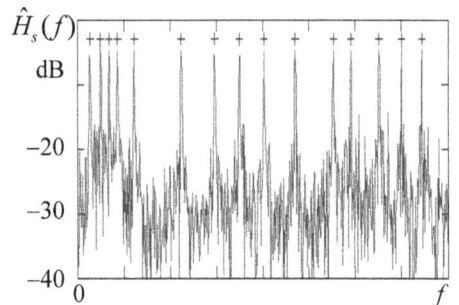

(a)   (b)

FIGURE 5.2.2   Periodogram estimates of HSDs: (a) $\hat{H}_p(f)$ (5.2.3); (b) $\hat{H}_s(f)$ (5.2.4)

In each interval $I^{(p)}(i)$, sample spectrum is independently calculated (see, for instance, [140, (5.35)]):

$$\hat{P}_x^{(p)}(f) = \frac{\Delta t}{L} \cdot \left| \sum_{i=0}^{N-1} x^{(p)}(i) \exp\left(-j 2\pi f \cdot \frac{i\Delta t}{2N}\right) \right|^2, \qquad (5.2.7)$$

where Δt is a sample interval of initial continuous stochastic process $x(t)$; f is discrete frequency parameter; $x^{(p)}(i)$ is the p-th interval of samples $\{x(i)\}$ determined by the relationship (5.2.6).

Further, at each frequency, N_I separate non-modified periodograms are averaged to obtain averaged Bartlett periodogram:

$$\hat{P}_B(f) = \frac{1}{N_I} \sum_{p=0}^{N_I-1} \hat{P}_x^{(p)}(f). \qquad (5.2.8)$$

A variance of averaged Bartlett periodogram $\hat{P}_B(f)$ is inversely proportional to a number of used intervals N_I.

Taking into account the relationships (5.2.3), (5.2.4), by the analogy with the estimator (5.2.8), one can obtain modified Bartlett periodogram estimators of HSDs $\hat{H}_B^p(f)$ and $\hat{H}_B^s(f)$ based on MSIs (1.4.10) and (1.4.11), respectively, that are defined by the following relationships:

$$\hat{H}_B^p(f) = \frac{1}{N_I} \sum_{p=0}^{N_I-1} \hat{H}_p^{(p)}(f); \qquad (5.2.9)$$

$$\hat{H}_p^{(p)}(f) = \Big| N_p(\text{Re}[x(i)], \cos F(f,i); I^{(p)}(i)) + N_p(\text{Im}[x(i)], \sin F(f,i); I^{(p)}(i))$$

$$+j \cdot N_p(\text{Im}[x(i)], \cos F(f,i); I^{(p)}(i))$$

$$-j \cdot N_p(\text{Re}[x(i)], \sin F(f,i); I^{(p)}(i)) \Big|^2; \qquad (5.2.9a)$$

$$N_p(u(i), v(i); z(i))$$

$$= \sum_{i=0}^{N-1} [1 - (\text{sgn}(u(i) \vee v(i)) - \text{sgn}(u(i) \wedge v(i)))] \cdot z(i); \qquad (5.2.9b)$$

$$\hat{H}_B^s(f) = \frac{1}{N_I} \sum_{p=0}^{N_I-1} \hat{H}_s^{(p)}(f); \qquad (5.2.10)$$

$$\hat{H}_s^{(p)}(f) = \Big| N_s(\text{Re}[x(i)], \cos F(f,i); I^{(p)}(i)) + N_s(\text{Im}[x(i)], \sin F(f,i); I^{(p)}(i))$$

$$+j \cdot N_s(\text{Im}[x(i)], \cos F(f,i); I^{(p)}(i))$$

$$-j \cdot N_s(\text{Re}[x(i)], \sin F(f,i); I^{(p)}(i)) \Big|^2; \qquad (5.2.10a)$$

$$N_s(u(i), v(i); z(i)) = \sum_{i=0}^{N-1} (|u(i)| + |v(i)|) \cdot z(i) \qquad (5.2.10b)$$

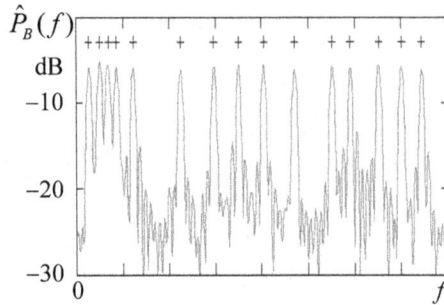

FIGURE 5.2.3   Bartlett periodogram PSD estimate $\hat{P}_B(f)$ (5.2.8)

$$-\sum_{i=0}^{N-1} |u(i) - v(i)| \cdot [\text{sgn}(u(i) \vee v(i)) - \text{sgn}(u(i) \wedge v(i))] \cdot z(i);$$

$$\text{cosF}(f, i) = \cos\left(\frac{\pi f \cdot (2i+1)\Delta t}{2N}\right); \ \text{sinF}(f, i) = \sin\left(\frac{\pi f \cdot (2i+1)\Delta t}{2N}\right), \quad (5.2.11)$$

where $\hat{H}_p^{(p)}(f)$, $\hat{H}_s^{(p)}(f)$ are HSD estimators in each interval $I^{(p)}(i)$ (5.2.6a); $N_p(u(i), v(i); z(i))$, $N_s(u(i), v(i); z(i))$ are sample MSIs (1.4.10) and (1.4.11), respectively; $z(i)$ is an arbitrary multiplier function that in this context is determined by the formula (5.2.6a).

Having obtained the algorithms (5.2.9) and (5.2.10) of modified Bartlett periodogram estimators of HSDs $\hat{H}_B^p(f)$ and $\hat{H}_B^s(f)$ based on MSIs (1.4.10) and (1.4.11), respectively, it is necessary to find out, first, how close are they with respect to modified periodogram estimator (5.2.8), and the second, could they be used for estimating spectral structure of a mixture of harmonic signals and noise? Algorithms (5.2.9) and (5.2.10) are illustrated by the following example.

Example 5.2.2. Let $x(i)$ be additive mixture of M complex discrete harmonic signals observed in the presence of complex quasi-white Gaussian noise $n(i)$ with zero mean (5.1.9).

Within the considered example, amplitudes of harmonics are chosen to be equal $A_l = A$, so that the relationships hold: $A^2/2 = D_n$, $F_n > 2\max_{l}\{f_l\}$, where D_n is a variance of noise $n(i)$; F_n is upper bound frequency of PSD of quasi-white Gaussian noise $n(i)$; $M = 15$. The number of samples N of stochastic process $x(i)$ used when forming the estimators (5.2.8), (5.2.9), (5.2.10) is equal to N=1024, and a number of non-overlapping intervals N_I with L samples in each is equal to N_I=4, so that L=256.

Fig. 5.2.3 depicts modified Bartlett periodogram PSD estimate $\hat{P}_B(f)$ of the mixture $x(i)$ (5.1.9) obtained as a result of calculating the relationship (5.2.8) for initial situation described above. True positions of the frequencies of harmonic signals from the mixture (5.1.9) are denoted by the symbols "+".

Fig. 5.2.4a and 5.2.4b illustrates modified Bartlett periodogram estimates of HSDs $\hat{H}_B^p(f)$ (5.2.9) and $\hat{H}_B^s(f)$ (5.2.10) of the signal $x(i)$ (5.1.9) obtained by

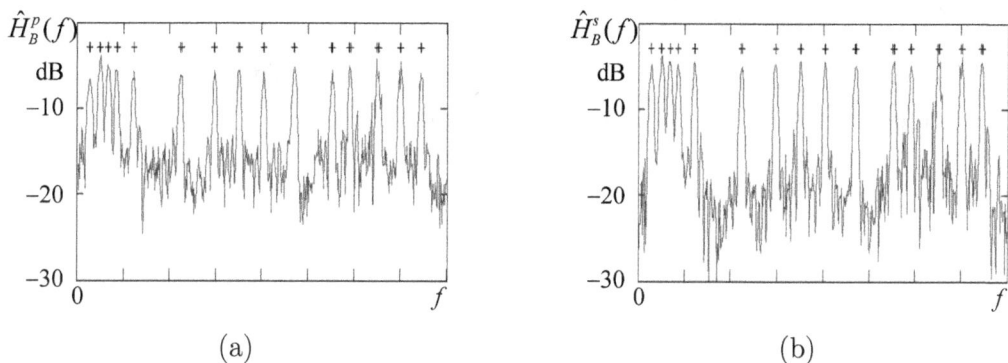

FIGURE 5.2.4   Bartlett periodogram estimates of HSDs: (a) $\hat{H}_B^p(f)$ (5.2.9); (b) $\hat{H}_B^s(f)$ (5.2.10)

Bartlett method based on MSIs (1.4.10) and (1.4.11), respectively, for initial situation described above. True positions of the frequencies of harmonic signals (5.1.9) are denoted by the signs "+". Correlation coefficients between the estimates of HSDs $\hat{H}_B^p(f)$ (5.2.9), $\hat{H}_B^s(f)$ (5.2.10) and modified periodogram estimate of PSD $\hat{P}_B(f)$ (5.2.8) are equal to: $r[\hat{H}_B^p(f), \hat{P}_B(f)]=0.967$ and $r[\hat{H}_B^s(f), \hat{P}_B(f)]=0.99$, respectively.

Comparing pairwise Figs. 5.2.1 and 5.2.3; Figs. 5.2.2a and 5.2.4a; Figs. 5.2.2b and 5.2.4b, we can see that resolution of the estimates $\hat{P}_B(f)$ (5.2.8), $\hat{H}_B^p(f)$ (5.2.9), and $\hat{H}_B^s(f)$ (5.2.10) is worse than resolution of the estimates $\hat{P}_x(f)$ (5.2.1), $\hat{H}_p(f)$ (5.2.3) and $\hat{H}_s(f)$ (5.2.4), respectively, that is explained by less (in $N_I = N/L=4$ times) number of samples used for estimation in each separate interval $I^{(p)}(i)$ (5.2.6a). Since $N = N_I \cdot L$, one can provide compromise relation between high resolution in frequency (maximizing L) and minimum variance of estimator (maximizing N_I). ▽

5.2.3   Periodogram Method of Spectral Estimation Based on Measure of Statistical Interrelation: Welch Approach

Welch approach lies in modifying Bartlett segmenting and averaging method by exploiting data window and using overlapping intervals. Before calculating periodogram in each interval, this interval is processed by window function. Using window function provide both side-lobes influence decreasing and estimator bias decreasing. Overlap of intervals provides increasing a number of averaged intervals on a given number of processed samples N that implies decreasing PSD estimator variance.

Sample spectrum smoothing based on Welch method supposes creating pseudo-ensemble of periodograms by partitioning a sequence of N data samples into N_I overlapping intervals with L samples in each and a shift S between adjacent intervals, so that $N_I = (N - L + S)/S$. Then, the processed p-th interval of samples $x^{(p)}(i)$ is determined by the relation:

$$x^{(p)}(i) = w(i) \cdot x(i) \cdot I^{(p)}(i);$$ (5.2.12)

$$I^{(p)}(i) = 1(i - pS) - 1(i - (pS + L)), \qquad (5.2.12a)$$

where $p = 0, \ldots, N_I - 1$; $i = 0, \ldots, N - 1$; $I^{(p)}(i)$ is interval function; $w(i)$ is a window function; $1(i)$ is Heaviside unit step function.

In each interval $I^{(p)}(i)$, sample spectrum is independently calculated (see, for instance, [140, (5.40)]):

$$\hat{P}_x^{(p)}(f) = \frac{\Delta t}{W \cdot L} \cdot \left| \sum_{i=0}^{N-1} x^{(p)}(i) \exp\left(-j2\pi f \cdot \frac{i\Delta t}{2N}\right) \right|^2; \qquad (5.2.13)$$

$$W = \sum_{i=0}^{L-1} w^2(i), \qquad (5.2.13a)$$

where Δt is a sampling interval of initial continuous stochastic process $x(t)$; f is discrete frequency parameter; $w(i)$ is a window function; $x^{(p)}(i)$ is the p-th interval of samples $\{x(i)\}$ determined by the relationship (5.2.12).

Then, at each frequency, N_I separate non-modified periodograms are averaged to obtain averaged Welch periodogram:

$$\hat{P}_W(f) = \frac{1}{N_I} \sum_{p=0}^{N_I-1} \hat{P}_x^{(p)}(f). \qquad (5.2.14)$$

Taking into account the relationships (5.2.9), (5.2.10), by the analogy with the estimator (5.2.14), one can obtain modified Welch periodogram estimators of HSDs $\hat{H}_W^p(f)$ and $\hat{H}_W^s(f)$ based on MSIs (1.4.10), (1.4.11) that are defined by the following relationships, respectively:

$$\hat{H}_W^p(f) = \frac{1}{N_I} \sum_{p=0}^{N_I-1} \hat{H}_p^{(p)}(f); \qquad (5.2.15)$$

$$\hat{H}_p^{(p)}(f) = \left| N_p(\text{Re}[x^{(p)}(i)], \cos F(f,i); I^{(p)}(i)) + N_p(\text{Im}[x^{(p)}(i)], \sin F(f,i); I^{(p)}(i)) \right.$$

$$+ j \cdot N_p(\text{Im}[x^{(p)}(i)], \cos F(f,i); I^{(p)}(i))$$

$$\left. - j \cdot N_p(\text{Re}[x^{(p)}(i)], \sin F(f,i); I^{(p)}(i)) \right|^2; \qquad (5.2.15a)$$

$$N_p(u(i), v(i); z(i))$$

$$= \sum_{i=0}^{N-1} \left[1 - (\text{sgn}(u(i) \vee v(i)) - \text{sgn}(u(i) \wedge v(i))) \right] \cdot z(i); \qquad (5.2.15b)$$

$$\hat{H}_W^s(f) = \frac{1}{N_I} \sum_{p=0}^{N_I-1} \hat{H}_s^{(p)}(f); \qquad (5.2.16)$$

$$\hat{H}_s^{(p)}(f) = \left| N_s(\text{Re}[x^{(p)}(i)], \cos F(f,i); I^{(p)}(i)) + N_s(\text{Im}[x^{(p)}(i)], \sin F(f,i); I^{(p)}(i)) \right.$$

$$+ j \cdot N_s(\text{Im}[x^{(p)}(i)], \cos F(f,i); I^{(p)}(i))$$

$$\left. - j \cdot N_s(\text{Re}[x^{(p)}(i)], \sin F(f,i); I^{(p)}(i)) \right|^2; \qquad (5.2.16a)$$

$$N_s(u(i), v(i); z(i)) = \sum_{i=0}^{N-1} (|u(i)| + |v(i)|) \cdot z(i)$$

$$- \sum_{i=0}^{N-1} |u(i) - v(i)| \cdot [\text{sgn}(u(i) \vee v(i)) - \text{sgn}(u(i) \wedge v(i))] \cdot z(i); \qquad (5.2.16b)$$

$$\cos F(f, i) = \cos\left(\frac{\pi f \cdot (2i+1)\Delta t}{2N}\right); \; \sin F(f, i) = \sin\left(\frac{\pi f \cdot (2i+1)\Delta t}{2N}\right), \quad (5.2.17)$$

where $\hat{H}_p^{(p)}(f)$, $\hat{H}_s^{(p)}(f)$ are HSD estimators in each interval $I^{(p)}(i)$ (5.2.12a); $N_p(u(i), v(i); z(i))$, $N_s(u(i), v(i); z(i))$ are sample MSIs (1.4.10) and (1.4.11), respectively; $x^{(p)}(i)$ is the p-th interval of the samples $\{x(i)\}$ determined by the relationship (5.2.12); $z(i)$ is an arbitrary multiplier function that in this context is determined by the formula (5.2.12a).

Having obtained the algorithms (5.2.15) and (5.2.16) of modified Welch periodogram estimators of HSDs $\hat{H}_W^p(f)$ and $\hat{H}_W^s(f)$ that are based on MSIs (1.4.10), (1.4.11), respectively, it is necessary to find out, first, how close are they with respect to modified periodogram estimator (5.2.14), and second, could they be used for estimating spectral structure of a mixture of harmonic signals and noise? Algorithms (5.2.15) and (5.2.16) are illustrated by the following example.

Example 5.2.3. Let $x(i)$ be additive mixture of M complex discrete harmonic signals observed in the presence of complex quasi-white Gaussian noise $n(i)$ with zero mean (5.1.9).

Within the considered example, amplitudes of harmonics are chosen to be equal $A_l = A$, so that the relationships hold: $A^2/2 = D_n$, $F_n > 2\max_l\{f_l\}$, where D_n is a variance of noise $n(i)$; F_n is upper bound frequency of PSD of quasi-white Gaussian noise $n(i)$; $M = 15$. The number of samples N of stochastic process $x(i)$ used when forming the estimators (5.2.14), (5.2.15), (5.2.16) is equal to $N=1024$, and a number of overlapping intervals N_I with L samples in each is equal to $N_I=7$, so that $L=256$; $S=128$.

Fig. 5.2.5 depicts modified Welch periodogram PSD estimate $\hat{P}_W(f)$ of the mixture $x(i)$ (5.1.9) obtained according to the relationship (5.2.14) for initial situation described above. True positions of the frequencies of harmonic signals from the mixture (5.1.9) are depicted by the signs "+".

Fig. 5.2.6a and 5.2.6b depict modified Welch peridogram estimates of HSDs $\hat{H}_W^p(f)$ (5.2.15) and $\hat{H}_W^s(f)$ (5.2.16) of the mixture $x(i)$ (5.1.9) based on MSIs (1.4.10) and (1.4.11), respectively, for initial situation described above. True positions of the frequencies of harmonic signals (5.1.9) are denoted by the signs "+". Correlation coefficients between the estimates of HSDs $\hat{H}_W^p(f)$ (5.2.15), $\hat{H}_W^s(f)$ (5.2.16) and modified periodogram PSD estimate $\hat{P}_W(f)$ (5.2.14) are equal to: $r[\hat{H}_W^p(f), \hat{P}_W(f)]=0.986$ and $r[\hat{H}_W^s(f), \hat{P}_W(f)]=0.996$, respectively.

Comparing pairwise Fig. 5.2.5 and 5.2.1; Fig. 5.2.6a and 5.2.2a; Fig. 5.2.6b and 5.2.2b, we can see that resolution of estimators $\hat{P}_W(f)$ (5.2.8), $\hat{H}_W^p(f)$ (5.2.15) and $\hat{H}_W^s(f)$ (5.2.16) is worse than resolution of estimators $\hat{P}_x(f)$ (5.2.1), $\hat{H}_p(f)$

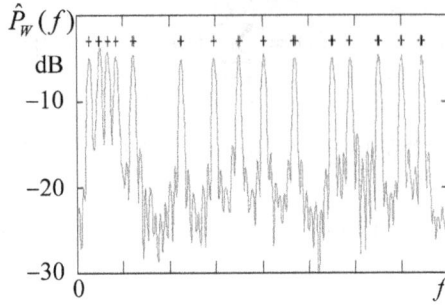

FIGURE 5.2.5   Welch periodogram PSD estimate $\hat{P}_W(f)$ (5.2.14)

(5.2.3) and $\hat{H}_s(f)$ (5.2.4), respectively, that is explained by less ($N/L=4$ times) number of samples used for estimation in each separate interval $I^{(p)}(i)$ (5.2.12a). When realizing Welch method by L-group algorithms, there is a strong impact of nonlinear processing on SNR worsening, especially in the upper part of frequency range, and also on a leakage effect over side-lobes owing to using relatively small number of the samples in a window function $w(i)$. $\triangledown$

5.3   Spectral Estimation Methods Based on Measure of Statistical Interrelation Matrix Estimator

Some algorithms of spectral estimation suppose forming an estimator $\hat{\mathbf{R}}_x$ of autocorrelation matrix (ACM) $\mathbf{R}_x$ with dimensionality $L \times L$ on the basis of an estimator $\hat{r}_x(m)$ of ACF $r_x(m)$ (5.1.2a). Sometimes, computational resources of a processing system are not sufficient to process all L samples of an estimator $\hat{r}_x(m)$ of ACF (5.1.2a). Then, the dimensionality of ACM estimator $\hat{\mathbf{R}}_x$ can be reduced down to

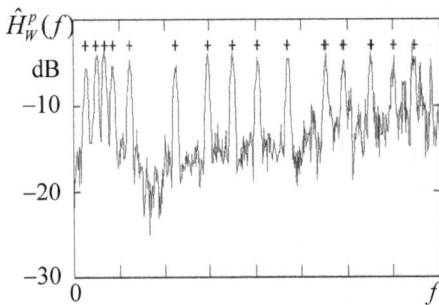

(a)

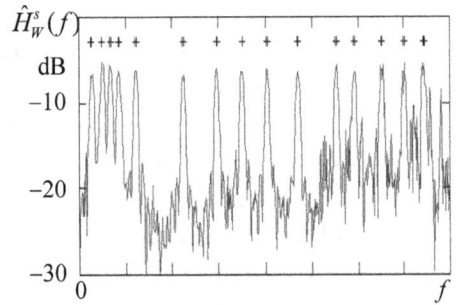

(b)

FIGURE 5.2.6   Welch peridogram estimates of HSDs: (a) $\hat{H}_W^p(f)$ (5.2.15); (b) $\hat{H}_W^s(f)$ (5.2.16).

some acceptable size $K \leq L$, $L = K \cdot \tau$, and ACM estimator $\hat{\mathbf{R}}_x$ takes the form:

$$\hat{\mathbf{R}}_x = \left\| R^x_{i,k} \right\| = \left\| \begin{matrix} \hat{r}_x(0) & \hat{r}^*_x(\tau) & \cdots & \hat{r}^*_x((K-1)\tau) \\ \hat{r}_x(\tau) & \hat{r}_x(0) & \cdots & \hat{r}^*_x((K-2)\tau) \\ \vdots & \vdots & \ddots & \vdots \\ \hat{r}_x((K-1)\tau) & \hat{r}_x((K-2)\tau) & \cdots & \hat{r}_x(0) \end{matrix} \right\|; \quad (5.3.1)$$

$$R^x_{i,k} = \begin{cases} \hat{r}_x(0), & i = k; \\ \hat{r}_x(|i-k|\tau), & i > k; \\ \hat{r}^*_x(|i-k|\tau), & i < k, \end{cases} \quad R^x_{i,k} = \bar{R}^x_{k,i}; \quad (5.3.1a)$$

$$i = 0, 1, \ldots, K-1; \; k = 0, 1, \ldots, K-1.$$

where, as before, $a^*(\tau)$ is a function conjugate to $a(\tau)$; $\bar{B}$ is a number conjugate to B.

In this section, we consider spectral estimation methods based on matrices of MSIs (MSI matrices) (1.4.10), (1.4.11), and (1.4.12), respectively. Since MSIs operate with real-valued vectors, remind a method of representation of complex stochastic process ACF based on ACFs and cross-correlation functions (CCF) of its real and imaginary components (see, for instance, [97, Theorem 15.1]).

Let $x(i) = u(i) + j \cdot v(i)$ be complex discrete stochastic signal, $i = 0, 1, \ldots, N-1$; $u(i) = \text{Re}[x(i)]$, $v(i) = \text{Im}[x(i)]$ its real and imaginary components. Then ACF estimator $\hat{r}_x(m)$ of complex discrete stochastic signal $x(i)$ is determined by the expression:

$$\hat{r}_x(m) = \hat{r}_u(m) + \hat{r}_v(m) - j[\hat{r}_{uv}(m) - \hat{r}_{vu}(m)], \quad (5.3.2)$$

where $\hat{r}_u(m)$, $\hat{r}_v(m)$ are ACF estimators of real $u(i) = \text{Re}[x(i)]$ and imaginary $v(i) = \text{Im}[x(i)]$ components of complex discrete stochastic signal $x(i)$ determined by the formula (5.1.2a); $\hat{r}_{uv}(m)$, $\hat{r}_{vu}(m)$ are CCF estimators of real $u(i)$ and imaginary $v(i)$ components of the signal $x(i)$ that are determined by the relationships:

$$\hat{r}_{uv}(m) = \frac{1}{N-m} \sum_{i=0}^{N-m-1} u(i+m)v^*(i); \quad (5.3.3a)$$

$$\hat{r}_{vu}(m) = \frac{1}{N-m} \sum_{i=0}^{N-m-1} v(i+m)u^*(i). \quad (5.3.3b)$$

Taking into account (5.3.2) and (5.3.3), ACM estimator $\hat{\mathbf{R}}_x$ has the following representation over ACM estimators $\hat{\mathbf{R}}_u$, $\hat{\mathbf{R}}_v$ of real $u(i)$ and imaginary $v(i)$ components of the signal $x(i)$, and also their cross-correlation matrix (CCM) estimators $\hat{\mathbf{R}}_{uv}$, $\hat{\mathbf{R}}_{vu}$:

$$\hat{\mathbf{R}}_x = \hat{\mathbf{R}}_u + \hat{\mathbf{R}}_v - j(\hat{\mathbf{R}}_{uv} - \hat{\mathbf{R}}_{vu}), \quad (5.3.4)$$

where ACM estimators $\hat{\mathbf{R}}_u$, $\hat{\mathbf{R}}_v$ are defined as symmetric matrices:

$$\hat{\mathbf{R}}_u = \left\| R^u_{i,k} \right\|; \quad \hat{\mathbf{R}}_v = \left\| R^v_{i,k} \right\|; \quad (5.3.5)$$

$$R^u_{i,k} = \begin{cases} \hat{r}_u(0), & i = k; \\ \hat{r}_u(|i-k|\tau, & i \neq k; \end{cases} \quad R^v_{i,k} = \begin{cases} \hat{r}_v(0), & i = k; \\ \hat{r}_v(|i-k|\tau, & i \neq k, \end{cases} \quad (5.3.5a)$$

and CCM estimators $\hat{\mathbf{R}}_{uv}$, $\hat{\mathbf{R}}_{vu}$ are defined as skew-symmetric matrices:

$$\hat{\mathbf{R}}_{uv} = \left\| R_{i,k}^{uv} \right\|; \quad \hat{\mathbf{R}}_{vu} = \left\| R_{i,k}^{vu} \right\|; \tag{5.3.6}$$

$$R_{i,k}^{uv} = \begin{cases} \hat{r}_{uv}(0), & i = k; \\ \hat{r}_{uv}(|i-k|\tau), & i > k; \\ -\hat{r}_{uv}(|i-k|\tau), & i < k; \end{cases} \tag{5.3.6a}$$

$$R_{i,k}^{vu} = \begin{cases} \hat{r}_{vu}(0), & i = k; \\ \hat{r}_{vu}(|i-k|\tau), & i > k; \\ -\hat{r}_{vu}(|i-k|\tau), & i < k; \end{cases} \tag{5.3.6b}$$

$$\tau = L/K; \ L << N; \ i = 0, 1, \dots, K-1; \ k = 0, 1, \dots, K-1.$$

If $x(i) = u(i) + j \cdot v(i)$ is complex discrete stochastic signal, $i = 0, 1, \dots, N-1$; $u(i) = \mathrm{Re}[x(i)]$, $v(i) = \mathrm{Im}[x(i)]$ are its real and imaginary components, then MSI estimator (1.4.13) $\hat{\mathrm{N}}_u^p(m)$ ($\hat{\mathrm{N}}_v^p(m)$) of its real $u(i)$ (imaginary $v(i)$) component is defined by the relationship:

$$\hat{\mathrm{N}}_u^p(m) = N - \frac{N}{N-m} \sum_{i=0}^{N-m-1} [\mathrm{sgn}(u(i+m) \vee u(i)) - \mathrm{sgn}(u(i+m) \wedge u(i))],$$
$$\tag{5.3.7}$$

and mutual MSI estimators (1.4.14a,b) $\hat{\mathrm{N}}_{uv}^p(m)$, $\hat{\mathrm{N}}_{vu}^p(m)$ of real $u(i)$ and imaginary $v(i)$ components of the signal $x(i)$ are defined by the relationships:

$$\hat{\mathrm{N}}_{uv}^p(m) = N - \frac{N}{N-m}$$
$$\times \sum_{i=0}^{N-m-1} [\mathrm{sgn}(u(i+m) \vee v(i)) - \mathrm{sgn}(u(i+m) \wedge v(i))]; \tag{5.3.8a}$$

$$\hat{\mathrm{N}}_{vu}^p(m) = N - \frac{N}{N-m}$$
$$\times \sum_{i=0}^{N-m-1} [\mathrm{sgn}(v(i+m) \vee u(i)) - \mathrm{sgn}(v(i+m) \wedge u(i))]. \tag{5.3.8b}$$

Quite similarly, MSI estimator (1.4.15) $\hat{\mathrm{N}}_u^s(m)$ ($\hat{\mathrm{N}}_v^s(m)$) of real $u(i)$ (imaginary $v(i)$) component of the signal $x(i)$ is defined by the relationship:

$$\hat{\mathrm{N}}_u^s(m) = \frac{1}{N-m} \left(\sum_{i=0}^{N-m-1} [|u(i+m)| + |u(i)|] + \sum_{i=0}^{N-m-1} |u(i+m) - u(i)| \right.$$
$$\left. \times [\mathrm{sgn}(u(i+m) \vee u(i)) - \mathrm{sgn}(u(i+m) \wedge u(i))] \right), \tag{5.3.9}$$

and mutual MSI estimators (1.4.16a,b) $\hat{\mathrm{N}}_{uv}^s(m)$, $\hat{\mathrm{N}}_{vu}^s(m)$ of real $u(i)$ and imaginary

$v(i)$ components of the signal $x(i)$ are defined by the relationships:

$$\hat{N}_{uv}^s(m) = \frac{1}{N-m} \left(\sum_{i=0}^{N-m-1} [|u(i+m)| + |v(i)|] + \sum_{i=0}^{N-m-1} |u(i+m) - v(i)| \right.$$
$$\left. \times [\text{sgn}(u(i+m) \vee v(i)) - \text{sgn}(u(i+m) \wedge v(i))] \right); \quad (5.3.10a)$$

$$\hat{N}_{vu}^s(m) = \frac{1}{N-m} \left(\sum_{i=0}^{N-m-1} [|v(i+m)| + |u(i)|] + \sum_{i=0}^{N-m-1} |v(i+m) - u(i)| \right.$$
$$\left. \times [\text{sgn}(v(i+m) \vee u(i)) - \text{sgn}(v(i+m) \wedge u(i))] \right). \quad (5.3.10b)$$

MSI estimator (1.4.17) $\hat{N}_u^{l_1}(m)$ $(\hat{N}_v^{l_1}(m))$ of real $u(i)$ (imaginary $v(i)$) component of the signal $x(i)$ is defined by the relationship:

$$\hat{N}_u^{l_1}(m) = \frac{1}{N-m} \sum_{i=0}^{N-m-1} [|u(i+m) + u(i)| - |u(i+m) - u(i)|], \quad (5.3.11)$$

and mutual MSI estimators (1.4.18a,b) $\hat{N}_{uv}^{l_1}(m)$, $\hat{N}_{vu}^{l_1}(m)$ of real $u(i)$ and imaginary $v(i)$ components of the signal $x(i)$ are defined by the relationships:

$$\hat{N}_{uv}^{l_1}(m) = \frac{1}{N-m} \sum_{i=0}^{N-m-1} [|u(i+m) + v(i)| - |u(i+m) - v(i)|]; \quad (5.3.12a)$$

$$\hat{N}_{vu}^{l_1}(m) = \frac{1}{N-m} \sum_{i=0}^{N-m-1} [|v(i+m) + u(i)| - |v(i+m) - u(i)|]. \quad (5.3.12b)$$

Here we notice that notations p, l_1, s, figuring in a subscript index in (1.4.10), (1.4.11), (1.4.12) and pointing at the relation between these measures and the corresponding generalized metrics, within the formulas (5.3.7)…(5.3.12), are, for convenience, used in a superscript index. Notice also, that maximal index L of time shift $m \leq L$, figuring in the estimators (5.3.7)…(5.3.12), as before, is chosen taking into account the relationship $L << N$.

Then, bearing in mind the relationships (5.3.7)…(5.3.12), the resulting MSI estimators $\hat{N}_x^p(m)$, $\hat{N}_x^s(m)$, and $\hat{N}_x^{l_1}(m)$ of complex discrete stochastic signal $x(i)$, by the analogy with (5.3.2), are defined by the following relationships, respectively:

$$\hat{N}_x^p(m) = \hat{N}_u^p(m) + \hat{N}_v^p(m) - j[\hat{N}_{uv}^p(m) - \hat{N}_{vu}^p(m)]; \quad (5.3.13a)$$

$$\hat{N}_x^s(m) = \hat{N}_u^s(m) + \hat{N}_v^s(m) - j[\hat{N}_{uv}^s(m) - \hat{N}_{vu}^s(m)]; \quad (5.3.13b)$$

$$\hat{N}_x^{l_1}(m) = \hat{N}_u^{l_1}(m) + \hat{N}_v^{l_1}(m) - j[\hat{N}_{uv}^{l_1}(m) - \hat{N}_{vu}^{l_1}(m)]. \quad (5.3.13c)$$

Taking into account the formulas (5.3.13a,b,c), MSI matrix estimators $\hat{\mathbf{N}}_{x,p}$, $\hat{\mathbf{N}}_{x,s}$, and $\hat{\mathbf{N}}_{x,l_1}$ of complex discrete stochastic signal $x(i)$ have the following representation over MSI matrix estimators $\hat{\mathbf{N}}_u^p$, $\hat{\mathbf{N}}_v^p$; $\hat{\mathbf{N}}_u^s$, $\hat{\mathbf{N}}_v^s$; $\hat{\mathbf{N}}_u^{l_1}$, $\hat{\mathbf{N}}_v^{l_1}$ of real $u(i)$ and imaginary $v(i)$ components of the signal $x(i)$, and also their mutual MSI matrix estimators $\hat{\mathbf{N}}_{uv}^p$, $\hat{\mathbf{N}}_{vu}^p$; $\hat{\mathbf{N}}_{uv}^s$, $\hat{\mathbf{N}}_{vu}^s$; $\hat{\mathbf{N}}_{uv}^{l_1}$, $\hat{\mathbf{N}}_{vu}^{l_1}$, respectively:

$$\hat{\mathbf{N}}_{x,p} = \hat{\mathbf{N}}_u^p + \hat{\mathbf{N}}_v^p - j(\hat{\mathbf{N}}_{uv}^p - \hat{\mathbf{N}}_{vu}^p); \tag{5.3.14a}$$

$$\hat{\mathbf{N}}_{x,s} = \hat{\mathbf{N}}_u^s + \hat{\mathbf{N}}_v^s - j(\hat{\mathbf{N}}_{uv}^s - \hat{\mathbf{N}}_{vu}^s); \tag{5.3.14b}$$

$$\hat{\mathbf{N}}_{x,l_1} = \hat{\mathbf{N}}_u^{l_1} + \hat{\mathbf{N}}_v^{l_1} - j(\hat{\mathbf{N}}_{uv}^{l_1} - \hat{\mathbf{N}}_{vu}^{l_1}), \tag{5.3.14c}$$

so that MSI matrix estimators $\hat{\mathbf{N}}_u^p$, $\hat{\mathbf{N}}_v^p$; $\hat{\mathbf{N}}_u^s$, $\hat{\mathbf{N}}_v^s$; $\hat{\mathbf{N}}_u^{l_1}$, $\hat{\mathbf{N}}_v^{l_1}$ are defined as symmetric matrices:

$$\hat{\mathbf{N}}_u^p = \left\| \mathrm{N}_{i,k}^{p,u} \right\|; \quad \hat{\mathbf{N}}_v^p = \left\| \mathrm{N}_{i,k}^{p,v} \right\|; \tag{5.3.15}$$

$$\mathrm{N}_{i,k}^{p,u} = \begin{cases} \hat{\mathrm{N}}_u^p(0), & i = k; \\ \hat{\mathrm{N}}_u^p(|i-k|\tau), & i \neq k; \end{cases} \quad \mathrm{N}_{i,k}^{p,v} = \begin{cases} \hat{\mathrm{N}}_v^p(0), & i = k; \\ \hat{\mathrm{N}}_v^p(|i-k|\tau), & i \neq k; \end{cases} \tag{5.3.15a}$$

$$\hat{\mathbf{N}}_u^s = \left\| \mathrm{N}_{i,k}^{s,u} \right\|; \quad \hat{\mathbf{N}}_v^s = \left\| \mathrm{N}_{i,k}^{s,v} \right\|; \tag{5.3.16}$$

$$\mathrm{N}_{i,k}^{s,u} = \begin{cases} \hat{\mathrm{N}}_u^s(0), & i = k; \\ \hat{\mathrm{N}}_u^s(|i-k|\tau), & i \neq k; \end{cases} \quad \mathrm{N}_{i,k}^{s,v} = \begin{cases} \hat{\mathrm{N}}_v^s(0), & i = k; \\ \hat{\mathrm{N}}_v^s(|i-k|\tau), & i \neq k, \end{cases} \tag{5.3.16a}$$

$$\hat{\mathbf{N}}_u^{l_1} = \left\| \mathrm{N}_{i,k}^{l_1,u} \right\|; \quad \hat{\mathbf{N}}_v^{l_1} = \left\| \mathrm{N}_{i,k}^{l_1,v} \right\|; \tag{5.3.17}$$

$$\mathrm{N}_{i,k}^{l_1,u} = \begin{cases} \hat{\mathrm{N}}_u^{l_1}(0), & i = k; \\ \hat{\mathrm{N}}_u^{l_1}(|i-k|\tau), & i \neq k; \end{cases} \quad \mathrm{N}_{i,k}^{l_1,v} = \begin{cases} \hat{\mathrm{N}}_v^{l_1}(0), & i = k; \\ \hat{\mathrm{N}}_v^{l_1}(|i-k|\tau), & i \neq k, \end{cases} \tag{5.3.17a}$$

and mutual MSI matrix estimators $\hat{\mathbf{N}}_{uv}^p$, $\hat{\mathbf{N}}_{vu}^p$; $\hat{\mathbf{N}}_{uv}^s$, $\hat{\mathbf{N}}_{vu}^s$; $\hat{\mathbf{N}}_{uv}^{l_1}$, $\hat{\mathbf{N}}_{vu}^{l_1}$ are defined as skew-symmetric matrices:

$$\hat{\mathbf{N}}_{uv}^p = \left\| \mathrm{N}_{i,k}^{p,uv} \right\|; \quad \hat{\mathbf{N}}_{vu}^p = \left\| \mathrm{N}_{i,k}^{p,vu} \right\|; \tag{5.3.18}$$

$$\mathrm{N}_{i,k}^{p,uv} = \begin{cases} \hat{\mathrm{N}}_{uv}^p(0), & i = k; \\ \hat{\mathrm{N}}_{uv}^p(|i-k|\tau), & i > k; \\ -\ \hat{\mathrm{N}}_{uv}^p(|i-k|\tau), & i < k; \end{cases} \tag{5.3.18a}$$

$$\mathrm{N}_{i,k}^{p,vu} = \begin{cases} \hat{\mathrm{N}}_{vu}^p(0), & i = k; \\ \hat{\mathrm{N}}_{vu}^p(|i-k|\tau), & i > k; \\ -\ \hat{\mathrm{N}}_{vu}^p(|i-k|\tau), & i < k; \end{cases} \tag{5.3.18b}$$

$$\hat{\mathbf{N}}_{uv}^s = \left\| \mathrm{N}_{i,k}^{s,uv} \right\|; \quad \hat{\mathbf{N}}_{vu}^s = \left\| \mathrm{N}_{i,k}^{s,vu} \right\|; \tag{5.3.19}$$

$$\mathrm{N}_{i,k}^{s,uv} = \begin{cases} \hat{\mathrm{N}}_{uv}^s(0), & i = k; \\ \hat{\mathrm{N}}_{uv}^s(|i-k|\tau), & i > k; \\ -\ \hat{\mathrm{N}}_{uv}^s(|i-k|\tau), & i < k; \end{cases} \tag{5.3.19a}$$

$$\mathrm{N}_{i,k}^{s,vu} = \begin{cases} \hat{\mathrm{N}}_{vu}^s(0), & i = k; \\ \hat{\mathrm{N}}_{vu}^s(|i-k|\tau), & i > k; \\ -\ \hat{\mathrm{N}}_{vu}^s(|i-k|\tau), & i < k; \end{cases} \tag{5.3.19b}$$

$$\hat{\mathbf{N}}_{uv}^{l_1} = \left\|\mathrm{N}_{i,k}^{l_1,uv}\right\|; \quad \hat{\mathbf{N}}_{vu}^{l_1} = \left\|\mathrm{N}_{i,k}^{l_1,vu}\right\|; \tag{5.3.20}$$

$$\mathrm{N}_{i,k}^{l_1,uv} = \begin{cases} \hat{\mathrm{N}}_{uv}^{l_1}(0), & i = k; \\ \hat{\mathrm{N}}_{uv}^{l_1}(|i-k|\tau), & i > k; \\ -\hat{\mathrm{N}}_{uv}^{l_1}(|i-k|\tau), & i < k; \end{cases} \tag{5.3.20a}$$

$$\mathrm{N}_{i,k}^{l_1,vu} = \begin{cases} \hat{\mathrm{N}}_{vu}^{l_1}(0), & i = k; \\ \hat{\mathrm{N}}_{vu}^{l_1}(|i-k|\tau), & i > k; \\ -\hat{\mathrm{N}}_{vu}^{l_1}(|i-k|\tau), & i < k; \end{cases} \tag{5.3.20b}$$

$$\tau = L/K; \; L << N; \; i = 0, 1, \ldots, K-1; \; k = 0, 1, \ldots, K-1.$$

The relationships $(5.3.7)\ldots(5.3.20)$ compose the basis for spectral estimation by L-group algorithms based on matrices of MSIs $(1.4.10)$, $(1.4.11)$, $(1.4.12)$. These relationships can be used when constructing other algorithms of signal processing in L-group, whose known analogues from linear signal space rely, in essential extent, on using correlation matrices of the observed signals.

5.3.1 Correlogram Method of Spectral Estimation Based on Measure of Statistical Interrelation Matrix Estimator

In first section, we found out that correlogram method of spectral estimation is based on substituting a finite set of values of the estimator $\hat{r}_x(m)$ of ACF $r_x(m)$ (correlogram) into the expression $(5.1.2)$. One of possible PSD estimators $\hat{P}_x(f)$ $(5.1.2)$ based on unbiased estimator $\hat{r}_x(m)$ can be written in the matrix form (see, for instance, $[140, (5.24)]$):

$$\hat{P}_x(f) = \Delta t \overline{\mathbf{e}(f)}^T \hat{\mathbf{R}}_x \mathbf{e}(f); \tag{5.3.21}$$

$$\mathbf{e}(f)^T = [1, \ldots, \exp(j2\pi f \cdot k\Delta t), \ldots, \exp(j2\pi f \cdot (K-1)\Delta t)], \tag{5.3.21a}$$

where Δt is a sampling interval of initial continuous stochastic process $x(t)$; f is discrete frequency parameter; $\hat{\mathbf{R}}_x$ is ACM estimator of complex discrete stochastic signal $x(i)$ $(5.3.1)$; $\mathbf{e}(f)$ is a vector of complex harmonics: $\mathbf{e}(f) = [\exp(j2\pi f \cdot k\Delta t)]$, $k = 0, 1, \ldots, K-1$; $j = \sqrt{-1}$; $\overline{\mathbf{e}(f)}^T$ is conjugate transpose of initial vector $\mathbf{e}(f)$.

Taking into account known relation between measure of statistical interrelation and scalar product of a pair of signals considered in Chapter 1, by the analogy with the estimator $(5.3.21)$, one can obtain correlogram estimators of hyperspectral densities (HSDs) $\hat{H}_p(f)$, $\hat{H}_s(f)$ and $\hat{H}_{l_1}(f)$ based on MSIs $(1.4.10)$, $(1.4.11)$, and $(1.4.12)$, respectively, that are defined by the following relationships:

$$\hat{H}_p(f) = \Delta t \overline{\mathbf{e}(f)}^T \hat{\mathbf{N}}_{x,p} \mathbf{e}(f); \tag{5.3.22a}$$

$$\hat{H}_s(f) = \Delta t \overline{\mathbf{e}(f)}^T \hat{\mathbf{N}}_{x,s} \mathbf{e}(f), \tag{5.3.22b}$$

$$\hat{H}_{l_1}(f) = \Delta t \overline{\mathbf{e}(f)}^T \hat{\mathbf{N}}_{x,l_1} \mathbf{e}(f), \tag{5.3.22c}$$

where $\hat{\mathbf{N}}_{x,p}$, $\hat{\mathbf{N}}_{x,s}$, and $\hat{\mathbf{N}}_{x,l_1}$ are MSI matrix estimators of complex discrete stochastic signal $x(i)$ determined by the relationships $(5.3.14a,b,c)$, correspondingly; $\mathbf{e}(f)$ is vector of complex harmonics defined by the formula $(5.3.21a)$.

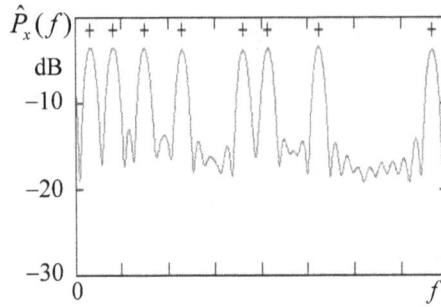

FIGURE 5.3.1   Correlogram PSD estimate $\hat{P}_x(f)$ (5.3.21)

Having obtained the algorithms (5.3.22) of correlogram estimators of HSDs $\hat{H}_p(f)$, $\hat{H}_s(f)$, and $\hat{H}_{l_1}(f)$ that are based on MSIs (1.4.10), (1.4.11), and (1.4.12), respectively, it is necessary to find out, first, how close are they with respect to classic correlogram estimator (5.3.21), and, second, could they be used for estimating spectral components of a mixture of harmonic signals and noise. Algorithms (5.3.22a,b) are illustrated by the following example.

Example 5.3.1. Let $x(i)$ be additive mixture of M complex discrete harmonic signals observed in the presence of complex quasi-white Gaussian noise $n(i)$ with zero mean (5.1.9).

Within the considered example, amplitudes of harmonics are chosen to be equal $A_l = A$, so that the relationships hold: $A^2/2 = D_n$, $F_n > 2\max_l\{f_l\}$, where D_n is a variance of noise $n(i)$; F_n is upper bound frequency of PSD of quasi-white Gaussian noise $n(i)$; $M = 8$. The number of samples N of stochastic process $x(i)$ used when forming ACF estimator (5.1.2a), and also estimators (5.3.7)...(5.3.10) is equal to N=1024. Maximal index L of time shift $m \leq L$, figuring in the ACF estimator (5.1.2a) and also in MSI estimators (5.3.7)...(5.3.10), is chosen to be equal L=128. Dimensionality $K \times K$ of MSI matrices $\hat{\mathbf{N}}_{x,p}$ (5.3.14a), $\hat{\mathbf{N}}_{x,s}$ (5.3.14b), and ACM $\hat{\mathbf{R}}_x$ (5.3.4) is chosen to be equal K=32.

Fig. 5.3.1 illustrates correlogram PSD estimate $\hat{P}_x(f)$ of a mixture $x(i)$ (5.1.9) obtained according to the formula (5.3.21) for initial situation described above. True positions of frequencies of harmonic signals from the mixture (5.1.9) are denoted by the signs "+".

Fig. 5.3.2a and 5.3.2b depict correlogram estimates of HSDs $\hat{H}_p(f)$ (5.3.22a) and $\hat{H}_s(f)$ (5.3.22b) of the signal $x(i)$ (5.1.9), respectively, for initial situation described above. True positions of the frequencies of harmonic signals (5.1.9) are denoted by the signs "+". Correlation coefficients between the estimates of HSDs $\hat{H}_p(f)$ (5.3.22a), $\hat{H}_s(f)$ (5.3.22b) and correlogram PSD estimate $\hat{P}_x(f)$ (5.3.21) are equal to: $r[\hat{H}_p(f), \hat{P}_x(f)]$=0.995 and $r[\hat{H}_s(f), \hat{P}_x(f)]$=0.999, respectively, whereas correlation coefficients between MSI matrix estimates $\hat{\mathbf{N}}_{x,p}$ (5.3.14a), $\hat{\mathbf{N}}_{x,s}$ (5.3.14b) and ACM estimate $\hat{\mathbf{R}}_x$ (5.3.4) are equal to: $r[\hat{\mathbf{N}}_{x,p}, \hat{\mathbf{R}}_x]$=0.975 and $r[\hat{\mathbf{N}}_{x,s}, \hat{\mathbf{R}}_x]$=0.999, respectively. This circumstance testifies to high extent correspondence between the results obtained in linear signal space (Fig. 5.3.1) and in signal space with L-group

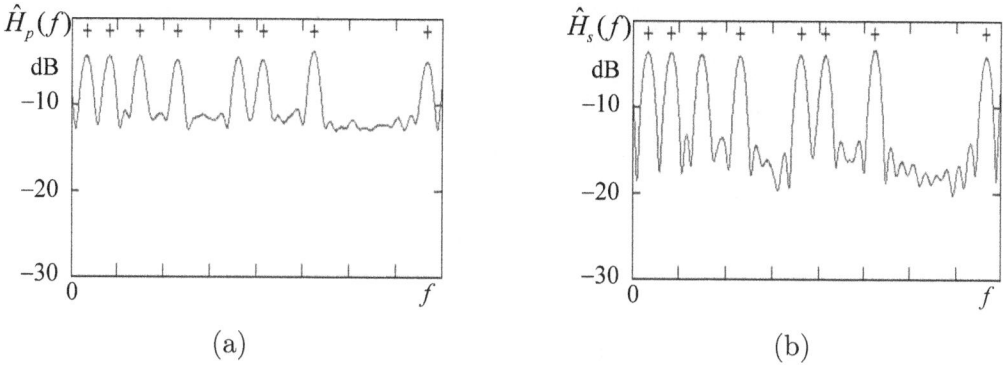

FIGURE 5.3.2 Correlogram estimates of HSDs: (a) $\hat{H}_p(f)$ (5.3.22a); (b) $\hat{H}_s(f)$ (5.3.22b)

properties (Fig. 5.3.2a,b), that is explained by high identities between MSI matrix estimators $\hat{\mathbf{N}}_{x,p}$ (5.3.14a), $\hat{\mathbf{N}}_{x,s}$ (5.3.14b) and ACM estimator $\hat{\mathbf{R}}_x$ (5.3.4).

$\triangledown$

5.3.2 Minimum Variance Spectral Estimation Method Based on Measure of Statistical Interrelation Matrix Estimator

Minimum variance (MV) spectral estimator was introduced by J. Capon [148] for solving the problem of spatial-temporal analysis of spatial signals in antenna arrays. MV estimator is not a true PSD function, inasmuch as it does not characterizes full power of the observed stochastic process. MV estimator is considered to be spectral estimator in the sense that it describes relative intensities of frequency spectrum components.

Spectral MV estimator $\hat{P}_{MV}(f)$ is defined by the relationship (see, for instance, [140, (12.1)]):

$$\hat{P}_{MV}(f) = \frac{\Delta t}{\overline{\mathbf{e}(f)}^T \hat{\mathbf{R}}_x^{-1} \mathbf{e}(f)}, \tag{5.3.23}$$

where Δt is a sampling interval of initial continuous stochastic process $x(t)$ (5.3.1); f is discrete frequency parameter; $\hat{\mathbf{R}}_x$ is ACM estimator; $\mathbf{e}(f)$ is vector of complex harmonics: $\mathbf{e}(f) = [\exp(j2\pi f \cdot k\Delta t)]$, $k = 0, 1, \ldots, K-1$; $j = \sqrt{-1}$; $\overline{\mathbf{e}(f)}^T$ is conjugate transpose of initial vector $\mathbf{e}(f)$.

By the analogy with MV estimator (5.3.23), one can obtain MV estimators of HSDs $\hat{H}_{MV}^p(f)$, $\hat{H}_{MV}^s(f)$, and $\hat{H}_{MV}^{l_1}(f)$ based on MSIs (1.4.10), (1.4.11), and (1.4.12), respectively, that are defined by the following relationships:

$$\hat{H}_{MV}^p(f) = \frac{\Delta t}{\overline{\mathbf{e}(f)}^T \hat{\mathbf{N}}_{x,p}^{-1} \mathbf{e}(f)}; \tag{5.3.24a}$$

$$\hat{H}_{MV}^s(f) = \frac{\Delta t}{\overline{\mathbf{e}(f)}^T \hat{\mathbf{N}}_{x,s}^{-1} \mathbf{e}(f)}, \tag{5.3.24b}$$

$$\hat{H}_{MV}^{l_1}(f) = \frac{\Delta t}{\overline{\mathbf{e}(f)}^T \hat{\mathbf{N}}_{x,l_1}^{-1} \mathbf{e}(f)}, \tag{5.3.24c}$$

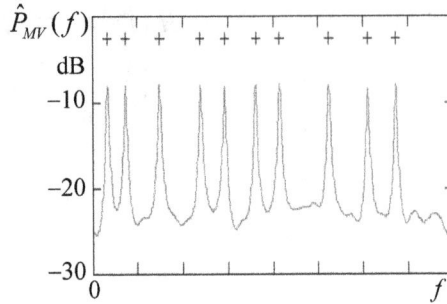

FIGURE 5.3.3   MV estimate of PSD $\hat{P}_{MV}(f)$ (5.3.23)

where $\hat{\mathbf{N}}_{x,p}$, $\hat{\mathbf{N}}_{x,s}$, $\hat{\mathbf{N}}_{x,l_1}$ are MSI matrix estimators of complex discrete stochastic signal $x(i)$ determined by the relationships (5.3.14a,b,c), respectively; $\mathbf{e}(f)$ is vector of complex harmonics determined by the formula (5.3.21a).

Having obtained the algorithms (5.3.24) of MV estimators of HSDs $\hat{H}^p_{MV}(f)$, $\hat{H}^s_{MV}(f)$, and $\hat{H}^{l_1}_{MV}(f)$ based on MSIs (1.4.10), (1.4.11), and (1.4.12), respectively, it is necessary to find out, first, how close are they with respect to MV estimator (5.3.23), and second, could they be used for estimating spectral components of a mixture of harmonic signals and noise. Algorithms (4.3.24a,b) are illustrated by the following example.

Example 5.3.2. Let $x(i)$ be additive mixture of M complex discrete harmonic signals observed in the presence of complex quasi-white Gaussian noise $n(i)$ with zero mean (5.1.9).

Within the considered example, amplitudes of harmonics are chosen to be equal $A_l = A$, so that the relationships hold: $A^2/2 = D_n$, $F_n > 2\max_l\{f_l\}$, where D_n is a variance of noise $n(i)$; F_n is upper bound frequency of PSD of quasi-white Gaussian noise $n(i)$; $M = 10$. The number of samples N of stochastic process $x(i)$ used when forming ACF estimator (5.1.2a), and also estimators (5.3.7)...(5.3.10) is equal to N=1024. Maximal index L of time shift $m \leq L$, figuring in the ACF estimator (5.1.2a) and also in MSI estimators (5.3.7)...(5.3.10), is chosen to be equal L=128. Dimensionality $K \times K$ of MSI matrices $\hat{\mathbf{N}}_{x,p}$ (5.3.14a), $\hat{\mathbf{N}}_{x,s}$ (5.3.14b), and ACM $\hat{\mathbf{R}}_x$ (5.3.4) is chosen to be equal K=32.

Fig. 5.3.3 illustrates MV estimate of PSD $\hat{P}_{MV}(f)$ of the mixture $x(i)$ (5.1.9) obtained according to the relationship (5.3.23) for initial situation described above. True positions of the frequencies of harmonic signals of the mixture (5.1.9) are denoted by the signs "+".

Fig. 5.3.4a and 5.3.4b depict MV estimates of HSDs $\hat{H}^p_{MV}(f)$ (5.3.24a) and $\hat{H}^s_{MV}(f)$ (5.3.24b) of the signal $x(i)$ (5.1.9), respectively, for initial situation described above. True positions of the frequencies of harmonic signals (5.1.9) are denoted by the symbols "+". Correlation coefficients between the estimators of HSDs $\hat{H}^p_{MV}(f)$ (5.3.23), $\hat{H}^s_{MV}(f)$ (5.3.24) and PSD MV estimator $\hat{P}_{MV}(f)$ (5.3.23) are equal to: $r[\hat{H}^p_{MV}(f), \hat{P}_{MV}(f)]$=0.906 and $r[\hat{H}^s_{MV}(f), \hat{P}_{MV}(f)]$=0.993, respectively, while correlation coefficients between MSI estimators $\hat{\mathbf{N}}_{x,p}$ (5.3.14a), $\hat{\mathbf{N}}_{x,s}$

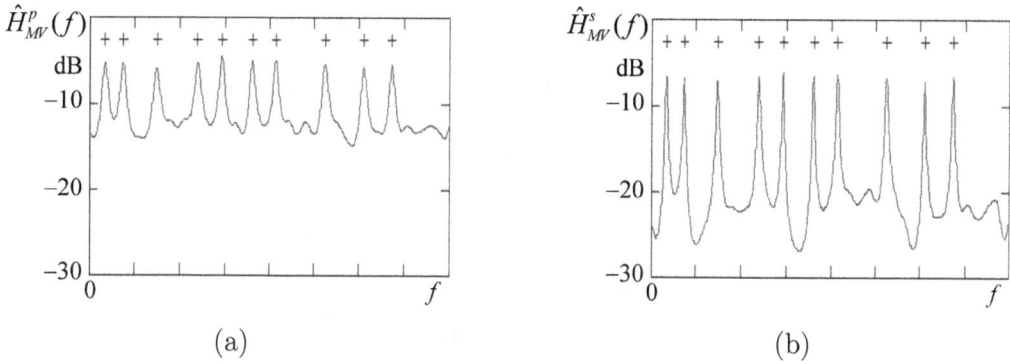

(a)   (b)

FIGURE 5.3.4   MV estimates of HSDs: (a) $\hat{H}^p_{MV}(f)$ (5.3.24a); (b) $\hat{H}^s_{MV}(f)$ (5.3.24b).

(5.3.14b) and ACM estimator $\hat{\mathbf{R}}_x$ (5.3.4) are equal to: $r[\hat{\mathbf{N}}_{x,p}, \hat{\mathbf{R}}_x]$=0.974 and $r[\hat{\mathbf{N}}_{x,s}, \hat{\mathbf{R}}_x]$=0.998, respectively, that testifies to high extent correspondence between the results obtained in linear signal space (Fig. 5.3.3) and in signal space with L-group properties (Fig. 5.3.4a,b), that is explained by high identities between MSI matrix estimators $\hat{\mathbf{N}}_{x,p}$ (5.3.14a), $\hat{\mathbf{N}}_{x,s}$ (5.3.14b) and ACM estimator $\hat{\mathbf{R}}_x$ (5.3.4).

$\triangledown$

5.4   Spectral Estimation Method Based on Eigenvectors of Measure of Statistical Interrelation Matrix Estimator

One of the classes of spectral estimation methods based on analysis of ACM eigenvectors is described in the literature as a group of methods providing better characteristics of resolution and frequency estimation than those of autoregressive methods and Prony's method [140]. A key idea in these methods is splitting information contained in ACM into two subspaces—signal subspace and noise subspace.

The class of spectral estimation methods based on analysis of ACM eigenvectors involves the following ones: eigenvector (EV) method [149], MUltiple SIgnal Classification (MUSIC) method [150], minimum norm (MN) method [151], and Estimation of Signal Parameters via Rotational Invariance Techniques (ESPRIT) method [152, 153]. The estimators obtained by these methods are not true PSD estimators, since they do not characterize a power of the observed stochastic signal, nevertheless, they allow obtaining PSD quasi-estimators with satisfactory quality indices of accuracy and resolution.

Return to the model of additive interaction of complex harmonics in the presence of complex quasi-white Gaussian noise (5.1.9) that can be written in vector form:

$$\mathbf{x} = \sum_{l=1}^{M} \mathbf{s}_l + \mathbf{n} \tag{5.4.1}$$

where vectors of signals are represented by their samples:

$$\mathbf{x} = [x(i)]; \quad \mathbf{s}_l = [s_l(i)] \quad \mathbf{n} = [n(i)]; \quad i = 0, 1, \ldots, N-1.$$

Then $K-1-M$ noise subspace eigenvectors $\mathbf{v}_M, \ldots, \mathbf{v}_{K-1}$ of ACM $\hat{\mathbf{R}}_x$ (5.3.1) are orthogonal to vectors of harmonic signals $\{\mathbf{s}_l\}$ at the corresponding frequencies, therefore the following linear combination is equal to zero:

$$\overline{\mathbf{e}(f_l)}^T \left(\sum_{k=M}^{K-1} \alpha_k \mathbf{v}_k \bar{\mathbf{v}}_k^T \right) \mathbf{e}(f_l) = 0; \tag{5.4.2}$$

$$\mathbf{e}(f)^T = [1, \ldots, \exp(j2\pi f \cdot k\Delta t), \ldots, \exp(j2\pi f \cdot (K-1)\Delta t)], \tag{5.4.2a}$$

where f_l is a frequency of a signal $s_l(i)$: $s_l(i) = A_l \exp[j2\pi f_l \cdot i + \varphi_l]$ (see (5.1.9)); $\{\alpha_k\}$ are arbitrary constants; $k = 0, \ldots, K-1$.

This means that the function of the estimator

$$\hat{P}(f) = \frac{1}{\overline{\mathbf{e}(f)}^T \left(\sum\limits_{k=M}^{K-1} \alpha_k \mathbf{v}_k \bar{\mathbf{v}}_k^T \right) \mathbf{e}(f)} \tag{5.4.3}$$

takes infinite value at the frequency $f = f_l$ of one of harmonic signals (5.1.9). In practice, due to noise influence, the function (5.4.3) takes finite values at the frequencies $f = f_l$ possessing rather thin peaks. The estimators (5.4.3) are not the estimators of true PSD, but are quasi-estimators that are useful for estimating frequencies of harmonic components of additive mixture (5.4.1).

5.4.1 Spectral Estimation by Eigenvector Method Based on Measure of Statistical Interrelation Matrix Estimator

Taking into account known decomposition of Hermitian matrix (ACM) $\hat{\mathbf{R}}_x$ (5.3.1) over eigenvectors $\{\mathbf{v}_k\}$ and eigenvalues $\{\lambda_k\}$ (see, for instance, [140, (3.86),(3.88)]):

$$\hat{\mathbf{R}}_x = \sum_{k=0}^{K-1} \lambda_k \mathbf{v}_k \bar{\mathbf{v}}_k^T; \quad \hat{\mathbf{R}}_x^{-1} = \sum_{k=0}^{K-1} \frac{1}{\lambda_k} \mathbf{v}_k \bar{\mathbf{v}}_k^T,$$

drawing a parallel with MV estimator $\hat{P}_{MV}(f)$ (5.3.23), and assuming in the formula (5.4.3) $\alpha_k = 1/\lambda_k$, we obtain spectral EV estimator based on eigenvectors:

$$\hat{P}_{EV}(f) = \frac{1}{\overline{\mathbf{e}(f)}^T \left(\sum\limits_{k=M}^{K-1} \frac{1}{\lambda_k} \mathbf{v}_k \bar{\mathbf{v}}_k^T \right) \mathbf{e}(f)}, \tag{5.4.4}$$

where, remind, $\mathbf{e}(f)$ is vector of complex harmonics determined by the formula (5.4.2a); M is a number of harmonic signals in the observed additive mixture (5.1.9); $K \times K$ is a dimensionality of ACM $\hat{\mathbf{R}}_x$.

By the analogy with EV estimator (5.4.4), one can obtain spectral EV estimators of HSDs $\hat{H}^p_{EV}(f)$, $\hat{H}^s_{EV}(f)$, $\hat{H}^{l_1}_{EV}(f)$ based on MSIs (1.4.10), (1.4.11), and (1.4.12), respectively, that are defined by the following relationships:

$$\hat{H}^p_{EV}(f) = \frac{1}{\overline{\mathbf{e}(f)}^T \left(\sum_{k=M}^{K-1} \frac{1}{\rho^p_k} \mathbf{u}^p_k \overline{\mathbf{u}^p_k}^T \right) \mathbf{e}(f)};$$ (5.4.5)

$$\hat{H}^s_{EV}(f) = \frac{1}{\overline{\mathbf{e}(f)}^T \left(\sum_{k=M}^{K-1} \frac{1}{\rho^s_k} \mathbf{u}^s_k \overline{\mathbf{u}^s_k}^T \right) \mathbf{e}(f)};$$ (5.4.6)

$$\hat{H}^{l_1}_{EV}(f) = \frac{1}{\overline{\mathbf{e}(f)}^T \left(\sum_{k=M}^{K-1} \frac{1}{\rho^{l_1}_k} \mathbf{u}^{l_1}_k \overline{\mathbf{u}^{l_1}_k}^T \right) \mathbf{e}(f)},$$ (5.4.7)

where $\{\mathbf{u}^p_k\}$, $\{\mathbf{u}^s_k\}$, $\{\mathbf{u}^{l_1}_k\}$ are eigenvectors of MSI matrix estimators $\hat{\mathbf{N}}_{x,p}$ (5.3.14a), $\hat{\mathbf{N}}_{x,s}$ (5.3.14b), $\hat{\mathbf{N}}_{x,l_1}$ (5.3.14c), respectively; $\{\rho^p_k\}$, $\{\rho^s_k\}$, $\{\rho^{l_1}_k\}$ are eigenvalues of MSI matrix estimators $\hat{\mathbf{N}}_{x,p}$ (5.3.14a), $\hat{\mathbf{N}}_{x,s}$ (5.3.14b), $\hat{\mathbf{N}}_{x,l_1}$ (5.3.14c), respectively; $\mathbf{e}(f)$ is vector of complex harmonics determined by the formula (5.4.2a); M is a number of harmonic signals in the observed additive mixture (5.1.9); $K \times K$ is a dimensionality of MSI matrices $\hat{\mathbf{N}}_{x,p}$ (5.3.14a), $\hat{\mathbf{N}}_{x,s}$ (5.3.14b), $\hat{\mathbf{N}}_{x,l_1}$ (5.3.14c).

Having obtained algorithms (5.4.5)...(5.4.7) of EV estimators of HSDs $\hat{H}^p_{EV}(f)$, $\hat{H}^s_{EV}(f)$, $\hat{H}^{l_1}_{EV}(f)$ based on MSIs (1.4.10), (1.4.11), and (1.4.12), respectively, it is necessary, first, to find out how close are they with respect to EV estimator (5.4.4), and second, could they be used for estimating spectral components of a mixture of harmonic signals and noise. Algorithms (5.4.5) and (5.4.6) are illustrated by the following example.

Example 5.4.1. Let $x(i)$ be additive mixture of M complex discrete harmonic signals observed in the presence of complex quasi-white Gaussian noise $n(i)$ with zero mean (5.1.9).

Within the considered example, amplitudes of harmonics are chosen to be equal $A_l = A$, so that the relationships hold: $A^2/2 = D_n$, $F_n > 2\max_l\{f_l\}$, where D_n is a variance of noise $n(i)$; F_n is upper bound frequency of PSD of quasi-white Gaussian noise $n(i)$; $M = 10$. The number of samples N of stochastic process $x(i)$ used when forming ACF estimator (5.1.2a) and also estimators (5.3.7)...(5.3.10) is equal to $N=1024$. Maximal index L of time shift $m \leq L$, figuring in the ACF estimator (5.1.2a) and also in MSI estimators (5.3.7)...(5.3.10), is chosen to be equal $L=128$. Dimensionality $K \times K$ of MSI matrices $\hat{\mathbf{N}}_{x,p}$ (5.3.14a), $\hat{\mathbf{N}}_{x,s}$ (5.3.14b), and ACM $\hat{\mathbf{R}}_x$ (5.3.4) is chosen to be equal $K=32$.

Fig. 5.4.1 depicts PSD EV estimate $\hat{P}_{EV}(f)$ of the mixture $x(i)$ (5.1.9) obtained according to the relationship (5.4.4) for situation described above. True positions of the frequencies of harmonic signals from the mixture (5.1.9) are denoted by the signs "+".

Fig. 5.4.2a and 5.4.2b illustrate EV estimates of HSDs $\hat{H}^p_{EV}(f)$ (5.4.5) and $\hat{H}^s_{EV}(f)$ (5.4.6) of the signal $x(i)$ (5.1.9), respectively, for initial situation

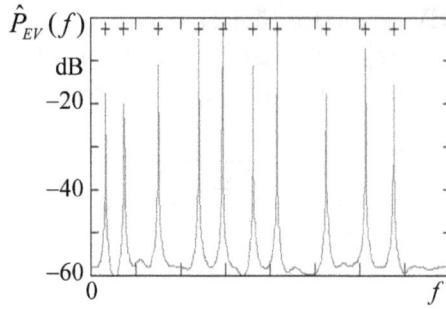

FIGURE 5.4.1　EV estimate of PSD $\hat{P}_{EV}(f)$ (5.4.4)

described above. True positions of the frequencies of harmonic signals (5.1.9) are denoted by the symbols "+". Correlation coefficients between EV estimators of HSDs $\hat{H}^p_{EV}(f)$ (5.4.5), $\hat{H}^s_{EV}(f)$ (5.4.6) and EV estimator of PSD $\hat{P}_{EV}(f)$ (5.4.4) are equal to: $r[\hat{H}^p_{EV}(f), \hat{P}_{EV}(f)]=0.524$ and $r[\hat{H}^s_{EV}(f), \hat{P}_{EV}(f)]=0.596$, that is explained by a slight level of linear statistical interrelations between the results of signal processing, while correlation coefficients between MSI estimators $\hat{\mathbf{N}}_{x,p}$ (5.3.14a), $\hat{\mathbf{N}}_{x,s}$ (5.3.14b) and ACM estimator $\hat{\mathbf{R}}_x$ (5.3.4) are equal to: $r[\hat{\mathbf{N}}_{x,p}, \hat{\mathbf{R}}_x]=0.974$ and $r[\hat{\mathbf{N}}_{x,s}, \hat{\mathbf{R}}_x]=0.998$, respectively. It should be noted that normalized MSIs (NMSIs) (1.4.6) and (1.4.7) between EV estimators of HSDs $\hat{H}^p_{EV}(f)$ (5.4.5), $\hat{H}^s_{EV}(f)$ (5.4.6) and EV estimator of PSD $\hat{P}_{EV}(f)$ (5.4.4) are, respectively, equal:

$$\nu_p[\hat{H}^s_{EV}(f) - \mathbf{M}(\hat{H}^s_{EV}(f)), \hat{P}_{EV}(f) - \mathbf{M}(\hat{P}_{EV}(f))] = 0.988;$$

$$\nu_p[\hat{H}^p_{EV}(f) - \mathbf{M}(\hat{H}^p_{EV}(f)), \hat{P}_{EV}(f) - \mathbf{M}(\hat{P}_{EV}(f))] = 0.922;$$

$$\nu_s[\hat{H}^s_{EV}(f) - \mathbf{M}(\hat{H}^s_{EV}(f)), \hat{P}_{EV}(f) - \mathbf{M}(\hat{P}_{EV}(f))] = 0.996;$$

$$\nu_s[\hat{H}^p_{EV}(f) - \mathbf{M}(\hat{H}^p_{EV}(f)), \hat{P}_{EV}(f) - \mathbf{M}(\hat{P}_{EV}(f))] = 0.966.$$

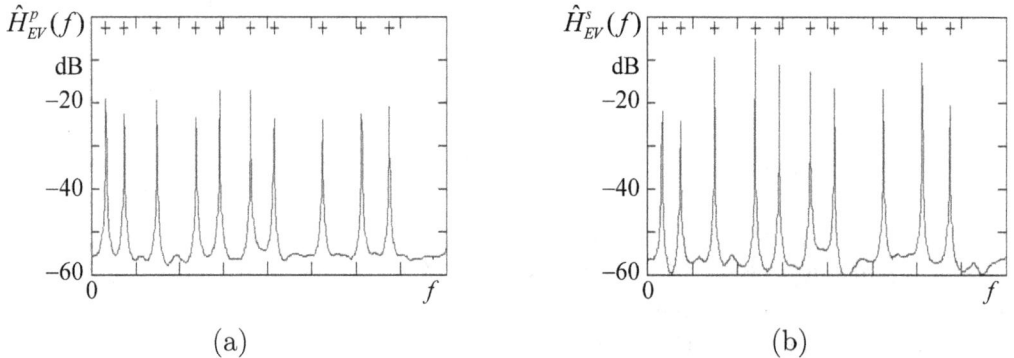

(a)　　　　　　　　　　　　　　　　　　　(b)

FIGURE 5.4.2　EV estimates of HSDs: (a) $\hat{H}^p_{EV}(f)$ (5.4.5); (b) $\hat{H}^s_{EV}(f)$ (5.4.6).

The last relationships testify to high extent correspondence between the results obtained in linear signal space (Fig. 5.4.1) and in signal space with L-group properties (Fig. 5.4.2a,b), that is explained by high identities between MSI matrix estimators $\hat{\mathbf{N}}_{x,p}$ (5.3.14a), $\hat{\mathbf{N}}_{x,s}$ (5.3.14b) and ACM estimator $\hat{\mathbf{R}}_x$ (5.3.4), and also by the fact that nonlinear statistical interrelations, unlike linear, are almost completely preserved. $\triangledown$

5.4.2   Spectral Estimation by MUSIC Method Based on Measure of Statistical Interrelation Matrix Estimator

Taking into account known decomposition of Hermitian matrix (ACM) $\hat{\mathbf{R}}_x$ (5.3.1) by eigenvectors $\{\mathbf{v}_k\}$ and eigenvalues $\{\lambda_k\}$ (see, for instance, [140, (3.86), (3.88)]), drawing a parallel with MV estimator $\hat{P}_{MV}(f)$ (5.3.23), and assuming in the formula (5.4.3) $\alpha_k = 1$ for $\forall k$, we obtain spectral MUSIC estimator $\hat{P}_{MUSIC}(f)$ based on eigenvectors of ACM $\hat{\mathbf{R}}_x$:

$$\hat{P}_{MUSIC}(f) = \frac{1}{\overline{\mathbf{e}(f)}^T \left(\sum_{k=M}^{K-1} \mathbf{v}_k \overline{\mathbf{v}}_k^T \right) \mathbf{e}(f)} = \frac{1}{\overline{\mathbf{e}(f)}^T \left(\mathbf{V}' \overline{\mathbf{V}'}^T \right) \mathbf{e}(f)}, \qquad (5.4.8)$$

where, remind, $\mathbf{e}(f)$ is vector of complex harmonics determined by the formula (5.4.2à); M is a number of harmonic signals in the observed additive mixture (5.1.9); $K \times K$ is a dimensionality of ACM $\hat{\mathbf{R}}_x$; $\mathbf{V}'$ is a matrix related to noise subspace and obtained by cutting off the matrix $\mathbf{V} = \|\mathbf{v}_0, \mathbf{v}_1, \ldots, \mathbf{v}_{K-1}\|$ with dimensionality $K \times K$ composed by eigenvectors $\{\mathbf{v}_k\}$ of ACM estimator $\hat{\mathbf{R}}_x$ (5.3.1) of additive mixture (5.1.9):

$$\mathbf{V}' = \text{submatrix}(\mathbf{V}, 0, K-1, M, K-1) = \|\mathbf{v}'_M, \mathbf{v}'_{M+1}, \ldots, \mathbf{v}'_{K-1}\|, \qquad (5.4.8a)$$

where submatrix$(\mathbf{A}, r_i, r_j, c_i, c_j)$ is a function that returns a submatrix of initial matrix $\mathbf{A}$ consisting of elements in rows r_i through r_j and columns c_i through c_j; eigenvector matrix $\mathbf{V} = \|\mathbf{V}_{i,k}\|$ has a dimensionality $K \times K$, $i = 0, \ldots, K-1$, $k = 0, \ldots, K-1$; matrix $\mathbf{V}' = \|\mathbf{v}'_M, \mathbf{v}'_{M+1}, \ldots, \mathbf{v}'_{K-1}\|$ has a dimensionality $K \times K - M$; M is expected number of harmonic signals in the mixture (5.1.9), $M < K$.

By the analogy with MUSIC estimator (5.4.8), one can obtain spectral MUSIC estimators of HSDs $\hat{H}^p_{MUSIC}(f)$, $\hat{H}^s_{MUSIC}(f)$, $\hat{H}^{l_1}_{MUSIC}(f)$ based on MSIs (1.4.10), (1.4.11), and (1.4.12), respectively, that are defined by the following relationships:

$$\hat{H}^p_{MUSIC}(f) = \frac{1}{\overline{\mathbf{e}(f)}^T \left(\sum_{k=M}^{K-1} \mathbf{u}_k^p \overline{\mathbf{u}_k^p}^T \right) \mathbf{e}(f)} = \frac{1}{\overline{\mathbf{e}(f)}^T \left(\mathbf{U}'_p \overline{\mathbf{U}'}_p^T \right) \mathbf{e}(f)}; \qquad (5.4.9)$$

$$\hat{H}^s_{MUSIC}(f) = \frac{1}{\overline{\mathbf{e}(f)}^T \left(\sum_{k=M}^{K-1} \mathbf{u}_k^s \overline{\mathbf{u}_k^s}^T \right) \mathbf{e}(f)} = \frac{1}{\overline{\mathbf{e}(f)}^T \left(\mathbf{U}'_s \overline{\mathbf{U}'}_s^T \right) \mathbf{e}(f)}; \qquad (5.4.10)$$

$$\hat{H}^{l_1}_{MUSIC}(f) = \frac{1}{\overline{\mathbf{e}(f)}^T \left(\sum_{k=M}^{K-1} \mathbf{u}_k^{l_1} \overline{\mathbf{u}_k^{l_1}}^T \right) \mathbf{e}(f)} = \frac{1}{\overline{\mathbf{e}(f)}^T \left(\mathbf{U}'_{l_1} \overline{\mathbf{U}'}_{l_1}^T \right) \mathbf{e}(f)}, \qquad (5.4.11)$$

where $\{\mathbf{u}_k^p\}$, $\{\mathbf{u}_k^s\}$, $\{\mathbf{u}_k^{l_1}\}$ are eigenvectors of MSI matrix estimators $\hat{\mathbf{N}}_{x,p}$ (5.3.14a), $\hat{\mathbf{N}}_{x,s}$ (5.3.14b), $\hat{\mathbf{N}}_{x,l_1}$ (5.3.14c), respectively; $\mathbf{e}(f)$ is vector of complex harmonics determined by the formula (5.4.2a); M is a number of harmonic signals in the observed additive mixture (5.1.9); $K \times K$ is a dimensionality of MSI matrices $\hat{\mathbf{N}}_{x,p}$ (5.3.14a), $\hat{\mathbf{N}}_{x,s}$ (5.3.14b), $\hat{\mathbf{N}}_{x,l_1}$ (5.3.14c); $\mathbf{U}'_p$, $\mathbf{U}'_s$, $\mathbf{U}'_{l_1}$ are matrices of noise subspace with dimensionality $K \times K - M$ obtained by cutting off the matrices $\mathbf{U}_p = \|\mathbf{u}_{p,0}, \mathbf{u}_{p,1}, \ldots, \mathbf{u}_{p,K-1}\|$, $\mathbf{U}_s = \|\mathbf{u}_{s,0}, \mathbf{u}_{s,1}, \ldots, \mathbf{u}_{s,K-1}\|$, $\mathbf{U}_{l_1} = \|\mathbf{u}_{l_1,0}, \mathbf{u}_{l_1,1}, \ldots, \mathbf{u}_{l_1,K-1}\|$ composed by eigenvectors of MSI matrix estimators $\hat{\mathbf{N}}_{x,p}$ (5.3.14a), $\hat{\mathbf{N}}_{x,s}$ (5.3.14b), $\hat{\mathbf{N}}_{x,l_1}$ (5.3.14c), respectively:

$$\mathbf{U}'_p = \text{submatrix}(\mathbf{U}_p, 0, K-1, M, K-1) = \|\mathbf{u}'_{p,M}, \mathbf{u}'_{p,M+1}, \ldots, \mathbf{u}'_{p,K-1}\| \,;$$

$$\mathbf{U}'_s = \text{submatrix}(\mathbf{U}_s, 0, K-1, M, K-1) = \|\mathbf{u}'_{s,M}, \mathbf{u}'_{s,M+1}, \ldots, \mathbf{u}'_{s,K-1}\| \,;$$

$$\mathbf{U}'_{l_1} = \text{submatrix}(\mathbf{U}_{l_1}, 0, K-1, M, K-1) = \|\mathbf{u}'_{l_1,M}, \mathbf{u}'_{l_1,M+1}, \ldots, \mathbf{u}'_{l-1,K-1}\| \,.$$

Having obtained the algorithms (5.4.9)...(5.4.11) of MUSIC estimators of HSDs $\hat{H}^p_{MUSIC}(f)$, $\hat{H}^s_{MUSIC}(f)$, $\hat{H}^{l_1}_{MUSIC}(f)$ based on MSIs (1.4.11), (1.4.11), and (1.4.12), respectively, it is necessary to find out, first, how close are they with respect to MUSIC estimator (5.4.8), and second, could they be used for estimating spectral components of a mixture of harmonic signals and noise. Algorithms (5.4.9) and (5.4.10) are illustrated by the following example.

Example 5.4.2. Let $x(i)$ be additive mixture of M complex discrete harmonic signals observed in the presence of complex quasi-white Gaussian noise $n(i)$ with zero mean (5.1.9).

Within the considered example, amplitudes of harmonics are chosen to be equal $A_l = A$, so that the following relationships hold: $A^2/2 = D_n$, $F_n > 2\max_l\{f_l\}$, where D_n is a variance of noise $n(i)$; F_n is upper bound frequency of PSD of quasi-white Gaussian noise $n(i)$; $M = 10$. The number of samples N of stochastic process $x(i)$ used when forming ACF estimator (5.1.2a), and also estimators (5.3.7)...(5.3.10) is equal to $N=1024$. Maximal index L of time shift $m \le L$, figuring in the ACF estimator (5.1.2a) and also in MSI estimators (5.3.7)...(5.3.10), is chosen to be equal $L=128$. Dimensionality $K \times K$ of MSI matrices $\hat{\mathbf{N}}_{x,p}$ (5.3.14a), $\hat{\mathbf{N}}_{x,s}$ (5.3.14b), and ACM $\hat{\mathbf{R}}_x$ (5.3.4) is chosen to be equal $K=32$.

Fig. 5.4.3 illustrates PSD MUSIC estimate $\hat{P}_{MUSIC}(f)$ of the mixture $x(i)$ (5.1.9) obtained according to the relationship (5.4.8) for initial situation described above. True positions of the frequencies of harmonic signals from the mixture (5.1.9) are denoted by the symbols "+".

Fig. 5.4.4a and 5.4.4b depict MUSIC estimates of HSDs $\hat{H}^p_{MUSIC}(f)$ (5.4.9) and $\hat{H}^s_{MUSIC}(f)$ (5.4.10) of the mixture $x(i)$ (5.1.9), respectively, for situation described above. True positions of the frequencies (5.1.9) are denoted by the signs "+". Correlation coefficients between MUSIC estimates of HSDs $\hat{H}^p_{MUSIC}(f)$ (5.4.9), $\hat{H}^s_{MUSIC}(f)$ (5.4.10) and PSD MUSIC estimate $\hat{P}_{MUSIC}(f)$ (5.4.8) are equal to:

$$r[\hat{H}^p_{MUSIC}(f), \hat{P}_{MUSIC}(f)] = 0.536; \quad r[\hat{H}^s_{MUSIC}(f), \hat{P}_{MUSIC}(f)] = 0.65,$$

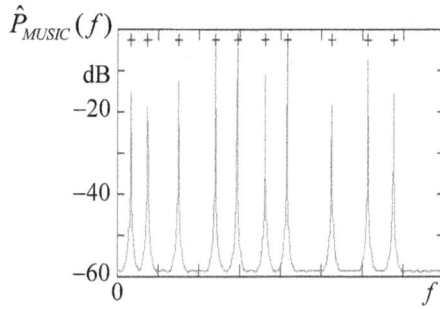

FIGURE 5.4.3   MUSIC estimate of PSD $\hat{P}_{MUSIC}(f)$ (5.4.8)

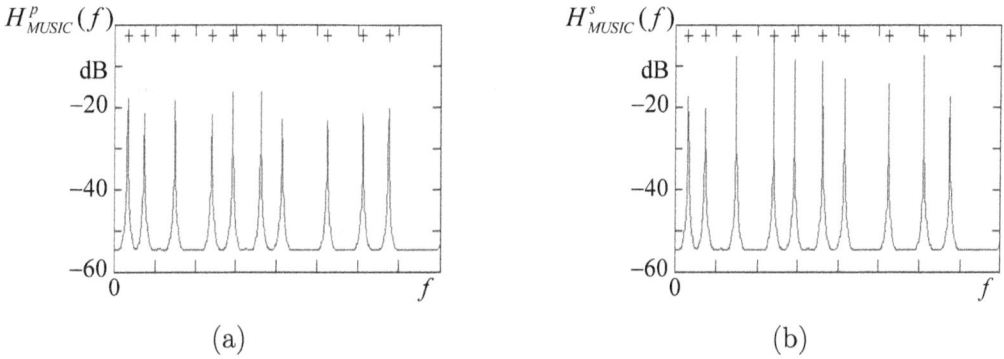

(a)   (b)

FIGURE 5.4.4   MUSIC estimates of HSDs: (a) $\hat{H}^p_{MUSIC}(f)$ (5.4.9); (b) $\hat{H}^s_{MUSIC}(f)$ (5.4.10)

that is explained by a slight level of linear statistical interrelations between the results of signal processing, whereas correlation coefficients between MSI matrix estimates $\hat{\mathbf{N}}_{x,p}$ (5.3.14a), $\hat{\mathbf{N}}_{x,s}$ (5.3.14b) and ACM estimate $\hat{\mathbf{R}}_x$ (5.3.4) are equal to $r[\hat{\mathbf{N}}_{x,p}, \hat{\mathbf{R}}_x]=0.974$ and $r[\hat{\mathbf{N}}_{x,s}, \hat{\mathbf{R}}_x]=0.998$, respectively. Notice, that NMSIs (1.4.6) and (1.4.7) between MUSIC estimates of HSDs $\hat{H}^p_{MUSIC}(f)$ (5.4.9), $\hat{H}^s_{MUSIC}(f)$ (5.4.10) and MUSIC estimate of PSD $\hat{P}_{MUSIC}(f)$ (5.4.8) are, respectively, equal:

$$\nu_p[\hat{H}^s_{MUSIC}(f) - \mathbf{M}(\hat{H}^s_{MUSIC}(f)), \hat{P}_{MUSIC}(f) - \mathbf{M}(\hat{P}_{MUSIC}(f))] = 0.99;$$

$$\nu_p[\hat{H}^p_{MUSIC}(f) - \mathbf{M}(\hat{H}^p_{MUSIC}(f)), \hat{P}_{MUSIC}(f) - \mathbf{M}(\hat{P}_{MUSIC}(f))] = 0.92;$$

$$\nu_s[\hat{H}^s_{MUSIC}(f) - \mathbf{M}(\hat{H}^s_{MUSIC}(f)), \hat{P}_{MUSIC}(f) - \mathbf{M}(\hat{P}_{MUSIC}(f))] = 0.997;$$

$$\nu_s[\hat{H}^p_{MUSIC}(f) - \mathbf{M}(\hat{H}^p_{MUSIC}(f)), \hat{P}_{MUSIC}(f) - \mathbf{M}(\hat{P}_{MUSIC}(f))] = 0.962.$$

The last relationships testify to a high extent correspondence of the results obtained in linear signal space (Fig. 5.4.3) and in signal space with L-group properties (Fig. 5.4.4a,b), that is explained by high identity between MSI matrix estimators $\hat{\mathbf{N}}_{x,p}$ (5.3.14a), $\hat{\mathbf{N}}_{x,s}$ (5.3.14b) and ACM estimator $\hat{\mathbf{R}}_x$ (5.3.4), and also by the fact that nonlinear statistical interrelations between these estimators, unlike linear, are almost completely preserved.

$\triangledown$

5.4.3 Spectral Estimation by Minimum Norm Method Based on Measure of Statistical Interrelation Matrix Estimator

Spectral estimation by minimum norm method [151], similarly as MUSIC method, supposes using information contained in the result $\mathbf{V}'$ of cutting off the matrix $\mathbf{V} = \|\mathbf{v}_0, \mathbf{v}_1, \ldots, \mathbf{v}_{K-1}\|$ with a dimensionality $K \times K$ composed by eigenvectors $\{\mathbf{v}_k\}$ of ACM estimator $\hat{\mathbf{R}}_x$ (5.3.1) of additive mixture (5.1.9):

$$\mathbf{V}' = \mathrm{submatrix}(\mathbf{V}, 0, K-1, M, K-1) = \|\mathbf{v}'_M, \mathbf{v}'_{M+1}, \ldots, \mathbf{v}'_{K-1}\|, \quad (5.4.12)$$

where $\mathrm{submatrix}(\mathbf{A}, r_i, r_j, c_i, c_j)$ is a function that returns a submatrix of initial matrix $\mathbf{A}$ consisting of elements in rows r_i through r_j and columns c_i through c_j; eigenvector matrix $\mathbf{V} = \|\mathbf{V}_{i,k}\|$ has a dimensionality $K \times K$, $i = 0, \ldots, K-1$, $k = 0, \ldots, K-1$; matrix $\mathbf{V}' = \|\mathbf{v}'_M, \mathbf{v}'_{M+1}, \ldots, \mathbf{v}'_{K-1}\|$ has a dimensionality $K \times K - M$; M is an expected number of harmonic signals in the mixture (5.1.9), $M < K$.

Matrix $\mathbf{V}'$ is related to noise subspace in which a vector $\mathbf{d}$, determined by the matrix $\mathbf{V}'$:

$$\mathbf{d} = \frac{\mathbf{V}' \cdot \bar{\mathbf{c}}^T}{\bar{\mathbf{c}}^T \mathbf{c}}, \quad (5.4.13)$$

where $\mathbf{c}$ is a vector composed by the upper row of the matrix $\mathbf{V}'$:

$$\mathbf{c} = [\mathrm{submatrix}(\mathbf{V}', 0, 0, 0, K-M-1)]^T, \quad (5.4.13a)$$

has the least l_2-norm: $\|\mathbf{d}\|^2 = \sum\limits_{i=0}^{K-M-1} d_i^2 \to \min.$

Then spectral estimator $\hat{P}_{MN}(f)$ obtained by minimum norm (MN) method is defined by the relationship:

$$\hat{P}_{MN}(f) = \frac{1}{\overline{\mathbf{e}(f)}^T \left(\mathbf{d} \cdot \bar{\mathbf{d}}^T\right) \mathbf{e}(f)}, \quad (5.4.14)$$

where, remind, $\mathbf{e}(f)$ is vector of complex harmonics that is similar to (5.4.2a), but has another dimensionality:

$$\mathbf{e}(f)^T = [1, \ldots, \exp(j2\pi f \cdot k\Delta t), \ldots, \exp(j2\pi f \cdot (K-M-1)\Delta t)], \quad (5.4.14a)$$

where M is a number of harmonic signals in the observed additive mixture (5.1.9); $K \times K$ is a dimensionality of ACM $\hat{\mathbf{R}}_x$ (5.3.1).

By the analogy with MN estimator (5.4.14), one can obtain MN estimators of HSDs $\hat{H}^p_{MN}(f)$, $\hat{H}^s_{MN}(f)$, $\hat{H}^{l_1}_{MN}(f)$ based on MSIs (1.4.10), (1.4.11), and (1.4.12), respectively, that are defined by the following similar relationships:

$$\hat{H}^p_{MN}(f) = \frac{1}{\overline{\mathbf{e}(f)}^T \left(\mathbf{g}_p \bar{\mathbf{g}}_p^T\right) \mathbf{e}(f)}; \quad (5.4.15)$$

$$\mathbf{g}_p = \frac{\mathbf{U}'_p \cdot \bar{\mathbf{a}}_p^T}{\bar{\mathbf{a}}_p^T \mathbf{a}_p}; \quad (5.4.15a)$$

$$\mathbf{a}_p = [\mathrm{submatrix}(\mathbf{U}'_p, 0, 0, 0, K-M-1)]^T; \quad (5.4.15b)$$

$$\mathbf{U}'_p = \mathrm{submatrix}(\mathbf{U}_p, 0, K-1, M, K-1) =$$
$$= \|\mathbf{u}'_{p,M}, \mathbf{u}'_{p,M+1}, \ldots, \mathbf{u}'_{p,K-1}\|; \quad (5.4.15c)$$

$$\hat{H}^s_{MN}(f) = \frac{1}{\overline{\mathbf{e}(f)}^T \left(\mathbf{g}_s \bar{\mathbf{g}}_s^T \right) \mathbf{e}(f)};\tag{5.4.16}$$

$$\mathbf{g}_s = \frac{\mathbf{U'}_s \cdot \bar{\mathbf{a}}_s^T}{\bar{\mathbf{a}}_s^T \mathbf{a}_s};\tag{5.4.16a}$$

$$\mathbf{a}_s = [\text{submatrix}(\mathbf{U'}_s, 0, 0, 0, K - M - 1)]^T;\tag{5.4.16b}$$

$$\mathbf{U'}_s = \text{submatrix}(\mathbf{U}_s, 0, K - 1, M, K - 1) =$$
$$= \|\mathbf{u'}_{s,M}, \mathbf{u'}_{s,M+1}, \ldots, \mathbf{u'}_{s,K-1}\|;\tag{5.4.16c}$$

$$\hat{H}^{l_1}_{MN}(f) = \frac{1}{\overline{\mathbf{e}(f)}^T \left(\mathbf{g}_{l_1} \bar{\mathbf{g}}_{l_1}^T \right) \mathbf{e}(f)};\tag{5.4.17}$$

$$\mathbf{g}_{l_1} = \frac{\mathbf{U'}_{l_1} \cdot \bar{\mathbf{a}}_{l_1}^T}{\bar{\mathbf{a}}_{l_1}^T \mathbf{a}_{l_1}};\tag{5.4.17a}$$

$$\mathbf{a}_{l_1} = [\text{submatrix}(\mathbf{U'}_{l_1}, 0, 0, 0, K - M - 1)]^T;\tag{5.4.17b}$$

$$\mathbf{U'}_{l_1} = \text{submatrix}(\mathbf{U}_{l_1}, 0, K - 1, M, K - 1) =$$
$$= \|\mathbf{u'}_{l_1,M}, \mathbf{u'}_{l_1,M+1}, \ldots, \mathbf{u'}_{l_1,K-1}\|;\tag{5.4.17c}$$

where

$$\mathbf{U}_p = \|\mathbf{u}_{p,0}, \mathbf{u}_{p,1}, \ldots, \mathbf{u}_{p,K-1}\|;$$

$$\mathbf{U}_s = \|\mathbf{u}_{s,0}, \mathbf{u}_{s,1}, \ldots, \mathbf{u}_{s,K-1}\|;$$

$$\mathbf{U}_{l_1} = \|\mathbf{u}_{l_1,0}, \mathbf{u}_{l_1,1}, \ldots, \mathbf{u}_{l_1,K-1}\|$$

are the matrices composed by eigenvectors of MSI matrix estimators $\hat{\mathbf{N}}_{x,p}$ (5.3.14a), $\hat{\mathbf{N}}_{x,s}$ (5.3.14b), $\hat{\mathbf{N}}_{x,l_1}$ (5.3.14c), respectively; $\mathbf{U'}_p$, $\mathbf{U'}_s$, $\mathbf{U'}_{l_1}$ are the matrices of *noise subspace* with a dimensionality $K \times K - M$ obtained by cutting off the matrices $\mathbf{U}_p$, $\mathbf{U}_s$, $\mathbf{U}_{l_1}$, respectively; $\mathbf{a}_p$, $\mathbf{a}_s$, $\mathbf{a}_{l_1}$ are vectors composed of the upper row of the matrices $\mathbf{U'}_p$, $\mathbf{U'}_s$, $\mathbf{U'}_{l_1}$, respectively; $\mathbf{g}_p$, $\mathbf{g}_s$, $\mathbf{g}_{l_1}$ are minimum norm vectors determined by the matrices $\mathbf{U}_p$, $\mathbf{U}_s$, $\mathbf{U}_{l_1}$, respectively; $\mathbf{e}(f)$ is vector of complex harmonics determined by the formula (5.4.14a); M is a number of harmonic signals in the observed additive mixture (5.1.9); $K \times K$ is a dimensionality of MSI matrices $\hat{\mathbf{N}}_{x,p}$, $\hat{\mathbf{N}}_{x,s}$, and $\hat{\mathbf{N}}_{x,l_1}$.

Having obtained the algorithms (5.4.15)... (5.4.17) of MN estimators of HSDs $\hat{H}^p_{MN}(f)$, $\hat{H}^s_{MN}(f)$, $\hat{H}^{l_1}_{MN}(f)$ based on MSIs (1.4.10), (1.4.11), and (1.4.12), respectively, it is necessary to find out, first, how close are they with respect to MN estimator (5.4.14), and second, could they be used for estimating spectral components of a mixture of harmonic signals and noise. Algorithms (5.4.15) and (5.4.16) are illustrated by the following example.

Example 5.4.3. Let $x(i)$ be additive mixture of M complex discrete harmonic signals observed in the presence of complex quasi-white Gaussian noise $n(i)$ with zero mean (5.1.9).

Within the considered example, amplitudes of harmonics are chosen to be equal $A_l = A$, so that the relationships hold: $A^2/2 = D_n$, $F_n > 2\max_l\{f_l\}$, where D_n is a

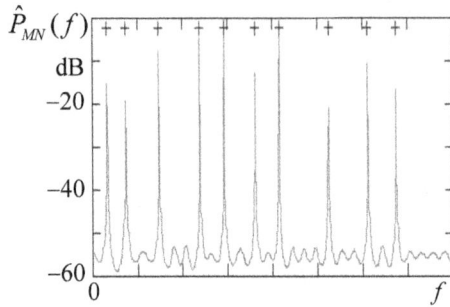

FIGURE 5.4.5   MN estimate of PSD $\hat{P}_{MN}(f)$ (5.4.14)

variance of noise $n(i)$; F_n is upper bound frequency of PSD of quasi-white Gaussian noise $n(i)$; $M = 10$. The number of samples N of stochastic process $x(i)$ used when forming ACF estimator (5.1.2a) and also estimators (5.3.7)…(5.3.10) is equal to N=1024. Maximal index L of time shift $m \leq L$, figuring in the ACF estimator (5.1.2a) and also in MSI estimators (5.3.7)…(5.3.10), is chosen to be equal L=128. Dimensionality $K \times K$ of MSI matrices $\hat{\mathbf{N}}_{x,p}$ (5.3.14a), $\hat{\mathbf{N}}_{x,s}$ (5.3.14b), and ACM $\hat{\mathbf{R}}_x$ (5.3.4) is chosen to be equal K=32.

Fig. 5.4.5 illustrates PSD MN estimate $\hat{P}_{MN}(f)$ of the mixture $x(i)$ (5.1.9) obtained according to the relationship (5.4.14) for initial situation described above. True positions of the frequencies of harmonic signals from the mixture (5.1.9) are denoted by the signs "+".

Fig. 5.4.6a and 5.4.6b depict MN estimates of HSDs $\hat{H}_{MN}^p(f)$ (5.4.15) and $\hat{H}_{MN}^s(f)$ (5.4.16) of the mixture $x(i)$ (5.1.9), respectively, for initial situation described above. True positions of the frequencies of harmonic signals (5.1.9) are denoted by the signs "+". Correlation coefficients between MN estimates of HSDs $\hat{H}_{MN}^p(f)$ (5.4.15), $\hat{H}_{MN}^s(f)$ (5.4.16) and MN estimate of PSD $\hat{P}_{MN}(f)$ (5.4.14) are equal to $r[\hat{H}_{MN}^p(f), \hat{P}_{MN}(f)]$=0.554 and $r[\hat{H}_{MN}^s(f), \hat{P}_{MN}(f)]$=0.148, that is explained by a slight level of linear statistical interrelations between the results of signal processing, while correlation coefficients between MSI matrix estimates $\hat{\mathbf{N}}_{x,p}$ (5.3.14a), $\hat{\mathbf{N}}_{x,s}$ (5.3.14b), and ACM estimate $\hat{\mathbf{R}}_x$ (5.3.4) are equal to: $r[\hat{\mathbf{N}}_{x,p}, \hat{\mathbf{R}}_x]$=0.974 and $r[\hat{\mathbf{N}}_{x,s}, \hat{\mathbf{R}}_x]$=0.998, respectively. Notice, that NMSIs (1.4.6) and (1.4.7) between MN estimates of HSDs $\hat{H}_{MN}^p(f)$ (5.4.14a), $\hat{H}_{MN}^s(f)$ (5.4.14b), and MN estimate of PSD $\hat{P}_{MN}(f)$ (5.4.14) are equal to 1:

$$\nu_p[\hat{H}_{MN}^s(f) - \mathbf{M}(\hat{H}_{MN}^s(f)), \hat{P}_{MN}(f) - \mathbf{M}(\hat{P}_{MN}(f))] = 0.996;$$

$$\nu_p[\hat{H}_{MN}^p(f) - \mathbf{M}(\hat{H}_{MN}^p(f)), \hat{P}_{MN}(f) - \mathbf{M}(\hat{P}_{MN}(f))] = 0.932;$$

$$\nu_s[\hat{H}_{MN}^s(f) - \mathbf{M}(\hat{H}_{MN}^s(f)), \hat{P}_{MN}(f) - \mathbf{M}(\hat{P}_{MN}(f))] = 0.999;$$

$$\nu_s[\hat{H}_{MN}^p(f) - \mathbf{M}(\hat{H}_{MN}^p(f)), \hat{P}_{MN}(f) - \mathbf{M}(\hat{P}_{MN}(f))] = 0.943.$$

The last relationships testify to high extent correspondence of the results obtained in linear signal space (Fig. 5.4.5) and in signal space with L-group properties (Fig. 5.4.6a,b), that is explained by high identities between MSI matrix estimators $\hat{\mathbf{N}}_{x,p}$ (5.3.14a), $\hat{\mathbf{N}}_{x,s}$ (5.3.14b) and ACM estimator $\hat{\mathbf{R}}_x$ (5.3.4), and also

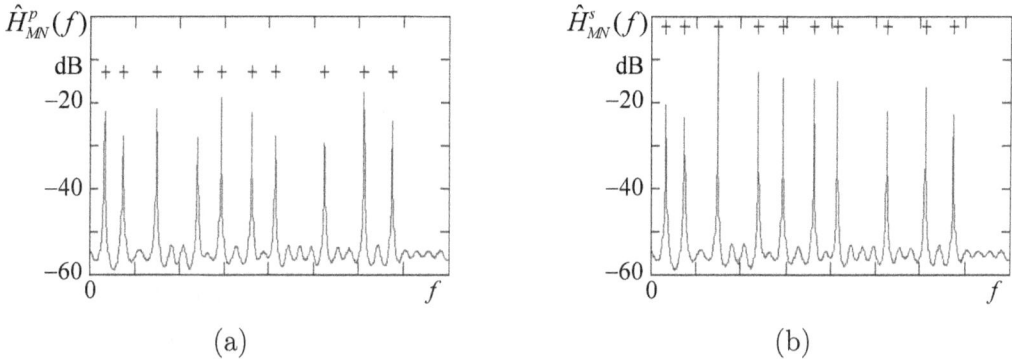

FIGURE 5.4.6   MN estimates of HSDs: (a) $\hat{H}^p_{MN}(f)$ (5.4.15); (b) $\hat{H}^s_{MN}(f)$ (5.4.16)

by the fact that nonlinear statistical interrelations between these estimators, unlike linear, are almost completely preserved.

$\triangledown$

5.4.4   Spectral Estimation by ESPRIT Method Based on Measure of Statistical Interrelation Matrix Estimator

Spectral estimation by Estimation of Signal Parameters via Rotational Invariance Techniques (ESPRIT) method [152, 153], as well as MUSIC method and MN method, supposes using information that is contained in the matrix $\mathbf{V} = \|\mathbf{v}_0, \mathbf{v}_1, \ldots, \mathbf{v}_{K-1}\|$ with dimensionality $K \times K$ composed by eigenvectors $\{\mathbf{v}_k\}$ of ACM estimator $\hat{\mathbf{R}}_x$ (5.3.4) (see decomposition $\hat{\mathbf{R}}_x$ over $\{\mathbf{v}_k\}$ in Subsection 5.4.1). However, unlike MUSIC method and MN method that realize signal processing in noise subspace based on the matrices (5.4.8a) and (5.4.12), respectively, ESPRIT method use information that is contained in the left part $\mathbf{V}''$ of the matrix $\mathbf{V}$ corresponding to so called *signal subspace*:

$$\mathbf{V}'' = \text{submatrix}(\mathbf{V}, 0, K-1, 0, M-1) = \|\mathbf{v}''_0, \mathbf{v}''_1, \ldots, \mathbf{v}''_{M-1}\|, \qquad (5.4.18)$$

where submatrix$(\mathbf{A}, r_i, r_j, c_i, c_j)$ is a function of extracting a submatrix from initial matrix $\mathbf{A}$ consisting of elements in rows r_i through r_j and columns c_i through c_j of $\mathbf{A}$; eigenvector matrix $\mathbf{V} = \|\mathbf{V}_{i,k}\|$ has a dimensionality $K \times K$, $i = 0, \ldots, K-1$, $k = 0, \ldots, K-1$; the matrix $\mathbf{V}'' = \|\mathbf{v}''_0, \mathbf{v}''_1, \ldots, \mathbf{v}''_{M-1}\|$ has a dimensionality $K \times M$; M is an expected number of harmonic signals in the mixture (5.1.9), $M < K$.

Then algorithm of forming ESPRIT estimator $\hat{\mathbf{f}} = [\hat{f}_l]$ of frequency vector $\mathbf{f} = [f_l]$ of harmonic signals in the mixture (5.1.9), $l = 0, \ldots, M-1$ takes the form:

$$\hat{f}_l = \frac{1}{2\pi} \text{arctg} \left[\frac{\text{Im}(\lambda_l^{\Phi})}{\text{Re}(\lambda_l^{\Phi})} \right], \qquad (5.4.19)$$

where λ_l^{Φ} are eigenvalues of the matrix $\mathbf{\Phi}$ with dimensionality $M \times M$:

$$\mathbf{\Phi} = \left(\bar{\mathbf{S}}_1^T \mathbf{S}_1 \right)^{-1} \bar{\mathbf{S}}_1^T \mathbf{S}_2, \qquad (5.4.20)$$

that is a solution of the equation:

$$\mathbf{S}_1\boldsymbol{\Phi} = \mathbf{S}_2; \tag{5.4.21}$$

$$\mathbf{S}_1 = \mathbf{G}_1\mathbf{V}''; \quad \mathbf{S}_2 = \mathbf{G}_2\mathbf{V}''; \tag{5.4.21a}$$

$$\mathbf{G}_1 = \|\mathbf{I}_{K-1}|\mathbf{0}\|; \quad \mathbf{G}_2 = \|\mathbf{0}|\mathbf{I}_{K-1}\|, \tag{5.4.21b}$$

where $\mathbf{G}_1$, $\mathbf{G}_2$ are selector matrices with dimensionality $K-1 \times K$; $\mathbf{V}''$ is truncated matrix of eigenvectors of ACM $\hat{\mathbf{R}}_x$ (5.3.1) corresponding to signal subspace that is determined by the relationship (5.4.18); $\mathbf{I}_{K-1}$ is $K-1 \times K-1$ identity matrix; $\mathbf{0}$ is zero column vector $\mathbf{0} = [0,\ldots,0]^T$ with a number of elements $K-1$; M is an expected number of harmonic signals in the mixture (5.1.9), $M < K$.

By the analogy with ESPRIT estimator (5.4.19)...(5.4.21), one can obtain ES-PRIT estimators $\hat{\mathbf{f}}^p = [\hat{f}_l^p]$, $\hat{\mathbf{f}}^s = [\hat{f}_l^s]$ and $\hat{\mathbf{f}}^{l_1} = [\hat{f}_l^{l_1}]$ of frequency vector $\mathbf{f} = [f_l]$ of harmonic signals from the mixture (5.1.9), $l = 0,\ldots,M-1$ that are based on MSIs (1.4.11), (1.4.11), and (1.4.12), respectively, and are defined by the following similar relationships:

$$\hat{f}_l^p = \frac{1}{2\pi}\operatorname{arctg}\left[\frac{\operatorname{Im}(\lambda_l^{\Phi^p})}{\operatorname{Re}(\lambda_l^{\Phi^p})}\right]; \tag{5.4.22}$$

$$\boldsymbol{\Phi}^p = \left(\bar{\mathbf{C}}_{p,1}^T\mathbf{C}_{p,1}\right)^{-1}\bar{\mathbf{C}}_{p,1}^T\mathbf{C}_{p,2}; \tag{5.4.22a}$$

$$\mathbf{C}_{p,1} = \mathbf{G}_1\mathbf{U}''_p; \quad \mathbf{C}_{p,2} = \mathbf{G}_2\mathbf{U}''_p; \tag{5.4.22b}$$

$$\mathbf{G}_1 = \|\mathbf{I}_{K-1}|\mathbf{0}\|; \quad \mathbf{G}_2 = \|\mathbf{0}|\mathbf{I}_{K-1}\|; \tag{5.4.22c}$$

$$\mathbf{U}''_p = \operatorname{submatrix}(\mathbf{U}_p, 0, K-1, 0, M-1)$$
$$= \|\mathbf{u}''_{p,0}, \mathbf{u}''_{p,1},\ldots,\mathbf{u}''_{p,M-1}\|; \tag{5.4.22d}$$

$$\hat{f}_l^s = \frac{1}{2\pi}\operatorname{arctg}\left[\frac{\operatorname{Im}(\lambda_l^{\Phi^s})}{\operatorname{Re}(\lambda_l^{\Phi^s})}\right]; \tag{5.4.23}$$

$$\boldsymbol{\Phi}^s = \left(\bar{\mathbf{C}}_{s,1}^T\mathbf{C}_{s,1}\right)^{-1}\bar{\mathbf{C}}_{s,1}^T\mathbf{C}_{s,2}; \tag{5.4.23a}$$

$$\mathbf{C}_{s,1} = \mathbf{G}_1\mathbf{U}''_s; \quad \mathbf{C}_{s,2} = \mathbf{G}_2\mathbf{U}''_s; \tag{5.4.23b}$$

$$\mathbf{G}_1 = \|\mathbf{I}_{K-1}|\mathbf{0}\|; \quad \mathbf{G}_2 = \|\mathbf{0}|\mathbf{I}_{K-1}\|; \tag{5.4.23c}$$

$$\mathbf{U}''_s = \operatorname{submatrix}(\mathbf{U}_s, 0, K-1, 0, M-1)$$
$$= \|\mathbf{u}''_{s,0}, \mathbf{u}''_{s,1},\ldots,\mathbf{u}''_{s,M-1}\|; \tag{5.4.23d}$$

$$\hat{f}_l^{l_1} = \frac{1}{2\pi}\operatorname{arctg}\left[\frac{\operatorname{Im}(\lambda_l^{\Phi^{l_1}})}{\operatorname{Re}(\lambda_l^{\Phi^{l_1}})}\right]; \tag{5.4.24}$$

$$\boldsymbol{\Phi}^{l_1} = \left(\bar{\mathbf{C}}_{l_1,1}^T\mathbf{C}_{l_1,1}\right)^{-1}\bar{\mathbf{C}}_{l_1,1}^T\mathbf{C}_{l_1,2}; \tag{5.4.24a}$$

$$\mathbf{C}_{l_1,1} = \mathbf{G}_1\mathbf{U}''_{l_1}; \quad \mathbf{C}_{l_1,2} = \mathbf{G}_2\mathbf{U}''_{l_1}; \tag{5.4.24b}$$

$$\mathbf{G}_1 = \|\mathbf{I}_{K-1}|\mathbf{0}\|; \quad \mathbf{G}_2 = \|\mathbf{0}|\mathbf{I}_{K-1}\|; \tag{5.4.24c}$$

$$\mathbf{U}''_{l_1} = \operatorname{submatrix}(\mathbf{U}_{l_1}, 0, K-1, 0, M-1)$$
$$= \|\mathbf{u}''_{l_1,0}, \mathbf{u}''_{l_1,1},\ldots,\mathbf{u}''_{l_1,M-1}\|, \tag{5.4.24d}$$

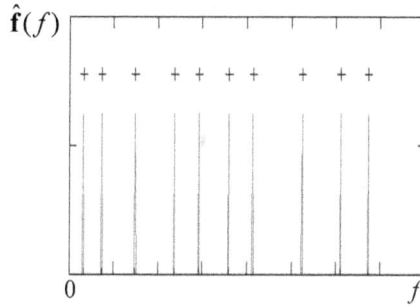

FIGURE 5.4.7   ESPRIT estimate $\hat{\mathbf{f}}(f)$ (5.4.19) of frequency vector $\mathbf{f} = [f_l]$ of harmonic signals

where $\lambda_l^{\Phi^p}$, $\lambda_l^{\Phi^s}$, $\lambda_l^{\Phi^{l_1}}$ are eigenvalues of matrices $\mathbf{\Phi}^p$, $\mathbf{\Phi}^s$, $\mathbf{\Phi}^{l_1}$ with dimensionality $M \times M$, respectively; $\mathbf{G}_1$, $\mathbf{G}_2$ are selector matrices with dimensionality $K - 1 \times K$; $\mathbf{U}_p = \|\mathbf{u}_{p,0}, \mathbf{u}_{p,1}, \ldots, \mathbf{u}_{p,K-1}\|$, $\mathbf{U}_s = \|\mathbf{u}_{s,0}, \mathbf{u}_{s,1}, \ldots, \mathbf{u}_{s,K-1}\|$, $\mathbf{U}_{l_1} = \|\mathbf{u}_{l_1,0}, \mathbf{u}_{l_1,1}, \ldots, \mathbf{u}_{l_1,K-1}\|$ are the matrices with dimensionality $K \times K$ composed by eigenvectors of MSI matrix estimators $\hat{\mathbf{N}}_{x,p}$ (5.3.14a), $\hat{\mathbf{N}}_{x,s}$ (5.3.14b), $\hat{\mathbf{N}}_{x,l_1}$ (5.3.14c), respectively; $\mathbf{U}''_p$, $\mathbf{U}''_s$, $\mathbf{U}''_{l_1}$ are the matrices of signal subspace with dimensionality $K \times M$ obtained by cutting off the matrices $\mathbf{U}_p$, $\mathbf{U}_s$, $\mathbf{U}_{l_1}$, respectively; $\mathbf{I}_{K-1}$ is $K - 1 \times K - 1$ identity matrix; $\mathbf{0}$ is zero column vector $\mathbf{0} = [0, \ldots, 0]^T$ with a number of elements $K - 1$; M is an expected number of harmonic signals in the mixture (5.1.9), $M < K$.

Having obtained the algorithms (5.4.22)...(5.4.24) of forming ESPRIT estimators $\hat{\mathbf{f}}^p = [\hat{f}_l^p]$, $\hat{\mathbf{f}}^s = [\hat{f}_l^s]$, $\hat{\mathbf{f}}^{l_1} = [\hat{f}_l^{l_1}]$ of frequency vector $\mathbf{f} = [f_l]$ of harmonic signals from the mixture (5.1.9), that are based on MSIs (1.4.10), (1.4.11), and (1.4.12), respectively, it is necessary to find out, first, how close are they with respect to ESPRIT estimator (5.4.19)...(5.4.21), and second, could they be used for estimating spectral components of a mixture of harmonic signals and noise. Algorithms (5.4.22) and (5.4.23) are illustrated by the following example.

Example 5.4.4. Let $x(i)$ be additive mixture of M complex discrete harmonic signals observed in the presence of complex quasi-white Gaussian noise $n(i)$ with zero mean (5.1.9).

Within the considered example, amplitudes of harmonics are chosen to be equal $A_l = A$, so that the relationships hold: $A^2/2 = D_n$, $F_n > 2\max_l\{f_l\}$, where D_n is a variance of noise $n(i)$; F_n is upper bound frequency of PSD of quasi-white Gaussian noise $n(i)$; $M = 10$. The number of samples N of stochastic process $x(i)$ used when forming ACF estimator (5.1.2a) and also estimators (5.3.7)...(5.3.10) is equal to $N=512$. Maximal index L of time shift $m \leq L$, figuring in the ACF estimator (5.1.2a) and also in MSI estimators (5.3.7)...(5.3.10), is chosen to be equal $L=64$. Dimensionality $K \times K$ of MSI matrices $\hat{\mathbf{N}}_{x,p}$ (5.3.14a), $\hat{\mathbf{N}}_{x,s}$ (5.3.14b), and ACM $\hat{\mathbf{R}}_x$ (5.3.4) is chosen to be equal $K = L=64$.

Fig. 5.4.7 illustrates ESPRIT estimate $\hat{\mathbf{f}} = [\hat{f}_l]$ of frequency vector $\mathbf{f} = [f_l]$ of harmonic signals from the mixture $x(i)$ (5.1.9) obtained according to the

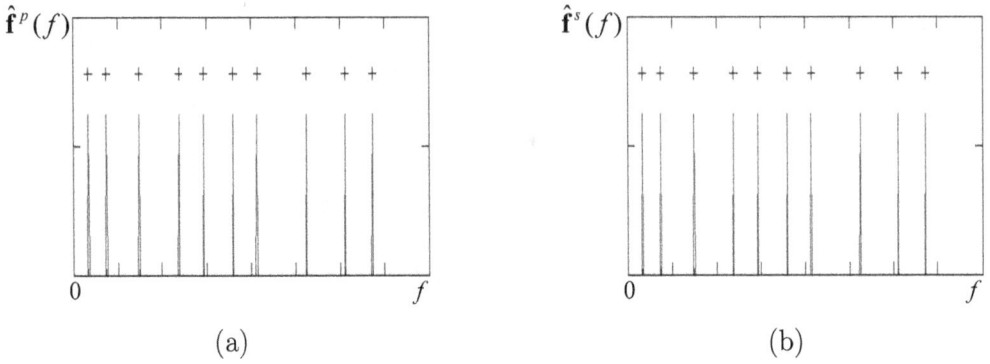

(a)   (b)

FIGURE 5.4.8   ESPRIT estimates of frequency vector $\mathbf{f} = [f_l]$: (a) $\hat{\mathbf{f}}^p(f)$; (b) $\hat{\mathbf{f}}^s(f)$

relationship (5.4.19) for initial situation described above in the form of a function $\hat{\mathbf{f}}(f)$:

$$\hat{\mathbf{f}}(f) = \sum_{l=0}^{M-1} \delta[f, \hat{f}_l],$$

where $\delta[a, b]$ is Kronecker symbol: $\delta[a, b]=1$, if $a = b$; $\delta[a, b]=0$, if $a \neq b$.

True positions of the frequencies of harmonic signals from the mixture (5.1.9) are denoted by the signs "+".

Fig. 5.4.8a and 5.4.8b depict ESPRIT estimates $\hat{\mathbf{f}}^p = [\hat{f}_l^p]$ (5.4.22) and $\hat{\mathbf{f}}^s = [\hat{f}_l^s]$ (5.4.23) of frequency vector $\mathbf{f} = [f_l]$ of harmonic signals from the mixture $x(i)$ (5.1.9), for initial situation described above, in the form of functions $\hat{\mathbf{f}}^p(f)$ and $\hat{\mathbf{f}}^s(f)$, respectively:

$$\hat{\mathbf{f}}^p(f) = \sum_{l=0}^{M-1} \delta[f, \hat{f}_l^p]; \ \hat{\mathbf{f}}^s(f) = \sum_{l=0}^{M-1} \delta[f, \hat{f}_l^s],$$

where $\delta[a, b]$ is Kronecker symbol.

True positions of the frequencies of harmonic signals from the mixture (5.1.9) are denoted by the signs "+".

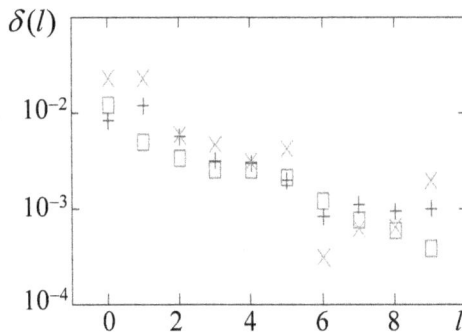

FIGURE 5.4.9   Realizations of normalized absolute errors $\delta(l)$, $\delta^p(l)$, and $\delta^s(l)$

Fig. 5.4.9 illustrates realizations of normalized absolute errors $\delta(l)$, $\delta^p(l)$, $\delta^s(l)$ of ESPRIT estimates $\hat{\mathbf{f}} = [\hat{f}_l]$ (5.4.19), $\hat{\mathbf{f}}^p = [\hat{f}_l^p]$ (5.4.22), $\hat{\mathbf{f}}^s = [\hat{f}_l^s]$ (5.4.23) of frequency vector $\mathbf{f} = [f_l]$, $l = 0, \ldots, M - 1$:

$$\delta(l) = |\hat{f}_l - f_l|/f_l; \ \delta^p(l) = |\hat{f}_l^p - f_l|/f_l; \ \delta^s(l) = |\hat{f}_l^s - f_l|/f_l.$$

Realizations of errors $\delta(l)$, $\delta^p(l)$, $\delta^s(l)$ are denoted by the symbols "□", "×", "+", respectively. As can be seen from the figure, estimators (5.4.22) and (5.4.23) based on *L*-group algorithms are just slightly inferior to classic ESPRIT estimator (5.4.19). ▽

6

Antenna Array Signal Processing Based on L–group Operations

6.1 Antenna Systems with Logical Signal Processing Based on L–group Operations

Two-channel antenna system with logical signal processing [154, Fig. 10.13], [155, Fig. 2.125], [131, Fig. 7.22] (or antenna system selecting the signals arriving within main lobe of field (power) pattern), involving directional (main) and omnidirectional (auxiliary) antennas, allow receiving useful signals within main lobe of field (power) pattern of main antenna, excluding interference signals, received within side lobes of main antenna, from processing. Possibility of using such a signal processing approach is based on the assumption that signal-to-interference ratio in main receiving channel is much higher than this in auxiliary receiving channel. In this section, we consider the basics of such systems formulating signal processing algorithms in terms of L-group operations for both coherent and incoherent (after-detector) signal processing.

6.1.1 Two-channel Antenna System with Logical Coherent Signal Processing Based on L–group Operations

Let $a(t)$ and $b(t)$ be some linearly independent functions of useful signal $s(t)$ and interference $n(t)$ in the inputs of processing unit T that come from two antennas, i.e. from directional antenna A of the main channel and omnidirectional antenna B of the auxiliary channel (see Fig. 6.1.1a,b). Useful signal $s(t)$ and interference $n(t)$ arrive in the antennas of two channels A and B (see Fig. 6.1.1a), so that, first, antennas A and B have a common phase center, second, antennas A and B have field patterns $F_A(\theta)$, $F_B(\theta)$, respectively, and $F_A(\theta_s) = G$; $F_A(\theta_n) = G_n$; $G > 1 > G_n$; $F_B(\theta)=1$. Complex frequency responses $\dot{K}_A(\omega)$, $\dot{K}_B(\omega)$ of antenna-feeder systems in the channels A and B are considered to be identical: $\dot{K}_A(\omega) = \dot{K}_B(\omega)$, so that in two processing channels (in the inputs of processing unit T) there appear the signals $a(t)$, $b(t)$:

$$\begin{cases} a(t) = G \cdot s(t) + G_n \cdot n(t); & (a) \\ b(t) = s(t) + n(t), & (b) \end{cases} \qquad (6.1.1)$$

where $G = F_A(\theta_s)$ is a gain of antenna A in the direction θ_s of arrival of the signal $s(t)$; $G_n = F_A(\theta_n)$ is a gain of antenna A in the direction θ_n of arrival of interference signal $n(t)$.

DOI: 10.1201/9781003275855-6

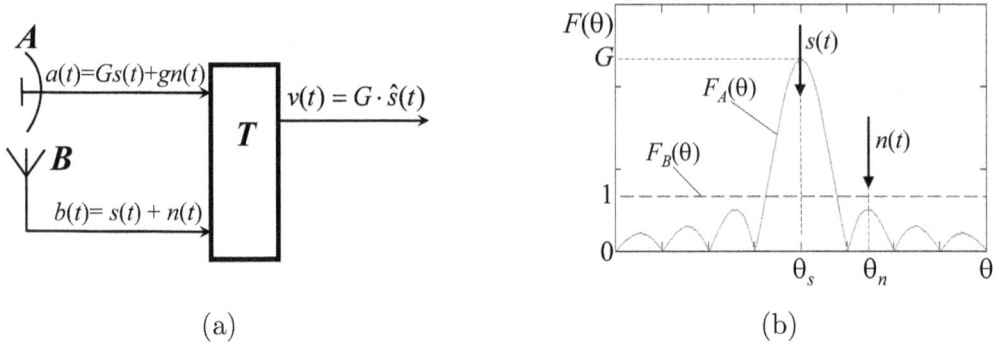

FIGURE 6.1.1   (a) Two-channel antenna system with logical signal processing; (b) field patterns $F_A(\theta)$, $F_B(\theta)$ of antennas A, B, respectively

We suppose that direction of signal arrival θ_s is known, and direction of interference signal arrival θ_n is unknown, so that any interference signal is received by side lobes of antenna pattern of main channel.

In the output of processing unit T, an estimator of the signal $v(t) = G \cdot \hat{s}(t)$ is formed according to the general processing algorithm:

$$v(t) = \operatorname*{med}_{t \in \tilde{T}}[y(t)]; \qquad (6.1.2)$$

$$y(t) = a(t) \cdot 1[|a(t)| - |b(t)|] = \begin{cases} a(t), & |a(t)| > |b(t)|; \quad (a) \\ 0, & |a(t)| \leqslant |b(t)|, \quad (b) \end{cases} \qquad (6.1.3)$$

where med[∗] is operation of calculating a median; $\tilde{T}$ is an interval in which the signal $y(t)$ is smoothed; $1(t)$ is Heaviside unit step function.

In terms of L-group operations, processing algorithm (6.1.3) is determined by the following relationship:

$$y(t) = a(t) \cdot [0 \vee (1 \wedge k \cdot [(a(t) \vee (-a(t))) - (b(t) \vee (-b(t)))])], \qquad (6.1.4)$$

where k=const, $k \gg 1$.

Fig. 6.1.2. . . 6.1.4 illustrate the results of simulating signal processing according to the relationships (6.1.2) and (6.1.4) in the presence of useful signal $s(t)$, arriving from the direction of main lobe, and interference signals $n_1(t)$, $n_2(t)$, $n_3(t)$, $n_4(t)$ received by side lobes of main channel antenna pattern (Fig. 6.1.2a,b):

$$\begin{cases} a(t) = G \cdot s(t) + \sum\limits_{i=1}^{4} G_i \cdot n_i(t); & (a) \\ b(t) = s(t) + \sum\limits_{i=1}^{4} n_i(t), & (b) \end{cases} \qquad (6.1.5)$$

where $G = F_A(\theta_s)$ is a gain of antenna A in the direction θ_s of arrival of the signal $s(t)$; $G_i = F_A(\theta_{n_i})$ is a gain of antenna A in the direction θ_{n_i} of arrival of interference $n_i(t)$, $G_i < F_B(\theta) = 1$.

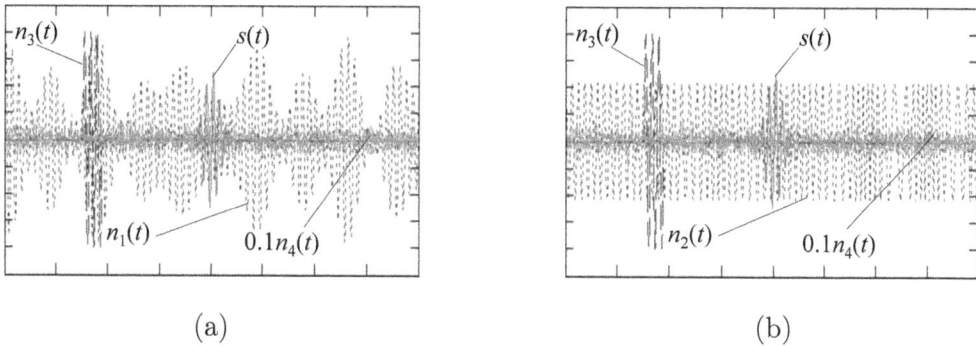

FIGURE 6.1.2   (a) Useful signal $s(t)$ and interference signals $n_1(t)$, $n_3(t)$, $n_4(t)$; (b) useful signal $s(t)$ and interference signals $n_2(t)$, $n_3(t)$, $n_4(t)$

Here useful signal $s(t)$ is radio frequency (RF) pulse with Gaussian (bell-shaped) envelope, and interference signals $n_1(t)$, $n_2(t)$, $n_3(t)$, $n_4(t)$ are amplitude-modulated interference, frequency-modulated interference, RF pulse with rectangular envelope, and quasi-white Gaussian noise, respectively, so that the ratio between upper bound frequency $f_{n,\max}$ of power spectral density (PSD) of Gaussian noise and carrier frequency f_0 of useful signal $s(t)$ is equal to $f_{n,\max}/f_0=16$. For convenience of visual perception, in Fig. 6.1.2a,b, the noise $n_4(t)$ is shown by scale decreasing in 100 times (in power). In auxiliary receiving channel, signal-to-k-th interference ratios are chosen to be equal to 1. Level of side lobes of antenna A in main receiving channel is within the interval $-17\ldots-22$ dB.

Spectra of realizations $S_s(f)$, $S_1(f)$, $S_2(f)$, $S_3(f)$, $S_4(f)$ of useful signal $s(t)$ and interference signals $n_1(t)$, $n_2(t)$, $n_3(t)$, $n_4(t)$, obtained in the result of their discrete cosine transform, are shown in Fig. 6.1.3a,b,c,d, respectively.

Fig. 6.1.4a illustrates realizations of the signals $a(t)$, $b(t)$ in antenna A of main channel and omnidirectional antenna B of auxiliary channel, respectively (see Fig. 6.1.1a), taking into account simultaneous impact of four interference signals $n_1(t)$, $n_2(t)$, $n_3(t)$, and $n_4(t)$ shown in Fig. 6.1.2a,b.

Fig. 6.1.4b depicts realization of stochastic process $y(t)$ (solid line) determined by the relationship (6.1.4) and useful signal $s(t)$ (dotted line). In the result of superposition of interference signals, the condition (b) of equation system (6.1.3) does not hold, that induces short interference overshoots. Fig. 6.1.4c,d illustrates realization of the signal $v(t)$ (solid line) determined by the relationship (6.1.2) and useful signal $s(t)$ (dotted line). In Fig. 6.1.4d, both signals are shown by scale changing along time axis with respect to Fig. 6.1.4c. Interference overshoots, that are present in stochastic process $y(t)$ (Fig. 6.1.4b) determined by the relationship (6.1.4), are removed by median filter performing operation of smoothing (6.1.2) of the process $y(t)$. During the experiment, correlation coefficient between useful signal $s(t)$ and its estimator $v(t) = G \cdot \hat{s}(t)$ is located within the interval $0.95\ldots0.96$.

Fig. 6.1.5 illustrates correlation coefficient $r_{sv}(q^2)$ between useful signal $s(t)$ and its estimator $v(t) = G \cdot \hat{s}(t)$ (6.1.2) as a dependence on signal-to-noise ratio (SNR) $q^2 = E/N_0$ (E is a signal energy, N_0 is noise PSD) in auxiliary channel for processing

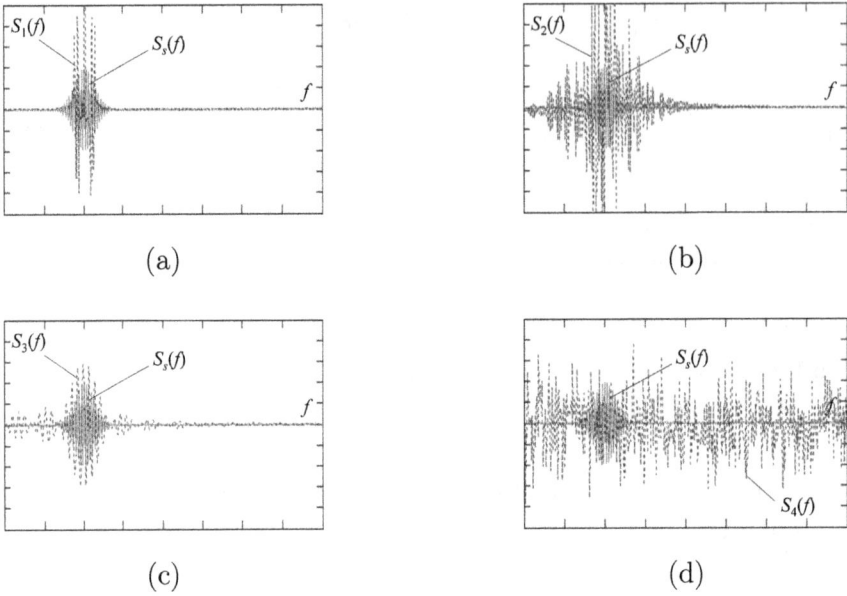

FIGURE 6.1.3 Spectra of realizations $S(f)$ of useful signal $s(t)$ and (a) amplitude-modulated interference signal $n_1(t)$; (b) frequency-modulated interference signal $n_2(t)$; (c) RF pulse interference signal $n_3(t)$; (d) noise interference signal $n_4(t)$

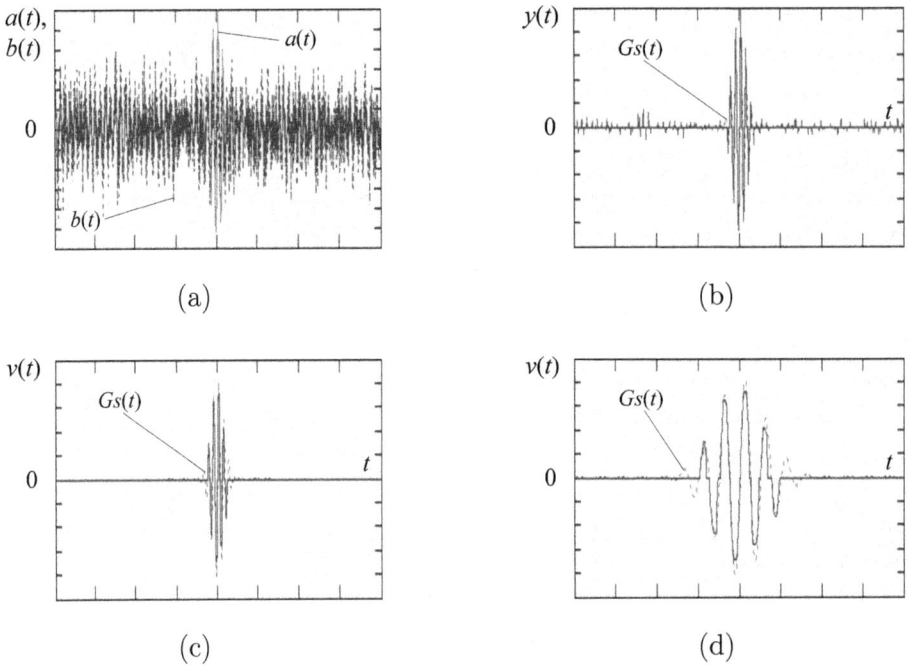

FIGURE 6.1.4 Realizations of the signals: (a) $a(t)$, $b(t)$ in antennas A and B; (b) $y(t)$ (6.1.4); (c), (d) $v(t)$ in the output of median filter (6.1.2)

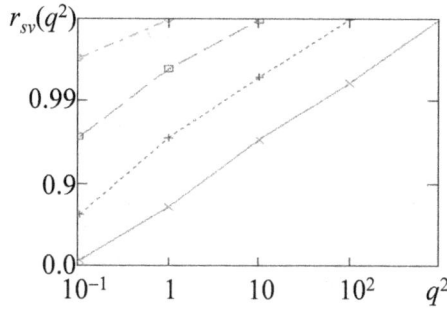

FIGURE 6.1.5   Dependence of correlation coefficient $r_{sv}(q^2)$ on SNR $q^2 = E/N_0$

scheme shown in the Fig. 6.1.1a,b. Scheme realizes signal processing algorithm that is determined by the relationships (6.1.2), (6.1.4). The sole interference $n(t)$ in the form of quasi-white Gaussian noise is present in antenna A from a direction of side lobes (see Fig. 6.1.1b). The dependences $r_{sv}(q^2)$ are represented for the gains G of antenna A that are equal to: $G=10^{0.5}$, 10, $10^{1.5}$, 10^2.

6.1.2   Two-channel Antenna System with Logical Incoherent Signal Processing Based on L–group Operations

Consider the same two-channel antenna system, shown in Fig. 6.1.1a, but with incoherent signal processing, when in the output of processing unit T, the estimator $v(t) = G \cdot \hat{E}_s(t)$ of envelope $E_s(t)$ of the signal $s(t)$ is formed according to a general processing algorithm that is slightly changed as compared with the relationships (6.1.2)...(6.1.4):

$$v(t) = \operatorname*{med}_{t \in \tilde{T}}[y(t)]; \tag{6.1.6}$$

$$y(t) = E_a(t) \cdot 1[E_a(t) - E_b(t)] = \begin{cases} E_a(t), & E_a(t) > E_b(t); \quad (a) \\ 0, & E_a(t) \leqslant E_b(t), \quad (b) \end{cases} \tag{6.1.7}$$

where med[∗] is an operation of calculating a median; $\tilde{T}$ is an interval in which a smoothing of the signal $y(t)$ is realized; $E_a(t)$, $E_b(t)$ are the envelopes of the signals $a(t)$ and $b(t)$ in antennas A and B of main and auxiliary channels, respectively (see Fig. 6.1.1a), $E_a(t) = \sqrt{a^2(t) + \tilde{a}^2(t)}$, $\tilde{a}(t) = -\frac{1}{\pi} \int\limits_{-\infty}^{\infty} \frac{a(\tau)}{\tau - t} \, d\tau$, $a(t) = \frac{1}{\pi} \int\limits_{-\infty}^{\infty} \frac{\tilde{a}(\tau)}{\tau - t} \, d\tau$; $1(t)$ is Heaviside unit step function.

In terms of L-group operations, processing algorithm (6.1.7) is determined by the following relationship:

$$y(t) = E_a(t) \cdot [0 \vee (1 \wedge k \cdot [E_a(t) - E_b(t)])], \tag{6.1.8}$$

where k=const, $k \gg 1$.

Fig. 6.1.6a,b,c,d illustrate the results of simulating signal processing according to the relationships (6.1.6), (6.1.8) for the case when in the inputs of main and auxiliary processing channels there are the useful signal $s(t)$, arriving from the

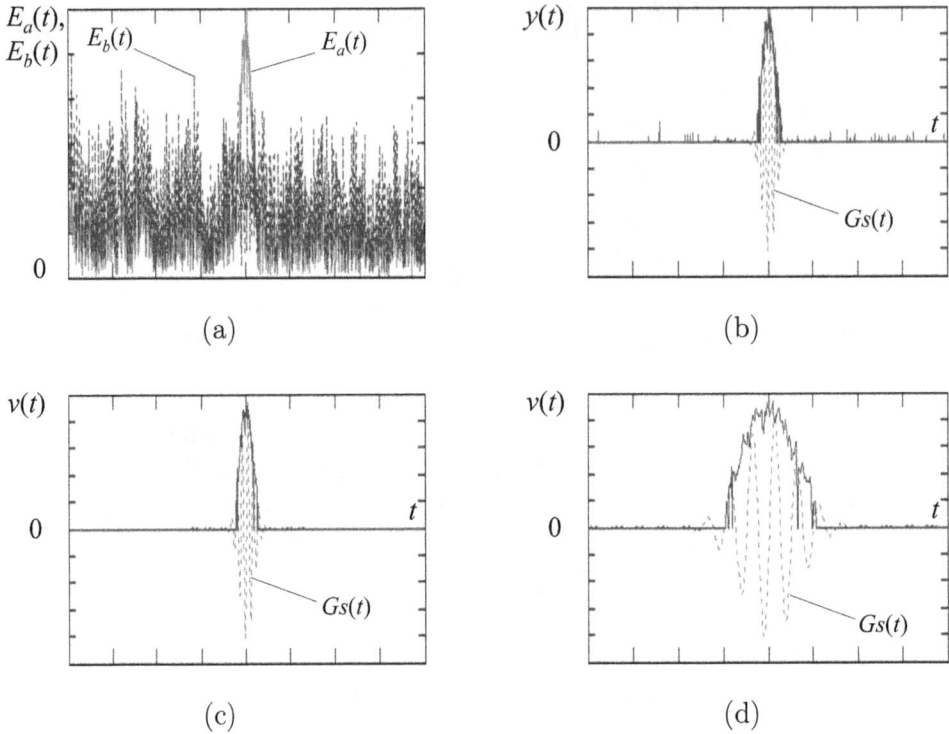

FIGURE 6.1.6   Realizations of (a) envelopes $E_a(t)$, $E_b(t)$ of the signals $a(t)$, $b(t)$; (b) stochastic process $y(t)$ (6.1.8); (c), (d) the signal $v(t)$ in the output of median filter (6.1.6)

direction of the main lobe, and also interference signals $n_1(t)$, $n_2(t)$, $n_3(t)$, $n_4(t)$ received by side lobes of main channel antenna and interacting with each other according to the relationships of the equation system (6.1.5) (see Fig. 6.1.2a,b; Fig. 6.1.3a,b,c,d). The conditions of simulation with respect to the received useful and interference signals are equivalent to those described in Subsection 6.1.1.

Fig. 6.1.6a shows realizations of envelopes $E_a(t)$, $E_b(t)$ of the signals $a(t)$, $b(t)$ in antenna A of main channel and omnidirectional antenna B of auxiliary channel, respectively (see Fig. 6.1.1a), taking into account simultaneous interaction of interference signals $n_1(t)$, $n_2(t)$, $n_3(t)$, $n_4(t)$ shown in Fig. 6.1.2a,b.

Fig. 6.1.6b illustrates realization of stochastic process $y(t)$ (solid line) determined by the relationship (6.1.8) and useful signal $s(t)$ (dotted line) received in main channel. In the result of superposition of interference signals, the condition (b) of the equation system (6.1.7) does not hold, that induces short overshoots.

Fig. 6.1.6c,d depict realizations of the signals $v(t)$ (solid line) determined by the relationship (6.1.6) and useful signal $s(t)$ (dotted line) received in main channel. In Fig. 6.1.6d, both signals are shown by scale changing along time axis with respect to Fig. 6.1.6c. Interference overshoots, presenting in stochastic process $y(t)$ (Fig. 6.1.6b) determined by the relationship (6.1.8), are removed by median filter performing operation of smoothing (6.1.6) of the process $y(t)$. During the

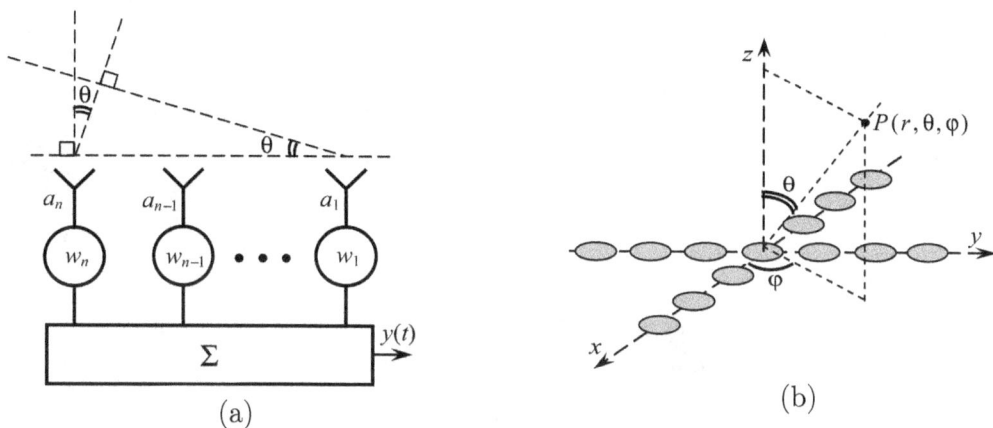

FIGURE 6.1.7   (a) Linear antenna array; (b) planar antenna array in the form of Mills cross

experiment, correlation coefficient between envelope $E_s(t)$ of useful signal $s(t)$ and estimator of signal envelope $v(t) = G \cdot \hat{E}_s(t)$ is located within the interval $0.97 \ldots 0.98$, that is quite satisfactory for most of practical applications.

6.1.3   Logical Signal Processing in Antenna Arrays Based on L–group Operations

In information transmitting systems, and also in electronic systems of information extracting, the problems of spatial-temporal signal processing are mainly solved by antenna arrays. Antenna array is a multichannel system consisting of a set of receiving elements and related to them signal processing unit that can change its own parameters according to information contained in receiving useful and interference signals.

The problem of optimal spatial-temporal signal processing is known for a long time and is well studied [102, 132, 156–159]. Optimal signal processing in antenna array is intended to improve quality of useful signal receiving in the presence of interference signals.

In this subsection, we discuss the results of comparative analysis of efficiency of spacial-temporal signal processing in linear equidistant antenna array (see Fig. 6.1.7a), and also in planar antenna array in the form of Mills cross (see Fig. 6.1.7b) for two variants of signal processing based on: (1) minimum noise variance criterion; (2) logical signal processing algorithm (6.1.2)...(6.1.4). Such an analysis is performed within the following constraints: useful signal is a narrow-band RF signal with Gaussian (bell-shaped) envelope that is received by main lobe of antenna pattern simultaneously with interference signals, so that antenna array is considered to be signal aligned; reception of useful an interference signals is realized in ideal conditions, i.e. signal propagation is realized in homogeneous, non-dispersive medium with nonrandom parameters, when signal wave front is plane, and the elements of antenna array do not induce any distortion.

6.1.3.1 Logical Signal Processing in Linear Antenna Array Based on L–group Operations

The resulting process $y(t)$ in the output of linear antenna array (see Fig. 6.1.7a) is determined by the expression:

$$y(t) = \sum_{i=1}^{n} \bar{w}_i a_i(t), \tag{6.1.9}$$

where $a_i(t)$ is the received signal in the input of i-th receiving element of antenna array; w_i is weight coefficient in i-th processing channel; $\bar{w}_i = \text{Re}(w_i) - j\,\text{Im}(w_i)$.

Antenna array operating on the basis of minimum noise variance criterion forms optimal solution for a vector of weight coefficients $\mathbf{w}$ in the following form [158, (3.132)], [102, (61.18)]:

$$\mathbf{w} = [w_1, \ldots, w_n]^T = \frac{\hat{\mathbf{R}}_n^{-1}\mathbf{1}}{\mathbf{1}^T \hat{\mathbf{R}}_n^{-1} \mathbf{1}}, \tag{6.1.10}$$

where $\hat{\mathbf{R}}_n$ is an estimator of correlation matrix of resulting interference, which, in the case of sufficiently small signal-to-interference ratio in antenna array receiving channels, is determined by correlation matrix $\mathbf{R}_a$ of the observed processes $\hat{\mathbf{R}}_n = \mathbf{R}_a$; $\mathbf{R}_a = \left\| r_{ij}\sqrt{D_i D_j} \right\|$, r_{ij} is correlation coefficient between the received signals $a_i(t)$, $a_j(t)$ in the inputs of i-th and j-th receiving elements of antenna array, respectively; D_i is a variance of the signal $a_i(t)$ in the input of i-th receiving element of antenna array; $\mathbf{1} = [1, 1, \ldots, 1]^T$; $\mathbf{w}^T \mathbf{1} = 1$.

Antenna array providing optimal solution (6.1.10) is considered to be signal aligned. Logical signal processing method determined by the algorithm (6.1.2), (6.1.4) can be also realized in antenna arrays.

Thus, for instance, in the output of processing unit T (see Fig. 6.1.1a), linear antenna array based on signal processing algorithm (6.1.2), (6.1.4) forms the signal estimator $v(t) = \hat{s}(t)$ according to general signal processing algorithm:

$$v(t) = \underset{t \in \tilde{T}}{\text{med}}[z(t)]; \tag{6.1.11}$$

$$z(t) = a(t) \cdot 1[|a(t)| - |b(t)|]$$
$$= a(t) \cdot [0 \vee (1 \wedge k \cdot [(a(t) \vee (-a(t))) - (b(t) \vee (-b(t)))])], \tag{6.1.12}$$

where $\text{med}[*]$ is operation of calculating a median; $\tilde{T}$ is an interval, in which a smoothing of a signal $y(t)$ is performed; $k=\text{const}$, $k \gg 1$; $1(t)$ is Heaviside unit step function.

Here we assume that the signals of main and auxiliary receiving channels $a(t)$ and $b(t)$ figuring in (6.1.12) are determined by the following relationships:

$$a(t) = \sum_{i=1}^{n} w_i a_i(t); \tag{6.1.13a}$$

$$b(t) = \bigwedge_{i=1}^{n} (w_i a_i(t)),$$ (6.1.13b)

$a_i(t)$ is a received signal in the input of i-th receiving element of antenna array; w_i is weight coefficient in i-th processing channel, $\mathbf{w}^T \mathbf{1} = 1$; $\bigwedge_{i=1}^{n} z_i(t) = z_1(t) \wedge \ldots \wedge z_n(t)$ is operation of meet between the functions $\{z_i(t)\}$; n is a number of receiving channels in linear antenna array.

Signal processing L-group algorithm (6.1.11)...(6.1.13) can be realized by data independent antenna arrays [102]. A simple case of this algorithm suppose exploiting a uniform array with equal weight coefficients $w_i = 1/n$. However, for some practically important applications, a level of side lobes provided by uniform arrays can be unsatisfactory. In this case, alternative variants of constructing data independent antenna arrays assume exploiting binomial arrays with weight coefficients w_i determined by the following relationship:

$$w_i = C_n^i / 2^n = \frac{n!}{2^n i!(n-i)!}; \quad \sum_{i=1}^{n} w_i = 1,$$

and also arrays with weight coefficients determined by Chebyshev polynomials (Dolph-Chebyshev arrays) [160–162]. However, practical realization of these methods is accompanied by known difficulties.

One can easily provide a given level of side lobes in data independent antenna arrays, choosing weight coefficients w_i based on some discrete distribution. Here, as an example, we consider weight coefficients w_i with Gaussian $p_G(i)$, Cauchy $p_C(i)$, and logistic $p_L(i)$ distributions, respectively:

$$w_{G,i} = p_G(i) = \frac{1}{\sqrt{2\pi \beta_G}} \exp\left[-\frac{(i-c)^2}{2\beta_G}\right];$$ (6.1.14a)

$$w_{C,i} = p_C(i) = \frac{1}{\pi} \frac{\beta_C}{\beta_C^2 + (i-c)^2};$$ (6.1.14b)

$$w_{L,i} = p_L(i) = \frac{\beta_L \exp[-\beta_L \cdot (i-c)]}{[1 + \exp(-\beta_L \cdot (i-c))]^2},$$ (6.1.14c)

where i is a number of array element, $i = 1, \ldots, n$; $c = (n+1)/2$ is a number of central element of antenna array with odd number of elements n; β_G, β_C, β_L are scale parameters of aforementioned distributions, respectively.

Fig. 6.1.8a illustrates weight coefficients w_i that are determined by these distributions. Fig. 6.1.8b depicts dependences $A(\theta)$ of normalized power patterns (in dB) of 13-element linear antenna arrays with: 1—uniform distribution $w_i = 1/n$ (dotted line); 2—Gaussian distribution (dashed line); 3—Cauchy distribution (dashed line); 4—logistic distribution (dash-dotted line) of weight coefficients. Ratio between interelement distance of antenna array d and wave length λ is equal to $d/\lambda = 0.5$.

As follows from Fig. 6.1.8b, weight coefficients with Gaussian (2) and logistic (4) distributions allow obtaining side lobes of normalized power pattern with a level that does not exceed a given one (in this case, it does not exceed –30 dB).

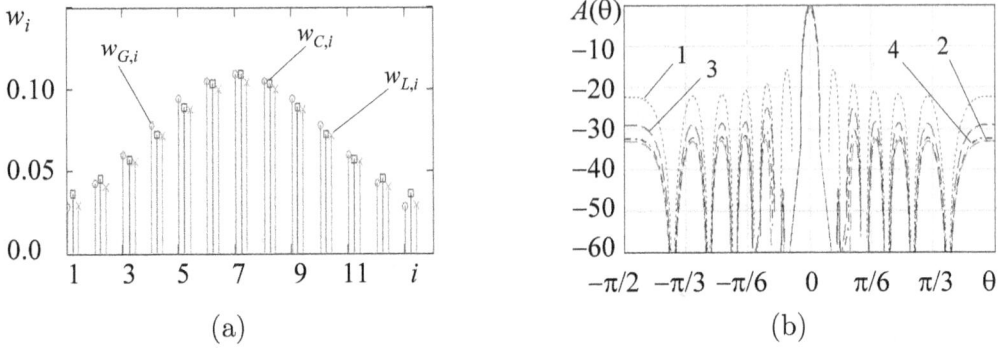

FIGURE 6.1.8   (a) Weight coefficients w_i with Gaussian, Cauchy, and logistic distributions, respectively; (b) dependences $A(\theta)$ of normalized power pattern of 13-element linear antenna array

Exploiting weight coefficients with Cauchy distribution provides a little bit worse level of side lobes.

Fig. 6.1.9a illustrates stochastic process $y(t)$ (solid line) in the output of 13-element linear equidistant antenna array ($n{=}13$) that realizes optimal signal processing according to the relationships (6.1.9), (6.1.10). Fig. 6.1.9b,c,d depicts stochastic processes $v(t)$ (solid line) in the outputs of 13-element linear equidistant antenna arrays ($n{=}13$) that provide signal processing based on L-group algorithm (6.1.11)...(6.1.13), so that weight coefficients are determined by Gaussian $p_G(i)$ (6.1.14a), Cauchy $p_C(i)$ (6.1.14b), and logistic $p_L(i)$ (6.1.14c) distributions, respectively.

Simulation of signal processing is provided in the presence of useful signal $s(t)$, arriving from main lobe direction ($\theta_s = 0$), and interference signals $n_1(t)$, $n_2(t)$, $n_3(t)$, $n_4(t)$, $n_5(t)$ arriving from the directions $\theta_1 = \pi/6$; $\theta_2 = -2\pi/9$; $\theta_3 = -5\pi/13$; $\theta_4 = 3\pi/7$; $\theta_5 = -2\pi/5$, respectively (see Fig. 6.1.2a,b; Fig. 6.1.3a,b,c,d). As before, interference signals $n_1(t)$, $n_2(t)$, $n_3(t)$ are amplitude-modulated oscillation, frequency-modulated oscillation, and also RF pulse with rectangular envelope, respectively; statistically independent interference signals $n_4(t)$, $n_5(t)$ are quasi-white Gaussian noise. Ratio between interelement distance of antenna array d and wave length λ is equal to $d/\lambda = 0.5$. Signal-to-k-th interference ratios in each separate i-th receiving channel of antenna array are closely equal to 0.25; 0.25; 1; 0.05; 0.05, respectively.

All the figures contain pulse interference $n_3(t)$ shown two times less than the original and useful signal $s(t)$ in order to mark their position along time axis. Correlation coefficients between useful signal $s(t)$ and output signals $y(t)$ and $v(t)$ take their values in the intervals: $r[y(t), s(t)]{=}0.96\ldots0.97$; for Gaussian and logistic distributions of weight coefficients $r[v(t), s(t)]{=}0.98\ldots0.985$, and for Cauchy distribution $r[v(t), s(t)]{=}0.97\ldots0.98$, respectively. Coefficients of interference suppression take the values: $K_y{=}45\ldots46$ dB; for Gaussian and logistic distributions of weight coefficients $K_v{=}47\ldots48$ dB, and for Cauchy distribution $K_v{=}46\ldots47$ dB, respectively. Here coefficients of interference suppression are defined over the

(a)

(b)

(c)

(d)

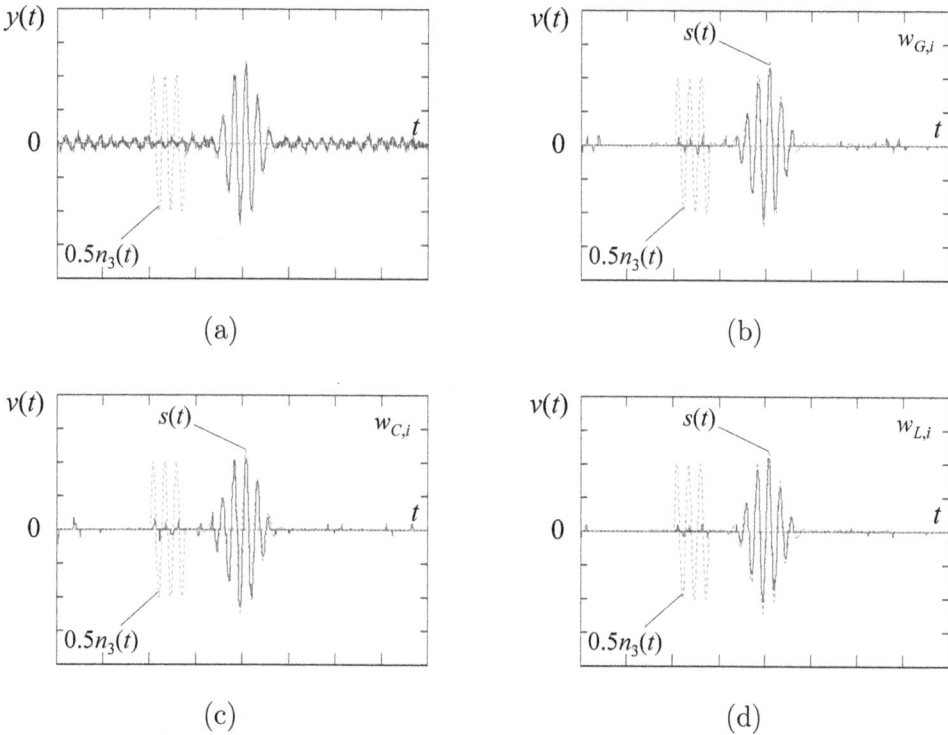

FIGURE 6.1.9   Stochastic processes in the output of antenna array: (a) $y(t)$ (6.1.9), (6.1.10); (b), (c), (d) $v(t)$ (6.1.11)...(6.1.13)

ratios: $K_y = 10 \lg(D/D[y(t) - s(t)])$, $K_v = 10 \lg(D/D[v(t) - s(t)])$, where D is a total variance of resulting interference, $D = \sum_{k=1}^{K} D_k$, D_k is a variance of k-th interference; $D[y(t) - s(t)]$, $D[v(t) - s(t)]$ are variances of noise components of the processes $y(t)$, $v(t)$ in the outputs of signal processing units.

To obtain more detailed envision on signal processing L-group algorithm (6.1.11)...(6.1.13), Fig. 6.1.10a, 6.1.11a, 6.1.12a illustrate the signal $a(t)$ of main receiving channel that is formed according to the formula (6.1.13a), so that weight coefficients are determined by Gaussian $p_G(i)$ (6.1.14a), Cauchy $p_C(i)$ (6.1.14b), and logistic $p_L(i)$ (6.1.14c) distributions, respectively.

Fig. 6.1.10b, 6.1.11b, 6.1.12b depict moduli of the signals of main $a(t)$ and auxiliary $b(t)$ receiving channels when forming weight coefficients with Gaussian $p_G(i)$ (6.1.14a), Cauchy $p_C(i)$ (6.1.14b), and logistic $p_L(i)$ (6.1.14c) distributions, respectively. As follows from the relationship (6.1.12), signal processing unit forms an output signal that is equal to the signal of main receiving channel $a(t)$ if and only if modulus of the signal $a(t)$ exceeds a value of modulus of the signal $b(t)$ in auxiliary receiving channel: $|a(t)| > |b(t)|$.

Correlation coefficients between useful signal $s(t)$ and output signals $a(t)$ take the values in the intervals: for Gaussian and logistic distributions

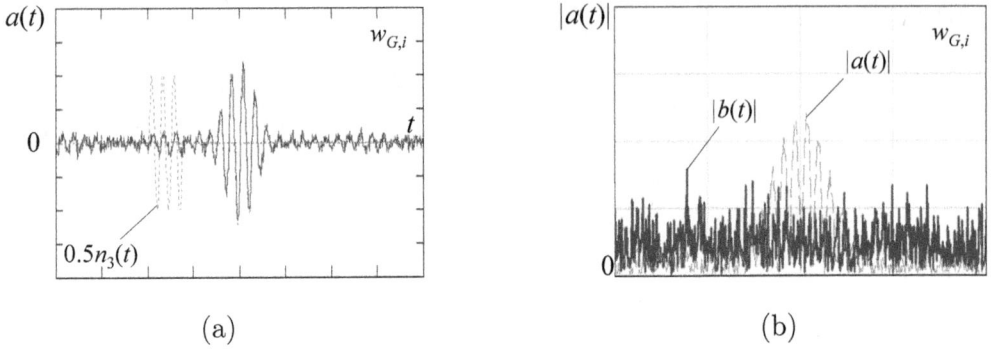

(a)   (b)

FIGURE 6.1.10   (a) The signal $a(t)$ in main receiving channel when forming weight coefficients according to Gaussian distribution; (b) moduli of the signals in main $a(t)$ and auxiliary $b(t)$ receiving channels

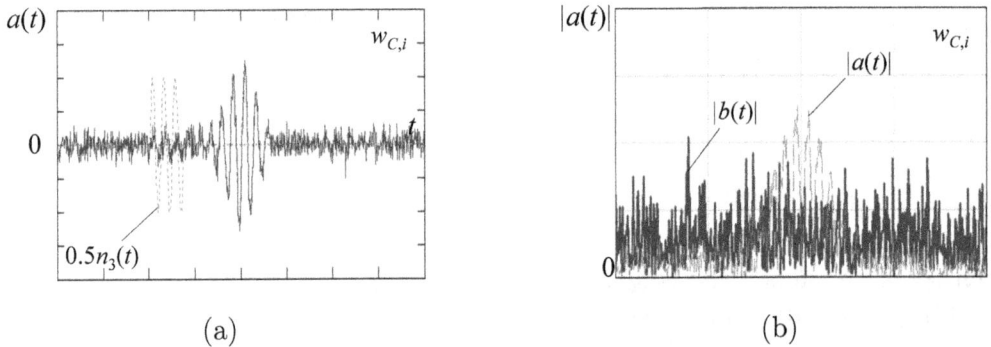

(a)   (b)

FIGURE 6.1.11   (a) The signal $a(t)$ in main receiving channel when forming weight coefficients according to Cauchy distribution; (b) moduli of the signals in main $a(t)$ and auxiliary $b(t)$ receiving channels

of weight coefficients $r[a(t), s(t)]$=0.94...0.97, and for Cauchy distribution $r[a(t), s(t)]$=0.87...0.88, respectively. Coefficients of interference suppression take the values: for Gaussian and logistic distributions of weight coefficients K_a=43...45 dB, and for Cauchy distribution K_a=39...40 dB, respectively. Comparing coefficients of interference suppression K_a and K_v, that characterize efficiency of signal processing before and after using the method of logical signal processing, one can conclude that Cauchy distributed weight coefficients of antenna array provide a gain ΔK in interference suppression coefficient $\Delta K = K_v - K_a \approx 6...7$ dB, whereas Gaussian and logistic distributed weight coefficients provide a smaller gain $\Delta K = K_v - K_a \approx 3...5$ dB.

Comparing Fig. 6.1.9b,c,d and Fig. 6.1.10a, 6.1.11a, 6.1.12a, respectively, one can see that meet (6.1.13b) of weighted received signals $a_i(t)$ in the input of i-th receiving elements of antenna array is quite satisfactory estimator of upper bound of field pattern side lobes of antenna array. Here we notice that other functions can also be used as such an estimator, for instance, mean function

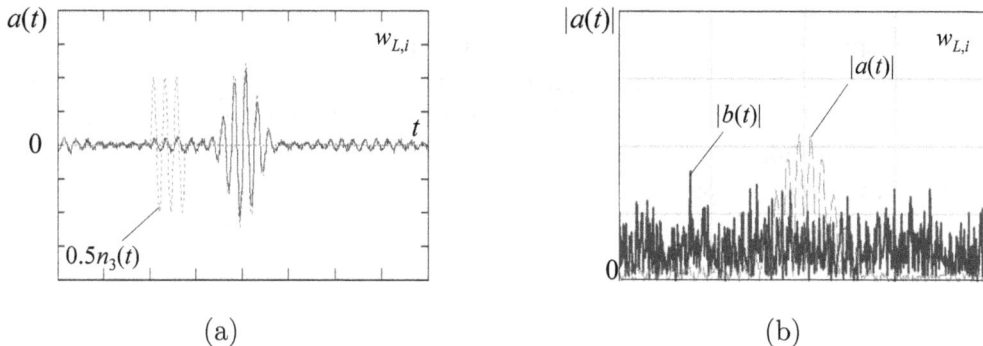

(a)   (b)

FIGURE 6.1.12   (a) The signal $a(t)$ in main receiving channel when forming weight coefficients according to logistic distribution; (b) moduli of the signals in main $a(t)$ and auxiliary $b(t)$ receiving channels

$$b'(t) = \left[\bigvee_{i=1}^{n} (w_i a_i(t)) + \bigwedge_{i=1}^{n} (w_i a_i(t)) \right] /2, \text{ or median } b''(t) = \underset{i=1,\ldots,n}{\text{med}} (w_i a_i(t)), \text{ that}$$

can however cause some worsening of signal processing quality as compared with the function (6.1.13b).

6.1.3.2   Logical Signal Processing in Mills Cross Antenna Array Based on L–group Operations

Planar antenna array in the form of Mills cross is a union of elements of two linear antenna arrays situated on two orthogonal axes Ox, Oy. The resulting process $y(t)$ in the output of antenna array in the form of Mills cross (see Fig. 6.1.7b) is determined by the expression:

$$y(t) = \sum_{i=1}^{n} \bar{w}_i^x a_i^x(t) + \sum_{k=1}^{n} \bar{w}_k^y a_k^y(t), \qquad (6.1.15)$$

where $a_i^x(t)$, $a_k^y(t)$ are received signals in the input i-th/k-th receiving elements of linear antenna arrays whose elements are situated along the axes Ox, Oy, respectively; w_i^x, w_k^y are weight coefficients in i-th/k-th processing channel of antenna array; $\bar{w}_i = \text{Re}(w_i) - j\,\text{Im}(w_i)$.

Planar antenna array in the form of Mills cross operating on the basis of minimum noise variance criterion provides optimal solution for vector of weight coefficients in the form [158, (3.132)], [102, (61.18)] that is similar to (6.1.10):

$$\mathbf{w} = [w_1^x, \ldots, w_n^x, w_1^y, \ldots, w_n^y]^T = \frac{\hat{\mathbf{R}}_n^{-1} \mathbf{1}}{\mathbf{1}^T \hat{\mathbf{R}}_n^{-1} \mathbf{1}}, \qquad (6.1.16)$$

where $\hat{\mathbf{R}}_n$ is an estimator of correlation matrix of resulting interference, which, on sufficiently small signal-to-interference ratio in antenna array receiving channels, is determined by correlation matrix $\mathbf{R}_a$ of the observed processes $\hat{\mathbf{R}}_n = \mathbf{R}_a$; $\mathbf{R}_a = \|r_{ij}\|$, r_{ij} is correlation coefficient between the received signals $a_i^{x,y}(t)$, $a_j^{x,y}(t)$ in the inputs of i-th and j-th receiving elements of antenna array that are situated along the axes Ox, Oy, respectively; $\mathbf{1} = [1,1,\ldots,1]^T$; $\mathbf{w}^T\mathbf{1} = 1$.

As before, antenna array providing optimal solution (6.1.16) is considered to be signal aligned. Planar antenna array in the form of Mills cross that performs initial signal processing algorithm (6.1.2)...(6.1.4) forms a signal estimator $v(t) = \hat{s}(t)$ according to general algorithm of signal processing:

$$v(t) = \operatorname*{med}_{t \in \tilde{T}}[z(t)]; \tag{6.1.17}$$

$$
\begin{aligned}
z(t) &= a(t) \cdot \mathbf{1}[|a(t)| - |b(t)|] \\
&= a(t) \cdot [0 \vee (1 \wedge k \cdot [(a(t) \vee (-a(t))) - (b(t) \vee (-b(t)))])], \tag{6.1.18}
\end{aligned}
$$

where med[$*$] is operation of calculating median; $\tilde{T}$ is an interval in which a smoothing of the signal $y(t)$ is realized; k=const, $k \gg 1$; $1(t)$ is Heaviside unit step function.

Here we assume that the signals of main $a(t)$ and auxiliary $b(t)$ receiving channels figuring in (6.1.18) are determined by the following relationships:

$$a(t) = \frac{1}{2}\left(\sum_{i=1}^{n} w_i^x a_i^x(t) + \sum_{k=1}^{n} w_k^y a_k^y(t)\right); \tag{6.1.19a}$$

$$b(t) = \left(\bigwedge_{i=1}^{(n-1)/2} w_i^x a_i^x(t)\right) \wedge \left(\bigwedge_{i=1}^{(n-1)/2} w_i^y a_i^y(t)\right)$$

$$\wedge\, 0.5\left(w_{\frac{(n+1)}{2}}^x a_{\frac{(n+1)}{2}}^x(t) + w_{\frac{(n+1)}{2}}^y a_{\frac{(n+1)}{2}}^y(t)\right), \tag{6.1.19b}$$

where $a_i^x(t)$, $a_k^y(t)$ are received signals in the inputs of i-th/k-th receiving elements of linear antenna arrays whose elements are situated along the axes Ox, Oy, respectively; w_i^x, w_k^y are weight coefficients in i-th/k-th processing channels of antenna array that are chosen on the basis of distributions (6.1.14a,b,c); $\bigwedge_{i=1}^{n} z_i(t) = z_1(t) \wedge \ldots \wedge z_n(t)$ is operation of meet between the functions $\{z_i(t)\}$; n is a number of receiving channels of a single linear antenna array.

Fig. 6.1.13a illustrates stochastic process $y(t)$ (solid line) in the output of Mills cross antenna array that is formed by two 13-element linear equidistant antenna arrays (n=13) and performs optimal signal processing according to the relationships (6.1.15), (6.1.16). Fig. 6.1.13b,c,d depicts stochastic processes $v(t)$ (solid line) in the output of Mills cross antenna array that is formed by two 13-element linear equidistant antenna arrays (n=13) and performs signal processing according to the relationships (6.1.17)...(6.1.19), so that weight coefficients are determined by Gaussian $p_G(i)$ (6.1.14a), Cauchy $p_C(i)$ (6.1.14b), and logistic $p_L(i)$ (6.1.14c) distributions, respectively.

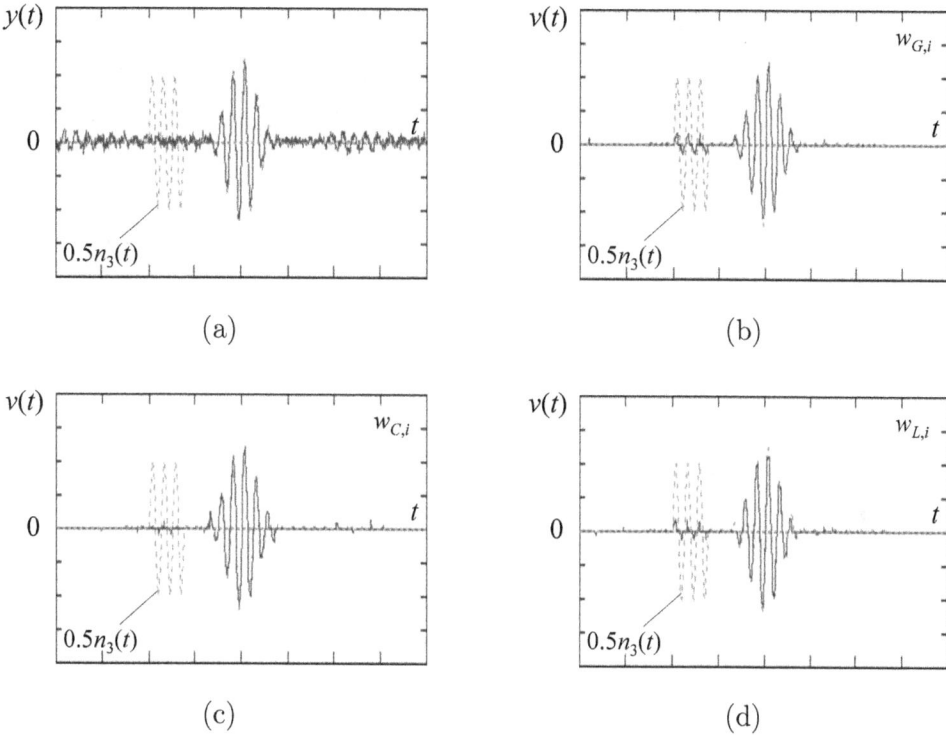

FIGURE 6.1.13   Stochastic processes in the output of Mills cross antenna array: (a) $y(t)$ (6.1.15), (6.1.16); (b), (c), (d) $v(t)$ (6.1.17)...(6.1.19) with weight coefficients chosen according to (6.1.14)

Simulating array signal processing is realized in the presence of useful signal $s(t)$ arriving from main lobe direction $(\theta_s, \varphi_s) = (0,0)$, and also interference signals $n_1(t)$, $n_2(t)$, $n_3(t)$, $n_4(t)$, $n_5(t)$ arriving from the directions $(\theta_1, \varphi_1) = (2\pi/9, \pi/6)$; $(\theta_2, \varphi_2) = (4\pi/13, 3\pi/5)$; $(\theta_3, \varphi_3) = (\pi/3, -4\pi/7)$; $(\theta_4, \varphi_4) = (3\pi/7, -2\pi/9)$; $(\theta_5, \varphi_5) = (2\pi/5, 2\pi/7)$, respectively (see Fig. 6.1.2a,b; Fig. 6.1.3a,b,c,d). Remind, that interference signals $n_1(t)$, $n_2(t)$, $n_3(t)$ are amplitude-modulated RF oscillation, frequency-modulated RF oscillation, and also RF pulse with rectangular envelope, respectively; statistically independent interference signals $n_4(t)$, $n_5(t)$ are quasi-white Gaussian noise. Ratio between interelement distance of antenna array d and wave length λ is equal to $d/\lambda = 0.5$. Signal-to-k-th interference ratios in each separate i-th receiving channel of antenna array are closely equal to 0.25; 0.25; 1; 0.05; 0.05, respectively.

Fig. 6.1.13a illustrates the signal $y(t)$ obtained according to optimal algorithm (6.1.15), (6.1.16), and Fig. 6.1.13b,c,d depicts the signals $v(t)$ obtained according to L-group algorithm (6.1.17)...(6.1.19). All the figures contain pulse interference $n_3(t)$ shown two times less than the original and useful signal $s(t)$ in order to mark their position along time axis. Correlation coefficients between useful signal $s(t)$ and the output signals $y(t)$ and $v(t)$ take their values in the intervals: $r[y(t), s(t)]=0.96...0.97$; $r[v(t), s(t)]=0.98...0.99$ (for Gaussian, Cauchy,

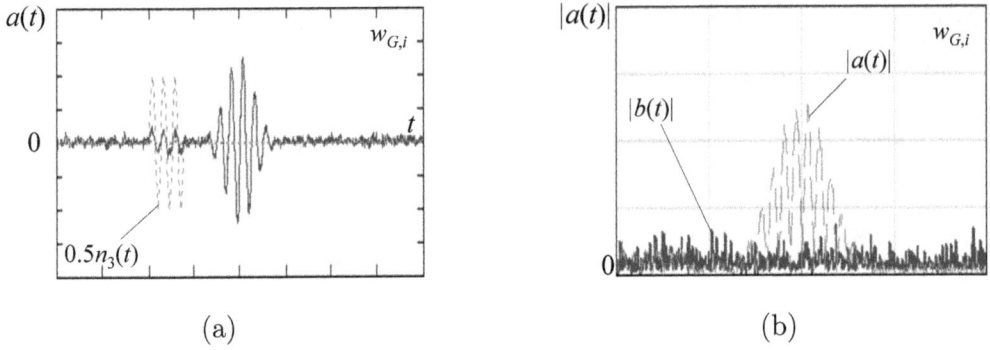

(a)   (b)

FIGURE 6.1.14   (a) The signal $a(t)$ in main receiving channel when forming weight co-efficients according to Gaussian distribution; (b) moduli of the signals in main $a(t)$ and auxiliary $b(t)$ receiving channels

and logistic distributions of weight coefficients), respectively. Coefficients of inter-ference suppression in the outputs of signal processing units take the values in the intervals: K_y=45...46 dB; and for Gaussian, Cauchy, and logistic distribu-tions of weight coefficients K_v=49.5...50.5 dB, respectively. Here coefficients of interference suppression are defined by the ratios: $K_y = 10\lg(D/D[y(t) - s(t)])$, $K_v = 10\lg(D/D[v(t) - s(t)])$, where D is total variance of resulting interference, $D = \sum_{k=1}^{K} D_k$, D_k is a variance of k-th interference; $D[y(t) - s(t)]$, $D[v(t) - s(t)]$ are variances of noise components of the output processes $y(t)$, $v(t)$, respectively.

To obtain an envision on signal processing L-group algorithm (6.1.17)...(6.1.19), Fig. 6.1.14a, 6.1.15a, 6.1.16a illustrate the signals $a(t)$ of main receiving channel that are formed according to the formula (6.1.19a), so that weight coefficients are determined by Gaussian $p_G(i)$ (6.1.14a), Cauchy $p_C(i)$ (6.1.14b), and logistic $p_L(i)$ (6.1.14c) distributions, respectively.

Fig. 6.1.14b, 6.1.15b, 6.1.16b illustrate moduli of signals of main $a(t)$ and aux-iliary $b(t)$ receiving channels when forming weight coefficients with Gaussian $p_G(i)$ (6.1.14a), Cauchy $p_C(i)$ (6.1.14b), and logistic $p_L(i)$ (6.1.14c) distributions, respec-tively. As follows from the relationship (6.1.18) (see also the relationship (6.1.3)), signal processing unit forms an output signal that is equal to main channel signal $a(t)$ if and only if modulus of the signal $a(t)$ exceeds modulus of the signal $b(t)$ in auxiliary receiving channel: $|a(t)| > |b(t)|$.

Correlation coefficients between useful signal $s(t)$ and output signal $a(t)$ take their values in the intervals: for Gaussian and logistic distributions of weight coefficients $r[a(t), s(t)]$=0.96...0.97, and for Cauchy distribution $r[a(t), s(t)]$=0.87...0.92, respectively. Coefficients of interference suppression take the values: for Gaussian and logistic distributions of weight coefficients K_a=45...47 dB, and for Cauchy distribution K_a=39...41 dB, respectively. In Fig. 6.1.14a, 6.1.15a, 6.1.16a, one can notice the remains of compensation of pulse interference $n_3(t)$ that is weakened during weight processing (6.1.19) ap-proximately up to 20 dB for all three types of weight coefficient distributions.

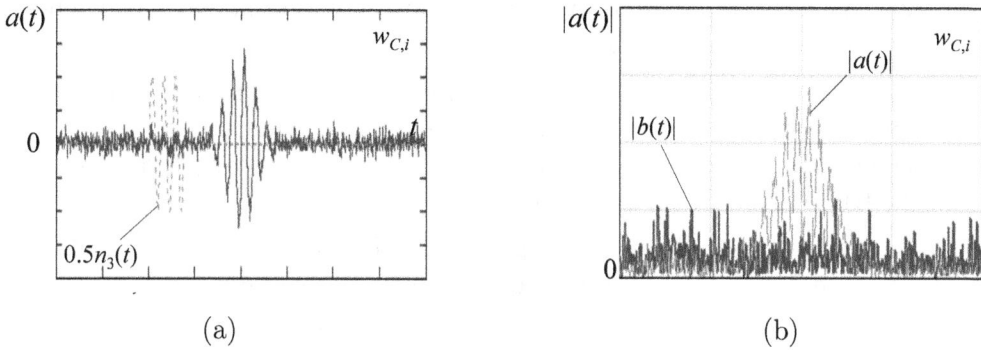

(a)   (b)

FIGURE 6.1.15   (a) The signal $a(t)$ in main receiving channel when forming weight coefficients according to Cauchy distribution; (b) moduli of the signals in main $a(t)$ and auxiliary $b(t)$ receiving channels

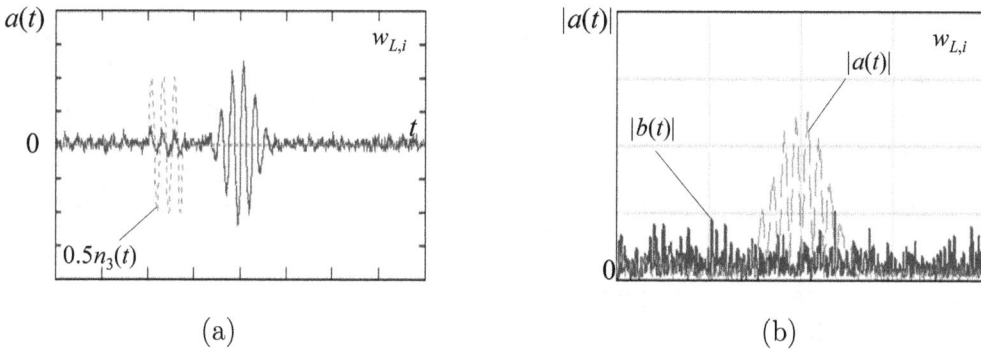

(a)   (b)

FIGURE 6.1.16   (a) The signal $a(t)$ in main receiving channel when forming weight coefficients according to logistic distribution; (b) moduli of the signals in main $a(t)$ and auxiliary $b(t)$ receiving channels

Comparing coefficients of interference suppression K_a and K_v characterizing efficiency of processing before and after using the method of logical signal processing, one can conclude that exploiting Cauchy distribution for weight coefficients of antenna array provides a gain ΔK in suppression coefficient that is equal $\Delta K = K_v - K_a \approx 10\ldots11$ dB, while exploiting Gaussian and logistic distributions provides a lesser gain $\Delta K = K_v - K_a \approx 3.5\ldots5$ dB.

Comparing Fig. 6.1.13b,c,d and Fig. 6.1.14a, 6.1.15a, 6.1.16a, one can see that meet (6.1.19b) of weighted received signals $a_i(t)$ in the inputs of i-th receiving elements of antenna array is quite satisfactory estimator of side lobe upper bound in antenna array. Here we notice that other functions can be also used as such estimator, that, however, can cause some worsening of processing quality with respect to the function (6.1.19b).

Two-channel and, in general, multichannel antenna systems with logical signal processing based on L-group operations possesses the following advantages.

1. Invariance with respect to nonparametric prior uncertainty conditions.

Efficiency of spatial signal processing does not depend on interference distribution.

2. Invariance with respect to a number of interference sources. This number can be arbitrary if phase centers of antennas in main and auxiliary receiving channels coincide. The last condition can be easily provided when constructing antenna system based on antenna array.

3. There is no need of exploiting a large number of receiving channels to eliminate ill effect of large number of interference sources.

4. There is no need of adaptation to signal receiving conditions.

5. Method of logical signal processing in antenna systems can be realized by different antenna types including antenna arrays.

6. Simplicity of realization of the method. There is no need of operation of correlation matrix inversion.

7. Efficiency of antenna array signal processing based on the method of logical L-group signal processing can be not worse, and under actual conditions it can be even better than efficiency provided by traditional methods of optimal signal processing in antenna arrays.

Main disadvantages of two-channel antenna systems with logical signal processing based on L-group operations are the following.

1. Noninvariance with respect to parametric prior uncertainty conditions (see Fig. 6.1.5).

2. Necessity of providing phase center matching in antennas of main and auxiliary receiving channels.

One can get rid of the first disadvantage by exploiting antennas with high directivity. The second disadvantage can be overcome by exploiting antenna arrays.

6.2   Spatial Filtering: Signal Space Mapping Method

Determining an estimator $\hat{s}(t)$ of a signal $s(t)$ as a functional $F_{\hat{s}}[\xi(t)]$ of an observed stochastic process $\xi(t)$ is called the problem of signal filtering (extracting) [97, 104, 106–108]:

$$\hat{s}(t) = F_{\hat{s}}[\xi(t)], \quad t \in T_s. \tag{6.2.1}$$

Along with stochastic processes, in the problems of *spatial-temporal signal processing* there exists a necessity of considering stochastic multivariate functions that are called *stochastic fields* or *multivariate signals*. The problem of spatial* filtering

---

*Hereinafter, we mean that spatial signal processing is, strictly speaking, spatial-temporal signal processing; it follows from considered below algorithms

(extracting) of signal $s(t)$ is obtaining an estimator $\hat{s}(t)$ in the form of functional F of the observed stochastic field $\xi(\mathbf{r}, t)$:

$$\hat{s}(t) = F_{\hat{s}}[\xi(\mathbf{r}, t)], \quad t \in T_s, \tag{6.2.2}$$

where $\mathbf{r}$ is m-dimensional vector with its own projections in a chosen coordinate system.

Subject of the following consideration is exploring antenna array spatial filtering algorithms based on L-group operations. In particular, we analyze efficiency of these algorithms as applied to linear equidistant antenna array (see Fig. 6.1.7a) and also to planar antenna array in the form of Mills cross (see Fig. 6.1.7b). These algorithms of signal processing are based on the method of mapping of linear signal space into signal space with lattice properties.

Remind that signal space with lattice properties $\mathcal{L}(\vee, \wedge)$ can be obtained by transforming linear space $\mathcal{LS}(+)$ in such a way that the results of interactions $x_\vee(t)$, $x_\wedge(t)$ between a signal $s(t)$ and interference $n(t)$ in signal space with lattice properties $\mathcal{L}(\vee, \wedge)$ and operations of join and meet $\vee, \wedge$ are realized according to the relationships [24, (7.7.1a,b)]:

$$x_\vee(t) = s(t) \vee n(t) = \{[s(t) + n(t)] + |s(t) - n(t)|\}/2; \tag{6.2.3a}$$

$$x_\wedge(t) = s(t) \wedge n(t) = \{[s(t) + n(t)] - |s(t) - n(t)|\}/2, \tag{6.2.3b}$$

that are consequence from known equations [29, § XIII.3;(14)], [29, § XIII.4;(22)].

Method of mapping of linear signal space into signal space with lattice properties that is determined by the relationships (6.2.3a,b) can be realized in antenna arrays. Further we consider some possible realizations of this method based on linear antenna arrays that differ from each other by a way of forming a vector of weight coefficients.

Before stating main material of the section, we formulate and prove the following lemma which is analogous to Lemma 3.2.1 and is used below.

Lemma 6.2.1. *In L-group $\mathcal{L}(+, \vee, \wedge)$, for any $a, b \in \mathcal{L}(+, \vee, \wedge)$ the following identity holds:*

$$F(a, b) = (a + |b|) \wedge (b + |a|) \vee 0 + (a - |b|) \vee (b - |a|) \wedge 0 = a + b$$

Proof. Introduce the following notations:

$$G = (a + |b|) \wedge (b + |a|); \quad H = (a - |b|) \vee (b - |a|); \tag{6.2.4a}$$

$$\Delta_+ = |a| + |b|; \quad \Delta_- = ||a| - |b||. \tag{6.2.4b}$$

Then expressions (6.2.4a) can be rewritten in the form:

$$G = (|a| \cdot \text{sgn}(a) + |b|) \wedge (|b| \cdot \text{sgn}(b) + |a|); \tag{6.2.5a}$$

$$H = (|a| \cdot \text{sgn}(a) - |b|) \vee (|b| \cdot \text{sgn}(b) - |a|). \tag{6.2.5b}$$

Fill the table of values for $F(a, b)$, G, H depending on signs of variables a, b, $a + b$.

TABLE 6.2.1   Values of $F(a,b)$, G, H depending on signs of variables a, b, $a+b$

$a+b$	a	b	G	H	$G \vee 0$	$H \wedge 0$	$F(a,b)$
$+$	$+$	$+$	$\Delta_+ \wedge \Delta_+$	$-\Delta_- \vee \Delta_-$	$\Delta_+ \vee 0$	0	$a+b$
$+$	$\pm$	$\mp$	$\Delta_\pm \wedge \Delta_\mp$	$\pm\Delta_\mp \vee \mp\Delta_\pm$	$\Delta_- \vee 0$	0	$a+b$
$-$	$-$	$-$	$-\Delta_- \wedge \Delta_-$	$-\Delta_+ \vee -\Delta_+$	0	$-\Delta_+ \wedge 0$	$a+b$
$-$	$\pm$	$\mp$	$\pm\Delta_\pm \wedge \mp\Delta_\mp$	$-\Delta_\mp \vee -\Delta_\pm$	0	$-\Delta_- \wedge 0$	$a+b$

The relationships in the last column of the table and signs of variables a, b imply that values of $F(a,b)$ are equal to $a+b$. $\qquad\qquad\qquad\qquad\qquad\qquad\square$

6.2.1   Spatial Filtering: Forming Vector of Weight Coefficients by Direct Inversion of Correlation Matrix

Consider signal aligned linear antenna array (see Fig. 6.1.7a) with odd number of channels, so that ratio between interelement distance of antenna array d and wave length λ is equal to $d/\lambda = 0.5$.

Resulting process $y(t)$ in the output of linear antenna array (see Fig. 6.1.7a) is determined by the expression:

$$y(t) = \sum_{i=1}^{n} \bar{w}_i a_i(t), \qquad\qquad (6.2.6)$$

where $a_i(t)$ is a received signal in the input of i-th receiving element of antenna array; w_i is weight coefficient in i-th processing channel; $\bar{w}_i = \text{Re}(w_i) - j\,\text{Im}(w_i)$.

Antenna array operating on the basis of minimum variance criterion forms optimal solution for vector of weight coefficients $\mathbf{w}$ according to the following relationship [158, (3.132)], [102, (61.18)]:

$$\mathbf{w} = [w_1, \ldots, w_n]^T = \frac{\hat{\mathbf{R}}_n^{-1}\mathbf{1}}{\mathbf{1}^T\hat{\mathbf{R}}_n^{-1}\mathbf{1}}, \qquad\qquad (6.2.7)$$

where $\hat{\mathbf{R}}_n$ is an estimator of interference correlation matrix that, in the case of sufficiently small signal-to-interference ratio in receiving channels, is considered to be equal to correlation matrix $\mathbf{R}_a$ of the observed processes $\hat{\mathbf{R}}_n = \mathbf{R}_a$; $\mathbf{R}_a = \left\|r_{ij}\sqrt{D_i D_j}\right\|$, r_{ij} is correlation coefficient between the received signals $a_i(t)$, $a_j(t)$ in the inputs of i-th and j-th receiving elements of antenna array, respectively; D_i is a variance of the signal $a_i(t)$ in the input of i-th receiving element of antenna array; $\mathbf{1} = [1, 1, \ldots, 1]^T$; $\mathbf{w}^T\mathbf{1} = 1$.

Linear antenna array based on L-group operations with forming vector of weight coefficients by direct inversion of correlation matrix is determined by general algorithm of pairwise forming the estimators $\hat{x}_\vee(t)$, $\hat{x}_\wedge(t)$ and $\hat{x}'_\vee(t)$, $\hat{x}'_\wedge(t)$ of join (6.2.3a) and meet (6.2.3b), respectively:

$$\hat{x}_\vee(t) = A_1(t) + |A_2(t)|; \qquad\qquad (6.2.8a)$$

$$\hat{x}_\wedge(t) = A_1(t) - |A_2(t)|; \qquad\qquad (6.2.8b)$$

$$\hat{x}'_\vee(t) = A_2(t) + |A_1(t)|; \tag{6.2.9a}$$

$$\hat{x}'_\wedge(t) = A_2(t) - |A_1(t)|; \tag{6.2.9b}$$

$$A_1(t) = \mathrm{Re}\left(\sum_{i=1}^{(n+1)/2} \bar{w}_{2i-1} \cdot a_{2i-1}(t) \right); \tag{6.2.10a}$$

$$A_2(t) = \mathrm{Re}\left(\sum_{i=2}^{(n+1)/2} \bar{w}_{2i-2} \cdot a_{2i-2}(t) \right), \tag{6.2.10b}$$

where $a_i(t)$ is received signal in the input of i-th receiving element of antenna array; w_i is weight coefficient in i-th processing channel; $\bar{w}_i = \mathrm{Re}(w_i) - j\,\mathrm{Im}(w_i)$; $A_1(t)$ and $A_2(t)$ are results of weight processing formed by the signals $\{a_i(t)\}$ in the outputs of odd and even channels of antenna array, respectively.

Weight coefficient in i-th processing channel is formed on the basis of optimal solution for vector of weight coefficients $\mathbf{w}$ determined by the relationship (6.2.7).

Two pairs of estimators (6.2.8a,b) and (6.2.9a,b) are necessary for the most complete exploiting information contained in a field of waves arriving at antenna array aperture.

Then the following theorem holds.

Theorem 6.2.1. *For antenna array operating according to signal processing L-group algorithm:*

$$z(t) = [\hat{x}_\vee(t) \wedge \hat{x}'_\vee(t)] \vee 0 + [\hat{x}_\wedge(t) \vee \hat{x}'_\wedge(t)] \wedge 0, \tag{6.2.11}$$

the latter is equivalent to optimal linear algorithm (6.2.6) in which vector of weight coefficients $\mathbf{w}(t)$ is determined by the relationship (6.2.7):

$$z(t) \equiv \mathrm{Re}[y(t)]. \tag{6.2.12}$$

Here the functions figuring in the right side of the identity (6.2.11) are determined by the expressions (6.2.8a,b), (6.2.9a,b), (6.2.10a,b).

Proof. Taking into account the equations (6.2.8a,b), (6.2.9a,b), rewrite the identity (6.2.11) in the following form:

$$z(t) = (A_2(t) + |A_1(t)|) \wedge (A_1(t) + |A_2(t)|) \vee 0$$
$$+ (A_2(t) - |A_1(t)|) \vee (A_1(t) - |A_2(t)|) \wedge 0. \tag{6.2.13}$$

Then basing on Lemma 6.2.1 and the expressions (6.2.10a,b), the relationship (6.2.13) can be rewritten in the form of the following identity:

$$z(t) = (A_2(t) + |A_1(t)|) \wedge (A_1(t) + |A_2(t)|) \vee 0$$
$$+ (A_2(t) - |A_1(t)|) \vee (A_1(t) - |A_2(t)|) \wedge 0$$
$$= A_1(t) + A_2(t) = \mathrm{Re}[y(t)].$$

$\square$

Distributivity of lattice $\mathcal{L}(\vee, \wedge)$ of L-group $\mathcal{L}(+, \vee, \wedge)$ implies the following corollary of Theorem 6.2.1.

Corollary 6.2.1. *For antenna array operating according to signal processing L-group algorithm:*

$$z'(t) = [\hat{x}_\vee(t) \vee 0] \wedge [\hat{x}'_\vee(t)) \vee 0] + [\hat{x}_\wedge(t) \wedge 0] \vee [\hat{x}'_\wedge(t)) \wedge 0], \qquad (6.2.14)$$

the latter is equivalent to algorithm (6.2.6) in which vector of weight coefficients $\mathbf{w}(t)$ is determined by the relationship (6.2.7):

$$z'(t) \equiv \mathrm{Re}[y(t)].$$

Theorem 6.2.1 contains the following important theoretical inferences.

1. Optimal signal processing algorithm providing optimal solution (6.2.7) is not a sole one.

2. Apparently, all optimal solutions known for antenna arrays have their own equivalent formulations written in terms of L-groups.

3. Efficiency of signal processing algorithms, based on L-group operations and defined by Theorem 6.2.1, is equivalent to efficiency of optimal signal processing algorithms synthesized for antenna arrays within linear signal space.

Theorem 6.2.1 is illustrated by the following example, within which we consider linear antenna array operating according to L-group algorithm defined by the relationships (6.2.7)...(6.2.11).

Example 6.2.1. Fig. 6.2.1a and 6.2.1b depict stochastic processes $a_1(t)$, $a_2(t)$ and $a_7(t)$, $a_8(t)$ in the inputs of the corresponding channels of 13-elements linear equidistant antenna array ($n=13$). Ratio between interelement distance of antenna array d and wave length λ is equal to $d/\lambda = 0.5$. The signals $a_1(t)$, $a_7(t)$ are shown by dotted line, the signals $a_2(t)$, $a_8(t)$ – by dashed line, and useful signal $s(t)$ amplified for convenience of visual perception is shown by solid line. Simulating signal processing is realized for the case that is similar to considered in Subsection 6.1.3, i.e. in the presence of useful signal $s(t)$ arriving from main lobe direction ($\theta_s = 0$) and interference signals $n_1(t)$, $n_2(t)$, $n_3(t)$, $n_4(t)$, $n_5(t)$ arriving from directions $\theta_1 = \pi/6$; $\theta_2 = -2\pi/9$; $\theta_3 = -5\pi/13$; $\theta_4 = 3\pi/7$; $\theta_5 = -2\pi/5$, respectively (see Fig. 6.1.2a,b; Fig. 6.1.3a,b,c,d). Remind that interference signals $n_1(t)$, $n_2(t)$, $n_3(t)$ are amplitude-modulated interference, frequency-modulated interference, and also RF pulse with rectangular envelope, respectively; and statistically independent interferences $n_4(t)$, $n_5(t)$ are quasi-white Gaussian noise. Signal-to-k-th interference ratios in each separate i-th receiving channel of antenna array are closely equal to 0.25; 0.25; 1; 0.05; 0.05, respectively.

Fig. 6.2.2a,b illustrates the estimates $\hat{x}_\vee(t)$, $\hat{x}_\wedge(t)$ of join (6.2.8a) and meet (6.2.8b), respectively (solid line), useful signal $s(t)$ and pulse interference $n_3(t)$ shown two times less than the original (dotted line). Fig. 6.2.3a,b depicts the

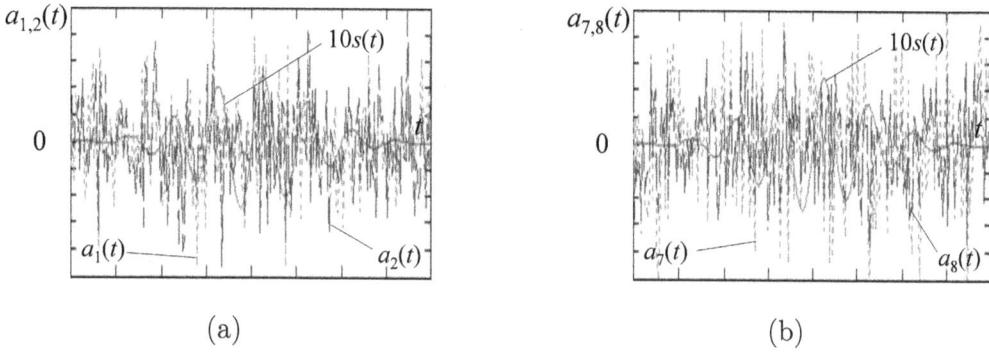

FIGURE 6.2.1 Stochastic processes in the inputs of corresponding channels of 13-element linear antenna array: (a) $a_1(t)$, $a_2(t)$; (b) $a_7(t)$, $a_8(t)$

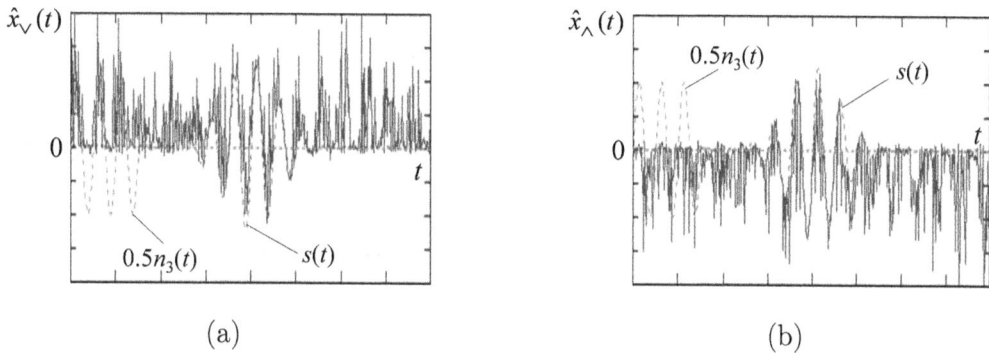

FIGURE 6.2.2 Estimates: (a) $\hat{x}_\vee(t)$ of join (6.2.8a) ; (b) $\hat{x}_\wedge(t)$ of meet (6.2.8b)

estimates $\hat{x}'_\vee(t)$, $\hat{x}'_\wedge(t)$ of join (6.2.9a) and meet (6.2.9b), respectively (solid line), useful signal $s(t)$ and pulse interference $n_3(t)$ shown two times less than the original (dotted line).

Fig. 6.2.4 illustrates the functions $g(t) = \hat{x}_\vee(t) \wedge \hat{x}'_\vee(t)$ (dashed line), $h(t) = \hat{x}_\wedge(t) \vee \hat{x}'_\wedge(t)$ (solid line) figuring in the equation (6.2.11) and formed by the estimators (6.2.8a,b); (6.2.9a,b), and also useful signal $s(t)$ and pulse interference $n_3(t)$ (dotted line). For convenience of visual perception, the functions $g(t)$ and $h(t)$ are shifted vertically at ± 4, respectively. Fig. 6.2.5 illustrates resulting signal $z(t)$ in the output of signal processing unit determined by the equation (6.2.11) and also pulse interference $n_3(t)$ shown two times less than the original little (dotted line). The signal $z(t)$ completely repeats the real part of the signal $y(t)$ (6.2.6) according to the identity (6.2.12) of Theorem 6.2.1.

Correlation coefficients between useful signal $s(t)$ and output signal $z(t)$ take their values in the interval: $r[z(t), s(t)]=0.96\ldots 0.985$. Coefficients of interference suppression take the values in the interval: $K_z=45\ldots 48$ dB. Here, coefficient of suppression is determined by the ratio: $K_z = 10\lg(D/D[z(t) - s(t)])$, where D is a total variance of resulting interference, $D = \sum_{k=1}^{K} D_k$; D_k is a variance of k-th

(a)   (b)

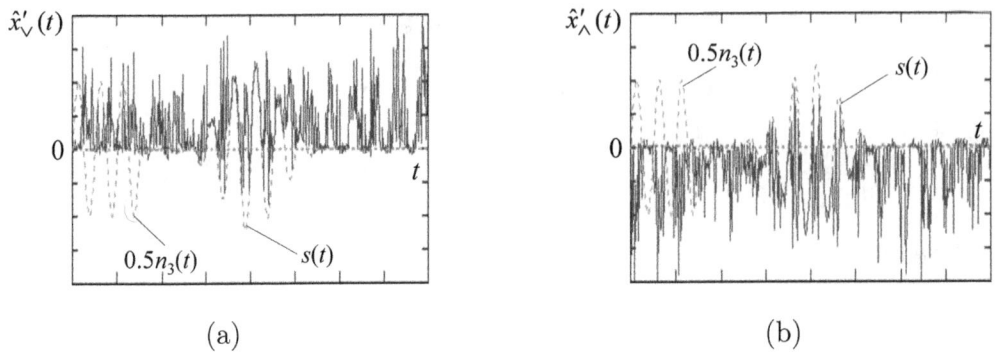

FIGURE 6.2.3   Estimates: (a) $\hat{x}'_\vee(t)$ of join (6.2.9a); (b) $\hat{x}'_\wedge(t)$ of meet (6.2.9b)

interference; $D[z(t) - s(t)]$ is a variance of noise component contained in the output process $z(t)$, respectively.   ▽

6.2.2 Spatial Filtering: Forming Vector of Weight Coefficients by Direct Inversion of Correlation Matrix with Decreasing Its Dimensionality

Signal processing rate in antenna arrays when using the method of direct inversion of correlation matrix depends essentially on its dimensionality. Thus, for instance, it is known that a common number of operations of multiplication and division that is necessary for inverting a matrix with dimensionality $m \times m$ (excluding operations that are necessary to form this correlation matrix) is equal to m^3 [163]. In this subsection we consider L-group algorithm of signal processing in linear antenna array with forming vector of weight coefficients by direct inversion of correlation matrix and decreasing its dimensionality. One can naturally decrease a dimensionality of correlation matrix when using signal processing algorithms based on L-group operations, inasmuch as these algorithms assume separate signal processing in odd and even receiving channels (see the formulas (6.2.10a,b)). Here, as before, we consider linear antenna array with odd number of receiving channels n.

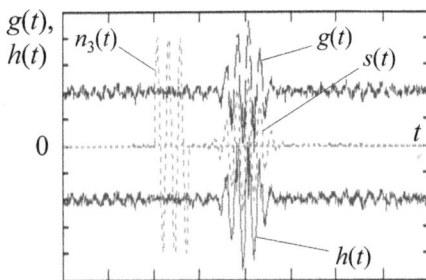

FIGURE 6.2.4   Functions $g(t) = \hat{x}_\vee(t) \wedge$
$\hat{x}'_\vee(t)$, $h(t) = \hat{x}_\wedge(t) \vee \hat{x}'_\wedge(t)$

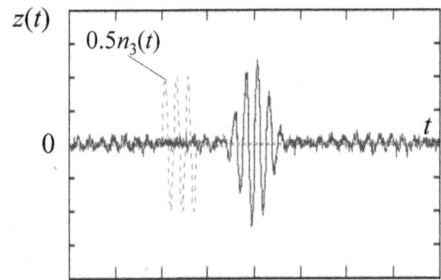

FIGURE 6.2.5   Resulting output signal $z(t)$
determined by the equation (6.2.11)

At first, find optimal solution for antenna array, whose output process $y'(t)$ (see Fig. 6.1.9a) is determined by linear combination of received signals $\{a_i(t)\}$ in the inputs of i-th receiving element of antenna array:

$$y'(t) = \sum_{i=1}^{n} \bar{w}_i a_i(t) = p_{\text{opt}} \cdot A_1'(t) + (1 - p_{\text{opt}}) \cdot A_2'(t), \qquad (6.2.15)$$

where w_i is weight coefficient in i-th processing channel; $\bar{w}_i = \text{Re}(w_i) - j\,\text{Im}(w_i)$; p is optimization parameter that is obtained by solving optimization equation:

$$p_{\text{opt}} = \arg\min_{p \in\,]0,1[} M\{y^2(t)\}, \qquad (6.2.16)$$

M is a symbol of mathematical expectation;

$$A_1'(t) = \sum_{i=1}^{(n+1)/2} \bar{w}_{2i-1} \cdot a_{2i-1}(t); \qquad (6.2.17a)$$

$$A_2'(t) = \sum_{i=2}^{(n+1)/2} \bar{w}_{2i-2} \cdot a_{2i-2}(t), \qquad (6.2.17b)$$

so that vectors of weight coefficients $\mathbf{w}_1$, $\mathbf{w}_2$ for odd and even channels are formed similarly as it is shown in (6.2.7):

$$\mathbf{w}_1 = [w_1, w_3, \ldots, w_n]^T = \frac{\hat{\mathbf{R}}_{1,n}^{-1}\mathbf{1}_1}{\mathbf{1}_1^T\hat{\mathbf{R}}_{1,n}^{-1}\mathbf{1}_1}; \qquad (6.2.18a)$$

$$\mathbf{w}_2 = [w_2, w_4, \ldots, w_{n-1}]^T = \frac{\hat{\mathbf{R}}_{2,n}^{-1}\mathbf{1}_2}{\mathbf{1}_2^T\hat{\mathbf{R}}_{2,n}^{-1}\mathbf{1}_2}, \qquad (6.2.18b)$$

where $\hat{\mathbf{R}}_{1,n}$, $\hat{\mathbf{R}}_{2,n}$ are estimators of interference correlation matrices for odd and even processing channels, so that in the case of small signal-to-interference ratios in receiving channels one can use correlation matrices $\mathbf{R}_{1,a}$, $\mathbf{R}_{2,a}$ of the observed processes instead of interference correlation matrices $\hat{\mathbf{R}}_{1,n}=\mathbf{R}_{1,a}$, $\hat{\mathbf{R}}_{2,n}=\mathbf{R}_{2,a}$; $\mathbf{R}_{1,2,a} = \left\|r_{ij}\sqrt{D_i D_j}\right\|$, r_{ij} is correlation coefficient between received signals $a_i(t)$, $a_j(t)$ in the inputs of i-th and j-th receiving elements of antenna array, respectively; D_i is a variance of the signal $a_i(t)$ in the input of i-th receiving element of antenna array; $\mathbf{1}_{1,2} = [1, 1, \ldots, 1]^T$; $\mathbf{w}_{1,2}^T \mathbf{1}_{1,2} = 1$.

The solution of optimization equation (6.2.16) is a quantity:

$$p_{\text{opt}} = \frac{M\{[A_2'(t)]^2\} - M\{A_1'(t)A_2'(t)\}}{M\{[A_1'(t)]^2\} + M\{[A_2'(t)]^2\} - 2\,M\{A_1'(t)A_2'(t)\}}. \qquad (6.2.19)$$

For linear antenna array with odd number of receiving channels n operating according to algorithm (6.2.15)...(6.2.18), a gain $K_{R^{-1}}$ in common number of multiplication and division, that is necessary to invert correlation matrix as compared with signal processing algorithm determined by the expressions (6.2.6),

(6.2.7), is determined by dimensionality of matrices $\hat{\mathbf{R}}_{1,n}$, $\hat{\mathbf{R}}_{2,n}$ (6.2.18a,b) and is equal to the quantity:

$$K_{R^{-1}} = 8n^3/[(n-1)^3 + (n+1)^3] = 8n^2/[2n^2 + 6].$$

The last relationship implies that when providing sufficiently large number of receiving channels n, $n \gg 1$ the gain $K_{R^{-1}}$ is approximately equal to 4: $K_{R^{-1}} \approx 4$ (for antenna arrays with even number n this gain is strictly equal to 4). Thus, algorithm determined by the relationships (6.2.15)...(6.2.18) can provide antenna array signal processing rate that is 4 times higher than algorithm determined by the relationships (6.2.6), (6.2.7). Necessary price to pay for higher processing rate is 0.5...3 dB smaller coefficient of interference suppression. Here, as before, coefficient of suppression is defined by the relationship: $K_y = 10 \lg(D/D[y(t) - s(t)])$, where D is a total variance of resulting interference, $D = \sum\limits_{k=1}^{K} D_k$, D_k is a variance of k-th interference; $D[y(t) - s(t)]$ is a variance of noise component contained in the output process $y(t)$; $s(t)$ is useful signal.

According to Theorem 6.2.1, optimal signal processing algorithm, which is based on L-group operations and is equivalent to optimal algorithm determined by the relationships (6.2.15)...(6.2.18), is defined by the following equations:

$$z(t) = [\hat{x}_\vee(t) \wedge \hat{x}'_\vee(t)] \vee 0 + [\hat{x}_\wedge(t) \vee \hat{x}'_\wedge(t)] \wedge 0. \qquad (6.2.20)$$

$$\hat{x}_\vee(t) = A_1(t) + |A_2(t)|; \qquad (6.2.21a)$$

$$\hat{x}_\wedge(t) = A_1(t) - |A_2(t)|; \qquad (6.2.21b)$$

$$\hat{x}'_\vee(t) = A_2(t) + |A_1(t)|; \qquad (6.2.22a)$$

$$\hat{x}'_\wedge(t) = A_2(t) - |A_1(t)|; \qquad (6.2.22b)$$

$$A_1(t) = \operatorname{Re}\left(p_{\text{opt}} \cdot \sum_{i=1}^{(n+1)/2} \bar{w}_{2i-1} \cdot a_{2i-1}(t)\right) = \operatorname{Re}[p_{\text{opt}} \cdot A'_1(t)]; \qquad (6.2.23a)$$

$$A_2(t) = \operatorname{Re}\left((1 - p_{\text{opt}}) \cdot \sum_{i=2}^{(n+1)/2} \bar{w}_{2i-2} \cdot a_{2i-2}(t)\right) = \operatorname{Re}[(1 - p_{\text{opt}}) \cdot A'_2(t)], \qquad (6.2.23b)$$

where vectors of weight coefficients $\mathbf{w}_1$, $\mathbf{w}_2$ for odd and even channels are formed according to the relationships (6.2.18a,b), respectively.

Efficiency of signal processing algorithm based on the relationships (6.2.20)... (6.2.23), (6.2.18) and L-group operations is equivalent to efficiency of algorithm determined by the relationships (6.2.15)...(6.2.18), so that in both cases signal processing rates are the same, and the gain (in a number of performed operations) with respect to optimal algorithm (6.2.6), (6.2.7) is approximately equal to $K_{R^{-1}} \approx 4$.

6.2.3 Spatial Filtering: Forming Vector of Weight Coefficients Based on Probability Distribution

Further direction of increasing signal processing rate in antenna arrays can be provided by refusing computation of correlation matrix inverse (see formula (6.2.7)) when exploiting data independent antenna arrays [102]. In previous section, we considered possibilities of providing a given level of side lobes in such antenna arrays by forming vector of weight coefficients on the basis of some symmetric discrete distributions (see Fig. 6.1.8a,b).

In this section, we investigate linear antenna array spatial filtering algorithm based on L-group operations and its analogue synthesized for linear signal space determined by the formulas (6.2.6), (6.2.7) in the form of linear combination of the received signals $a_i(t)$ in the inputs of i-th receiving elements of antenna array:

$$y(t) = \sum_{i=1}^{n} w_i a_i(t), \tag{6.2.24}$$

where w_i is weight coefficient in i-th processing channel taking its values according to one of distributions (6.1.14a,b,c).

Here we notice that unlike optimal algorithm (6.2.6), (6.2.7), weight coefficients w_i figuring in the equation (6.2.24) accept real values.

According to Theorem 6.2.1, signal processing L-group algorithm based on signal space mapping method, that is equivalent to algorithm defined by the relationships (6.2.24) and (6.1.14), is determined by the equations:

$$z(t) = [\hat{x}_\vee(t) \wedge \hat{x}'_\vee(t)] \vee 0 + [\hat{x}_\wedge(t) \vee \hat{x}'_\wedge(t)] \wedge 0. \tag{6.2.25}$$

$$\hat{x}_\vee(t) = A_1(t) + |A_2(t)|; \tag{6.2.26a}$$
$$\hat{x}_\wedge(t) = A_1(t) - |A_2(t)|; \tag{6.2.26b}$$

$$\hat{x}'_\vee(t) = A_2(t) + |A_1(t)|; \tag{6.2.27a}$$
$$\hat{x}'_\wedge(t) = A_2(t) - |A_1(t)|; \tag{6.2.27b}$$

$$A_1(t) = \sum_{i=1}^{(n+1)/2} w_{2i-1} \cdot a_{2i-1}(t); \tag{6.2.28a}$$

$$A_2(t) = \sum_{i=2}^{(n+1)/2} w_{2i-2} \cdot a_{2i-2}(t), \tag{6.2.28b}$$

where w_i is weight coefficient in i-th processing channel accepting its values according to one of distributions (6.1.14a,b,c), $w_i \in \mathbf{w}$, $\mathbf{w}^T \mathbf{1} = 1$, $\mathbf{1} = [1, 1, \ldots, 1]^T$.

Signal processing L-group algorithm determined by the relationships (6.2.25)... (6.2.28), (6.1.14) is illustrated by the following example.

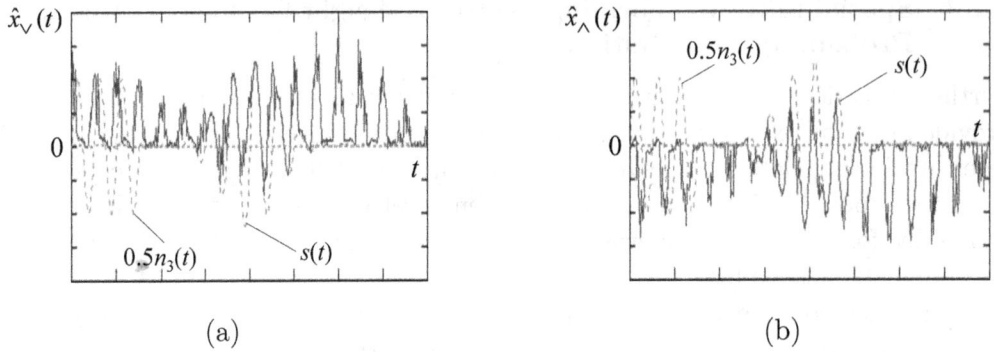

(a)   (b)

FIGURE 6.2.6   Estimates: (a) of join $\hat{x}_\vee(t)$ (6.2.26a); (b) of meet $\hat{x}_\wedge(t)$ (6.2.26b)

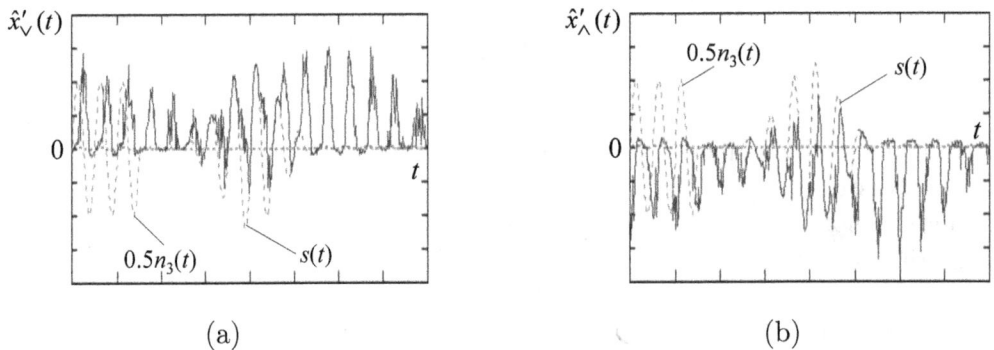

(a)   (b)

FIGURE 6.2.7   Estimates: (a) of join $\hat{x}'_\vee(t)$ (6.2.27a); (b) of meet $\hat{x}'_\wedge(t)$ (6.2.27b)

Example 6.2.2. As before, we consider 13-element linear antenna array and interferences $n_1(t)$, $n_2(t)$, $n_3(t)$, $n_4(t)$, $n_5(t)$ described in Example 6.2.1 arriving at array aperture with signal-to-k-th interference ratios in each separate i-th receiving channel that are closely equal to 0.25; 0.25; 1; 0.05; 0.05, respectively. Weight coefficients w_i figuring in the formulas (6.2.28a,b) are described by Cauchy distribution (6.1.14b). Ratio between interelement distance of antenna array d and wave length λ is equal to $d/\lambda = 0.5$.

Fig. 6.2.6a,b depicts the estimates $\hat{x}_\vee(t)$, $\hat{x}_\wedge(t)$ of join (6.2.26a) and meet (6.2.26b), respectively (solid line), useful signal $s(t)$ and pulse interference $n_3(t)$ shown two times less than the original (dotted line). Fig. 6.2.7a,b illustrates the estimates $\hat{x}'_\vee(t)$, $\hat{x}'_\wedge(t)$ of join (6.2.27a) and meet (6.2.27b), respectively (solid line), useful signal $s(t)$ and pulse interference $n_3(t)$ shown two times less than the original (dotted line).

Fig. 6.2.8 illustrates the functions $g(t) = \hat{x}_\vee(t) \wedge \hat{x}'_\vee(t)$ (dashed line), $h(t) = \hat{x}_\wedge(t) \vee \hat{x}'_\wedge(t)$ (solid line), figuring in the equation (6.2.25) and based on the estimators (6.2.26a,b), (6.2.27a,b), and also useful signal $s(t)$ and pulse interference $n_3(t)$ (dotted line). For convenience of visual perception, the functions $g(t)$ and $h(t)$ are shifted vertically at ± 4, respectively. Fig. 6.2.9 depicts resulting signal $z(t)$ defined by the equation (6.2.25) and also pulse interference $n_3(t)$ two times less than the

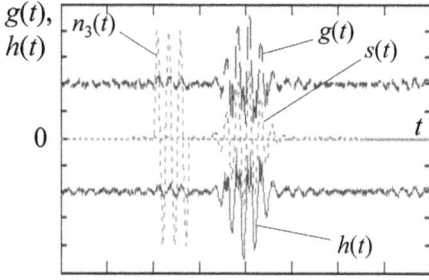

FIGURE 6.2.8   Functions $g(t) = \hat{x}_\vee(t) \wedge \hat{x}'_\vee(t)$, $h(t) = \hat{x}_\wedge(t) \vee \hat{x}'_\wedge(t)$

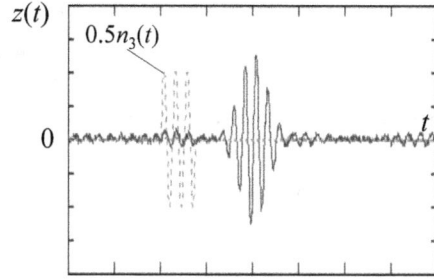

FIGURE 6.2.9   Resulting output signal $z(t)$ defined by the equation (6.2.25)

original (dotted line). The signal $z(t)$ completely repeats the signal $y(t)$ (6.2.24) according to the identity (6.2.12) of Theorem 6.2.1.

Correlation coefficients between useful signal $s(t)$ and the output signal $z(t)$ take their values in the interval $r[z(t), s(t)]$=0.96...0.985. Coefficients of interference suppression take the values: K_z=45...48 dB. As before, coefficient of suppression is defined by the relationship $K_z = 10\lg(D/D[z(t) - s(t)])$, where D is a total variance of resulting interference, $D = \sum\limits_{k=1}^{K} D_k$, D_k is a variance of k-th interference; $D[z(t) - s(t)]$ is a variance of noise component of the output process $z(t)$. Comparing the values $r[z(t), s(t)]$ and K_z, L-group algorithm defined by the relationships (6.2.25)...(6.2.28), (6.1.14) is not inferior to optimal algorithm determined by the expressions (6.2.6), (6.2.7) in efficiency. $\triangledown$

6.2.4   Spatial Filtering Suboptimal L-group Algorithms: Forming Vector of Weight Coefficients by Direct Inversion of Correlation Matrix

In Subsection 6.2.1, we found out that for linear antenna array based on L-group operations with forming vector of weight coefficients by direct inversion of correlation matrix, general algorithm of pairwise forming the estimators $\hat{x}_\vee(t)$, $\hat{x}_\wedge(t)$ and $\hat{x}'_\vee(t)$, $\hat{x}'_\wedge(t)$ of join (6.2.3a) and meet (6.2.3b), respectively, take the form:

$$\hat{x}_\vee(t) = A_1(t) + |A_2(t)|; \tag{6.2.29a}$$

$$\hat{x}_\wedge(t) = A_1(t) - |A_2(t)|; \tag{6.2.29b}$$

$$\hat{x}'_\vee(t) = A_2(t) + |A_1(t)|; \tag{6.2.30a}$$

$$\hat{x}'_\wedge(t) = A_2(t) - |A_1(t)|; \tag{6.2.30b}$$

$$A_1(t) = \text{Re}\left(\sum_{i=1}^{(n+1)/2} \bar{w}_{2i-1} \cdot a_{2i-1}(t)\right); \tag{6.2.31a}$$

$$A_2(t) = \operatorname{Re}\left(\sum_{i=2}^{(n+1)/2} \bar{w}_{2i-2} \cdot a_{2i-2}(t)\right), \qquad (6.2.31b)$$

where w_i is weight coefficient in i-th processing channel; $\bar{w}_i = \operatorname{Re}(w_i) - j\operatorname{Im}(w_i)$.

Here weight coefficient w_i in i-th processing channel is calculated basing on optimal solution for vector of weight coefficients $\mathbf{w}$ determined by the equation (6.2.7).

For antenna array spatial filtering suboptimal algorithms based on L-group operations, subsequent processing is performed according to signal filtering algorithm obtained by synthesis for spaces with L-group properties [24, § 7.3]:

$$v(t) = \operatorname*{med}_{t_k \in \tilde{T}}\{w(t_k)\}; \quad v'(t) = \operatorname*{med}_{t_k \in \tilde{T}}\{w'(t_k)\}; \qquad (6.2.32)$$

$$w(t) = y_+(t) + y_-(t); \quad w'(t) = y'_+(t) + y'_-(t); \qquad (6.2.33)$$

$$y_+(t) = y_\wedge(t) \vee 0; \quad y'_+(t) = y'_\wedge(t) \vee 0; \qquad (6.2.34a)$$

$$y_-(t) = y_\vee(t) \wedge 0; \quad y'_-(t) = y'_\vee(t) \wedge 0; \qquad (6.2.34b)$$

$$y_\wedge(t) = \bigwedge_{j=0}^{N-1} \hat{x}_\vee(t_j) = \bigwedge_{j=0}^{N-1} \hat{x}_\vee(t - j\Delta t); \qquad (6.2.35a)$$

$$y'_\wedge(t) = \bigwedge_{j=0}^{N-1} \hat{x}'_\vee(t_j) = \bigwedge_{j=0}^{N-1} \hat{x}'_\vee(t - j\Delta t); \qquad (6.2.35b)$$

$$y_\vee(t) = \bigvee_{j=0}^{N-1} \hat{x}_\wedge(t_j) = \bigvee_{j=0}^{N-1} \hat{x}_\wedge(t - j\Delta t); \qquad (6.2.35c)$$

$$y'_\vee(t) = \bigvee_{j=0}^{N-1} \hat{x}'_\wedge(t_j) = \bigvee_{j=0}^{N-1} \hat{x}'_\wedge(t - j\Delta t), \qquad (6.2.35d)$$

where med$\{*\}$ is a sample median of a set of samples $\{w(t_k)\}$ of stochastic process $w(t)$; $t_k = t - \frac{k}{M}\Delta\tilde{T}$, $k = 0, 1, \ldots, M - 1$; $t_k \in \tilde{T} =]t - \Delta\tilde{T}, t]$; $\tilde{T}$ is an interval (window) in which smoothing of stochastic process $w(t)$ is performed; $t_j = t - j\Delta t$, $j = 0, 1, \ldots, N - 1$, $t_j \in T^*$; T^* is interval of processing: $T^* = [t - (N - 1)\Delta t, t]$; $N \in \mathbf{N}$, $\mathbf{N}$ is set of natural numbers.

The results of processing $v(t)$, $v'(t)$ (6.2.32) are united according to the following relationships to obtain output signal $u(t)$:

$$u(t) = v_+(t) \wedge v'_+(t) + v_-(t) \vee v'_-(t); \qquad (6.2.36)$$

$$v_+(t) = v(t) \vee 0; \quad v'_+(t) = v'(t) \vee 0; \qquad (6.2.37a)$$

$$v_-(t) = v(t) \wedge 0; \quad v'_-(t) = v'(t) \wedge 0. \qquad (6.2.37b)$$

For coherent processing systems, when a condition of coherence between received useful signal $s(t)$ and some reference signal $s_0(t)$ holds:

$$r\{\operatorname{sgn}[s(t)], \operatorname{sgn}[s_0(t)]\} = 1, \qquad (6.2.38)$$

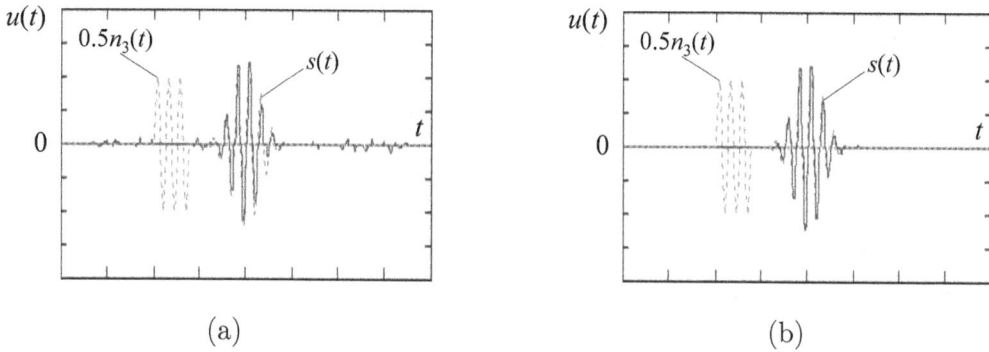

FIGURE 6.2.10 Output signals $u(t)$: (a) incoherent system; (b) coherent system

where r is correlation coefficient,
the relationships (6.2.34a,b) take the form:

$$y_+(t) = [y_\wedge(t) \vee 0] \wedge [s_0(t) \vee 0]; \quad y'_+(t) = [y'_\wedge(t) \vee 0] \wedge [s_0(t) \vee 0]; \quad (6.2.39a)$$

$$y_-(t) = [y_\vee(t) \wedge 0] \vee [s_0(t) \wedge 0]; \quad y'_-(t) = [y'_\vee(t) \wedge 0] \vee [s_0(t) \wedge 0]. \quad (6.2.39b)$$

Signal processing algorithm determined by the relationships (6.2.29)...(6.2.31); (6.2.32)...(6.2.35); (6.2.36), (6.2.37) is illustrated by the following example.

Example 6.2.3. As before, we consider 13-element linear antenna array and interferences $n_1(t)$, $n_2(t)$, $n_3(t)$, $n_4(t)$, $n_5(t)$ described in Example 6.2.1 arriving at array aperture with signal-to-k-th interference ratios in each separate i-th receiving channel that are closely equal to 0.25; 0.25; 1; 0.05; 0.05, respectively. Ratio between interelement distance of antenna array d and wave length λ is equal to $d/\lambda = 0.5$.

Fig. 6.2.10a depicts the resulting signal $u(t)$ in the output of incoherent system determined by the relationships (6.2.36), (6.2.37a,b). Correlation coefficients between useful signal $s(t)$ and output signal $u(t)$ take their values in the interval $r[u(t), s(t)]=0.88...0.96$. Coefficient of interference suppression takes the values in the interval $K_u=39...45$ dB that is $3...6$ dB less than coefficient of suppression K_y provided by antenna array based on minimum variance criterion and determined by the relationships (6.2.6), (6.2.7).

Fig. 6.2.10b illustrates the resulting signal $u(t)$ in the output of coherent system (see formulas (6.2.39a,b)) determined by the relationships (6.2.36), (6.2.37a,b). Correlation coefficients between useful signal $s(t)$ and output signal $u(t)$ take their values in the interval $r[u(t), s(t)]=0.94...0.97$. Coefficient of interference suppression takes the values in the interval $K_u=43...46$ dB that is $0...2.5$ dB less than coefficient of suppression K_y provided by antenna array based on minimum variance criterion and determined by the relationships (6.2.6), (6.2.7). $\triangledown$

6.2.5 Spatial Filtering Suboptimal L-group Algorithms: Forming Vector of Weight Coefficients Based on Probability Distribution

In Subsection 6.2.3 we found out that for linear antenna array based on L-group operations with forming vector of weight coefficients based on probability distribution, general algorithm of pairwise forming the estimators $\hat{x}_\vee(t)$, $\hat{x}_\wedge(t)$ and $\hat{x}'_\vee(t)$, $\hat{x}'_\wedge(t)$ of join (6.2.3a) and meet (6.2.3b), respectively, take the form:

$$\hat{x}_\vee(t) = A_1(t) + |A_2(t)|; \tag{6.2.40a}$$

$$\hat{x}_\wedge(t) = A_1(t) - |A_2(t)|; \tag{6.2.40b}$$

$$\hat{x}'_\vee(t) = A_2(t) + |A_1(t)|; \tag{6.2.41a}$$

$$\hat{x}'_\wedge(t) = A_2(t) - |A_1(t)|; \tag{6.2.41b}$$

$$A_1(t) = \sum_{i=1}^{(n+1)/2} w_{2i-1} \cdot a_{2i-1}(t); \tag{6.2.42a}$$

$$A_2(t) = \sum_{i=2}^{(n+1)/2} w_{2i-2} \cdot a_{2i-2}(t), \tag{6.2.42b}$$

where w_i is weight coefficient in i-th processing channel taking its values according to one of distributions (6.1.14a,b,c), $w_i \in \mathbf{w}$, $\mathbf{w}^T \mathbf{1} = 1$, $\mathbf{1} = [1,1,\ldots,1]^T$.

For suboptimal antenna array spatial filtering algorithms based on L-group operations, subsequent processing is performed according to known signal filtering algorithm, obtained by synthesis for spaces with L-group properties [24, § 7.3], the relationships (6.2.32)... (6.2.37), and, besides, for coherent processing systems, according to the relationships (6.2.39a,b).

Signal processing algorithm determined by the relationships (6.2.40)... (6.2.42); (6.2.32)... (6.2.35); (6.2.36), (6.2.37), is illustrated by the following example.

Example 6.2.4. As before, we consider 13-element linear antenna array and interferences $n_1(t)$, $n_2(t)$, $n_3(t)$, $n_4(t)$, $n_5(t)$ described in Example 6.2.1 arriving at array aperture with signal-to-k-th interference ratios in each separate i-th receiving channel that are closely equal to 0.25; 0.25; 1; 0.05; 0.05, respectively. Weight coefficients w_i figuring in the formulas (6.2.42a,b) are determined by Cauchy distribution (6.1.14b). Ratio between interelement distance of antenna array d and wave length λ is equal to $d/\lambda = 0.5$.

Fig. 6.2.11a depicts the resulting signal $u(t)$ in the output of processing unit of incoherent system determined by the relationships (6.2.36), (6.2.37a,b). Correlation coefficients between useful signal $s(t)$ and output signal $u(t)$ take their values in the interval $r[u(t), s(t)]=0.88\ldots0.94$. Coefficient of interference suppression takes the values in the interval $K_u=39\ldots44$ dB that is $4\ldots6$ dB lesser than coefficient of suppression K_y provided by antenna array based on minimum variance criterion and the relationships (6.2.6), (6.2.7).

Fig. 6.2.11b illustrates the resulting signal $u(t)$ in the output of processing unit of coherent system (see the formulas (6.2.39a,b)) determined by the relationships (6.2.36), (6.2.37a,b). Correlation coefficients between useful signal $s(t)$ and output signal $u(t)$ take their values in the interval $r[u(t), s(t)]=0.94\ldots0.96$. Coefficient of interference suppression takes the values in the interval $K_u=43\ldots45$ dB that is

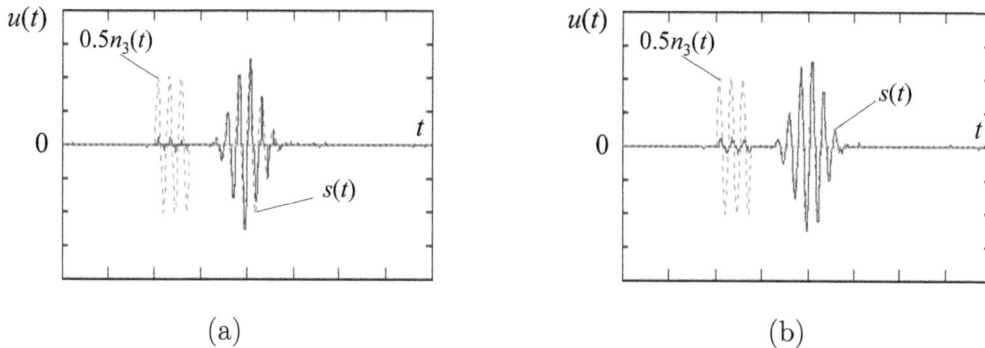

(a)   (b)

FIGURE 6.2.11   Output signals $u(t)$: (a) incoherent system; (b) coherent system

$0 \ldots 3$ dB lesser than coefficient of suppression K_y provided by antenna array based on minimum variance criterion and the relationships (6.2.6), (6.2.7). ∇

6.2.6   Spatial Filtering Suboptimal L-group Algorithms: Nonlinearities of Amplitude Characteristics of Antenna Array Receiving Channels

Providing a desired dynamic range of multichannel receiving and processing systems of various functionality is rather difficult technology problem. One of acceptable variants of its solution is based on exploiting receiving systems with nonlinear amplitude characteristics. Using multichannel receiving and processing systems with nonlinear amplitude characteristics causes destroying linear statistical relations (correlation) between the signals in receiving channels. This worsens an efficiency of signal processing in multichannel systems.

In this subsection, we consider the results of comparative analysis of efficiency of suboptimal spatial filtering algorithms, based on L-group operations discussed in Subsections 6.2.4 and 6.2.5, and also algorithm synthesized on the basis of minimum variance criterion (6.2.6), (6.2.7) in the presence of nonlinearities of receiving channel amplitude characteristics in antenna array.

Comparative analysis is performed under conditions that are similar to those described in Examples $6.2.1 \ldots 6.2.4$, i.e. in the presence of five interferences at the aperture of 13-elements of linear antenna array with signal-to-k-th interference ratios in each separate i-th receiving channel that are closely equal to 0.25; 0.25; 1; 0.05; 0.05, respectively. Ratio between interelement distance of antenna array d and wave length λ is equal to $d/\lambda = 0.5$. Influence of receiving channel nonlinearity on signal processing efficiency in linear antenna array is evaluated for amplitude characteristics of two types shown in Fig. 6.2.12a,b, respectively:

$$g_1(x) = k \cdot \tanh[x/(4l)]; \tag{6.2.43a}$$

$$g_2(x) = 2k \cdot \arctan[x/(2l)]/\pi, \tag{6.2.43b}$$

where l is a quantity determining a linear part of amplitude characteristic; k is some constant, k=const.

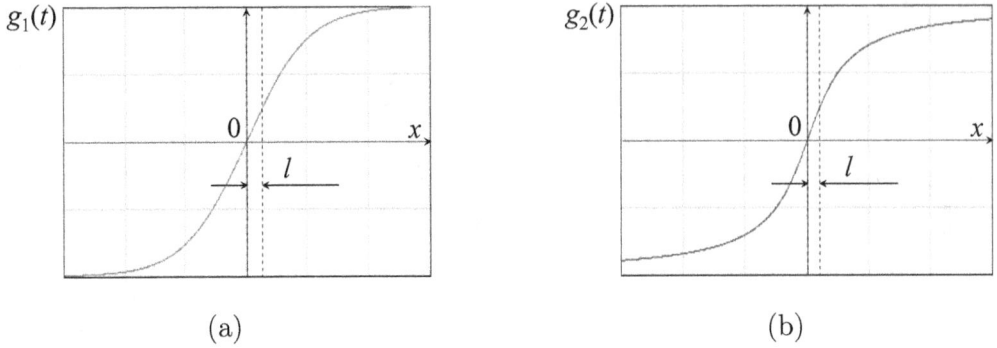

(a)   (b)

FIGURE 6.2.12   Amplitude characteristics of receiving channels: (a) $g_1(x)$ (6.2.43a); (b) $g_2(x)$ (6.2.43b)

Fig. 6.2.13a, 6.2.14a illustrate dependences of coefficient of interference suppression $K_u(b)$ for suboptimal spatial filtering algorithms based on L-group operations described in Subsections 6.2.4 and 6.2.5 on the ratio $b = l/\sqrt{D}$ characterizing nonlinearity degree, where D is a total variance of resulting interference, $D = \sum\limits_{k=1}^{K} D_k$, D_k is a variance of k-th interference. Dependences $K_u(b)$ characterizing L-group signal processing algorithms based on forming vector of weight coefficients determined by Cauchy distribution are shown by dashed line; so that for coherent processing system these dependences are shown by circles, and for incoherent processing system – by squares. Dependences $K_u(b)$, characterizing L-group signal processing algorithms based on forming vector of weight coefficients by direct inversion of correlation matrix, are shown by solid line; so that for coherent processing system these dependences are shown by circles, and for incoherent processing system – by squares. To make a comparison, for optimal linear signal processing algorithm determined by the relationships (6.2.6) and (6.2.7), dependence $K_y(b)$ is shown by dotted line with crosses.

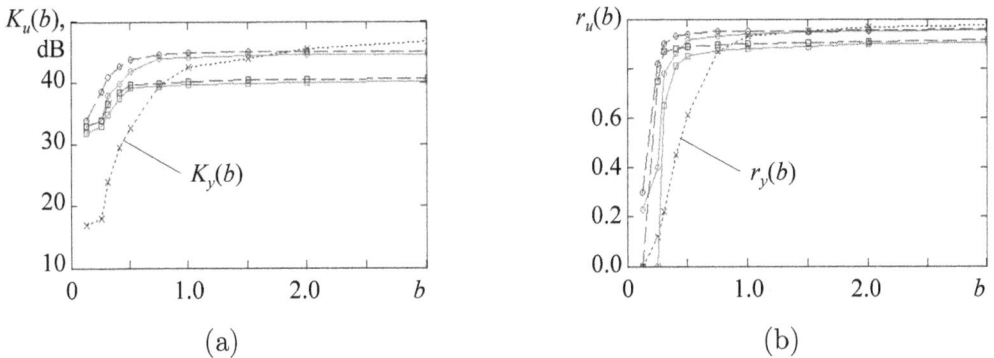

(a)   (b)

FIGURE 6.2.13   Dependences corresponding to amplitude characteristic $g_1(x)$ (6.2.43a): (a) coefficient of interference suppression $K_u(b)$; (b) correlation coefficient $r_u(b)$

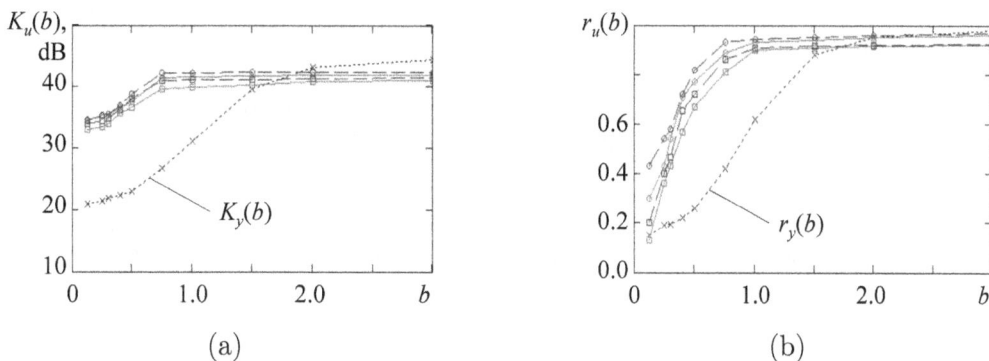

FIGURE 6.2.14   Dependences corresponding to amplitude characteristic $g_2(x)$ (6.2.43b): (a) coefficient of interference suppression $K_u(b)$; (b) correlation coefficient $r_u(b)$

Fig. 6.2.13b, 6.2.14b depict dependences of correlation coefficient $r_u(b)$ between useful signal $s(t)$ and its estimator $u(t)$ ($r_u(b){=}r[u(t), s(t); b]$) in the outputs of signal processing units performing suboptimal L-group spatial filtering algorithms discussed in Subsections 6.2.4 and 6.2.5 on the ratio $b = l/\sqrt{D}$ determining nonlinearity degree.

Dependences $r_u(b)$ characterizing L-group signal processing algorithms with forming vector of weight coefficients determined by Cauchy distribution are shown by dashed line; so that for coherent processing system the dependence is shown by circles, and for incoherent processing system—by squares. Dependences $r_u(b)$ characterizing L-group signal processing algorithms with forming vector of weight coefficients by direct inversion of correlation matrix are shown by solid line; so that for coherent processing system the dependence is shown by circles, and for incoherent processing system—by squares. To provide a comparison, dependence $r_y(b){=}r[y(t), s(t); b]$ characterizing signal processing algorithm determined by the relationships (6.2.6) and (6.2.7) is shown by crosses.

Analyzing Figs. 6.2.13a,b and 6.2.14a,b, notice the following. Despite essential difference in provided coefficients of suppression $K_u(b)$ when $b \leq 1.0$ (see Figs. 6.2.13a and 6.2.14a), operating part of abscissa axis corresponds to the interval in which the inequality holds $r_u(b) \geq 0.65$. Thus, when amplitude characteristic of antenna array receiving channels is determined by (6.2.43a), suboptimal spatial filtering L-group algorithms discussed in Subsections 6.2.4 and 6.2.5 possess better efficiency in the intervals $b \in [0.3, 0.8]$ and $b \in [0.3, 1.5]$, correspondingly, than optimal signal processing algorithm determined by the relationships (6.2.6) and (6.2.7). When amplitude characteristic of antenna array receiving channels is determined by (6.2.43b), suboptimal spatial filtering L-group algorithms discussed in Subsections 6.2.4 and 6.2.5 possess better efficiency in the intervals $b \in [0.5, 1.6]$ and $b \in [0.5, 2.0]$, respectively, than optimal signal processing algorithm determined by the relationships (6.2.6) and (6.2.7). In the interval $b \in [2.5, \infty[$ providing linearity of amplitude characteristics, efficiency of signal processing algorithms for different amplitude characteristics (see Figs. 6.2.13b and 6.2.14b) is approximately the same.

Summarizing the aforementioned, one can formulate the following conclusions.

1. Optimal signal processing algorithm corresponding to optimal solution (6.2.7) is not a sole one.

2. Apparently, for all optimal solutions known for antenna arrays, there exist equivalent formulations that can be written in terms of L-groups.

3. Efficiency of antenna array signal processing L-group algorithms determined by Theorem 6.2.1 is equivalent to efficiency of corresponding optimal algorithms synthesized within a framework of linear signal space.

4. Under conditions of nonlinearities of amplitude characteristics of antenna array receiving channels, suboptimal spatial filtering L-group algorithms discussed in Subsections 6.2.4 and 6.2.5 can provide higher efficiency than optimal algorithm based on minimum variance criterion and determined by the relationships (6.2.6) and (6.2.7).

6.3 Spatial Filtering: Method Based on Measure of Statistical Interrelation

Consider the problem of spatial filtering of narrowband RF useful signal $s(t)$ that is solved by antenna array with arbitrary configuration.

Let $a_i(t)$ be additive mixture of useful signal $s(t)$, M complex narrowband interference signals $u_l(t)$ (each arriving from l-th point source) and complex quasi-white Gaussian noise $n(t)$ with zero mean in the input of i-th receiving channel of linear antenna array:

$$a_i(t) = s_i(t) + \sum_{l=1}^{M} u_{l,i}(t) + n_i(t). \tag{6.3.1}$$

A solution of spatial filtering problem assumes forming the estimator $\hat{s}(t)$ of useful signal $s(t)$ that must be the best from the standpoint of some chosen criterion. Here, direction of arrival θ_s of the signal $s(t)$ is considered to be known.

Spatial filtering algorithms suppose forming an estimator of autocorrelation matrix (ACM) $\hat{\mathbf{R}}_a$ with dimensionality $n \times n$ calculated on the basis of scalar products between the signals $a_i(t)$ and $a_k(t)$, $i = 1, \ldots, n$; $k = 1, \ldots, n$ observed in i-th and k-th channels of linear antenna array, respectively:

$$\hat{\mathbf{R}}_a = \left\| R_{i,k}^a \right\| = \frac{1}{N} \begin{Vmatrix} \mathbf{a}_1^T \bar{\mathbf{a}}_1 & \mathbf{a}_1^T \bar{\mathbf{a}}_2 & \cdots & \mathbf{a}_1^T \bar{\mathbf{a}}_n \\ \mathbf{a}_2^T \bar{\mathbf{a}}_1 & \mathbf{a}_2^T \bar{\mathbf{a}}_2 & \cdots & \mathbf{a}_2^T \bar{\mathbf{a}}_n \\ \vdots & \vdots & \ddots & \vdots \\ \mathbf{a}_n^T \bar{\mathbf{a}}_1 & \mathbf{a}_n^T \bar{\mathbf{a}}_2 & \cdots & \mathbf{a}_n^T \bar{\mathbf{a}}_n \end{Vmatrix}, \tag{6.3.2}$$

where $\mathbf{a}_i = [a_i(t_0), a_i(t_1), \ldots, a_i(t_{N-1})]^T$ is a vector composed of samples of the signal $a_i(t)$ (6.3.1) observed in the input of i-th element of antenna array; N is a number of samples of the signal $a_i(t)$ used in processing.

In this section, we consider method of spatial filtering in antenna arrays based on matrix of measure of statistical interrelation (MSI) (1.4.11) in signal space with semimetric (1.3.26). MSI operates with real values, thought spatial filtering problem statement assumes processing complex signals (see the formula (6.3.1)), therefore we remind method of autocorrelation function (ACF) representation based on ACFs and cross-correlation functions (CCF) of real and imaginary components of complex stochastic process (see, for instance, [97, Theorem 15.1]).

Let $\mathbf{a}_i = \mathbf{u}_i + j \cdot \mathbf{v}_i$ be complex random vector:

$$\mathbf{a}_i = [a_i(t_0), a_i(t_1), \ldots, a_i(t_{N-1})]^T,$$

composed of the samples of stochastic process $a(t_m)$, $t = t_m = t_0 + m \cdot \Delta t$; $m = 0, 1, \ldots, N - 1$:

$$a(t_m) = u(t_m) + j \cdot v(t_m),$$

where Δt is a sampling interval; $\mathbf{u}_i = \text{Re}[\mathbf{a}_i]$, $\mathbf{v}_i = \text{Im}[\mathbf{a}_i]$ are real and imaginary components of random vector $\mathbf{a}_i = \mathbf{u}_i + j \cdot \mathbf{v}_i$, $u(t_m) = \text{Re}[a(t_m)]$, $v(t_m) = \text{Im}[a(t_m)]$.

Then, taking into account (6.3.2), ACM estimator $\hat{\mathbf{R}}_a$ based on vectors $\{\mathbf{a}_i\}$ has the following representation over estimators of ACMs $\hat{\mathbf{R}}_u$, $\hat{\mathbf{R}}_v$ of real $\mathbf{u}_i = \text{Re}[\mathbf{a}_i]$ and imaginary $\mathbf{v}_i = \text{Im}[\mathbf{a}_i]$ components of the vector $\mathbf{a}_i = \mathbf{u}_i + j \cdot \mathbf{v}_i$ and also estimators of cross-correlation matrices (CCMs) $\hat{\mathbf{R}}_{uv}$, $\hat{\mathbf{R}}_{vu}$:

$$\hat{\mathbf{R}}_a = \hat{\mathbf{R}}_u + \hat{\mathbf{R}}_v - j(\hat{\mathbf{R}}_{uv} - \hat{\mathbf{R}}_{vu}), \tag{6.3.3}$$

where estimators of ACMs $\hat{\mathbf{R}}_u$, $\hat{\mathbf{R}}_v$ of vectors $\{\mathbf{u}_i\}$, $\{\mathbf{v}_i\}$ are defined as symmetric matrices:

$$\hat{\mathbf{R}}_u = \left\| R_{i,k}^u \right\| = \frac{1}{N} \left\| \mathbf{u}_i^T \mathbf{u}_k \right\|; \quad \hat{\mathbf{R}}_v = \left\| R_{i,k}^v \right\| = \frac{1}{N} \left\| \mathbf{v}_i^T \mathbf{v}_k \right\|, \tag{6.3.4}$$

and estimators of CCMs $\hat{\mathbf{R}}_{uv}$, $\hat{\mathbf{R}}_{vu}$ of vectors $\{\mathbf{u}_i\}$, $\{\mathbf{v}_i\}$ are defined as skew-symmetric matrices:

$$\hat{\mathbf{R}}_{uv} = \left\| R_{i,k}^{uv} \right\| = \frac{1}{N} \left\| \mathbf{u}_i^T \mathbf{v}_k \right\|; \quad \hat{\mathbf{R}}_{vu} = \left\| R_{i,k}^{vu} \right\| = \frac{1}{N} \left\| \mathbf{v}_i^T \mathbf{u}_k \right\|; \tag{6.3.5}$$

$$R_{i,k}^{uv} = -R_{k,i}^{uv}; \quad R_{i,k}^{vu} = -R_{k,i}^{vu};$$

$$i = 0, 1, \ldots, n - 1; \quad k = 0, 1, \ldots, n - 1.$$

Estimator of MSI matrix $\hat{\mathbf{N}}_{a,s}$ based on vectors $\{\mathbf{a}_i\}$ of complex stochastic signal $a_i(t)$ (6.3.1) has the following representation over MSI matrices $\hat{\mathbf{N}}_u^s$, $\hat{\mathbf{N}}_v^s$ of real $\mathbf{u}_i = \text{Re}[\mathbf{a}_i]$ and imaginary $\mathbf{v}_i = \text{Im}[\mathbf{a}_i]$ components of vector $\mathbf{a}_i = \mathbf{u}_i + j \cdot \mathbf{v}_i$, and also their mutual MSI matrices $\hat{\mathbf{N}}_{uv}^s$, $\hat{\mathbf{N}}_{vu}^s$, respectively:

$$\hat{\mathbf{N}}_{a,s} = \hat{\mathbf{N}}_u^s + \hat{\mathbf{N}}_v^s - j(\hat{\mathbf{N}}_{uv}^s - \hat{\mathbf{N}}_{vu}^s), \tag{6.3.6}$$

so that MSI matrices $\hat{\mathbf{N}}_u^s$, $\hat{\mathbf{N}}_v^s$ based on vectors $\{\mathbf{u}_i\}$, $\{\mathbf{v}_i\}$ are defined as symmetric matrices:

$$\hat{\mathbf{N}}_u^s = \left\| N_{i,k}^u \right\| = \left\| \mathbf{N}_s(\mathbf{u}_i, \mathbf{u}_k) \right\|; \quad \hat{\mathbf{N}}_v^s = \left\| N_{i,k}^v \right\| = \left\| \mathbf{N}_s(\mathbf{v}_i, \mathbf{v}_k) \right\|, \tag{6.3.7}$$

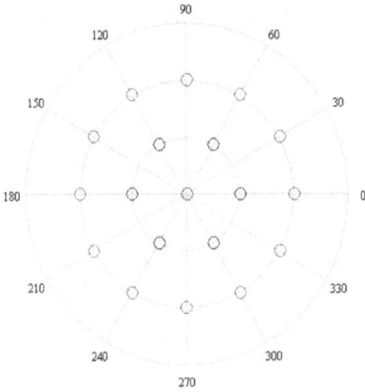

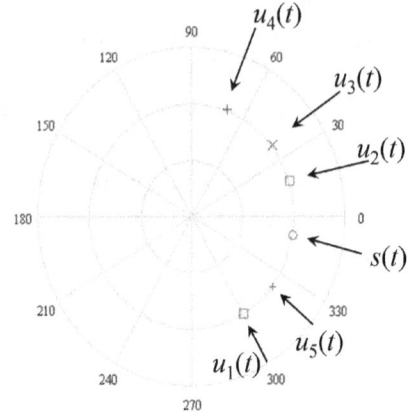

FIGURE 6.3.1 Configuration of circular FIGURE 6.3.2 Directions of arrivals of use-
19-element antenna array   ful and interference signals

and mutual MSI matrices $\hat{\mathbf{N}}_{uv}^s$, $\hat{\mathbf{N}}_{vu}^s$ based on vectors $\{\mathbf{u}_i\}$, $\{\mathbf{v}_i\}$ are defined as skew-symmetric matrices:

$$\hat{\mathbf{N}}_{uv}^s = \left\|N_{i,k}^{uv}\right\| = \|\mathbf{N}_s(\mathbf{u}_i, \mathbf{v}_k)\| ; \quad \hat{\mathbf{N}}_{vu}^s = \left\|N_{i,k}^{vu}\right\| = \|\mathbf{N}_s(\mathbf{v}_i, \mathbf{u}_k)\| ; \qquad (6.3.8)$$

$$N_{i,k}^{uv} = -N_{k,i}^{uv}; \quad N_{i,k}^{vu} = -N_{k,i}^{vu};$$

$$i = 0, 1, \ldots, n-1; \quad k = 0, 1, \ldots, n-1,$$

where sample MSIs $\mathbf{N}_s(\mathbf{u}_i, \mathbf{u}_k)$, $\mathbf{N}_s(\mathbf{v}_i, \mathbf{v}_k)$, $\mathbf{N}_s(\mathbf{u}_i, \mathbf{v}_k)$, $\mathbf{N}_s(\mathbf{v}_i, \mathbf{u}_k)$ are defined according to the relationship (1.4.11):

$$\mathbf{N}_s(a, b) = \sum_{m=0}^{N-1} (|a_m| + |b_m|) - \sum_{m=0}^{N-1} |a_m - b_m| \cdot [\operatorname{sgn}(a_m \vee b_m) - \operatorname{sgn}(a_m \wedge b_m)]; \ (6.3.9)$$

$$a = [a_0, a_1, \ldots, a_{N-1}]^T, \quad b = [b_0, b_1, \ldots, b_{N-1}]^T.$$

The relationships (6.3.6)...(6.3.9) create a foundation for signal processing in antenna arrays (including spatial filtering) exploiting L-group algorithms based on MSI (1.4.11) in signal space with semimetric (1.3.26). Further we consider spatial filtering algorithms that use estimator of MSI matrix (6.3.6). This discussion is illustrated by examples within which we simulate operation of linear 19-element antenna array (see Fig. 6.1.7a) and also circular 19-element antenna array whose configuration is shown in Fig. 6.3.1. Here we suppose that radiating sources are situated in the same plane with the directions of arrivals shown in Fig. 6.3.2.

6.3.1 Spatial Filtering in Linear Antenna Array: L-group Algorithm

Consider spatial filtering problem solved by linear equidistant antenna array (see Fig. 6.1.7a).

Let $a_i(t)$ be additive mixture of useful signal $s(t)$, M complex narrowband interference signals $u_l(t)$ (each arriving from l-th point source) and complex quasi-white Gaussian noise $n(t)$ with zero mean in the input of i-th receiving channel of linear antenna array:

$$a_i(t) = s_i(t) + \sum_{l=1}^{M} u_{l,i}(t) + n_i(t)$$

$$= s(t, \omega_s) \exp[-j\omega_s \tfrac{d}{c} \sin\theta_s (i-1)]$$

$$+ \sum_{l=1}^{M} u_l(t, \omega_l) \exp[-j\omega_l \tfrac{d}{c} \sin\theta_l (i-1)] + n_i(t), \quad (6.3.10)$$

where $t = t_m = t_0 + m \cdot \Delta t$; $m = 0, 1, \ldots, N-1$; Δt is a sampling interval; $s_i(t)$ is useful signal $s(t)$ observed in i-th channel of linear antenna array; $\omega_s = 2\pi f_s = $ const, f_s is a carrier frequency of useful signal $s(t)$; θ_s is a direction of arrival of the signal $s(t)$; $\omega_l = 2\pi f_l = $ const, f_l is a carrier frequency of interference signal $u_{l,i}(t)$ observed in i-th channel of linear antenna array; θ_l is a direction of arrival of interference signal $u_l(t)$; frequencies $\{f_l\}$ of interference signals $\{u_l(t)\}$ are different and uniformly distributed in a narrow interval around some central frequency f_0 coinciding with carrier frequency f_s of useful signal $s(t)$: $f_0 = f_s$; $f_l \in [f_0 - \frac{\Delta F}{2}, f_0 + \frac{\Delta F}{2}]$, $\Delta F << f_0$; $n_i(t) = n_i^c(t) + j \cdot n_i^s(t)$ is noise observed in i-th channel of linear antenna array, $j = \sqrt{-1}$; $\mathbf{M}\{u_l(t)u_k(t)\} = 0, l \neq k$; $\mathbf{M}\{n_i(t)n_j(t)\} = 0, i \neq j$; $\mathbf{M}\{*\}$ is a symbol of mathematical expectation; $\mathbf{M}\{(n_i^c(t))^2\} = \mathbf{M}\{(n_i^s(t))^2\} = D_n/2$, D_n is a variance of complex quasi-white Gaussian noise $n_i(t)$ with zero mean; d is interelement distance of antenna array; c is a velocity of electromagnetic waves propagation.

The resulting process $y(t)$ in the output of signal processing unit of linear antenna array (see Fig. 6.1.7a) is determined by the expression:

$$y(t) = \sum_{i=1}^{n} \bar{w}_i a_i(t), \quad (6.3.11)$$

where $a_i(t)$ is received signal in the input of i-th receiving element of antenna array; w_i is weight coefficient in i-th processing channel; $\bar{w}_i = \mathrm{Re}(w_i) - j\,\mathrm{Im}(w_i)$.

Antenna array synthesized on the basis of minimum variance criterion forms optimal solution for vector of weight coefficients in the following form [158, (3.132)], [102, (61.18)]:

$$\mathbf{w} = [w_1, \ldots, w_n]^T = \frac{\hat{\mathbf{R}}_a^{-1} \mathbf{e}(\theta_s, f_s)}{\mathbf{e}(\theta_s, f_s)^T \hat{\mathbf{R}}_a^{-1} \mathbf{e}(\theta_s, f_s)}; \quad (6.3.12)$$

$$\mathbf{e}(\theta_s, f_s) = [\, 1 \ e^{(-j2\pi f_s d \sin(\theta_s)/c)} \ \ldots \ e^{(-j2\pi f_s d(n-1)\sin(\theta_s)/c)} \,]^T, \quad (6.3.12a)$$

where $\hat{\mathbf{R}}_a$ is an estimator of ACM $\mathbf{R}_a$ of the observed processes (5.3.2); $\mathbf{R}_a = \left\| r_{ij}\sqrt{D_i D_j} \right\|$, r_{ij} is correlation coefficient between received signals $a_i(t)$, $a_j(t)$ in the inputs of i-th and j-th receiving elements of antenna array, respectively; D_i

is a variance of the signal $a_i(t)$ in the input of i-th receiving element of antenna array; $\mathbf{e}(\theta_s, f_s)$ is vector of complex spatial harmonics (steering vector) as a function of direction of arrival θ_s and carrier frequency f_s of the signal $s(t)$; $\overline{\mathbf{e}(\theta_s, f_s)^T}$ is conjugate transpose of initial vector $\mathbf{e}(\theta_s, f_s)$; c is velocity of electromagnetic waves propagation; d is interelement distance of linear antenna array.

Taking into account known analogy between MSI and correlation coefficient (scalar product) of a pair of signals discussed in Chapter 1, also by the analogy with relationship (6.3.12), one can form vector of weight coefficients $\mathbf{w}'_L$ determined by MSI (1.4.11) and the following relationship:

$$\mathbf{w}'_L = [w'_{L,1}, \ldots, w'_{L,n}]^T = \frac{\hat{\mathbf{N}}_{a,s}^{-1} \mathbf{e}(\theta_s, f_s)}{\mathbf{e}(\theta_s, f_s)^T \hat{\mathbf{N}}_{a,s}^{-1} \mathbf{e}(\theta_s, f_s)}, \qquad (6.3.13)$$

where $\hat{\mathbf{N}}_{a,s}$ is estimator of MSI matrix of vectors $\{\mathbf{a}_i\}$ composed from the samples of signals $a_i(t)$ (6.3.10) and determined by the relationship (6.3.6); $\mathbf{e}(\theta_s, f_s)$ is vector of complex spatial harmonics (steering vector) determined by the formula (6.3.12a).

Since the estimator of MSI matrix $\hat{\mathbf{N}}_{a,s}$ is determined by semimetric, direct use of formula (6.3.13) for obtaining an estimator of useful signal $s(t)$ can be accompanied by worsening its quality because of possible negative eigenvalues of matrix $\hat{\mathbf{N}}_{a,s}$. Here we use known decomposition of Hermitian matrix over eigenvectors and eigenvalues (see, for instance, [140, (3.88)]):

$$\hat{\mathbf{N}}_{a,s}^{-1} = \sum_{i=1}^{n} \frac{1}{\rho_i^s} \mathbf{u}_i^s \overline{\mathbf{u}_i^s}^T,$$

where $\{\mathbf{u}_i^s\}$, $\{\rho_i^s\}$ are eigenvectors and eigenvalues of estimator of MSI matrix $\hat{\mathbf{N}}_{a,s}$ (6.3.6).

Then modified spatial filtering algorithm based on MSI matrix takes the form:

$$z(t) = \sum_{i=1}^{n} \bar{w}_{L,i} a_i(t); \qquad (6.3.14)$$

$$\mathbf{w}_L = [w_{L,1}, \ldots, w_{L,n}]^T = \frac{\hat{\mathbf{N}}_a^{-1} \mathbf{e}(\theta_s, f_s)}{\mathbf{e}(\theta_s, f_s)^T \hat{\mathbf{N}}_a^{-1} \mathbf{e}(\theta_s, f_s)} \qquad (6.3.14a)$$

$$\hat{\mathbf{N}}_a^{-1} = \sum_{i=1}^{n} \frac{1}{|\rho_i^s|} \mathbf{u}_i^s \overline{\mathbf{u}_i^s}^T, \qquad (6.3.14b)$$

where $t = t_m = t_0 + m \cdot \Delta t$; $m = 0, 1, \ldots, N-1$; $\hat{\mathbf{N}}_a$ is modified according to (6.3.14b) estimator of MSI matrix $\hat{\mathbf{N}}_{a,s}$; $\mathbf{e}(\theta_s, f_s)$ is vector of complex spatial harmonics (steering vector) determined by (6.3.12a).

Stated above algorithm is illustrated by the following example.

Example 6.3.1. Consider 19-elements linear antenna array (see Fig. 6.1.7a) that receives additive mixture of signals (6.3.10). The latter contains: useful signal $s(t)$ arriving from the direction $\theta_s = -\pi/18$, interference signals $u_1(t)$, $u_2(t)$, $u_3(t)$,

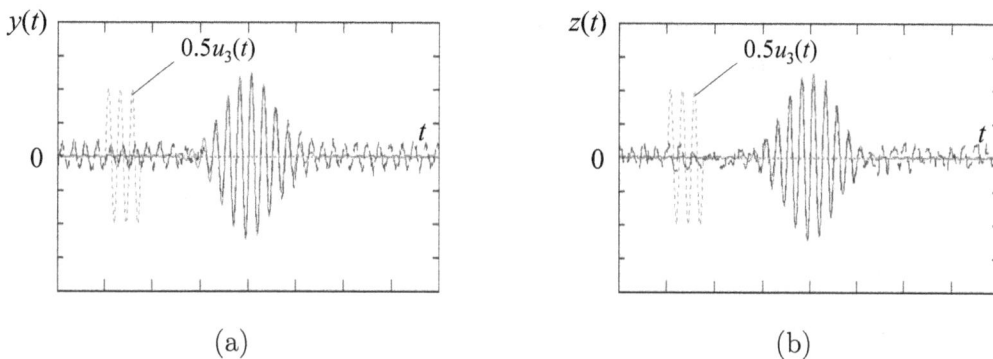

(a)   (b)

FIGURE 6.3.3 Estimates of useful signal $s(t)$ formed according to spatial filtering algorithms: (a) $y(t) = \hat{s}(t)$ (6.3.11), (6.3.12) (classic optimal algorithm); (b) $z(t) = \hat{s}(t)$ (6.3.14) (L-group algorithm)

$u_4(t)$, $u_5(t)$ arriving from the directions $\theta_1 = -\pi/3$; $\theta_2 = \pi/10$; $\theta_3 = 3\pi/14$; $\theta_4 = 7\pi/18$; $\theta_5 = -3\pi/14$, respectively (see Fig. 6.3.2; 6.1.2, 6.1.3), and also quasi-white Gaussian noise $n(t)$. Interference signals $u_1(t)$, $u_2(t)$, $u_3(t)$ are amplitude-modulated interference, frequency-modulated interference, and also RF pulse with rectangular envelope, respectively; and statistically independent interferences $u_4(t)$, $u_5(t)$ are quasi-white Gaussian noises. Signal-to-k-th interference ratios in each separate i-th receiving channel of antenna array take the values that are closely equal to 0.25; 0.25; 1; 0.05; 0.05, respectively. Signal-to-noise ratio in receiving channels of antenna array is equal to 10^3. Number of samples N of signals $a_i(t)$ used in processing (6.3.10) is chosen to be equal $N=1024$. Interelement distance d of linear antenna array is equal $d = 0.5\lambda_0$, where λ_0 is wave length of useful signal $s(t)$.

Fig. 6.3.3a,b illustrate the estimates $y(t) = \hat{s}(t)$, $z(t) = \hat{s}(t)$ of useful signal $s(t)$ that are formed in linear antenna array according to classic optimal algorithm (6.3.11), (6.3.12), and also L-group algorithm (6.3.14) based on the estimator of MSI matrix (6.3.6), respectively. Useful signal $s(t)$ is shown by solid line, the results of filtering $y(t)$, $z(t)$ are shown by dashed line, pulse interference $u_3(t)$ is shown two times less than the original by dotted line. Correlation coefficients between useful signal $s(t)$ and the output signals $y(t)$, $z(t)$ take their values in the intervals $r[y(t), s(t)]=0.92\ldots0.97$ and $r[z(t), s(t)]=0.9\ldots0.95$, respectively. Coefficients of interference suppression take the values in the intervals $K_y=39\ldots46$ dB and $K_z=38\ldots45$ dB, correspondingly. Here coefficient of suppression is defined as a ratio $K_z = 10\lg(D/D[z(t) - s(t)])$, where D is a total variance of resulting interference, $D = \sum_{k=1}^{K} D_k$, D_k is a variance of k-th interference; $D[z(t) - s(t)]$ is a variance of residual noise component of the output signal $z(t)$. $\triangledown$

6.3.2 Spatial Filtering in Circular Antenna Array: L-group Algorithm

Consider spatial filtering problem solved by circular antenna array (see Fig. 6.3.1).

Let $a_i(t)$ be additive mixture of useful signal $s(t)$, M complex narrowband interference signals $u_l(t)$ (each arriving from l-th point source) and complex quasi-white Gaussian noise $n(t)$ with zero mean in the input of i-th receiving channel:

$$a_0(t) = s(t, \omega_s) + \sum_{l=1}^{M} u_l(t, \omega_l) + n_0(t); \qquad (6.3.15a)$$

$$a_{i_I}(t) = s_{i_I}(t) + \sum_{l=1}^{M} u_{l,i_I}(t) + n_{i_I}(t)$$

$$= s(t, \omega_s) \exp\left[-j\omega_s \tfrac{d}{c} C_I (\cos(\Delta\theta_I \cdot i_I) \cos(\theta_s) + \sin(\Delta\theta_I \cdot i_I) \sin(\theta_s))\right]$$

$$+ \sum_{l=1}^{M} u_l(t, \omega_l) \exp\left[-j\omega_l \tfrac{d}{c} C_I (\cos(\Delta\theta_I \cdot i_I) \cos(\theta_l) + \sin(\Delta\theta_I \cdot i_I) \sin(\theta_l))\right]$$

$$+ n_{i_I}(t); \quad (6.3.15b)$$

$$a_{i_{II}}(t) = s_{i_{II}}(t) + \sum_{l=1}^{M} u_{l,i_{II}}(t) + n_{i_{II}}(t)$$

$$= s(t, \omega_s) \exp\left[-j\omega_s \tfrac{d}{c} C_{II} (\cos(\Delta\theta_{II} \cdot i_{II}) \cos(\theta_s) + \sin(\Delta\theta_{II} \cdot i_{II}) \sin(\theta_s))\right]$$

$$+ \sum_{l=1}^{M} u_l(t, \omega_l) \exp\left[-j\omega_l \tfrac{d}{c} C_{II} (\cos(\Delta\theta_{II} \cdot i_{II}) \cos(\theta_l) + \sin(\Delta\theta_{II} \cdot i_{II}) \sin(\theta_l))\right]$$

$$+ n_{i_{II}}(t), \quad (6.3.15c)$$

where $i_I = 0, 1, \ldots, 5$; $i_{II} = 0, 1, \ldots, 11$ are indexes of the elements of internal and external circles of circular antenna array, respectively; C_I, C_{II} are coefficients determining radius of internal and external circles of circular antenna array, respectively: $C_I = 1$, $C_{II} = 2$; $\Delta\theta_I$, $\Delta\theta_{II}$ are angular measures between the adjacent elements of internal and external circles of circular antenna array, respectively: $\Delta\theta_I = \pi/3$, $\Delta\theta_{II} = \pi/6$; $t = t_m = t_0 + m \cdot \Delta t$; $m = 0, 1, \ldots, N-1$; Δt is a sampling interval; $s_i(t)$ is useful signal $s(t)$ observed in i-th channel of circular antenna array; $\omega_s = 2\pi f_s = \text{const}$, f_s is a carrier frequency of useful signal $s(t)$; θ_s is a direction of arrival of the signal $s(t)$; $\omega_l = 2\pi f_l = \text{const}$, f_l is a carrier frequency of interference signal $u_{l,i}(t)$ observed in i-th channel of circular antenna array; θ_l is a direction of arrival of interference signal $u_l(t)$; frequencies $\{f_l\}$ of interference signals $\{u_l(t)\}$ are different and uniformly distributed in a narrow interval around some central frequency f_0 coinciding with a carrier frequency f_s of useful signal $s(t)$: $f_0 = f_s$; $f_l \in [f_0 - \frac{\Delta F}{2}, f_0 + \frac{\Delta F}{2}]$, $\Delta F \ll f_0$; $n_i(t) = n_i^c(t) + j \cdot n_i^s(t)$ is noise observed in i-th channel of circular antenna array, $j = \sqrt{-1}$; $\mathbf{M}\{u_l(t)u_k(t)\} = 0$, $l \neq k$; $\mathbf{M}\{n_i(t)n_j(t)\} = 0$, $i \neq j$; $\mathbf{M}\{*\}$ is a symbol of mathematical expectation; $\mathbf{M}\{(n_i^c(t))^2\} = \mathbf{M}\{(n_i^s(t))^2\} = D_n/2$, D_n is a variance of complex quasi-white Gaussian noise $n_i(t)$ with zero mean; d is interelement distance of antenna array; c is velocity of electromagnetic waves propagation.

According to spatial filtering algorithm (6.3.14), the observed processes $a_i(t)$ in the inputs of i-th elements of antenna array (6.3.15a,b,c) are used for forming a vector $\mathbf{a}_i = [a_i(t_0), a_i(t_1), \ldots, a_i(t_{N-1})]^T$ composed of signal samples $a_i(t_j)$ (N is a number of signal samples $a_i(t_j)$ used in processing), so that the following relationships hold:

$$\mathbf{a}_i \equiv \mathbf{a}_0 = [a_0(t_0), a_0(t_1), \ldots, a_0(t_{N-1})]^T, \ i = 1; \qquad (6.3.16a)$$

$$\mathbf{a}_i \equiv \mathbf{a}_{i_I} = [a_{i_I}(t_0), a_{i_I}(t_1), \ldots, a_{i_I}(t_{N-1})]^T; \qquad (6.3.16b)$$

$$i = 2, 3, \ldots, 7; \ i_I = i - 2;$$

$$\mathbf{a}_i \equiv \mathbf{a}_{i_{II}} = [a_{i_{II}}(t_0), a_{i_{II}}(t_1), \ldots, a_{i_{II}}(t_{N-1})]^T; \qquad (6.3.16c)$$

$$i = 8, 9, \ldots, 19; \ i_{II} = i - 8.$$

Vector $\mathbf{e}(\theta_s, f_s) = [\ e_1(\theta_s, f_s) \quad e_2(\theta_s, f_s) \quad \cdots \quad e_n(\theta_s, f_s)\ ]^T$ of complex spatial harmonics (steering vector) for a circular antenna array is formed according to the observed signals $a_i(t)$ in the input of i-th element of array (6.3.15):

$$e_i(\theta_s, f_s) = 1, \ i = 1; \qquad (6.3.17a)$$

$$e_i(\theta_s, f_s) = e^{-j2\pi f_s \frac{d}{c} C_I(\cos(\Delta\theta_I \cdot (i-2))\cos(\theta_s) + \sin(\Delta\theta_I \cdot (i-2))\sin(\theta_s))}, \qquad (6.3.17b)$$

$$i = 2, 3, \ldots, 7;$$

$$e_i(\theta_s, f_s) = e^{-j2\pi f_s \frac{d}{c} C_{II}(\cos(\Delta\theta_{II} \cdot (i-8))\cos(\theta_s) + \sin(\Delta\theta_{II} \cdot (i-8))\sin(\theta_s))}, \qquad (6.3.17c)$$

$$i = 8, 9, \ldots, 19.$$

Taking into account aforementioned features of forming observation vector $\{\mathbf{a}_i\}$ (6.3.16) and steering vector $\mathbf{e}(\theta_s, f_s)$ (6.3.17) for the circular antenna array with configuration shown in Fig. 6.3.1, spatial filtering algorithm (6.3.14) based on MSI matrix estimator (6.3.6) stays unchanged:

$$z(t) = \sum_{i=1}^{n} \bar{w}_{L,i} a_i(t); \qquad (6.3.18)$$

$$\mathbf{w}_L = [w_{L,1}, \ldots, w_{L,n}]^T = \frac{\hat{\mathbf{N}}_a^{-1} \mathbf{e}(\theta_s, f_s)}{\mathbf{e}(\theta_s, f_s)^T \hat{\mathbf{N}}_a^{-1} \mathbf{e}(\theta_s, f_s)} \qquad (6.3.18a)$$

$$\hat{\mathbf{N}}_a^{-1} = \sum_{i=1}^{n} \frac{1}{|\rho_i^s|} \mathbf{u}_i^s \overline{\mathbf{u}_i^s}^T, \qquad (6.3.18b)$$

where $t = t_m = t_0 + m \cdot \Delta t$; $m = 0, 1, \ldots, N - 1$; $\hat{\mathbf{N}}_a$ is MSI matrix estimator $\hat{\mathbf{N}}_{a,s}$ modified according to (6.3.14b); $\mathbf{e}(\theta_s, f_s)$ is vector of complex spatial harmonics (steering vector) defined by the relationships (6.3.17).

Spatial filtering algorithm (6.3.18) is illustrated by the following example.

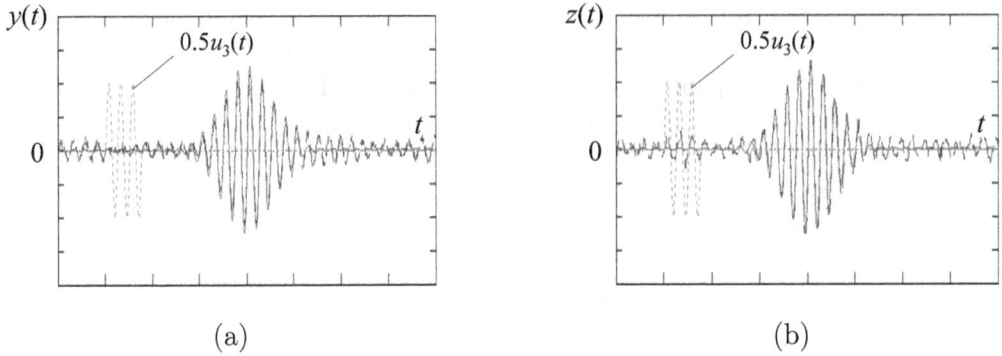

(a)   (b)

FIGURE 6.3.4 Estimates of useful signal $s(t)$ formed in circular antenna array according to spatial filtering algorithms: (a) $y(t) = \hat{s}(t)$ (6.3.11), (6.3.12) (classic optimal algorithm); (b) $z(t) = \hat{s}(t)$ (6.3.18) (*L*-group algorithm)

Example 6.3.2. Consider 19-elements circular antenna array (see Fig. 6.3.1) that receives additive mixture of signals (6.3.15). The latter contains: useful signal $s(t)$ arriving from the direction $\theta_s = -\pi/18$, interference signals $u_1(t)$, $u_2(t)$, $u_3(t)$, $u_4(t)$, $u_5(t)$ arriving from the directions $\theta_1 = -\pi/3$; $\theta_2 = \pi/10$; $\theta_3 = 3\pi/14$; $\theta_4 = 7\pi/18$; $\theta_5 = -3\pi/14$, respectively (see Fig. 6.3.2; 6.1.2, 6.1.3), and also quasi-white Gaussian noise $n(t)$. Interference signals $u_1(t)$, $u_2(t)$, $u_3(t)$ are amplitude-modulated interference, frequency-modulated interference, and also RF pulse with rectangular envelope, respectively; and statistically independent interferences $u_4(t)$, $u_5(t)$ are quasi-white Gaussian noises. Signal-to-k-th interference ratios in each separate i-th receiving channel of antenna array take the values that are closely equal to 0.25; 0.25; 1; 0.05; 0.05, respectively. Signal-to-noise ratio in receiving channels is equal to 10^3. Number of samples N of signals $a_i(t)$ used in processing (6.3.14) is chosen to be equal $N=1024$. Interelement distance d of linear antenna array is equal $d = 0.5\lambda_0$, where λ_0 is wave length of useful signal $s(t)$.

Fig. 6.3.4a,b illustrate the estimates $y(t) = \hat{s}(t)$, $z(t) = \hat{s}(t)$ of useful signal $s(t)$ formed in circular antenna array according to classic optimal algorithm (6.3.11), (6.3.12) and also L-group algorithm (6.3.18) based on MSI matrix estimator (6.3.6), respectively. Useful signal $s(t)$ is shown by solid line, the filtering results $y(t)$, $z(t)$ are shown by dashed line, pulse interference $u_3(t)$ is shown two times less than the original by dotted line. Correlation coefficients between useful signal $s(t)$ and the output signals $y(t)$, $z(t)$ take their values the intervals $r[y(t), s(t)]$=0.9...0.97 and $r[z(t), s(t)]$=0.88...0.95, respectively. Coefficients of interference suppression take the values in the intervals K_y=39...46 dB and K_z=38...45 dB, correspondingly. Here coefficient of suppression is defined as a ratio $K_z = 10\lg(D/D[z(t) - s(t)])$, where D is a total variance of resulting interference, $D = \sum_{k=1}^{K} D_k$, D_k is a variance of k-th interference; $D[z(t) - s(t)]$ is a variance of residual noise component of the output signal $z(t)$. $\bigtriangledown$

6.4 Direction of Arrival Estimation Based on *L*-group Operations

Known methods of analysis and estimation of spatial spectrum are well stated in numerous publications [141, 148–153, 162, 164–172]. An energy radiated from wave process sources could have different origin, including electromagnetic, acoustic, etc. Antenna system transforms obtained wave energy into electric signals. Such processes take place in radar, sonar, information transmitting systems, and also in radio astronomy, seismology, etc. The problems of detection and classification within some set of signals in both time, frequency, and spatial domains are directly related with estimation problem in the corresponding domain of observations. In this section we consider alternative spatial spectrum estimation algorithms based on *L*-group operations, that being analogues of known algorithms possess similar frequency resolution and are realized by lesser computational resources. Multichannel antenna systems (arrays) provide spatial and temporal sampling of spatial-temporal signals for their further processing. All methods of spatial spectrum estimation based on *L*-groups discussed in this section are stated in comparison with their known linear analogues based on known methods considered in Chapter 5 with respect to discrete time sequences (discrete signals). The considered signal processing algorithms are based on measure of statistical interrelation (MSI) (1.4.11) in sample space with semimetric (1.3.26).

Let $a_i(t)$ be additive mixture of M complex narrowband signals $s_l(t)$, each arriving from *l*-th point source, and complex quasi-white Gaussian noise $n(t)$ with zero mean in the input of *i*-th receiving channel of linear antenna array (see Fig. 6.1.7a):

$$a_i(t) = \sum_{l=1}^{M} s_{l,i}(t) + n_i(t)$$

$$= \sum_{l=1}^{M} A_l \exp[j\omega_l t + \varphi_l] \exp\left(-j\omega_l \frac{d}{c} \sin\theta_l (i-1)\right) + n_i(t), \quad (6.4.1)$$

where $i = 1,\ldots,n$ is an index of receiving channel of antenna array; $t = t_m = t_0 + m \cdot \Delta t$; $m = 0, 1, \ldots, N-1$; Δt is a sampling interval; $A_l = \text{const}$, $\omega_l = 2\pi f_l = \text{const}$, A_l, f_l are amplitude and carrier frequency of a signal $s_{l,i}(t)$ observed in *i*-th channel of linear antenna array; frequencies $\{f_l\}$ of the signals $s_l(t)$ are different and distributed in a narrow interval around some central frequency f_c: $f_l \in [f_c - \frac{\Delta F}{2}, f_c + \frac{\Delta F}{2}]$, $\Delta F \ll f_c$; φ_l is unknown initial phase belonging to the interval $\varphi_l \in [0, 2\pi]$; $n_i(t) = n_i^c(t) + j \cdot n_i^s(t)$ is noise observed in *i*-th channel of linear antenna array, $j = \sqrt{-1}$; $\mathbf{M}\{u_l(t)u_k(t)\} = 0, l \neq k$; $\mathbf{M}\{n_i(t)n_j(t)\} = 0, i \neq j$; $\mathbf{M}\{*\}$ is a symbol of mathematical expectation; $\mathbf{M}\{(n_i^c(t))^2\} = \mathbf{M}\{(n_i^s(t))^2\} = D_n/2$, D_n is a variance of complex quasi-white Gaussian noise $n_i(t)$ with zero mean; θ_l is a direction of arrival of the signal $s_l(t)$; d is interelement distance of antenna array; c is velocity of electromagnetic wave propagation.

In this section, we consider spatial spectrum estimation methods based on matrix of MSI (1.4.11) concerned with semimetric (1.3.26).

Relationships (6.3.6)...(6.3.9) create the basis for antenna array signal processing (including spatial spectrum estimation) using L-group algorithms based on MSI (1.4.11). In this section we use them to formulate corresponding signal processing algorithms.

6.4.1 Minimum Variance Direction of Arrival Estimation Based on Measure of Statistical Interrelation Matrix Estimator

Minimum variance (MV) estimator of spatial spectrum was introduced by J. Capon [148] for solving the problem of spatial-temporal analysis of multivariable signals in antenna arrays. Spectral MV estimator $\hat{P}_{MV}(\theta)$ is defined by the expression (see, for instance, [140, (12.1)]):

$$\hat{P}_{MV}(\theta) = \frac{1}{\overline{e(\theta)^T} \hat{R}_a^{-1} e(\theta)};\qquad (6.4.2)$$

$$\mathbf{e}(\theta) = [1, \exp[-j\omega_c \tfrac{d}{c} \sin\theta], \ldots, \exp[-j\omega_c \tfrac{d}{c}(K-1)\sin\theta]]^T, \qquad (6.4.2a)$$

where θ is a discrete parameter of angle of signal arrival; $\hat{R}_a$ is autocorrelation matrix (ACM) estimator (6.3.2) of vectors $\{\mathbf{a}_i\}$ composed of signal samples $a_i(t)$ (6.4.1); $\mathbf{e}(\theta)$ is vector of complex spatial harmonics; $\overline{\mathbf{e}(\theta)^T}$ is conjugate transpose of initial vector $\mathbf{e}(\theta)$; $\omega_c = 2\pi f_c = \text{const}$, f_c is a central frequency around which frequencies of the signals $s_{l,i}(t)$ are distributed; $l = 1, \ldots, M$; $i = 1, \ldots, n$; d is interelement distance of antenna array; c is velocity of electromagnetic wave propagation; $K \times K$ is a dimensionality of matrix $\hat{R}_a$, $K = n$; n is a number of antenna array elements (receiving channels).

Using known decomposition of Hermitian matrix (ACM) $\hat{R}_a$ over eigenvectors $\{\mathbf{v}_k\}$ and eigenvalues $\{\lambda_k\}$ (see, for instance, [140, (3.86), (3.88)]):

$$\hat{R}_a = \sum_{i=1}^{K} \lambda_i \mathbf{v}_i \overline{\mathbf{v}_i^T};\qquad \hat{R}_a^{-1} = \sum_{i=1}^{K} \frac{1}{\lambda_i} \mathbf{v}_i \overline{\mathbf{v}_i^T}, \qquad (6.4.3)$$

MV estimator (6.4.2) can be represented in the following form:

$$\hat{P}_{MV}(\theta) = \frac{1}{\overline{\mathbf{e}(\theta)^T} \left(\sum\limits_{i=1}^{K} \frac{1}{\lambda_i} \mathbf{v}_i \overline{\mathbf{v}_i^T} \right) \mathbf{e}(\theta)}.$$

Taking into account known relation between normalized MSI (or MSI) and correlation coefficient (or scalar product, respectively) of a pair of signals considered in Chapter 1, by the analogy, one can obtain MV estimator of spatial hyperspectrum[†] $\hat{H}_{MV}^s(\theta)$ (MV L-group estimator) based on MSI (1.4.11) that is defined by the following relationship:

$$\hat{H}_{MV}^s(\theta) = \frac{1}{\overline{\mathbf{e}(\theta)^T} \hat{N}_{a,s}^{-1} \mathbf{e}(\theta)}, \qquad (6.4.4)$$

[†]Hereinafter, we use term "hyperspectrum" instead of "spectrum" to put an emphasis on non-linearity of signal processing and spatial nonlinear characteristics of signals

where $\hat{\mathbf{N}}_{a,s}$ is MSI matrix estimator of vectors $\{\mathbf{a}_i\}$ composed of signal samples $a_i(t)$ (6.4.1), which is determined by the relationship (6.3.6); $\mathbf{e}(\theta)$ is vector of complex spatial harmonics determined by the formula (6.4.2a).

Since MSI matrix estimator $\hat{\mathbf{N}}_{a,s}$ is based on generalized metric (1.3.26) (namely, semimetric), direct using the formula (6.4.4) for obtaining spatial hyperspectrum estimator $\hat{H}_{MV}^s(\theta)$ can be accompanied by a considerable number of false peaks due to possible negative eigenvalues of matrix $\hat{\mathbf{N}}_{a,s}$. Thus, using decomposition (6.4.3), modified algorithm of MV estimator (6.4.4) takes the following form:

$$\hat{H}_{MV}^s(\theta) = \frac{1}{\overline{\mathbf{e}(\theta)^T} \left(\sum_{i=1}^{K} \frac{1}{|\rho_i^s|} \mathbf{u}_i^s \overline{\mathbf{u}_i^s}^T \right) \mathbf{e}(\theta)}, \tag{6.4.5}$$

where $\{\mathbf{u}_i^s\}$, $\{\rho_i^s\}$ are eigenvectors and eigenvalues of MSI matrix estimator $\hat{\mathbf{N}}_{a,s}$ (6.3.6), respectively; $\mathbf{e}(\theta)$ is vector of complex spatial harmonics determined by the formula (6.4.2a); M is a number of narrowband signals in the observed additive mixture (6.4.1); $K \times K$ is a dimensionality of the matrix $\hat{\mathbf{R}}_a$, $K = n$.

Having obtained algorithm (6.4.5) of calculating MV estimator of spatial hyperspectrum $\hat{H}_{MV}^s(\theta)$ that is based on MSI (1.4.11), it is necessary, first, to find out how closely this algorithm is with respect to classic MV estimator (6.4.2), and second, is it possible to use it for joint signal detection and DOA estimation. Algorithms (6.4.2) and (6.4.5) are illustrated by the following example.

Example 6.4.1. Let $a_i(t)$ be additive mixture (6.4.1) of M complex narrowband signals $s_l(t)$ (M=8) (each of them arrives from l-th point source), and complex quasi-white Gaussian noise $n(t)$ with zero mean in the input of i-th receiving channel of linear antenna array (see Fig. 6.1.7a).

In the considered example, amplitudes of narrowband signals $s_l(t)$ are chosen to be equal to each other $A_l = A$, so that signal-to-noise ratios $A^2/(2D_n)$ in each i-th channel of antenna array are equal to 20 dB, D_n is a variance of noise $n_i(t)$; $F_n > 2\max_l\{f_l\}$, where F_n is upper bound frequency of PSD of quasi-white Gaussian noise $n_i(t)$. The number of samples N of stochastic process $a_i(t)$ (6.4.1) used when forming ACM estimator $\hat{\mathbf{R}}_a$ (6.3.2) and MSI matrix estimator $\hat{\mathbf{N}}_{a,s}$ (6.3.6) is equal to N=2048. The number n of antenna array elements is equal to $n = 31$, and correspondingly, dimensionality $K \times K$ of MSI matrix $\hat{\mathbf{N}}_{a,s}$ (6.3.6) and ACM $\hat{\mathbf{R}}_a$ (6.3.2) are equal to $K = n = 31$. The signals $s_l(t)$ arrive on the aperture of antenna array from the directions: $\theta_1 = -32°$, $\theta_2 = -30°$; $\theta_3 = -15°$, $\theta_4 = -10°$; $\theta_5 = 13.85°$, $\theta_6 = 10.85°$; $\theta_7 = 26°$, $\theta_8 = 30°$.

Fig. 6.4.1a illustrates spatial spectrum MV estimate $\hat{P}_{MV}(\theta)$ of the mixture $a_i(t)$ (6.4.1) obtained as a result of calculation (6.4.2) for initial situation described above. True positions of angles $\{\theta_l\}$ of arrival of the signals $\{s_l(t)\}$ from the mixture $a_i(t)$ (6.4.1) are shown by the signs "+". Fig. 6.4.1b depicts spatial spectrum MV estimate $\hat{H}_{MV}^s(\theta)$ (6.4.5) of the mixture $a_i(t)$ (6.4.1) for the situation described above. Correlation coefficient between MV estimators $\hat{H}_{MV}^s(\theta)$ (6.4.5) and $\hat{P}_{MV}(\theta)$ (6.4.2) takes the value $r[\hat{H}_{MV}^s(\theta), \hat{P}_{MV}(\theta)]$=0.996, while correlation coefficient between MSI matrix estimator $\hat{\mathbf{N}}_{a,s}$ (6.3.6) and ACM estimator $\hat{\mathbf{R}}_a$ (6.3.2) accept the

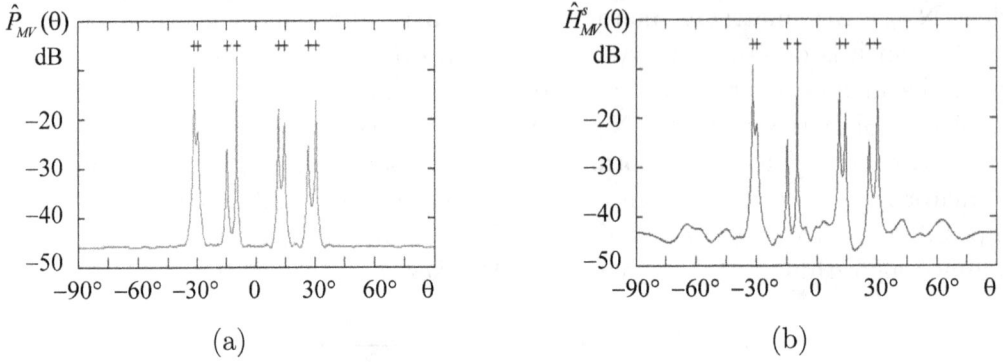

FIGURE 6.4.1   MV estimates of spatial spectrum: (a) $\hat{P}_{MV}(\theta)$ (6.4.2); (b) $\hat{H}^s_{MV}(\theta)$ (6.4.5)

value $r[\hat{\mathbf{N}}_{a,s}, \hat{\mathbf{R}}_a]=0.996$ that indicates high degree of the correspondence between the results obtained in both linear signal space (Fig. 6.4.1a) and signal space with L-group properties (Fig. 6.4.1b). Notice, that normalized MSI (1.4.6) and (1.4.7) between MV estimators $\hat{H}^s_{MV}(\theta)$ (6.4.5) and $\hat{P}_{MV}(\theta)$ (6.4.2) are, respectively, equal:

$$\nu_p[\hat{H}^s_{MV}(\theta) - \mathbf{M}[\hat{H}^s_{MV}(\theta)], \hat{P}_{MV}(\theta) - \mathbf{M}[\hat{P}_{MV}(\theta)]] = 0.99;$$

$$\nu_s[\hat{H}^s_{MV}(\theta) - \mathbf{M}[\hat{H}^s_{MV}(\theta)], \hat{P}_{MV}(\theta) - \mathbf{M}[\hat{P}_{MV}(\theta)]] = 0.999,$$

where $\mathbf{M}[*]$ is sample mean function. $\triangledown$

6.4.2   Direction of Arrival Estimation Methods Based on Eigenvectors of Measure of Statistical Interrelation Matrix Estimator

Eigenvector method [149], MUSIC method [150], minimum norm method [151], and ESPRIT method [152, 153] can be referred to the class of spatial spectrum estimation methods based on analysis of ACM estimator eigenvectors. Return to the model of additive interaction of M narrowband signals $a_i(t)$ in the presence of complex quasi-white Gaussian noise (6.4.1) that can be rewritten in the following vector form:

$$\mathbf{a}_i = \sum_{l=1}^{M} \mathbf{s}_{l,i} + \mathbf{n}_i, \tag{6.4.6}$$

where vectors of the signals are represented by their samples:

$$\mathbf{a}_i = [a_i(t_0), a_i(t_1), \ldots, a_i(t_{N-1})]^T;$$

$$\mathbf{s}_{l,i} = [s_{l,i}(t_0), s_{l,i}(t_1), \ldots, s_{l,i}(t_{N-1})]^T;$$

$$\mathbf{n}_i = [n_i(t_0), n_i(t_1), \ldots, n_i(t_{N-1})]^T;$$

$$i = 1, 2, \ldots, n;\ l = 1, 2, \ldots, M;\ m = 0, 1, \ldots, N-1.$$

Then $K - 1 - M$ eigenvectors $\mathbf{v}_M, \ldots, \mathbf{v}_{K-1}$ of *noise subspace* of ACM $\hat{\mathbf{R}}_a$

(6.3.2) are orthogonal with respect to vectors of narrowband signals $\{\mathbf{s}_{l,i}\}$ at the corresponding angles of arrival, therefore their linear combination is equal to zero:

$$\overline{\mathbf{e}(\theta_l)^T} \left(\sum_{i=M+1}^{K} \alpha_i \mathbf{v}_i \overline{\mathbf{v}_i^T} \right) \mathbf{e}(\theta_l) = 0; \tag{6.4.7}$$

$$\mathbf{e}(\theta) = [1, \exp[-j\omega_c \tfrac{d}{c} \sin \theta], \dots, \exp[-j\omega_c \tfrac{d}{c}(K-1) \sin \theta]]^T, \tag{6.4.7a}$$

where θ is a discrete parameter of angle of signal arrival; $\mathbf{e}(\theta)$ is vector of complex spatial harmonics; $\overline{\mathbf{e}(\theta)^T}$ is conjugate transpose of initial vector $\mathbf{e}(\theta)$; $\omega_c = 2\pi f_c = $ const, f_c is a central frequency around which frequencies of the signals $s_{l,i}(t)$ are distributed; $l = 1, \dots, M$; $i = 1, \dots, n$; d is interelement distance of antenna array; c is velocity of electromagnetic wave propagation; θ_l is a direction of arrival of the signal $s_l(t)$; $\{\alpha_k\}$ are arbitrary constants; $K \times K$ is a dimensionality of ACM $\hat{\mathbf{R}}_a$, $K = n$.

This means that the following estimator function

$$\hat{P}(\theta) = \frac{1}{\overline{\mathbf{e}(\theta)}^T \left(\sum_{i=M+1}^{K} \alpha_i \mathbf{v}_i \overline{\mathbf{v}_i^T} \right) \mathbf{e}(\theta)} \tag{6.4.8}$$

takes infinite value at the angle $\theta = \theta_l$ of arrival of the signal $s_l(t)$ (6.4.1). In practice, owing to the presence of noise, the function (6.4.8) takes the finite values at the angles $\theta = \theta_l$ possessing rather sharp peaks.

6.4.2.1   Direction of Arrival Estimation by Eigenvector Method Based on Measure of Statistical Interrelation Matrix Estimator

Using known decomposition of Hermitian matrix (ACM) $\hat{\mathbf{R}}_a$ over eigenvectors $\{\mathbf{v}_k\}$ and eigenvalues $\{\lambda_k\}$ (see, for instance, [140, (3.86), (3.88)]):

$$\hat{\mathbf{R}}_a = \sum_{i=1}^{K} \lambda_i \mathbf{v}_i \overline{\mathbf{v}_i^T}; \quad \hat{\mathbf{R}}_a^{-1} = \sum_{i=1}^{K} \frac{1}{\lambda_i} \mathbf{v}_i \overline{\mathbf{v}_i^T},$$

also drawing a parallel with MV estimator $\hat{P}_{MV}(\theta)$ (6.4.2) and assuming in the formula (6.4.8) $\alpha_i = 1/\lambda_i$, we obtain spatial spectrum eigenvector (EV) estimator based on eigenvectors taking into account eigenvalues of ACM estimator $\hat{\mathbf{R}}_a$ (6.3.2):

$$\hat{P}_{EV}(\theta) = \frac{1}{\overline{\mathbf{e}(\theta)^T} \left(\sum_{i=M+1}^{K} \frac{1}{\lambda_i} \mathbf{v}_i \overline{\mathbf{v}_i^T} \right) \mathbf{e}(\theta)}, \tag{6.4.9}$$

where, remind, $\mathbf{e}(\theta)$ is vector of complex spatial harmonics determined by the formula (6.4.2a); M is a number of narrowband signals in the observed additive mixture (6.4.1); $K \times K$ is a dimensionality of ACM estimator $\hat{\mathbf{R}}_a$, $K = n$.

Using known relation between normalized MSI (or MSI) and correlation coefficient (or scalar product) of a pair of signals considered in Chapter 1, by the

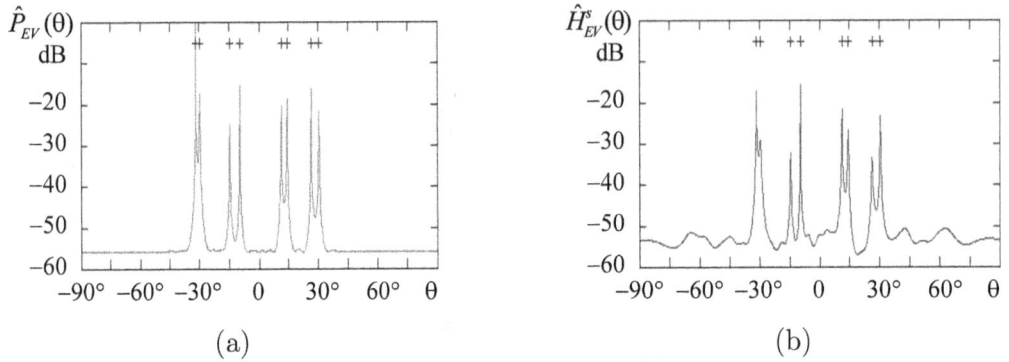

FIGURE 6.4.2   EV estimates of spatial spectrum: (a) $\hat{P}_{EV}(\theta)$ (6.4.9); (b) $\hat{H}_{EV}^s(\theta)$ (6.4.10)

analogy with EV estimator (6.4.9), one can obtain spatial hyperspectrum EV estimator $\hat{H}_{EV}^s(\theta)$ based on MSI (1.4.11) that is defined by the following relationship:

$$\hat{H}_{EV}^s(\theta) = \frac{1}{\mathbf{e}(\theta)^T \left(\sum_{i=M+1}^{K} \frac{1}{|\rho_i^s|} \mathbf{u}_i^s \overline{\mathbf{u}_i^s}^T \right) \mathbf{e}(\theta)}, \qquad (6.4.10)$$

where $\{\mathbf{u}_i^s\}$, $\{\rho_i^s\}$ are eigenvectors and eigenvalues of MSI matrix estimator $\hat{\mathbf{N}}_{a,s}$ (6.3.6), respectively; $\mathbf{e}(\theta)$ is vector of complex spacial harmonics determined by the formula (6.4.7a); M is a number of narrowband signals in the observed additive mixture (6.4.1); $K \times K$ is a dimensionality of MSI matrix estimator $\hat{\mathbf{N}}_{a,s}$, $K = n$.

Having obtained the algorithm (6.4.10) of calculating spatial hyperspectrum EV estimator $\hat{H}_{EV}^s(\theta)$ based on MSI (1.4.11), it is necessary, first, to find out, how close this estimator is with respect to EV estimator (6.4.9), and second, can it be used for estimating spatial spectrum of a mixture between narrowband signals and noise. Operation of the algorithms (6.4.9) and (6.4.10) is illustrated by the following example.

Example 6.4.2. Let $a_i(t)$ be additive mixture (6.4.1) of M complex narrowband signals $s_l(t)$ ($M{=}8$) (each arriving from l-th point source) and complex quasi-white Gaussian noise $n_i(t)$ with zero mean (6.4.1) in the input of i-th receiving channel of linear antenna array (see Fig. 6.1.7a), so that conditions pointed in preamble of Example 6.4.1 hold.

Fig. 6.4.2a depicts EV estimate of spatial spectrum $\hat{P}_{EV}(\theta)$ of the mixture $a_i(t)$ (6.4.1) obtained as a result of calculating the relationship (6.4.9) for initial situation described above. True positions of angles $\{\theta_l\}$ of arrival of the signals $\{s_l(t)\}$ from the mixture $a_i(t)$ (6.4.1) are denoted by the signs "+". Fig. 6.4.2b illustrates EV estimate of spatial hyperspectrum $\hat{H}_{EV}^s(\theta)$ (6.4.10) of the mixture $a_i(t)$ (6.4.1) for initial situation described above. Correlation coefficient between EV estimators $\hat{H}_{EV}^s(\theta)$ (6.4.10) and $\hat{P}_{EV}(\theta)$ (6.4.9) takes the value $r[\hat{H}_{EV}^s(\theta), \hat{P}_{EV}(\theta)]{=}0.578$ that can be explained by insignificant level of linear statistical interrelations between the results of processing, whereas correlation coefficient between MSI matrix estimator $\hat{\mathbf{N}}_{a,s}$ (6.3.6) and ACM estimator $\hat{\mathbf{R}}_a$ (6.3.2) takes the value $r[\hat{\mathbf{N}}_{a,s}, \hat{\mathbf{R}}_a]{=}0.996$.

Notice, that normalized MSI (1.4.6) and (1.4.7) between EV estimators $\hat{H}^s_{EV}(\theta)$ (6.4.10) and $\hat{P}_{EV}(\theta)$ (6.4.9) are, respectively, equal to:

$$\nu_p[\hat{H}^s_{EV}(\theta) - \mathbf{M}[\hat{H}^s_{EV}(\theta)], \hat{P}_{EV}(\theta) - \mathbf{M}[\hat{P}_{EV}(\theta)]] = 0.912;$$

$$\nu_s[\hat{H}^s_{EV}(\theta) - \mathbf{M}[\hat{H}^s_{EV}(\theta)], \hat{P}_{EV}(\theta) - \mathbf{M}[\hat{P}_{EV}(\theta)]] = 0.921,$$

where $\mathbf{M}[*]$ is sample mean function.

The last relationships indicate high degree correspondence between the results obtained in linear space and signal space with L-group properties (Fig. 6.4.2a,b), that can be explained by high identity of MSI matrix estimator $\hat{\mathbf{N}}_{a,s}$ (6.3.6) and ACM estimator $\hat{\mathbf{R}}_a$ (6.3.2), and also by the fact that nonlinear statistical interrelations between these estimators, unlike linear statistical interrelations, preserve more completely.   ▽

6.4.2.2 Direction of Arrival Estimation by MUSIC Method

Using known decomposition of Hermitian matrix (ACM) $\hat{\mathbf{R}}_a$ over eigenvectors $\{\mathbf{v}_i\}$ and eigenvalues $\{\lambda_i\}$ (see, for instance, [140, (3.86), (3.88)]), also drawing a parallel with EV estimator $\hat{P}_{EV}(\theta)$ (6.4.9) and assuming in the formula (6.4.8) $\alpha_i = 1$ for $\forall i$, we obtain spatial spectrum MUSIC estimator $\hat{P}_{MUSIC}(\theta)$ based on eigenvectors of ACM estimator $\hat{\mathbf{R}}_a$ (6.3.2):

$$\hat{P}_{MUSIC}(\theta) = \frac{1}{\overline{\mathbf{e}(\theta)^T} \left(\sum_{i=M+1}^{K} \mathbf{v}_i \overline{\mathbf{v}_i^T} \right) \mathbf{e}(\theta)} = \frac{1}{\overline{\mathbf{e}(\theta)^T} \left(\mathbf{V}' \overline{\mathbf{V}'^T} \right) \mathbf{e}(\theta)}, \qquad (6.4.11)$$

where, repeat, $\mathbf{e}(\theta)$ is vector of complex spatial harmonics determined by the formula (6.4.2a); M is a number of narrowband signals in the observed additive mixture (6.4.1); $K \times K$ is a dimensionality of ACM estimator $\hat{\mathbf{R}}_a$, $K = n$; $\mathbf{V}'$ is a matrix related to a noise subspace obtained as the result of truncating the matrix $\mathbf{V} = \|\mathbf{v}_1, \mathbf{v}_2, \dots, \mathbf{v}_K\|$ of dimensionality $K \times K$ composed of eigenvectors $\{\mathbf{v}_i\}$ of ACM estimator $\hat{\mathbf{R}}_a$ (6.3.2):

$$\mathbf{V}' = \text{submatrix}(\mathbf{V}, 1, K, M+1, K) = \|\mathbf{v}'_{M+1}, \mathbf{v}'_{M+2}, \dots, \mathbf{v}'_K\|,$$

where $\text{submatrix}(\mathbf{A}, r_i, r_j, c_i, c_j)$ is operation of extracting a submatrix containing the elements in rows r_i trough r_j and columns c_i trough c_j from the initial matrix $\mathbf{A}$; eigenvector matrix $\mathbf{V} = \|\mathbf{V}_{i,k}\|$ has a dimensionality $K \times K$, $i = 1, \dots, K$, $j = 1, \dots, K$; matrix $\mathbf{V}' = \|\mathbf{v}'_{M+1}, \mathbf{v}'_{M+2}, \dots, \mathbf{v}'_K\|$ has a dimensionality $K \times (K-M)$; M is expected number of narrowband signals in the mixture (6.4.1), $M < K$.

Taking into account known relation between normalized MSI (or MSI) and correlation coefficient (or scalar product) of a pair of signals considered in Chapter 1, by the analogy with MUSIC estimator (6.4.11), one can obtain spatial spectrum MUSIC estimator $\hat{H}^s_{MUSIC}(\theta)$ based on MSI (1.4.11) (spatial hyperspectrum MUSIC estimator based on MSI) that is defined by the following relationship:

$$\hat{H}^s_{MUSIC}(\theta) = \frac{1}{\overline{\mathbf{e}(\theta)^T} \left(\sum_{i=M+1}^{K} \mathbf{u}^s_i \overline{\mathbf{u}^s_i}^T \right) \mathbf{e}(\theta)} = \frac{1}{\overline{\mathbf{e}(\theta)^T} \left(\mathbf{U}'_s \overline{\mathbf{U}'^T_s} \right) \mathbf{e}(\theta)}, \qquad (6.4.12)$$

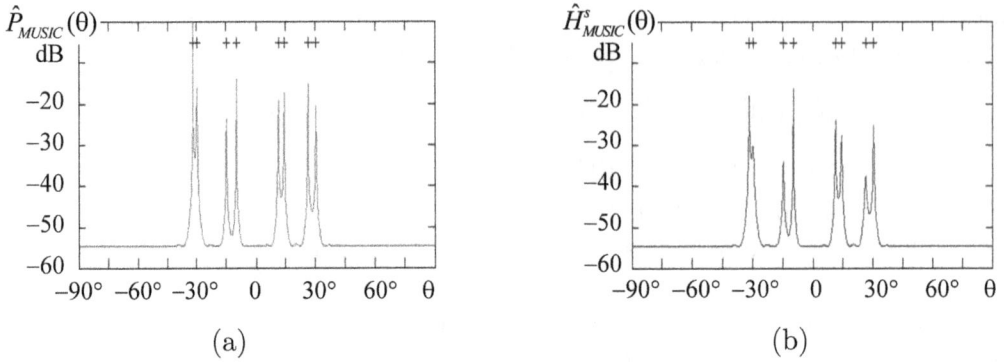

(a)

(b)

FIGURE 6.4.3   MUSIC estimates of spatial spectrum: (a) $\hat{P}_{MUSIC}(\theta)$ (6.4.11); (b) $\hat{H}_{MUSIC}^s(\theta)$ (6.4.12)

where $\{\mathbf{u}_i^s\}$ are eigenvectors of MSI matrix estimator $\hat{\mathbf{N}}_{a,s}$ (6.3.6); $\mathbf{e}(\theta)$ is vector of complex spatial harmonics determined by the formula (6.4.2a); M is a number of narrowband signals in the observed additive mixture (6.4.1); $K \times K$ is a dimensionality of MSI matrix estimator $\hat{\mathbf{N}}_{a,s}$; $\mathbf{U}'_s$ is a noise subspace matrix with a dimensionality $K \times (K - M)$ obtained by truncating the matrix $\mathbf{U}_s = \|\mathbf{u}_{s,1}, \mathbf{u}_{s,2}, \ldots, \mathbf{u}_{s,K}\|$ composed of eigenvectors of MSI matrix estimator $\hat{\mathbf{N}}_{a,s}$ (6.3.6):

$$\mathbf{U}'_s = \text{submatrix}(\mathbf{U}_s, 1, K, M + 1, K) = \|\mathbf{u}'_{s,M+1}, \mathbf{u}'_{s,M+2}, \ldots, \mathbf{u}'_{s,K}\|.$$

Having obtained the algorithm (6.4.12) of calculating MUSIC estimator of spatial hyperpectrum $\hat{H}_{MUSIC}^s(\theta)$ based on MSI (1.4.11), it is necessary, first, to find out, how close this estimator is with respect to MUSIC estimator (6.4.11), and second, could it be used for estimating spatial spectrum of the mixture of narrowband signals and noise. Algorithms (6.4.11) and (6.4.12) are illustrated by the following example.

Example 6.4.3. Let $a_i(t)$ be additive mixture (6.4.1) of M complex narrowband signals $s_l(t)$ (M=8) (each arriving from l-th point source) and complex quasi-white Gaussian noise $n_i(t)$ with zero mean in the input of i-th receiving channel of linear antenna array (see Fig. 6.1.7a), so that conditions pointed in preamble of Example 6.4.1 hold.

Fig. 6.4.3a illustrates MUSIC estimate of spatial spectrum $\hat{P}_{MUSIC}(\theta)$ of the mixture $a_i(t)$ (6.4.1) obtained by algorithm (6.4.11) for initial situation described above. True positions of the angles $\{\theta_l\}$ of arrival of the signals $\{s_l(t)\}$ from the mixture $a_i(t)$ (6.4.1) are shown by the signs "+". Fig. 6.4.3b depicts MUSIC estimate of spatial hyperspectrum $\hat{H}_{MUSIC}^s(\theta)$ (6.4.12) of the mixture $a_i(t)$ (6.4.1) for initial situation described above. Correlation coefficient between MUSIC estimates $\hat{H}_{MUSIC}^s(\theta)$ (6.4.12) and $\hat{P}_{MUSIC}(\theta)$ (6.4.11) takes the value $r[\hat{H}_{EV}^s(\theta), \hat{P}_{EV}(\theta)]$=0.541, that can be explained by insignificant level of linear statistical interrelations between the results of processing, whereas correlation coefficient between MSI matrix estimate $\hat{\mathbf{N}}_{a,s}$ (6.3.6) and ACM estimate $\hat{\mathbf{R}}_a$ (6.3.2) takes the value $r[\hat{\mathbf{N}}_{a,s}, \hat{\mathbf{R}}_a]$=0.996. Notice, that normalized MSIs (1.4.6) and (1.4.7) between MUSIC estimates $\hat{H}^s_{MUSIC}(\theta)$ (6.4.12) and $\hat{P}_{MUSIC}(\theta)$ (6.4.11) are, respectively, equal to:

$$\nu_p[\hat{H}^s_{MUSIC}(\theta) - \mathbf{M}[\hat{H}^s_{MUSIC}(\theta)], \hat{P}_{MUSIC}(\theta) - \mathbf{M}[\hat{P}_{MUSIC}(\theta)]] = 0.914;$$

$$\nu_s[\hat{H}^s_{MUSIC}(\theta) - \mathbf{M}[\hat{H}^s_{MUSIC}(\theta)], \hat{P}_{MUSIC}(\theta) - \mathbf{M}[\hat{P}_{MUSIC}(\theta)]] = 0.964,$$

where $\mathbf{M}[*]$ is sample mean function.

The last relationships indicate high degree correspondence between the results obtained in linear space and in signal space with L-group properties (Fig. 6.4.3a,b) that is explained by high degree identity between MSI matrix estimate $\hat{\mathbf{N}}_{a,s}$ (6.3.6) and ACM estimate $\hat{\mathbf{R}}_a$ (6.3.2), and also by a fact that nonlinear statistical interrelations between these estimators, unlike linear statistical interrelations, preserve more completely. $\quad\triangledown$

6.4.2.3 Direction of Arrival Estimation by Minimum Norm Method

Spatial spectrum estimation by minimum norm method [151], as well as MUSIC method, assumes using information contained in the result $\mathbf{V}'$ of truncating the matrix $\mathbf{V} = \|\mathbf{v}_1, \mathbf{v}_2, \ldots, \mathbf{v}_K\|$ with dimensionality $K \times K$ composed by eigenvectors $\{\mathbf{v}_i\}$ of ACM estimator $\hat{\mathbf{R}}_a$ (6.3.2) based on the additive mixture (6.4.1):

$$\mathbf{V}' = \text{submatrix}(\mathbf{V}, 1, K, M+1, K) = \|\mathbf{v}'_{M+1}, \mathbf{v}'_{M+2}, \ldots, \mathbf{v}'_K\|, \qquad (6.4.13)$$

where submatrix$(\mathbf{A}, r_i, r_j, c_i, c_j)$ is a function of extracting submatrix containing the elements in rows r_i trough r_j and columns c_i trough c_j from initial matrix $\mathbf{A}$; eigenvectors matrix $\mathbf{V} = \|\mathbf{V}_{i,k}\|$ has a dimensionality $K \times K$, $i = 1, \ldots, K$, $j = 1, \ldots, K$; matrix $\mathbf{V}' = \|\mathbf{v}'_{M+1}, \mathbf{v}'_{M+2}, \ldots, \mathbf{v}'_K\|$ has a dimensionality $K \times (K-M)$; M is an expected number of narrowband signals in the mixture (6.4.1), $M < K$.

Matrix $\mathbf{V}'$ relates to noise subspace in which vector $\mathbf{d}$ based on a matrix $\mathbf{V}'$

$$\mathbf{d} = \frac{\mathbf{V}' \cdot \overline{\mathbf{c}^T}}{\overline{\mathbf{c}^T}\mathbf{c}}, \qquad (6.4.14)$$

where $\mathbf{c}$ is vector composed of upper row of the matrix $\mathbf{V}'$:

$$\mathbf{c} = [\text{submatrix}(\mathbf{V}', 1, 1, 1, K - M)]^T,$$

has the least norm: $\|\mathbf{d}\|^2 = \sum_{i=1}^{K-M} d_i^2 \to \min.$

Then spatial spectrum estimator $\hat{P}_{MN}(\theta)$ obtained by minimum norm method has the form:

$$\hat{P}_{MN}(\theta) = \frac{1}{\mathbf{e}(\theta)^T \left(\mathbf{d} \cdot \overline{\mathbf{d}^T}\right) \mathbf{e}(\theta)}; \qquad (6.4.15)$$

$$\mathbf{e}(\theta) = [1, \exp[-j\omega_c \tfrac{d}{c} \sin\theta], \ldots, \exp[-j\omega_c \tfrac{d}{c}(K - M - 1)\sin\theta]]^T, \qquad (6.4.15a)$$

where, remind, $\mathbf{e}(\theta)$ is vector of complex spatial harmonics that is similar to (6.4.7a) but has another dimensionality; θ is a discrete parameter of angle of signal arrival; $\overline{\mathbf{e}(\theta)^T}$ is conjugate transpose of initial vector $\mathbf{e}(\theta)$; $\omega_c = 2\pi f_c = \text{const}$, f_c is a central frequency around which frequencies of the signals $s_{l,i}(t)$ are distributed; $l = 1, \ldots, M; i = 1, \ldots, n; d$ is interelement distance of antenna array; c is a velocity of electromagnetic wave propagation; θ_l is a DOA of the signal $s_l(t)$; $K \times K$ is a dimensionality of the matrix $\hat{\mathbf{R}}_a$, $K = n$.

Taking into account known relation between normalized MSI (or MSI) and correlation coefficient (or scalar product) of a pair of signals considered in Chapter 1, by the analogy with MN-estimator (6.4.15), one can obtain spatial hyperspectrum MN estimator $\hat{H}_{MN}^s(\theta)$ based on MSI (1.4.11) that is defined by the following relationship:

$$\hat{H}_{MN}^s(\theta) = \frac{1}{\overline{\mathbf{e}(\theta)^T}\left(\mathbf{g}_s\overline{\mathbf{g}_s^T}\right)\mathbf{e}(\theta)}; \tag{6.4.16}$$

$$\mathbf{g}_s = \frac{\mathbf{U}' \cdot \overline{\mathbf{a}_s^T}}{\mathbf{a}_s^T\mathbf{a}_s}; \tag{6.4.16a}$$

$$\mathbf{a}_s = [\text{submatrix}(\mathbf{U}'_s, 1, 1, 1, K - M)]^T; \tag{6.4.16b}$$

$$\mathbf{U}'_s = \text{submatrix}(\mathbf{U}_s, 1, K, M + 1, K) = \|\mathbf{u}'_{s,M+1}, \mathbf{u}'_{s,M+2}, \ldots, \mathbf{u}'_{s,K}\|, \tag{6.4.16c}$$

where $\mathbf{U}_s = \|\mathbf{u}_{s,1}, \mathbf{u}_{s,2}, \ldots, \mathbf{u}_{s,K}\|$ is a matrix composed of eigenvectors of MSI matrix estimator $\hat{\mathbf{N}}_{a,s}$ (6.3.6); $\mathbf{U}'_s$ is a matrix of noise subspace with dimensionality $K \times K - M$ obtained by truncating the matrix $\mathbf{U}_s$; $\mathbf{a}_s$ is a vector composed of upper row of the matrix $\mathbf{U}'_s$; $\mathbf{g}_s$ is a minimum norm vector based on the matrix $\mathbf{U}'_s$; $\mathbf{e}(\theta)$ is vector of complex spatial harmonics determined by the formula (6.4.15a); M is a number of narrowband signals in the observed additive mixture (6.4.1); $K \times K$ is a dimensionality of MSI matrix estimator $\hat{\mathbf{N}}_{a,s}$, $K = n$.

Having obtained the algorithm (6.4.16) of calculating MN-estimator of spatial hyperspectrum $\hat{H}_{MN}^s(\theta)$ that is based on MSI (1.4.11), it is necessary, first, to find out how close this estimator is with respect to MN-estimator (6.4.15), and second, could it be used for estimating spacial spectrum of the mixture of narrowband signals and noise. Algorithms (6.4.15) and (6.4.16) are illustrated by the following example.

Example 6.4.4. Let $a_i(t)$ be additive mixture (6.4.1) of M complex narrowband signals $s_l(t)$ ($M=8$) (each arriving from l-th point source) and complex quasi-white Gaussian noise $n_i(t)$ with zero mean in the input of i-th receiving channel of linear antenna array (see Fig. 6.1.7a), so that conditions pointed in preamble of Example 6.4.1 hold.

Fig. 6.4.4a illustrates MN-estimate of spatial spectrum $\hat{P}_{MN}(\theta)$ of the mixture $a_i(t)$ (6.4.1) obtained as a result of calculating the relationship (6.4.15) for initial situation described above. True positions of angles $\{\theta_l\}$ of arrival of the signals $\{s_l(t)\}$ from the mixture $a_i(t)$ (6.4.1) are shown by the signs "+". Fig. 6.4.4b depicts MN-estimate of spatial hyperspectrum $\hat{H}_{MUSIC}^s(\theta)$ (6.4.16) of the mixture $a_i(t)$ (6.4.1) for initial situation described above. Correlation coefficient between MN-estimates $\hat{H}_{MN}^s(\theta)$ (6.4.16) and $\hat{P}_{MN}(\theta)$ (6.4.15) takes the value $r[\hat{H}_{MN}^s(\theta), \hat{P}_{MN}(\theta)]=0.6$,

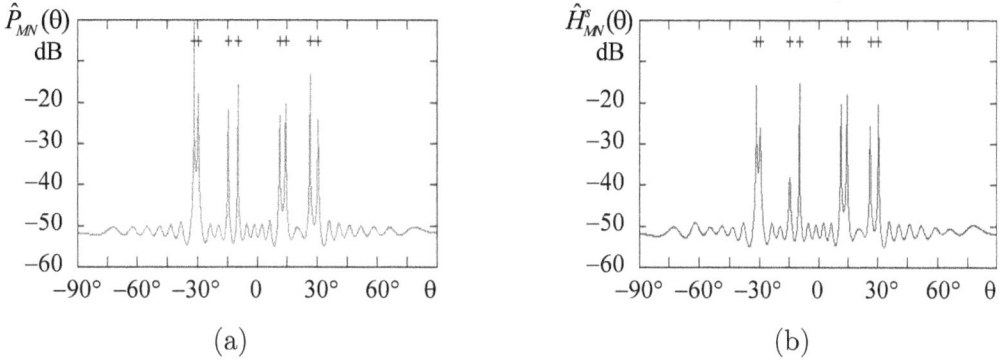

FIGURE 6.4.4 MN estimates of spatial spectrum: (a) $\hat{P}_{MN}(\theta)$ (6.4.15); (b) $\hat{H}^s_{MN}(\theta)$ (6.4.16)

that is explained by insignificant level of linear statistical interrelations between the processing results, while correlation coefficient between MSI matrix estimate $\hat{\mathbf{N}}_{a,s}$ (6.3.6) and ACM estimate $\hat{\mathbf{R}}_a$ (6.3.2) takes the value $r[\hat{\mathbf{N}}_{a,s}, \hat{\mathbf{R}}_a]=0.996$. Notice, that normalized MSIs (1.4.6) and (1.4.7) between MN-estimates $\hat{H}^s_{MN}(\theta)$ (6.4.16) and $\hat{P}_{MN}(\theta)$ (6.4.15) are, respectively, equal to:

$$\nu_p[\hat{H}^s_{MN}(\theta) - \mathbf{M}[\hat{H}^s_{MN}(\theta)], \hat{P}_{MN}(\theta) - \mathbf{M}[\hat{P}_{MN}(\theta)]] = 0.943;$$

$$\nu_s[\hat{H}^s_{MN}(\theta) - \mathbf{M}[\hat{H}^s_{MN}(\theta)], \hat{P}_{MN}(\theta) - \mathbf{M}[\hat{P}_{MN}(\theta)]] = 0.974,$$

where $\mathbf{M}[*]$ is a sample mean function.

The last relationships indicate high degree correspondence between the results obtained in linear signal space and signal space with L-group properties (Fig. 6.4.4a,b), that is explained by high degree identity between MSI matrix estimate $\hat{\mathbf{N}}_{a,s}$ (6.3.6) and ACM estimate $\hat{\mathbf{R}}_a$ (6.3.2), and also by the fact that nonlinear statistical interrelations between these estimates, unlike linear statistical interrelations, preserve more completely. $\triangledown$

6.4.2.4 Direction of Arrival Estimation by ESPRIT Method

Spatial spectrum estimation by ESPRIT method [152], as well as both MUSIC and minimum norm methods, assumes using information contained in a matrix $\mathbf{V} = \|\mathbf{v}_1, \mathbf{v}_2, \ldots, \mathbf{v}_K\|$ with dimensionality $K \times K$ composed by eigenvectors $\{\mathbf{v}_i\}$ of ACM estimator $\hat{\mathbf{R}}_a$ (6.3.2). However, unlike MUSIC and minimum norm methods realizing signal processing in noise subspace based on the matrix $\mathbf{V}'$ (6.4.13), ESPRIT method uses information contained in the left part $\mathbf{V}''$ of the matrix $\mathbf{V}$ corresponding to so called *signal subspace*:

$$\mathbf{V}'' = \text{submatrix}(\mathbf{V}, 1, K, 1, M) = \|\mathbf{v}''_1, \mathbf{v}''_2, \ldots, \mathbf{v}''_M\|, \qquad (6.4.17)$$

where $\text{submatrix}(\mathbf{A}, r_i, r_j, c_i, c_j)$ is a function of extracting a submatrix containing the elements in rows r_i trough r_j and columns c_i trough c_j from the initial matrix $\mathbf{A}$; matrix $\mathbf{V}'' = \|\mathbf{v}''_1, \mathbf{v}''_2, \ldots, \mathbf{v}''_M\|$ has a dimensionality $K \times M$; M is an expected number of narrowband signals in the mixture (6.4.1), $M < K$.

Then algorithm of forming ESPRIT estimator $\hat{\theta} = [\hat{\theta}_l]$ of a vector of DOAs $\theta = [\theta_l]$ of narrowband signals from the mixture (6.4.1), $l = 1, \ldots, M$ is defined by the following relationships:

$$\hat{\theta}_l = \arcsin\left[\frac{-\arg(\lambda_l^{\Phi})}{2\pi f_c d/c}\right],\tag{6.4.18}$$

where λ_l^{Φ} are eigenvalues of the matrix $\mathbf{\Phi}$ with dimensionality $M \times M$:

$$\mathbf{\Phi} = \left(\bar{\mathbf{S}}_1^T \mathbf{S}_1\right)^{-1} \bar{\mathbf{S}}_1^T \mathbf{S}_2,\tag{6.4.19}$$

that is a solution of the equation:

$$\mathbf{S}_1 \mathbf{\Phi} = \mathbf{S}_2;\tag{6.4.20}$$

$$\mathbf{S}_1 = \mathbf{G}_1 \mathbf{V}'', \quad \mathbf{S}_2 = \mathbf{G}_2 \mathbf{V}'';\tag{6.4.20a}$$

$$\mathbf{G}_1 = \|\mathbf{I}_{K-1}|\mathbf{0}\|, \quad \mathbf{G}_2 = \|\mathbf{0}|\mathbf{I}_{K-1}\|,\tag{6.4.20b}$$

where $\mathbf{G}_1$, $\mathbf{G}_2$ are selector matrices with dimensionality $(K-1) \times K$; $\mathbf{V}''$ is a truncated matrix of eigenvectors of ACM $\hat{\mathbf{R}}_a$ (6.3.2) corresponding to signal subspace that is determined by the relationship (6.4.17); $\mathbf{I}_{K-1}$ is the identity matrix with dimensionality $(K-1) \times (K-1)$; $\mathbf{0}$ is zero vector-column $\mathbf{0} = [0, \ldots, 0]^T$ with a number of element equal to $K-1$; M is an expected number of narrowband signals in the mixture (6.4.1), $M < K$.

Taking into account known relation between normalized MSI (or MSI) and correlation coefficient (or scalar product) of a pair of signals considered in Chapter 1, by the analogy with ESPRIT estimator (6.4.18)...(6.4.20), one can obtain ESPRIT estimator $\hat{\theta}_s = [\hat{\theta}_{s,l}]$ of DOAs $\theta = [\theta_l]$ of narrowband signals from the mixture (6.4.1), $l = 1, \ldots, M$ based on MSI (1.4.11) that is defined by the following relationships:

$$\hat{\theta}_{s,l} = \arcsin\left[\frac{-\arg(\lambda_l^{\Phi_s})}{2\pi f_c d/c}\right];\tag{6.4.21}$$

$$\mathbf{\Phi}_s = \left(\bar{\mathbf{C}}_1^T \mathbf{C}_1\right)^{-1} \bar{\mathbf{C}}_1^T \mathbf{C}_2;\tag{6.4.21a}$$

$$\mathbf{C}_1 = \mathbf{G}_1 \mathbf{U}''_s, \quad \mathbf{C}_2 = \mathbf{G}_2 \mathbf{U}''_s;\tag{6.4.21b}$$

$$\mathbf{G}_1 = \|\mathbf{I}_{K-1}|\mathbf{0}\|, \quad \mathbf{G}_2 = \|\mathbf{0}|\mathbf{I}_{K-1}\|;\tag{6.4.21c}$$

$$\mathbf{U}''_s = \text{submatrix}(\mathbf{U}_s, 1, K, 1, M) = \|\mathbf{u}''_{s,1}, \mathbf{u}''_{s,2}, \ldots, \mathbf{u}''_{s,M}\|,\tag{6.4.21d}$$

where $\lambda_l^{\Phi_s}$ are eigenvalues of the matrix $\mathbf{\Phi}_s$ with dimensionality $M \times M$; $\mathbf{G}_1$, $\mathbf{G}_2$ are selector matrices with dimensionality $(K-1) \times K$; $\mathbf{U}_s = \|\mathbf{u}_{s,1}, \mathbf{u}_{s,2}, \ldots, \mathbf{u}_{s,K}\|$ are matrices with dimensionality $K \times K$ composed of eigenvectors of MSI matrix estimator $\hat{\mathbf{N}}_{a,s}$ (6.3.6); $\mathbf{U}'_s$ is a signal subspace matrix with dimensionality $K \times M$ obtained by truncating the matrix $\mathbf{U}_s$; $\mathbf{I}_{K-1}$ is the identity matrix with dimensionality $(K-1) \times (K-1)$; $\mathbf{0}$ is zero vector-column $\mathbf{0} = [0, \ldots, 0]^T$ with a number of elements $K-1$; M is an expected number of narrowband signals in the mixture (6.4.1), $M < K$.

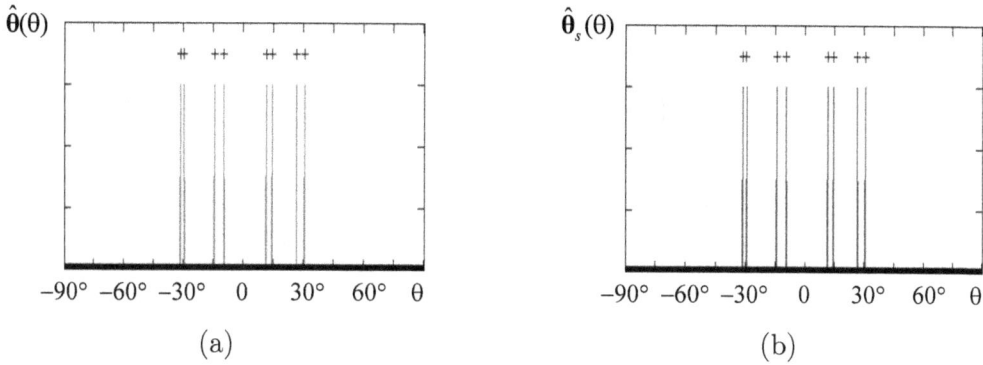

FIGURE 6.4.5   ESPRIT estimates of spatial spectrum: (a) $\hat{\theta} = [\hat{\theta}_l]$ (6.4.18); (b) $\hat{\theta}_s = [\hat{\theta}_{s,l}]$ (6.4.21)

Having obtained the algorithm (6.4.21) of calculating ESPRIT estimator $\hat{\theta}_s = [\hat{\theta}_{s,l}]$ of vector of DOAs $\theta = [\theta_l]$ of narrowband signals from the mixture (6.4.1) that is based on MSI (1.4.11), it is necessary, first, to find out how close this estimator is with respect to ESPRIT estimator (6.4.18)...(6.4.20), and second, could it be used for estimating spatial spectrum of a mixture of narrowband signals and noise. Algorithms (6.4.18)...(6.4.20) and (6.4.21) are illustrated by the following example.

Example 6.4.5. Let $a_i(t)$ be additive mixture (6.4.1) of M complex narrowband signals $s_l(t)$ ($M{=}8$) (each arriving from l-th point source) and complex quasi-white Gaussian noise $n_i(t)$ with zero mean in the input of i-th receiving channel of linear antenna array (see Fig. 6.1.7a), so that conditions pointed in preamble of Example 6.4.1 hold.

Fig. 6.4.5a depicts ESPRIT estimate of spatial spectrum $\hat{\theta} = [\hat{\theta}_l]$ of the mixture $a_i(t)$ (6.4.1) obtained as a result of calculating the relationship (6.4.18) for initial situation described above:

$$\hat{\theta}(\theta) = \sum_{l=1}^{M} \delta[\theta, \hat{\theta}_l],$$

where $\delta[a, b]$ is Kronecker function: $\delta[a, b] = 1$, if $a = b$; $\delta[a, b] = 0$, if $a \neq b$.

Actual positions of DOAs $\{\theta_l\}$ of the signals $\{s_l(t)\}$ from the mixture $a_i(t)$ (6.4.1) are shown by the signs "+". Fig. 6.4.5b illustrates L-group ESPRIT estimate of spatial hyperspectrum $\hat{\theta}_s = [\hat{\theta}_{s,l}]$ (6.4.21) of the mixture $a_i(t)$ (6.4.1) for initial situation described above:

$$\hat{\theta}_s(\theta) = \sum_{l=1}^{M} \delta[\theta, \hat{\theta}_{s,l}],$$

where $\delta[a, b]$ is Kronecker function.

Fig. 6.4.6 depicts realizations of absolute errors $\Delta(l)$, $\Delta_s(l)$ of ESPRIT estimates $\hat{\theta} = [\hat{\theta}_l]$ (6.4.18), $\hat{\theta}_s = [\hat{\theta}_{s,l}]$ (6.4.21) of DOAs $\theta = [\theta_l]$, $l = 1, \ldots, M$ of narrowband signals from the mixture (6.4.1):

$$\Delta(l) = |\hat{\theta}_l - \theta_l|;$$

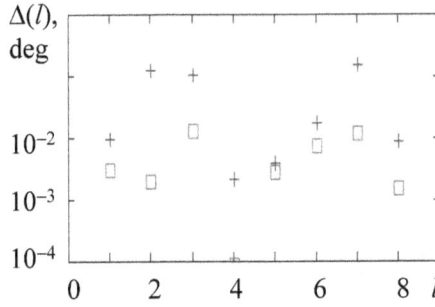

FIGURE 6.4.6 Realizations of absolute errors $\Delta(l)$ ($\square$—$\square$), $\Delta_s(l)$ (+—+) of ESPRIT estimates $\hat{\theta} = [\hat{\theta}_l]$ (6.4.18), $\hat{\theta}_s = [\hat{\theta}_{s,l}]$ (6.4.21), respectively

$$\Delta_s(l) = |\hat{\theta}_{s,l} - \theta_l|.$$

Realizations of absolute errors $\Delta(l)$, $\Delta_s(l)$ are denoted by the symbols $\square$, +, respectively. As follows from the figure, the ESPRIT estimator (6.4.21) based on L-group operations is incidentally inferior to classic ESPRIT estimator (6.4.18) in estimation accuracy. $\triangledown$

6.5 Wideband Antenna Array Signal Processing Based on L-group Operations

In this section, we consider the features of signal processing algorithms in wideband antenna arrays. Here, to the class of wideband arrays we refer linear antenna arrays with ratios $f_{\max}/f_{\min}$ of maximum to minimum frequencies of processed signals that belong to the interval $f_{\max}/f_{\min} \in [1.01, 2.0]$, and also circular antenna arrays with ratios $f_{\max}/f_{\min} > 2$. The signals $u_l(t, f_l)$ processed in antenna array are considered to be narrowband if their PSDs (or spectra) are concentrated in rather narrow frequency band around some central frequency f_l: $\Delta F_l/f_l << 1$, where ΔF_l is a bandwidth of PSD (or spectrum) of a signal.

6.5.1 Spatial Filtering in Wideband Linear Antenna Array: L-group Algorithm

Consider the problem of spatial filtering of narrowband useful signal that is solved by wideband linear equidistant antenna array (see Fig. 6.1.7a). Here we consider linear antenna arrays with the ratios $f_{\max}/f_{\min}$ of maximum to minimum frequencies of processed signals that belong to the interval $f_{\max}/f_{\min} \in [1.01, 2.0]$.

Let $a_i(t)$ be additive mixture of useful signal $s(t)$, complex narrowband interference signals $u_l(t)$ (each arriving from l-th point source), and quasi-white Gaussian noise $n(t)$ with zero mean observed in the input of i-th element of linear antenna array (see Fig. 6.1.7a):

$$a_i(t) = s_i(t) + \sum_{l=1}^{M} u_{l,i}(t) + n_i(t) =$$

$$= s(t, \omega_s) \exp\left(-j2\pi \frac{f_s}{f_0} \frac{d}{\lambda_0} \sin\theta_s (i-1)\right)$$

$$+ \sum_{l=1}^{M} u_l(t, \omega_l) \exp\left(-j2\pi \frac{f_l}{f_0} \frac{d}{\lambda_0} \sin\theta_l (i-1)\right) + n_i(t) \quad (6.5.1)$$

$t = t_m = t_0 + m \cdot \Delta t$; $m = 0, 1, \ldots, N-1$; Δt is a sampling interval; $s_i(t)$ is useful signal $s(t)$ observed in i-th channel of linear antenna array; $\omega_s = 2\pi f_s = \text{const}$, f_s is a carrier frequency of useful signal $s(t)$; θ_s is DOA of the signal $s(t)$; $\omega_l = 2\pi f_l = \text{const}$, f_l is a frequency of interference signal $u_{l,i}(t)$ observed in i-th channel of linear antenna array; θ_l is DOA of the interference signal $u_l(t)$; frequencies $\{f_l\}$ of interference signals $\{u_l(t)\}$ and carrier frequency f_s of useful signal $s(t)$ are different and distributed in some interval: $f_l, f_s \in [f_{\min}, f_{\max}]$, $f_0 \leqslant f_{\min}$, $f_{\max}/f_{\min} \in [1.01, 2.0]$; $n_i(t) = n_i^c(t) + j \cdot n_i^s(t)$ is complex quasi-white Gaussian noise observed in i-th channel of linear antenna array, $j = \sqrt{-1}$; $\mathbf{M}\{u_l(t)u_k(t)\} = 0$, $l \neq k$; $\mathbf{M}\{n_i(t)n_j(t)\} = 0$, $i \neq j$; $\mathbf{M}\{*\}$ is a symbol of mathematical expectation; $\mathbf{M}\{(n_i^c(t))^2\} = \mathbf{M}\{(n_i^s(t))^2\} = D_n/2$, D_n is a variance of complex quasi-white Gaussian noise $n_i(t)$ with zero mean; d is interelement distance of antenna array, $d = 0.5\lambda_0$; λ_0 is a wave length that corresponds to minimum frequency of the processed signals $f_0 \leqslant f_{\min}$.

The resulting process $y(t)$ in the output of wideband linear antenna array (see Fig. 6.1.7a) is determined by weighted sum of delayed copies of signals in the receiving channels $a_i(t)$ [158, (4.170)], [102, (61.2)]:

$$y(t) = \sum_{i=1}^{n} \sum_{k=0}^{K-1} \bar{w}_{i,k} \cdot a_i(t - \Delta_k), \quad (6.5.2)$$

where $t = t_m = t_0 + m \cdot \Delta t$; $m = 0, 1, \ldots, N-1$; Δt is a sampling interval; $a_i(t)$ is received signal in the input of i-th receiving element of antenna array; $w_{i,k}$ is weight coefficient in i-th processing channel in the output of k-th delay element; $\bar{w}_{i,k} = \text{Re}(w_{i,k}) - j \text{Im}(w_{i,k})$; Δ_k is a value of time delay of a signal in the output of k-th delay element, $\Delta_{k=0} = 0$, $\Delta_k < \Delta_{k+1}$; K is a number of delay elements in each channel.

According to spatial filtering algorithm (6.5.2), basing on observed signals $a_i(t)$ in the input of i-th element of antenna array (6.5.1), it is necessary to form vectors $\mathbf{a}_i = [a_i(t_0), a_i(t_1), \ldots, a_i(t_{N-1})]^T$ composed of the signals $a_i(t)$ and their samples, and also their copies delayed at Δ_k, so that the resulting matrix $\mathbf{A}$ formed by these vectors takes the following form:

$$\mathbf{A} = \|A_{m,i}\| = \text{augment}(\mathbf{A}_0, \mathbf{A}_1, \ldots \mathbf{A}_{K-1}); \quad (6.5.3)$$

$$\mathbf{A}_k = \|a_{m,i}\|, \ a_{m,i} = a_i(t_m - \Delta_k), \quad (6.5.3a)$$

where augment$(\mathbf{A}, \mathbf{B}, \mathbf{C}, \ldots)$ is a function returning a matrix formed by placing the matrices $\mathbf{A}, \mathbf{B}, \mathbf{C}, \ldots$ left to right; $t = t_m = t_0 + m \cdot \Delta t$; $m = 0, 1, \ldots, N - 1$; $l = 1, \ldots, n \cdot K$; $i = 1, 2, \ldots, n$; $k = 0, 1, \ldots, K - 1$; $\Delta_{k=0} = 0$, $\Delta_k < \Delta_{k+1}$; N is a number of samples used in signal processing.

Thus, matrix $\mathbf{A}$ has a dimensionality $N \times (n \cdot K)$, where n is a number of elements of linear antenna array; K is a number of delay elements in each channel. Then estimator $\hat{\mathbf{R}}_a$ of autocorrelation matrix (ACM) $\mathbf{R}_a$ (6.3.2) formed in the basis of the observed processes (6.5.3) has a dimensionality $(n \cdot K) \times (n \cdot K)$:

$$\hat{\mathbf{R}}_a = \left\| R_{i,l}^a \right\| = \frac{1}{N} \left\| \left(\mathbf{A}^{<i>} \right)^T \overline{\mathbf{A}^{<l>}} \right\|, \tag{6.5.4}$$

where $i = 1, \ldots, n \cdot K$; $l = 1, \ldots, n \cdot K$; $\mathbf{A}^{<l>}$ is a vector formed by l-th column of the matrix $\mathbf{A}$ (6.5.3).

Antenna array, operating on the basis of minimum variance criterion, forms optimal solution for vector of weight coefficients according to the following relationships [158, (3.132)], [102, (61.18)]:

$$\mathbf{w} = [w_1, \ldots, w_{n \cdot K}]^T = \frac{\hat{\mathbf{R}}_a^{-1} \mathbf{E}(\theta_s, f_s)}{\mathbf{E}(\theta_s, f_s)^T \hat{\mathbf{R}}_a^{-1} \mathbf{E}(\theta_s, f_s)}; \tag{6.5.5}$$

$$\mathbf{E}(\theta_s, f_s) =$$
$$= \text{stack} \left[\mathbf{e}(\theta_s, f_s), \mathbf{e}(\theta_s, f_s) \cdot e^{-j2\pi f_s \cdot \Delta_1}, \ldots, \mathbf{e}(\theta_s, f_s) \cdot e^{-j2\pi f_s \cdot \Delta_{K-1}} \right]; \tag{6.5.5a}$$

$$\mathbf{e}(\theta_s, f_s) = \left[1 \; e^{-j2\pi \frac{f_s}{f_0} \frac{d}{\lambda_0} \sin \theta_s} \; \ldots \; e^{-j2\pi \frac{f_s}{f_0} \frac{d}{\lambda_0} \sin \theta_s (n-1)} \right]^T, \tag{6.5.5b}$$

where $\hat{\mathbf{R}}_a$ is an estimator of ACM $\mathbf{R}_a$ of the observed processed (6.5.4); $\mathbf{e}(\theta_s, f_s)$ is vector of complex spatial harmonics (steering vector) as a function of DOA θ_s and carrier frequency f_s of the signal $s(t)$; $\mathbf{E}(\theta_s, f_s)$ is a modified vector of complex spatial harmonics (steering vector) that assumes processing vectors $\mathbf{a}_i = [a_i(t_0), a_i(t_1), \ldots, a_i(t_{N-1})]^T$ composed of the samples of signals $a_i(t)$, and also their copies (6.5.3) delayed at Δ_k; stack$(\mathbf{A}, \mathbf{B}, \mathbf{C}, \ldots)$ is a function returning a matrix (vector) formed by placing the matrices (vectors) $\mathbf{A}, \mathbf{B}, \mathbf{C}, \ldots$ top to bottom.

Then, by the analogy with the algorithms (6.3.14) and (6.5.5), L-group algorithm of spatial filtering based on MSI matrix is defined by the following relationship:

$$z(t) = \sum_{i=1}^{n} \sum_{k=0}^{K-1} \bar{w}_{L;i,k} \cdot a_i(t - \Delta_k); \tag{6.5.6}$$

$$\mathbf{w}_L = [w_{L,1}, \ldots, w_{L,n \cdot K}]^T = \frac{\hat{\mathbf{N}}_a^{-1} \mathbf{E}(\theta_s, f_s)}{\mathbf{E}(\theta_s, f_s)^T \hat{\mathbf{N}}_a^{-1} \mathbf{E}(\theta_s, f_s)}; \tag{6.5.6a}$$

$$\hat{\mathbf{N}}_a^{-1} = \sum_{i=1}^{n \cdot K} \frac{1}{|\rho_i^s|} \mathbf{u}_i^s \overline{\mathbf{u}_i^s}^T; \tag{6.5.6b}$$

$$\hat{\mathbf{N}}_{a,s} = \hat{\mathbf{N}}_u^s + \hat{\mathbf{N}}_v^s - j(\hat{\mathbf{N}}_{uv}^s - \hat{\mathbf{N}}_{vu}^s); \tag{6.5.6c}$$

$$\hat{\mathbf{N}}_u^s = \left\| \mathrm{N}_s(\mathbf{u}^{<i>}, \mathbf{u}^{<k>}) \right\|; \quad \hat{\mathbf{N}}_v^s = \left\| \mathrm{N}_s(\mathbf{v}^{<i>}, \mathbf{v}^{<k>}) \right\|; \tag{6.5.6d}$$

$$\hat{\mathbf{N}}_{uv}^s = \left\| \mathrm{N}_s(\mathbf{u}^{<i>}, \mathbf{v}^{<k>}) \right\|; \quad \hat{\mathbf{N}}_{vu}^s = \left\| \mathrm{N}_s(\mathbf{v}^{<i>}, \mathbf{u}^{<k>}) \right\|; \tag{6.5.6e}$$

$$\mathbf{u} = \mathrm{Re}[\mathbf{A}]; \quad \mathbf{v} = \mathrm{Im}[\mathbf{A}], \tag{6.5.6f}$$

where $t = t_m = t_0 + m \cdot \Delta t$; $m = 0, 1, \ldots, N - 1$; $\hat{\mathbf{N}}_a$ is MSI matrix estimator $\hat{\mathbf{N}}_{a,s}$ (6.3.6) that is modified according to the relationship (6.5.6b) based on vectors-columns (see explanation in Subsection 6.3.1); $\{\mathbf{u}_i^s\}$, $\{\rho_i^s\}$ are eigenvectors and eigenvalues of MSI matrix estimator $\hat{\mathbf{N}}_{a,s}$ (6.3.6), respectively; $\mathbf{E}(\theta_s, f_s)$ is modified vector of complex spatial harmonics (steering vector) determined by the formula (6.5.5a); $\Delta_{k=0} = 0$, $\Delta_k < \Delta_{k+1}$; $\mathbf{u}^{<i>}$, $\mathbf{v}^{<k>}$ are vectors composed of i-th and k-th columns of the matrices $\mathbf{u}$, $\mathbf{v}$, respectively; the matrix $\mathbf{A}$ is defined by the formula (6.5.3).

Algorithm (6.5.6) is illustrated by the following example.

Example 6.5.1. Consider 19-elements linear antenna array (see Fig. 6.1.7a) that receives additive mixture of signals (6.5.1). The latter contains: useful signal $s(t)$ arriving from the direction $\theta_s = -\pi/18$, interference signals $u_1(t)$, $u_2(t)$, $u_3(t)$, $u_4(t)$, $u_5(t)$ arriving from the directions $\theta_1 = -\pi/3$; $\theta_2 = \pi/10$; $\theta_3 = 3\pi/14$; $\theta_4 = 7\pi/18$; $\theta_5 = -3\pi/14$, respectively (see Fig. 6.3.2; 6.1.2, 6.1.3), and also quasi-white Gaussian noise $n(t)$. Interference signals $u_1(t)$, $u_2(t)$, $u_3(t)$ are amplitude-modulated interference, frequency-modulated interference, and also RF pulse with rectangular envelope, respectively; and statistically independent interferences $u_4(t)$, $u_5(t)$ are quasi-white Gaussian noises. Signal-to-k-th interference ratios in each separate i-th receiving channel of antenna array take the values that are closely equal to 0.25; 0.25; 1; 0.05; 0.05, respectively. Signal-to-noise ratio in receiving channels is equal to 10^3. Frequencies of received interference and useful signals are distributed in some interval $[f_{\min}, f_{\max}]$, $f_0 \leqslant f_{\min}$, $f_{\max}/f_{\min} \approx 1.9$. Interelement distance d of linear antenna array is equal $d = 0.5\lambda_0$, where λ_0 is a wave length that corresponds to minimum frequency of the processed signals $f_0 \leqslant f_{\min}$. Values of time delays Δ_k in receiving channels (6.5.6) are chosen to be equal to $\Delta_k = 0; 2\Delta t; 4\Delta t; 5\Delta t; 8\Delta t$, $\Delta t = 1/(2F_{\max})$; $F_{\max}$ is maximum frequency of PSD of quasi-white Gaussian noise. Number of samples N of signals $a_i(t)$ used in processing (6.5.6) is chosen to be equal $N{=}1024$.

Fig. 6.5.1a,b illustrates the estimates $y(t) = \hat{s}(t)$, $z(t) = \hat{s}(t)$ of useful signal $s(t)$ that are formed in wideband linear antenna array according to classic algorithm (6.5.2), (6.5.5), and also L-group algorithm (6.5.6) based on MSI matrix estimator (6.3.6), respectively. Useful signal $s(t)$ is shown by solid line, filtering results $y(t)$, $z(t)$ are shown by dashed line, pulse interference $u_3(t)$ is shown two times less than the original by dotted line. Correlation coefficients between useful signal $s(t)$ and the output signals $y(t)$, $z(t)$ take their values in the intervals $r[y(t), s(t)]{=}0.92\ldots0.95$ and $r[z(t), s(t)]{=}0.89\ldots0.92$, respectively. Coefficients of interference suppression take the values in the intervals $K_y{=}40\ldots43$ dB and $K_z{=}39\ldots42$ dB, correspondingly. Here, coefficient of suppression is defined as a

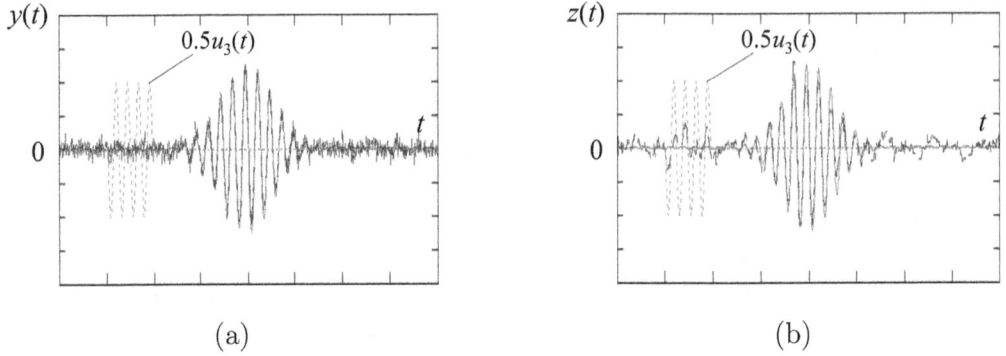

(a)   (b)

FIGURE 6.5.1   Estimates of useful signal $s(t)$ obtained in wideband linear antenna array according to algorithms of spatial filtering: (a) $y(t)$ (6.5.2), (6.5.5) (classic optimal algorithm); (b) $z(t) = $ (6.5.6) (L-group algorithm)

ratio $K_z = 10\lg(D/D[z(t) - s(t)])$, where D is a total variance of a resulting interference, $D = \sum_{k=1}^{K} D_k$, D_k is a variance of k-th interference; $D[z(t) - s(t)]$ is a variance of residual noise component of the process $z(t)$ in the output of antenna array. $\triangledown$

6.5.2   Spatial Filtering in Wideband Circular Antenna Array: L-group Algorithm

Consider the problem of signal filtering of narrowband useful signal $s(t)$ solved by wideband circular antenna array (see Fig. 6.3.1). We consider wideband circular antenna arrays with a ratio $f_{\max}/f_{\min}$ of maximum to minimum frequencies of processed signals that satisfies the inequality $f_{\max}/f_{\min} > 2$.

Let $a_i(t)$ be additive mixture of useful signal $s(t)$, complex narrowband interference signals $u_l(t)$ (each arriving from l-th point source), and quasi-white Gaussian noise $n(t)$ with zero mean observed in the input of i-th element of circular antenna array (see Fig. 6.3.1):

$$a_0(t) = s(t, \omega_s) + \sum_{l=1}^{M} u_l(t, \omega_l) + n_0(t); \qquad (6.5.7a)$$

$$a_{i_I}(t) = s_{i_I}(t) + \sum_{l=1}^{M} u_{l,i_I}(t) + n_{i_I}(t)$$

$$= s(t, \omega_s) \exp\left[-j2\pi \frac{f_s}{f_0}\frac{d}{\lambda_0}C_I(\cos(\Delta\theta_I \cdot i_I)\cos(\theta_s) + \sin(\Delta\theta_I \cdot i_I)\sin(\theta_s))\right]$$

$$+ \sum_{l=1}^{M} u_l(t, \omega_l) \exp\left[-j2\pi \frac{f_l}{f_0}\frac{d}{\lambda_0}C_I(\cos(\Delta\theta_I \cdot i_I)\cos(\theta_l) + \sin(\Delta\theta_I \cdot i_I)\sin(\theta_l))\right]$$

$$+ n_{i_I}(t); \quad (6.5.7b)$$

$$a_{i_{II}}(t) = s_{i_{II}}(t) + \sum_{l=1}^{M} u_{l,i_{II}}(t) + n_{i_{II}}(t)$$

$$= s(t, \omega_s) \exp\left[-j2\pi\frac{f_s}{f_0}\frac{d}{\lambda_0}C_{II}(\cos(\Delta\theta_{II} \cdot i_{II})\cos(\theta_s) + \sin(\Delta\theta_{II} \cdot i_{II})\sin(\theta_s))\right]$$

$$+ \sum_{l=1}^{M} u_l(t, \omega_l) \exp\left[-j2\pi\frac{f_l}{f_0}\frac{d}{\lambda_0}C_{II}(\cos(\Delta\theta_{II} \cdot i_{II})\cos(\theta_l) + \sin(\Delta\theta_{II} \cdot i_{II})\sin(\theta_l))\right]$$

$$+ n_{i_{II}}(t), \quad (6.5.7c)$$

where the equations (6.5.7a,b,c) define the signals $a_i(t)$ in the inputs of central element, elements of internal and external circles, respectively; $i_I = 0, 1, \ldots, 5$; $i_{II} = 0, 1, \ldots, 11$ are indexes of the elements of internal and external circles of circular antenna array, respectively; C_I, C_{II} are coefficients determining radii of internal and external circles of array, respectively: $C_I = 1$, $C_{II} = 2$; $\Delta\theta_I$, $\Delta\theta_{II}$ are angle distances between the adjacent elements of internal and external circles of array, respectively: $\Delta\theta_I = \pi/3$, $\Delta\theta_{II} = \pi/6$; $t = t_m = t_0 + m \cdot \Delta t$; $m = 0, 1, \ldots, N - 1$; Δt is a sample interval; $s_i(t)$ is useful signal $s(t)$ observed in i-th channel of circular antenna array; $\omega_s = 2\pi f_s = $ const, f_s is a carrier frequency of useful signal $s(t)$; θ_s is DOA of the signal $s(t)$; $\omega_l = 2\pi f_l = $ const, f_l is a frequency of interference signal $u_{l,i}(t)$ observed in i-th channel of circular antenna array; θ_l is DOA of interference signal $u_l(t)$; frequencies $\{f_l\}$ of interference signals $\{u_l(t)\}$ and carrier frequency f_s of useful signal $s(t)$ are different and distributed in some interval: $f_l, f_s \in [f_{\min}, f_{\max}]$, $f_0 \leqslant f_{\min}$, $f_{\max}/f_{\min} > 2$; $n_i(t) = n_i^c(t) + j \cdot n_i^s(t)$ is complex quasi-white Gaussian noise noise observed in i-th channel of circular antenna array, $j = \sqrt{-1}$; $\mathbf{M}\{u_l(t)u_k(t)\} = 0, l \neq k$; $\mathbf{M}\{n_i(t)n_j(t)\} = 0, i \neq j$; $\mathbf{M}\{*\}$ is a symbol of mathematical expectation; $\mathbf{M}\{(n_i^c(t))^2\} = \mathbf{M}\{(n_i^s(t))^2\} = D_n/2$, D_n is a variance of complex quasi-white Gaussian noise $n_i(t)$ with zero mean; d is interelement distance of antenna array in the internal circle, $d = 0.5\lambda_0$; λ_0 is wave length corresponding to minimum frequency of processed signals $f_0 \leqslant f_{\min}$.

According to spatial filtering algorithm (6.5.2), basing on the observed signal $a_i(t)$ in the input of i-th element of antenna array (6.5.7a,b,c), it is necessary to form vectors $\mathbf{a}_i = [a_i(t_0), a_i(t_1), \ldots, a_i(t_{N-1})]^T$ composed of samples of signals $a_i(t)$, so that the following relationships hold:

$$\mathbf{a}_i \equiv \mathbf{a}_0 = [a_0(t_0), a_0(t_1), \ldots, a_0(t_{N-1})]^T, \ i = 1; \quad (6.5.8a)$$

$$\mathbf{a}_i \equiv \mathbf{a}_{i_I} = [a_{i_I}(t_0), a_{i_I}(t_1), \ldots, a_{i_I}(t_{N-1})]^T; \quad (6.5.8b)$$

$$i = 2, 3, \ldots, 7; \ i_I = i - 2;$$

$$\mathbf{a}_i \equiv \mathbf{a}_{i_{II}} = [a_{i_{II}}(t_0), a_{i_{II}}(t_1), \ldots, a_{i_{II}}(t_{N-1})]^T; \quad (6.5.8c)$$

$$i = 8, 9, \ldots, 19; \ i_{II} = i - 8.$$

Vector $\mathbf{e}(\theta_s, f_s) = [\ e_1(\theta_s, f_s)\ \ e_2(\theta_s, f_s)\ \ \cdots\ \ e_n(\theta_s, f_s)\ ]^T$ of complex spatial harmonics (steering vector) for wideband circular antenna array is formed according to the following relationships:

$$e_i(\theta_s, f_s) = 1, \ i = 1; \quad (6.5.9a)$$

$$e_i(\theta_s, f_s) = e^{-j2\pi \frac{f_s}{f_0} \frac{d}{\lambda_0} C_I (\cos(\Delta\theta_I \cdot (i-2)) \cos(\theta_s) + \sin(\Delta\theta_I \cdot (i-2)) \sin(\theta_s))}, \qquad (6.5.9b)$$

$$i = 2, 3, \ldots, 7;$$

$$e_i(\theta_s, f_s) = e^{-j2\pi \frac{f_s}{f_0} \frac{d}{\lambda_0} C_{II} (\cos(\Delta\theta_{II} \cdot (i-8)) \cos(\theta_s) + \sin(\Delta\theta_{II} \cdot (i-8)) \sin(\theta_s))}, \qquad (6.5.9c)$$

$$i = 8, 9, \ldots, 19.$$

Taking into account the aforementioned features of forming observation vectors $\{\mathbf{a}_i\}$ (6.5.8) and steering vector $\mathbf{e}(\theta_s, f_s)$ (6.5.9) for wideband circular antenna array with configuration depicted in Fig. 6.3.1, spatial filtering L-group algorithm (6.5.6) based on MSI matrix estimator (6.3.6) is defined by the following relationships:

$$z(t) = \sum_{i=1}^{n} \sum_{k=0}^{K-1} \bar{w}_{L;i,k} \cdot a_i(t - \Delta_k); \qquad (6.5.10)$$

$$\mathbf{w}_L = [w_{L,1}, \ldots, w_{L,n \cdot K}]^T = \frac{\hat{\mathbf{N}}_a^{-1} \mathbf{E}(\theta_s, f_s)}{\mathbf{E}(\theta_s, f_s)^T \hat{\mathbf{N}}_a^{-1} \mathbf{E}(\theta_s, f_s)}; \qquad (6.5.10a)$$

$$\hat{\mathbf{N}}_a^{-1} = \sum_{i=1}^{n \cdot K} \frac{1}{|\rho_i^s|} \mathbf{u}_i^s \overline{\mathbf{u}}_i^{s\,T}, \qquad (6.5.10b)$$

where variables have the same meaning like these variables from (6.5.6).

The said above is illustrated by the following example.

Example 6.5.2. Consider 19-elements circular antenna array (see Fig. 6.3.1) that receives additive mixture of signals (6.3.15). The latter contains: useful signal $s(t)$ arriving from the direction $\theta_s = -\pi/18$, interference signals $u_1(t)$, $u_2(t)$, $u_3(t)$, $u_4(t)$, $u_5(t)$ arriving from the directions $\theta_1 = -\pi/3$; $\theta_2 = \pi/10$; $\theta_3 = 3\pi/14$; $\theta_4 = 7\pi/18$; $\theta_5 = -3\pi/14$, respectively (see Fig. 6.3.2; 6.1.2, 6.1.3), and also quasi-white Gaussian noise $n(t)$. Interference signals $u_1(t)$, $u_2(t)$, $u_3(t)$ are amplitude-modulated interference, frequency-modulated interference, and also RF pulse with rectangular envelope, respectively; and statistically independent interferences $u_4(t)$, $u_5(t)$ are quasi-white Gaussian noise. Signal-to-k-th interference ratios in each separate i-th receiving channel of antenna array take the values that are closely equal to 0.25; 0.25; 1; 0.05; 0.05, respectively. Signal-to-noise ratio in receiving channels is equal to 10^3. Frequencies of received interference and useful signals are distributed in the interval $[f_{\min}, f_{\max}]$, $f_0 \leqslant f_{\min}$, $f_{\max}/f_{\min} \approx 10.5$. Interelement distance d of antenna array is equal $d = 0.5\lambda_0$, where λ_0 is a wave length that corresponds to minimum frequency of the processed signals $f_0 \leqslant f_{\min}$. Values of time delays Δ_k in receiving channels (6.5.10) are chosen to be equal to $\Delta_k = 0; 2\Delta t; 4\Delta t; 5\Delta t; 8\Delta t$, $\Delta t = 1/(2F_{\max})$; $F_{\max}$ is maximum frequency of PSD of quasi-white Gaussian noise. Number of samples N of signals $a_i(t)$ used in processing (6.5.10) is chosen to be equal $N=1024$.

Fig. 6.5.2a,b illustrates the estimates $y(t) = \hat{s}(t)$, $z(t) = \hat{s}(t)$ of useful signal $s(t)$ obtained in wideband circular antenna array according to classic optimal algorithm (6.5.2), (6.5.5), and also L-group algorithm (6.5.10) based on MSI matrix estimator (6.3.6), respectively. Useful signal $s(t)$ is shown by solid line, filtering results $y(t)$,

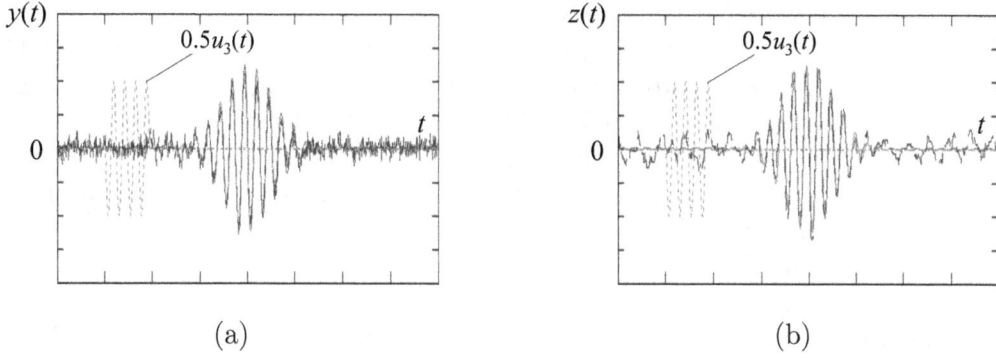

(a)   (b)

FIGURE 6.5.2   Estimates of useful signal $s(t)$ obtained by wideband circular antenna array according to spatial filtering algorithms: (a) $y(t)$ (6.5.2), (6.5.5) (classic optimal algorithm); (b) $z(t)$ (6.5.10) (*L*-group algorithm)

$z(t)$ are shown by dashed line, pulse interference $u_3(t)$ is shown two times less than the original by dotted line. Correlation coefficients between useful signal $s(t)$ and the output signals $y(t)$, $z(t)$ take their values in the intervals $r[y(t), s(t)]=0.88\ldots0.95$ and $r[z(t), s(t)]=0.85\ldots0.92$, respectively. Coefficients of interference suppression take the values in the intervals $K_y=36\ldots43$ dB and $K_z=35\ldots42$ dB, correspondingly. $\triangledown$

6.5.3   Direction of Arrival Estimation in Wideband Linear Antenna Array: *L*-group Algorithm

Consider the problem of direction of arrival estimation solved by wideband linear equidistant antenna array (see Fig. 6.1.7a). We consider wideband linear antenna array with ratios $f_{\max}/f_{\min}$ of maximum to minimum frequencies of processed signals that belong to the interval $f_{\max}/f_{\min} \in [1.01, 2.0]$.

Let $a_i(t)$ be additive mixture of M complex narrowband RF signals $s_l(t)$ (each arriving from l-th point source) and quasi-white Gaussian noise $n(t)$ with zero mean observed in the input of i-th element of linear antenna array (see Fig. 6.1.7a):

$$a_i(t) = \sum_{l=1}^{M} s_{l,i}(t) + n_i(t)$$

$$= \sum_{l=1}^{M} A_l \exp[j\omega_l t + \varphi_l] \exp\left(-j2\pi \frac{f_l}{f_0} \frac{d}{\lambda_0} \sin\theta_l(i-1)\right) + n_i(t), \quad (6.5.11)$$

where $i = 1, \ldots, n$ is an index of received channel of antenna array; $t = t_m = t_0 + m \cdot \Delta t$; $m = 0, 1, \ldots, N-1$; Δt is a sampling interval; $A_l = \text{const}, \omega_l = 2\pi f_l = \text{const}, A_l$, f_l are amplitude and carrier frequency of a signal $s_{l,i}(t)$ observed in i-th channel of linear antenna array; frequencies $\{f_l\}$ of signals $s_l(t)$ are different and distributed in some interval: $f_l, f_s \in [f_{\min}, f_{\max}]$, $f_0 \leqslant f_{\min}$, $f_{\max}/f_{\min} \in [1.01, 2.0]$; φ_l is unknown initial phase belonging to the interval $\varphi_l \in [0, 2\pi]$; $n_i(t) = n_i^c(t) + j \cdot n_i^s(t)$ is complex quasi-white Gaussian noise observed in i-th channel of linear antenna array,

$j = \sqrt{-1}$; $\mathbf{M}\{n_i(t)n_j(t)\} = 0$, $i \neq j$; $\mathbf{M}\{*\}$ is a symbol of mathematical expectation; $\mathbf{M}\{(n_i^c(t))^2\} = \mathbf{M}\{(n_i^s(t))^2\} = D_n/2$, D_n is a variance of complex quasi-white Gaussian noise $n_i(t)$ with zero mean; θ_l is DOA of a signal $s_l(t)$; d is interelement distance of antenna array, $d = 0.5\lambda_0$, where λ_0 is wave length corresponding to minimum frequency of processed signals $f_0 = c/\lambda_0 \leqslant f_{\min}$.

In this subsection, we consider the features of spatial spectrum estimation in wideband linear antenna array taking as an example MUSIC method based on MSI matrix estimator (6.3.6) that for the case of narrowband antenna array is considered in Subsubsection 6.4.2.2.

In the presence of prior information on frequencies $\{f_l\}$ of signals $\{s_l(t)\}$, MUSIC estimator of spatial hyperspectrum $\hat{H}_L(\theta, f_l)$ based on MSI (1.4.11) is determined by the relationship (6.4.12):

$$\hat{H}_L(\theta, f_l) = \frac{1}{\mathbf{e}(\theta, f_l)^T \left(\sum\limits_{i=M+1}^{n} \mathbf{u}_i \overline{\mathbf{u}_i^T} \right) \mathbf{e}(\theta, f_l)} \qquad (6.5.12)$$

$$= \frac{1}{\mathbf{e}(\theta, f_l)^T \left(\mathbf{U'}_s \overline{\mathbf{U'}_s^T} \right) \mathbf{e}(\theta, f_l)};$$

$$\mathbf{e}(\theta, f_l) = \left[1, e^{-j2\pi \frac{f_l}{f_0} \frac{d}{\lambda_0} \sin\theta}, \ldots, e^{-j2\pi \frac{f_l}{f_0} \frac{d}{\lambda_0} (n-1)\sin\theta} \right]^T, \qquad (6.5.12a)$$

where $\{\mathbf{u}_i\}$ are eigenvectors of MSI matrix estimator $\hat{\mathbf{N}}_{a,s}$ (6.3.6); $\mathbf{e}(\theta, f_l)$ is vector of complex spatial harmonics; M is a number of narrowband signals in the observed mixture (6.5.11); $n \times n$ is a dimensionality of MSI matrix estimator $\hat{\mathbf{N}}_{a,s}$; $\mathbf{U'}_s$ is noise subspace matrix with dimensionality $n \times (n - M)$ obtained by truncating the matrix $\mathbf{U}_s = \|\mathbf{u}_1, \mathbf{u}_2, \ldots, \mathbf{u}_n\|$ composed of eigenvectors of MSI matrix estimator $\hat{\mathbf{N}}_{a,s}$ (6.3.6):

$$\mathbf{U'}_s = \text{submatrix}(\mathbf{U}_s, 1, n, M+1, n) = \|\mathbf{u'}_{M+1}, \mathbf{u'}_{M+2}, \ldots, \mathbf{u'}_n\|.$$

Direct using MUSIC algorithm (6.5.12) for calculating spatial hyperspectrum L-group estimator $\hat{H}_L(\theta, f_l)$ in the presence of information on the frequencies $\{f_l\}$ of signals $\{s_l(t)\}$ from the mixture (6.5.11) causes false peaks at the angles θ_l^k determined by the formula:

$$\theta_l^k = \arcsin\left(\frac{f_k \sin(\theta_k)}{f_l} \right), k \neq l, \qquad (6.5.13)$$

that is explained by dispersion phenomena in wideband linear antenna array.

One can overcome negative impact of dispersion and get rid of ambiguity among the obtained estimators in the following way. Direct using MUSIC algorithm (6.5.12) puts at our disposal a totality of estimates $\{\hat{H}_L(\theta, f_l)\}$ for known frequencies $\{f_l\}$ of signals $\{s_l(t)\}$ from the mixture (6.5.11) that allow obtaining complete totality $\{\hat{\theta}_l^k\}$ of DOA estimates containing among them M estimates $\hat{\theta}_l^k = \hat{\theta}_l$, $k = l$ corresponding to real DOAs θ_l, and also no more than $M^2 - M$ false estimates $\hat{\theta}_l^k$, $k \neq l$ corresponding to values that determined by the formula (6.5.13).

Basing on such totality $\{\hat{\theta}_l^k\}$ and using enumeration of possibilities to solve the equation (6.5.13), one can obtain a complete composition of estimates $\hat{\theta}_l$, $l = 1, \ldots, M$ corresponding to real DOAs θ_l excluding false estimates $\hat{\theta}_l^k$, $k \neq l$ from processing according to the following relationships:

$$H(\theta) = H'(\theta) \vee [\varepsilon \cdot \text{mean}(H_{\sup}(\theta)) + (1 - \varepsilon) \cdot \text{median}(H_{\sup}(\theta))]; \quad (6.5.14)$$

$$H_{\sup}(\theta) = \bigvee_{l=1}^{M} \hat{H}_L(\theta, f_l); \quad (6.5.14a)$$

$$H'(\theta) = \bigvee_{l=1}^{M} H'_l(\theta); \quad (6.5.14b)$$

$$H'_l(\theta) = \hat{H}_L(\theta, f_l) \cdot \left[1 - \sum_{k \neq l} [1(\theta - (\hat{\theta}_l^k - \Delta\theta)) - 1(\theta - (\hat{\theta}_l^k + \Delta\theta))] \right], \quad (6.5.14c)$$

where $H(\theta)$ is a resulting function determining a sought set of estimates $\hat{\theta}_l$, $l = 1, \ldots, M$ corresponding to real DOAs θ_l; $H_{\sup}(\theta)$ is a supremum (least upper bound) of MUSIC estimates of spatial spectrum $\hat{H}_L(\theta, f_l)$ (6.5.12) for known frequencies $\{f_l\}$ of signals $\{s_l(t)\}$ from the mixture (6.5.11); $\vee$ is operation of join; $\text{mean}(\mathbf{a})$, $\text{median}(\mathbf{a})$ are operations of sample mean and sample median of a vector $\mathbf{a}$, respectively; $\varepsilon = \text{const}$ is sufficiently small positive quantity: $\varepsilon << 1$, $\varepsilon > 0$; $H'(\theta)$, $H'_l(\theta)$ are the functions determining intermediate results; $\Delta\theta = \text{const}$ is a constant determining a window size.

Complete DOA estimation L-group algorithm defined by the relationships (6.5.12) and (6.5.14) is illustrated by the following example.

Example 6.5.3. Let $a_i(t)$ be additive mixture (6.5.11) of $M{=}4$ complex narrowband signals $s_l(t)$ (each arriving from l-th point source) and complex quasi-white Gaussian noise $n_i(t)$ with zero mean observed in the input of i-th element of 19-element linear antenna array (see Fig. 6.1.7a). Frequencies of received signals $s_l(t)$ are distributed in some interval: $[f_{\min}, f_{\max}]$, $f_0 \leqslant f_{\min}$, $f_{\max}/f_{\min} \approx 1.9$; interelement distance of antenna array is equal to $d = 0.5\lambda_0$, where λ_0 is a wave length corresponding to minimum frequency of processed signals $f_0 \leqslant f_{\min}$. Signal-to-noise ratios in receiving channels of wideband linear antenna array are equal to 20 dB. Number N of samples $a_i(t_m)$, $m = 0, 1, \ldots, N - 1$ of signals $a_i(t)$ (6.5.11), that are used in processing, is chosen to be equal to $N{=}512$.

Fig. 6.5.3a depicts MUSIC estimate of spatial hyperspectrum $\{\hat{H}_L(\theta, f_l)\}$ for known frequencies $\{f_l\}$ of four signals $\{s_l(t)\}$ from the mixture (6.5.11) obtained by calculating the relationship (6.5.12) for initial situation described above. True positions of DOAs $\{\theta_l\}$ of signals $\{s_l(t)\}$ from the mixture (6.5.11) are shown by the sign "+". False estimates $\hat{\theta}_l^k$, $k \neq l$, that correspond to the values determined by the formula (6.5.13) and must be excluded from further processing, are shown by symbols "o". Fig. 6.5.3b depicts a resulting function $H(\theta)$ (solid line) determining the sought set of estimates $\hat{\theta}_l$, $l = 1, \ldots, M$ corresponding to real DOAs θ_l, supremum $H_{\sup}(\theta)$ (dashed line) of MUSIC estimate of spatial hyperspectrum $\hat{H}_L(\theta, f_l)$ (6.5.12), and also levels of $\text{mean}(H_{\sup}(\theta))$ (dotted line) and $\text{median}(H_{\sup}(\theta))$ (dash-dotted line) from the formula (6.5.14).

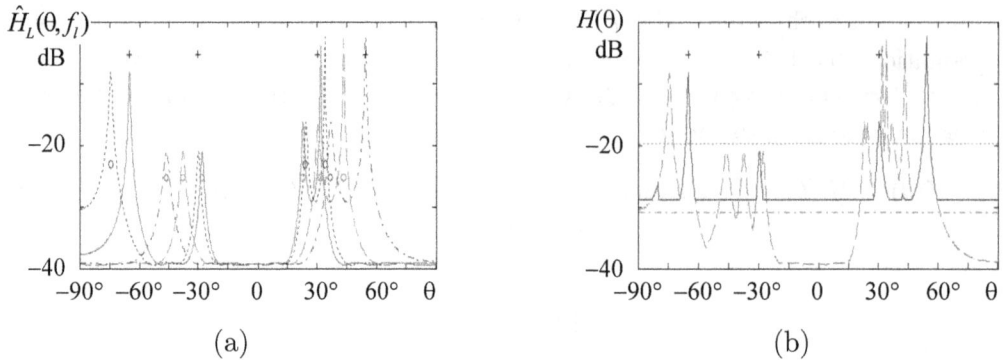

FIGURE 6.5.3   (a) MUSIC estimate of spatial hyperspectrum $\{\hat{H}_L(\theta, f_l)\}$ (6.5.12); (b) functions $H(\theta)$ (6.5.14) (solid line) and $H_{\sup}(\theta)$ (6.5.14a) (dashed line)

$\triangledown$

6.5.4   Direction of Arrival Estimation in Wideband Circular Antenna Array: L-group Algorithm

Consider the problem of direction of arrival estimation solved by wideband circular antenna array (see Fig. 6.3.1). We consider wideband circular antenna array with ratios $f_{\max}/f_{\min}$ of maximum to minimum frequencies of processed signals that satisfies the inequality $f_{\max}/f_{\min} > 2$.

Let $a_i(t)$ be additive mixture of M complex narrowband RF signals $s_l(t, \omega_l)$, $s_l(t, \omega_l) = A_l \exp[j\omega_l t + \varphi_l]$ (each arriving from l-th point source) and complex quasi-white Gaussian noise $n_i(t)$ with zero mean observed in the input of i-th element of circular antenna array (see Fig. 6.3.1):

$$a_i(t) = \sum_{l=1}^{M} s_{l,i}(t) + n_i(t);$$

$$a_0(t) = \sum_{l=1}^{M} s_l(t, \omega_l) + n_0(t); \tag{6.5.15a}$$

$$a_{i_I}(t) = \sum_{l=1}^{M} s_l(t, \omega_l)$$
$$\times e^{-j2\pi \frac{f_l}{f_0} \frac{d}{\lambda_0} C_I (\cos(\Delta\theta_I \cdot i_I) \cos(\theta_l) + \sin(\Delta\theta_I \cdot i_I) \sin(\theta_l))} + n_{i_I}(t); \tag{6.5.15b}$$

$$a_{i_{II}}(t) = \sum_{l=1}^{M} s_l(t, \omega_l)$$
$$\times e^{-j2\pi \frac{f_l}{f_0} \frac{d}{\lambda_0} C_{II} (\cos(\Delta\theta_{II} \cdot i_{II}) \cos(\theta_l) + \sin(\Delta\theta_{II} \cdot i_{II}) \sin(\theta_l))} + n_{i_{II}}(t), \tag{6.5.15c}$$

where the equations (6.5.15a,b,c) define the signals $a_i(t)$ from the mixture in the inputs of the central element, the elements of internal and external circles, respectively; $i_I = 0, 1, \ldots, 5$; $i_{II} = 0, 1, \ldots, 11$ are indexes of the elements of internal and external circles of circular antenna array, respectively; C_I, C_{II} are coefficients determining radii of internal and external circles of array, respectively: $C_I = 1$, $C_{II} = 2$; $\Delta\theta_I$, $\Delta\theta_{II}$ are angle distances between the adjacent elements of internal and external circles of array, respectively: $\Delta\theta_I = \pi/3$, $\Delta\theta_{II} = \pi/6$; $t = t_m = t_0 + m \cdot \Delta t$; $m = 0, 1, \ldots, N - 1$; Δt is a sampling interval; $\omega_s = 2\pi f_s = $ const, $A_l = $ const, $\omega_l = 2\pi f_l = $ const, A_l, f_l are amplitude and carrier frequency of the signal $s_{l,i}(t)$ observed in i-th channel of circular antenna array; θ_l is a DOA of signal $s_l(t)$; frequencies $\{f_l\}$ of signals $\{s_l(t)\}$ are different and distributed in the interval: $f_l \in [f_{\min}, f_{\max}]$, $f_0 \leqslant f_{\min}$, $f_{\max}/f_{\min} > 2$; φ_l is unknown initial phase belonging to the interval $\varphi_l \in [0, 2\pi]$; $n_i(t) = n_i^c(t) + j \cdot n_i^s(t)$ is complex quasi-white Gaussian noise observed in i-th channel of circular antenna array, $j = \sqrt{-1}$; $\mathbf{M}\{n_i(t)n_j(t)\} = 0$, $i \neq j$; $\mathbf{M}\{*\}$ is a symbol of mathematical expectation; $\mathbf{M}\{(n_i^c(t))^2\} = \mathbf{M}\{(n_i^s(t))^2\} = D_n/2$, D_n is a variance of complex quasi-white Gaussian noise $n_i(t)$ with zero mean; d is interelement distance of antenna array in the internal circle, $d = 0.5\lambda_0$; λ_0 is a wave length corresponding to minimum frequency of processed signals $f_0 \leqslant f_{\min}$.

In this subsection, we consider the features of spatial spectrum estimation in wideband circular antenna array taking as an example MUSIC method based on MSI matrix estimator (6.3.6) that, for the case of narrowband antenna array, is considered in Subsubsection 6.4.2.2.

In the presence of prior information on the frequencies $\{f_l\}$ of the signals $\{s_l(t)\}$, MUSIC estimator of spatial hyperspectrum $\hat{H}_L(\theta, f_l)$ based on MSI (1.4.11) is determined by the relationship (6.4.20):

$$\hat{H}_L(\theta, f_l) = \cfrac{1}{\mathbf{e}(\theta, f_l)^T \left(\sum_{i=M+1}^{n} \mathbf{u}_i \overline{\mathbf{u}_i^T} \right) \mathbf{e}(\theta, f_l)}$$

$$= \cfrac{1}{\mathbf{e}(\theta, f_l)^T \left(\mathbf{U'}_s \overline{\mathbf{U'}_s^T} \right) \mathbf{e}(\theta, f_l)}, \quad (6.5.16)$$

where $\{\mathbf{u}_i\}$ are eigenvectors of MSI matrix estimator $\hat{\mathbf{N}}_{a,s}$ (6.3.6); $\mathbf{e}(\theta, f_l)$ is vector of complex spatial harmonics; M is a number of narrowband signals in the observed additive mixture (6.5.15); $n \times n$ is a dimensionality of MSI matrix estimator $\hat{\mathbf{N}}_{a,s}$; $\mathbf{U'}_s$ is noise subspace matrix with dimensionality $n \times (n - M)$ obtained by truncating the matrix $\mathbf{U}_s = \|\mathbf{u}_1, \mathbf{u}_2, \ldots, \mathbf{u}_n\|$ composed of eigenvectors of MSI matrix estimator $\hat{\mathbf{N}}_{a,s}$ (6.3.6):

$$\mathbf{U'}_s = \mathrm{submatrix}(\mathbf{U}_s, 1, n, M + 1, n) = \|\mathbf{u'}_{M+1}, \mathbf{u'}_{M+2}, \ldots, \mathbf{u'}_n\|.$$

It should be noted, that vector

$$\mathbf{e}(\theta_s, f_l) = [\, e_1(\theta_s, f_l) \quad e_2(\theta_s, f_l) \quad \ldots \quad e_n(\theta_s, f_l) \,]^T$$

of complex spatial harmonics (steering vector) for wideband circular antenna array figuring in the formula (6.5.16) is formed according to the observed signals $a_i(t)$ in the input of i-th element of antenna array (6.5.15):

$$e_i(\theta_s, f_s) = 1, \ i = 1; \tag{6.5.17a}$$

$$e_i(\theta_s, f_l) = e^{-j2\pi \frac{f_l}{f_0} \frac{d}{\lambda_0} C_I (\cos(\Delta\theta_I \cdot (i-2)) \cos(\theta_l) + \sin(\Delta\theta_I \cdot (i-2)) \sin(\theta_l))}, \tag{6.5.17b}$$

$$i = 2, 3, \ldots, 7;$$

$$e_i(\theta_s, f_l) = e^{-j2\pi \frac{f_l}{f_0} \frac{d}{\lambda_0} C_{II} (\cos(\Delta\theta_{II} \cdot (i-8)) \cos(\theta_l) + \sin(\Delta\theta_{II} \cdot (i-8)) \sin(\theta_l))}, \tag{6.5.17c}$$

$$i = 8, 9, \ldots, 19.$$

Direct using MUSIC algorithm (6.5.16) for calculating spatial hyperspectrum L-group estimator $\hat{H}_L(\theta, f_l)$ in the presence of information on the frequencies $\{f_l\}$ of signals $\{s_l(t)\}$ from the mixture (6.5.15) causes false peaks. One can overcome negative impact of dispersion and get rid of ambiguity among the obtained estimators in the following way. Direct using MUSIC algorithm (6.5.16) puts at our disposal a totality of estimates $\{\hat{H}_L(\theta, f_l)\}$ for known frequencies $\{f_l\}$ of signals $\{s_l(t)\}$ from the mixture (6.5.15) that allow obtaining complete totality $\{\hat{\theta}_l^k\}$ of DOA estimates containing among them M estimates $\hat{\theta}_l^k = \hat{\theta}_l$, $k = l$ corresponding to real DOAs θ_l, and also some quantity of false estimates $\hat{\theta}_l^k$, $k \neq l$.

Basing on such totality $\{\hat{\theta}_l^k\}$, one can obtain a complete composition of estimates $\hat{\theta}_l$, $l = 1, \ldots, M$ corresponding to real DOAs θ_l excluding false estimates $\hat{\theta}_l^k$, $k \neq l$ from processing according to the following relationships:

$$H(\theta) = H'(\theta) \vee [\varepsilon \cdot \text{mean}(H_{\sup}(\theta)) + (1 - \varepsilon) \cdot \text{median}(H_{\sup}(\theta))]; \tag{6.5.18}$$

$$H_{\sup}(\theta) = \bigvee_{l=1}^{M} \hat{H}_L(\theta, f_l); \tag{6.5.18a}$$

$$H'(\theta) = \bigvee_{l=1}^{M} H'_l(\theta); \tag{6.5.18b}$$

$$H'_l(\theta) = \hat{H}_L(\theta, f_l) \cdot [1(\theta - (\hat{\theta}_{l,\max} - \Delta\theta)) - 1(\theta - (\hat{\theta}_{l,\max} + \Delta\theta))]; \tag{6.5.18c}$$

$$\hat{\theta}_{l,\max} = \text{match}(\max_{\theta}(\hat{H}_L(\theta, f_l)), \hat{H}_L(\theta, f_l)), \tag{6.5.18d}$$

where $H(\theta)$ is a resulting function determining the sought set of estimates $\hat{\theta}_l$, $l = 1, \ldots, M$ corresponding to real DOAs θ_l; $H_{\sup}(\theta)$ is supremum (least upper bound) of MUSIC estimates of spatial hyperspectrum $\hat{H}_L(\theta, f_l)$ (6.5.16) for known frequencies $\{f_l\}$ of signals $\{s_l(t)\}$ from the mixture (6.5.15); $\vee$ is operation of join; mean($\mathbf{a}$), median($\mathbf{a}$) are operations of sample mean and sample median of a vector $\mathbf{a}$, respectively; $\varepsilon = \text{const}$ is a sufficiently small positive quantity: $\varepsilon \ll 1$, $\varepsilon > 0$; $H'(\theta)$, $H'_l(\theta)$ are the functions determining intermediate results; $\Delta\theta = \text{const}$ is a constant determining a window size; $\hat{\theta}_{l,\max}$ are estimated values of DOAs θ_l determining maximum value of the estimate $\hat{H}_L(\theta, f_l)$ (6.5.16); match($z, \mathbf{a}$) is a function returning an index of an element of vector $\mathbf{a}$ that is equal to a given value z.

Complete DOA estimation L-group algorithm defined by the relationships (6.5.12) and (6.5.14) is illustrated by the following example.

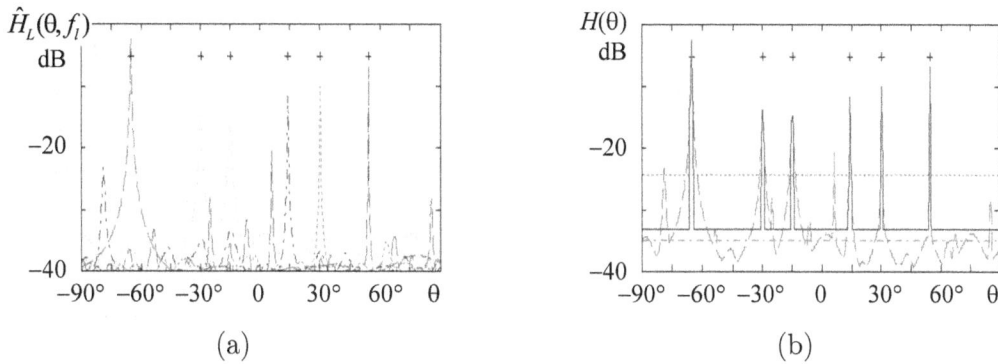

FIGURE 6.5.4 (a) MUSIC estimate of spatial hyperspectrum $\{\hat{H}_L(\theta, f_l)\}$ (6.5.16); (b) functions $H(\theta)$ (6.5.18) (solid line) and $H_{\mathrm{sup}}(\theta)$ (6.5.18a) (dashed line)

Example 6.5.4. Let $a_i(t)$ be additive mixture (6.5.15) of $M=6$ complex narrow-band RF signals $s_l(t)$ (each arriving from l-th point source) and complex quasi-white Gaussian noise $n_i(t)$ with zero mean observed in the input of i-th element of 19-element circular antenna array (see Fig. 6.3.1). Frequencies of received signals $s_l(t)$ are distributed in the interval: $[f_{\mathrm{min}}, f_{\mathrm{max}}]$, $f_0 \leqslant f_{\mathrm{min}}$, $f_{\mathrm{max}}/f_{\mathrm{min}} \approx 10.5$; interelement distance of antenna array is equal to $d = 0.5\lambda_0$, where λ_0 is a wave length corresponding to minimum frequency of processed signals $f_0 \leqslant f_{\mathrm{min}}$. Signal-to-noise ratios in receiving channels of wideband circular antenna array are equal to 20 dB. Number N of samples $a_i(t_m)$, $m = 0, 1, \ldots, N-1$ of signals $a_i(t)$ (6.5.15), that are used in processing, is chosen to be equal to $N=1024$.

Fig. 6.5.4a depicts MUSIC estimate of spatial hyperspectrum $\{\hat{H}_L(\theta, f_l)\}$ for known frequencies $\{f_l\}$ of six signals $\{s_l(t)\}$ from the mixture (6.5.15) obtained by calculating the relationship (6.5.16) for the situation described above. True positions of DOAs $\{\theta_l\}$ of signals $\{s_l(t)\}$ from the mixture (6.5.15) are shown by the signs "+". Fig. 6.5.4b depicts the resulting function $H(\theta)$ (6.5.18) (solid line) determining the sought set of estimates $\hat{\theta}_l$, $l = 1, \ldots, M$ corresponding to real DOAs θ_l, supremum $H_{\mathrm{sup}}(\theta)$ (6.5.18a) (dashed line) of MUSIC estimate of spatial hyperspectrum $\hat{H}_L(\theta, f_l)$ (6.5.16), and also levels of $\mathrm{mean}(H_{\mathrm{sup}}(\theta))$ (dotted line) and $\mathrm{median}(H_{\mathrm{sup}}(\theta))$ (dash-dotted line) from the formula (6.5.18).
▽

6.6 Adaptive Algorithms of Spatial Filtering Based on L–group Operations

The subject of further consideration is exploring adaptive algorithms of spatial filtering based on L-group operations. Adaptive systems and algorithms of signal processing are, as a rule, used when: (a) conditions of signal receiving (for instance, PSD of interference signal) are unknown and/ or changed in time; (b) computational resources of a system are extremely confined and do not allow performing algorithms assuming direct inversion of correlation matrix; etc. [92, 94, 95, 100–102, 158]. In particular, we consider adaptive algorithms of signal filtering based on known variants of least squares algorithms (recursive least squares (RLS), least mean squares (LMS)), and also on recursive method of forming autocorrelation matrix (ACM) estimator. On the other hand, we consider adaptive algorithms of spatial filtering based on L-group operations. These algorithms of signal processing are based on either method of mapping of linear signal space into signal space with lattice properties or recursive method of forming MSI matrix estimator (6.3.6).

Consider the problem of spatial filtering solved by linear equidistant antenna array (see Fig. 6.1.7a).

Let $a_i(t)$ be additive mixture of useful signal $s(t)$, complex narrowband interference signals $u_l(t)$ (each arriving from l-th point source), and complex quasi-white Gaussian noise $n_i(t)$ with zero mean observed in the input of i-th element of linear antenna array (see Fig. 6.1.7a):

$$\mathbf{a}(t) = \mathbf{s}(t) + \sum_{l=1}^{M} \mathbf{u}_l(t) + \mathbf{n}(t) \Leftrightarrow a_i(t) = s_i(t) + \sum_{l=1}^{M} u_{l,i}(t) + n_i(t)$$

$$= s(t, \omega_s) \exp[-j\omega_s \tfrac{d}{c} \sin \theta_s (i-1)]$$

$$+ \sum_{l=1}^{M} u_l(t, \omega_l) \exp[-j\omega_l \tfrac{d}{c} \sin \theta_l (i-1)] + n_i(t), \quad (6.6.1)$$

where $t = t_m = t_0 + m \cdot \Delta t$; $m = 0, 1, \dots, N-1$; Δt is a sampling interval; $s_i(t)$ is useful signal $s(t)$ observed in i-th channel of linear antenna array; $\omega_s = 2\pi f_s = \text{const}$, f_s is a carrier frequency of useful signal $s(t)$; θ_s is DOA of signal $s(t)$; $\omega_l = 2\pi f_l = \text{const}$, f_l is a carrier frequency of interference signal $u_l(t)$ observed in i-th channel of linear antenna array; θ_l is DOA of interference signal $u_l(t)$; frequencies $\{f_l\}$ of interference signals $\{u_l(t)\}$ are different and distributed in a narrow interval around some central frequency f_0 that coincides with a carrier frequency f_s of useful signal $s(t)$: $f_s = f_0$; $f_l \in [f_0 - \tfrac{\Delta F}{2}, f_0 + \tfrac{\Delta F}{2}]$, $\Delta F \ll f_0$; $n_i(t) = n_i^c(t) + j \cdot n_i^s(t)$ is complex quasi-white Gaussian noise observed in i-th channel of linear antenna array, $j = \sqrt{-1}$; $\mathbf{M}\{u_l(t)u_k(t)\} = 0$, $l \neq k$; $\mathbf{M}\{n_i(t)n_j(t)\} = 0$, $i \neq j$; $\mathbf{M}\{*\}$ is a symbol of mathematical expectation; $\mathbf{M}\{(n_i^c(t))^2\} = \mathbf{M}\{(n_i^s(t))^2\} = D_n/2$, D_n is a variance of complex quasi-white Gaussian noise $n_i(t)$ with zero mean; d is interelement distance of antenna array; c is velocity of electromagnetic wave propagation.

The resulting process $y(t)$ in the output of linear antenna array (see Fig. 6.1.7a) is determined by the expression:

$$y(t) = \sum_{i=1}^{n} \overline{w_i(t)} a_i(t) = \overline{\mathbf{w}^T(t)} \mathbf{a}(t), \quad (6.6.2)$$

where $a_i(t)$ is a received signal in the input of i-th receiving element of antenna array; $w_i(t)$ is weight coefficient in i-th processing channel; $\overline{w_i(t)} = \text{Re}(w_i(t)) - j\,\text{Im}(w_i(t))$; $t = t_m = t_0 + m \cdot \Delta t$; $m = 0, 1, \dots, N-1$; Δt is a sample interval.

Remind, that signal space with lattice properties $\mathcal{L}(\vee, \wedge)$ can be realized by mapping of linear signal space $\mathcal{LS}(+)$ in such a way that the results of interactions $x_\vee(t)$, $x_\wedge(t)$ between the signal $s(t)$ and interference $n(t)$ in $\mathcal{L}(\vee, \wedge)$ with operations of join and meet are determined by the following relationships [24, (7.7.1a,b)]:

$$x_\vee(t) = s(t) \vee n(t) = \{[s(t) + n(t)] + |s(t) - n(t)|\}/2; \tag{6.6.3a}$$

$$x_\wedge(t) = s(t) \wedge n(t) = \{[s(t) + n(t)] - |s(t) - n(t)|\}/2, \tag{6.6.3b}$$

that follows from known equations [29, § XIII.3;(14)], [29, § XIII.4;(22)].

Method of mapping of linear signal space into signal space with lattice properties, that is defined by the relationships (6.6.3a,b), can be realized on the basis of known adaptive filters. Further we consider some possible variants of practical realization of this method using known adaptive algorithms of filtering.

When stating the main material of the section we use Theorem 6.2.1.

6.6.1   Least Mean Squares Algorithm of Spatial Filtering Based on L-group Operations

Least mean squares (LMS) algorithm of spatial filtering is based on steepest descent method [101] without direct inversion of correlation matrix and forms optimal solution for vector of weight coefficients $\mathbf{w}(t)$ that is defined by the following relationships [158, (4.22)], [102, (61.28)], [95, (7.102)], [100, (10.20)]:

$$\mathbf{w}(t_m) = [w_1(t_m), \dots, w_n(t_m)]^T = \mathbf{w}(t_{m-1}) + \mu \cdot \overline{e(t_{m-1})\mathbf{a}(t_{m-1})}; \tag{6.6.4}$$

$$e(t_m) = d(t_m) - \overline{\mathbf{w}^T(t_m)\mathbf{a}(t_m)}, \tag{6.6.4a}$$

where $t = t_m = t_0 + m \cdot \Delta t$; $m = 0, 1, \dots, N - 1$; Δt is a sampling interval; N is a number of samples used in processing; μ is a size of adaptation step that is chosen depending on the following parameters of correlation matrix $\mathbf{R}_a$: $0 < \mu < 1/(3 \operatorname{tr}(\mathbf{R}_a))$ or $0 < \mu < 1/\lambda_{\max}$, where $\operatorname{tr}(\mathbf{R}_a)$ is a trace of ACM $\mathbf{R}_a$; $\lambda_{\max}$ is a maximum eigenvalue of ACM $\mathbf{R}_a$; $\mathbf{a}(t_m) = [a_1(t_m), \dots, a_n(t_m)]^T$ is vector (6.6.1) of the observed process $\mathbf{a}(t)$; $e(t)$ is an error signal; $d(t)$ is a desired signal.

Here we notice that in order to perform LMS algorithm, one must form an error signal determined by the relationship (6.6.4a). In its turn, to obtain such an error signal, one must have a desired signal $d(t)$ characterizing the received useful signal $s(t)$. In information transmitting systems, where useful signal is usually present, this requirement can be achieved by forming a signal that is close to received useful signal.

LMS algorithm of spatial filtering based on L-group operations with pairwise forming the estimators $x_\vee(t)$, $x_\wedge(t)$ and $x'_\vee(t)$, $x'_\wedge(t)$ of join (6.6.3a) and meet (6.6.3b) is defined by the relationships (6.2.11), (6.2.8)...(6.2.10), respectively:

$$z(t) = [\hat{x}_\vee(t) \wedge \hat{x}'_\vee(t)] \vee 0 + [\hat{x}_\wedge(t) \vee \hat{x}'_\wedge(t)] \wedge 0, \tag{6.6.5}$$

$$\hat{x}_\vee(t) = A_1(t) + |A_2(t)|; \tag{6.6.6a}$$

$$\hat{x}_\wedge(t) = A_1(t) - |A_2(t)|; \tag{6.6.6b}$$

$$\hat{x}'_\vee(t) = A_2(t) + |A_1(t)|; \tag{6.6.7a}$$

$$\hat{x}'_\wedge(t) = A_2(t) - |A_1(t)|; \tag{6.6.7b}$$

$$A_1(t) = \mathrm{Re}\left(\sum_{i=1}^{(n+1)/2} \overline{w_{2i-1}(t)} \cdot a_{2i-1}(t)\right); \tag{6.6.8a}$$

$$A_2(t) = \mathrm{Re}\left(\sum_{i=1}^{(n-1)/2} \overline{w_{2i}(t)} \cdot a_{2i}(t)\right); \tag{6.6.8b}$$

where $w_i(t)$ is weight coefficient in i-th processing channel; n is a number of channels in linear antenna array; $A_1(t)$ and $A_2(t)$ are results of weight processing of the observed stochastic process $a_i(t)$ (6.6.1) that are formed on the basis of signals in the outputs of odd and even channels of linear antenna array, respectively.

Weight coefficient $w_i(t)$ in i-th processing channel is formed as a result of optimal solution for the corresponding vector $\mathbf{w}(t)$ determined by the relationship (6.6.4).

As for L-group algorithm (6.6.5)…(6.6.8), Theorem 6.2.1 holds only with respect to real part of stochastic process $y(t)$ (6.6.2) (i.e. $z(t) \equiv \mathrm{Re}[y(t)]$), where $y(t)$ is the resulting signal in the output of adaptive LMS spatial filter and is determined by the relationships (6.6.2), (6.6.4)), inasmuch as the formulas (6.6.8a,b) take into account just real part of the result $y(t)$. Operation of adaptive LMS algorithm (6.6.4)…(6.6.8) based on L-group operations is illustrated by the following example.

Example 6.6.1. Consider 13-elements linear antenna array (see Fig. 6.1.7a) that receives additive mixture of signals (6.6.1). The latter contains: amplitude modulated useful signal $s(t) = A(t)\cos(\omega_s t + \varphi_s)$ arriving from the direction $\theta_s = -\pi/18$, interference signals $u_1(t)$, $u_2(t)$, $u_3(t)$, $u_4(t)$, $u_5(t)$ arriving from the directions $\theta_1 = -\pi/3$; $\theta_2 = \pi/10$; $\theta_3 = 3\pi/14$; $\theta_4 = 7\pi/18$; $\theta_5 = -3\pi/14$, respectively (see Fig. 6.3.2; 6.1.2, 6.1.3), and also quasi-white Gaussian noise $n(t)$. Interference signals $u_1(t)$, $u_2(t)$, $u_3(t)$ are amplitude-modulated oscillation, frequency-modulated oscillation, and also RF pulse with rectangular envelope, respectively; and statistically independent interferences $u_4(t)$, $u_5(t)$ are quasi-white Gaussian noises. Signal-to-k-th interference ratios in each separate i-th receiving channel of antenna array take the values that are closely equal to 0.25; 0.25; 1; 0.05; 0.05, respectively. Signal-to-noise ratio in receiving channels is equal to 500. Number of samples N of signals $a_i(t)$ used in processing (6.6.1) is chosen to be equal $N=1024$. Interelement distance d of linear antenna array is equal $d = 0.5\lambda_0$, where λ_0 is a wave length of useful signal $s(t)$. Scatter of eigenvalues $\lambda_{\max}/\lambda_{\min}$ of ACM $\mathbf{R}_a$ takes the value about $\lambda_{\max}/\lambda_{\min} \approx 3.7 \cdot 10^4$. As a desired signal $d(t)$ figuring in the formula (6.6.4a) we use unmodulated coherent oscillation $d(t) = A\cos(\omega_s t + \varphi_s)$, where $A = \mathrm{const}$.

(a)   (b)

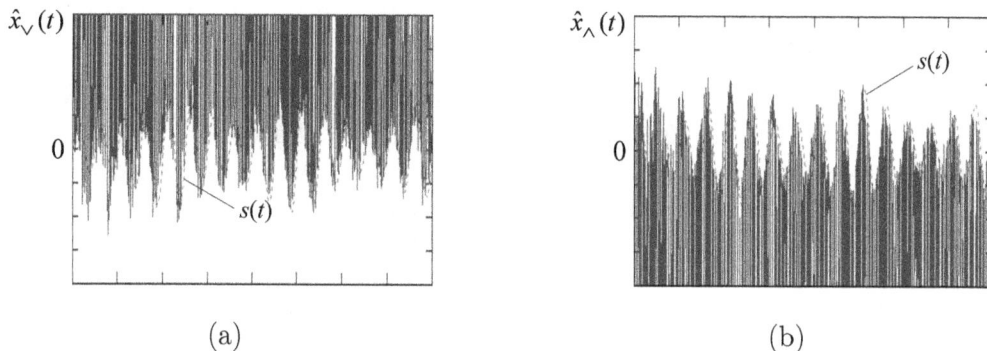

FIGURE 6.6.1   Estimates: (a) $\hat{x}_\vee(t)$ of join (6.6.5a); (b) $\hat{x}_\wedge(t)$ of meet (6.6.5b)

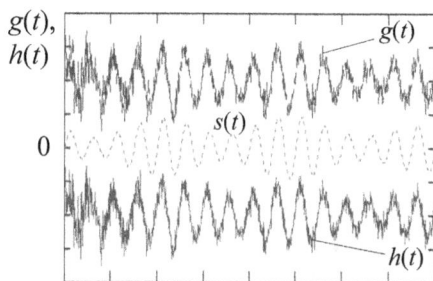

FIGURE 6.6.2   Functions $g(t) = \hat{x}_\vee(t) \wedge$ FIGURE 6.6.3   Output signal $z(t)$ deter-
$\hat{x}'_\vee(t)$, $h(t) = \hat{x}_\wedge(t) \vee \hat{x}'_\wedge(t)$ mined by the equation (6.6.8)

Fig. 6.6.1a,b depicts the estimates $\hat{x}_\vee(t)$, $\hat{x}_\wedge(t)$ of join (6.6.6a) and meet (6.6.6b) (solid line) that are formed by adaptive LMS algorithm based on L-group operations, respectively, and also useful signal $s(t)$ (dotted line).

Fig. 6.6.2 depicts the functions $g(t) = \hat{x}_\vee(t) \wedge \hat{x}'_\vee(t)$ (dashed line), $h(t) = \hat{x}_\wedge(t) \vee \hat{x}'_\wedge(t)$ (solid line) that figure in the equation (6.6.5) and are formed over the estimates (6.6.6a,b); (6.6.7a,b), and also useful signal $s(t)$ (dotted line). For convenience of visual perception, the functions $g(t)$ and $h(t)$ are shifted vertically at ± 2, respectively. Fig. 6.6.3 illustrates the resulting signal $z(t)$ in the output of antenna array (solid line) determined by the equation (6.6.5), useful signal $s(t)$ (dotted line), and also pulse interference $u_3(t)$ (dashed line) scaled down with coefficient 0.75. Correlation coefficients between useful signal $s(t)$ and the signal $z(t)$ in the output of LMS spatial filter based on L-group operations according to algorithm (6.6.5), (6.6.6)...(6.6.8) take their values in the interval $r[z(t), s(t)]=0.72...0.79$. Using auxiliary filtering of the signal $z(t)$, quality of final processing can be improved, however this aspect is considered in Subsection 6.6.3. According to Theorem 6.2.1, the signal (6.6.5) completely repeats the real part of the signal $y(t)$ in the output of LMS spatial filter realizing classic LMS algorithm (6.6.2), (6.6.4): $z(t) \equiv \text{Re}[y(t)]$.
▽

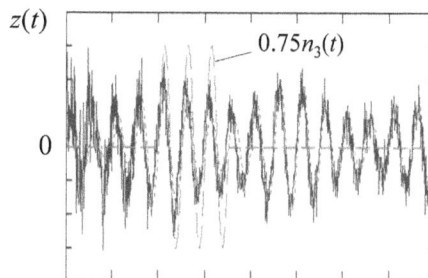

6.6.2   Recursive Least Squares Algorithm of Spatial Filtering Based on L-group Operations

Recursive least squares (RLS) algorithm of spatial filtering is a variant of Wiener spatial filter (see Formula (2.2.20)) in which estimation of a vector of weight coefficients is recursively calculated without direct inversion of correlation matrix, and optimal solution is determined by the following relationships [158, (7.15)], [102, (21.37)... (21.39)], [95, (7.78)... (7.81)], [100, (10.32)]:

$$\mathbf{w}(t_m) = [w_1(t_m), \ldots, w_n(t_m)]^T = \mathbf{w}(t_{m-1}) + \overline{e(t_m)}\mathbf{k}(t_m); \qquad (6.6.9)$$

$$e(t_m) = d(t_m) - \overline{\mathbf{w}^T(t_m)}\mathbf{a}(t_m); \qquad (6.6.9a)$$

$$\mathbf{k}(t_m) = \frac{\lambda^{-1}\mathbf{\Phi}_a(t_{m-1})\mathbf{a}(t_m)}{1 + \lambda^{-1}\overline{\mathbf{a}^T(t_m)}\mathbf{\Phi}_a(t_{m-1})\mathbf{a}(t_m)}; \qquad (6.6.9b)$$

$$\mathbf{\Phi}_a(t_m) = \lambda^{-1}\mathbf{\Phi}_a(t_{m-1}) - \lambda^{-1}\mathbf{k}(t_m)\overline{\mathbf{a}^T(t_m)}\mathbf{\Phi}_a(t_{m-1}); \qquad (6.6.9c)$$

$$\mathbf{\Phi}_a(t_0) = \hat{\mathbf{R}}_a^{-1}(t_0) = c \cdot \mathbf{I}, \ \mathbf{I} = [\delta_{ik}]; \qquad (6.6.9d)$$

$$\mathbf{w}(t_0) = [1, 1, \ldots, 1]^T, \qquad (6.6.9e)$$

where $t = t_m = t_0 + m \cdot \Delta t$; $m = 0, 1, \ldots, N-1$; Δt is a sample interval; N is a number of samples used in processing; $\mathbf{a}(t_m) = [a_1(t_m), \ldots, a_n(t_m)]^T$ is vector (6.6.1) of the observed process $\mathbf{a}(t)$; $e(t)$ is an error signal; $d(t)$ is a desired signal; $\mathbf{\Phi}_a(t) = \hat{\mathbf{R}}_a^{-1}(t)$; $\hat{\mathbf{R}}_a(t)$ is an estimator of correlation matrix of vector $\mathbf{a}(t_m) = [a_1(t_m), \ldots, a_n(t_m)]^T$; λ is forgetting coefficient; the relationship (6.6.9b) determines gain vector with dimensionality $n \times 1$; the relationship (6.6.9c) determines iterative modification of matrix $\mathbf{\Phi}_a(t) = \hat{\mathbf{R}}_a^{-1}(t)$; the relationships (6.6.9d), (6.6.9e) determine initial conditions for correlation matrix and vector of weight coefficients, respectively; $\mathbf{I}$ is identity matrix; δ_{ik} is Kronecker symbol: $\delta_{ik} = 1$, if $i = k$; $\delta_{ik} = 0$, if $i \neq k$; c=const.

RLS algorithm of spatial filtering based on L-group operations is defined by the relationship (6.6.5):

$$z(t) = [\hat{x}_\vee(t) \wedge \hat{x}'_\vee(t)] \vee 0 + [\hat{x}_\wedge(t) \vee \hat{x}'_\wedge(t)] \wedge 0, \qquad (6.6.10)$$

so that pairwise forming of the estimators $\hat{x}_\vee(t)$, $\hat{x}_\wedge(t)$ and $\hat{x}'_\vee(t)$, $\hat{x}'_\wedge(t)$ of join and meet is performed according to the relationships (6.6.6)... (6.6.8), and weight coefficient $w_i(t)$ in i-th processing channel is formed according to optimal solution for vector of weight coefficients $\mathbf{w}(t)$ determined by the relationship (6.6.9).

As for algorithm (6.6.10), (6.6.9), (6.6.6)... (6.6.8), Theorem 6.2.1 holds only with respect to real part of stochastic process $y(t)$ (6.6.2) (i.e. $z(t) \equiv \text{Re}[y(t)]$), where $y(t)$ is the resulting signal in the output of adaptive RLS spatial filter and is determined by the relationships (6.6.2), (6.6.9)), since the formulas (6.6.8a,b) take into account just real part of weight processing. Operation of adaptive RLS algorithm (6.6.10), (6.6.9), (6.6.6)... (6.6.8) based on L-group operations is illustrated by the following example.

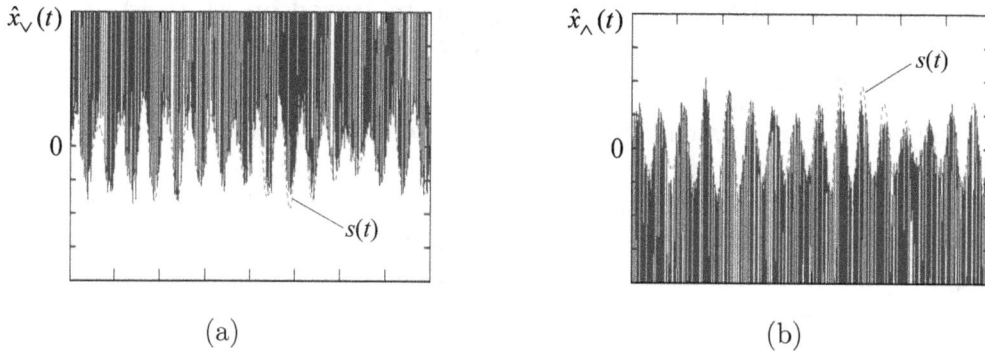

(a)

(b)

FIGURE 6.6.4   Estimates: (a) $\hat{x}_\vee(t)$ of join (6.6.6a); (b) $\hat{x}_\wedge(t)$ of meet (6.6.6b)

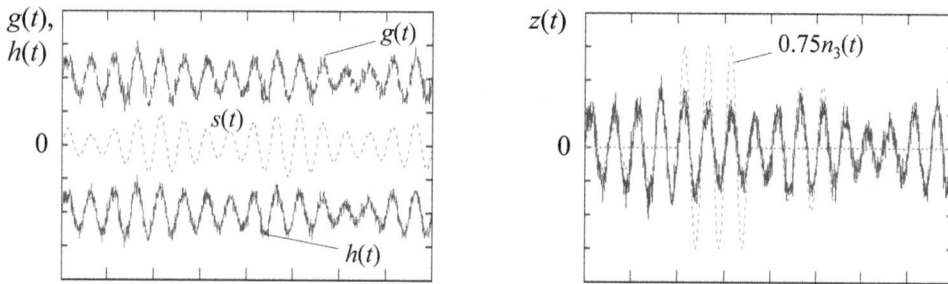

FIGURE 6.6.5   Functions $g(t) = \hat{x}_\vee(t) \wedge \hat{x}'_\vee(t)$, $h(t) = \hat{x}_\wedge(t) \vee \hat{x}'_\wedge(t)$

FIGURE 6.6.6   Output signal $z(t)$ determined by the equation (6.6.10)

Example 6.6.2. Consider again 13-elements linear antenna array (see Fig. 6.1.7a) that receives additive mixture of signals (6.6.1). The receiving conditions are similar as shown in Example 6.6.1.

Fig. 6.6.4a, b depicts the estimates $\hat{x}_\vee(t)$, $\hat{x}_\wedge(t)$ of join (6.6.6a) and meet (6.6.6b), respectively, (solid line), formed by adaptive RLS algorithm based on L-group operations, and also useful signal $s(t)$ (dotted line).

Fig. 6.6.5 illustrates the functions $g(t) = \hat{x}_\vee(t) \wedge \hat{x}'_\vee(t)$ (dashed line), $h(t) = \hat{x}_\wedge(t) \vee \hat{x}'_\wedge(t)$ (solid line), that figure in the equation (6.6.10) and are formed over the estimates (6.6.6a,b); (6.6.7a,b), and also useful signal $s(t)$ (dotted line). For convenience of visual perception, the functions $g(t)$ and $h(t)$ are shifted vertically at ± 2, respectively. Fig. 6.6.6 depict the resulting signal $z(t)$ in the output of antenna array (solid line) determined by the equation (6.6.10), useful signal $s(t)$ (dotted line), and also pulse interference $u_3(t)$ (dashed line) scaled down with coefficient 0.75. Correlation coefficients between useful signal $s(t)$ and the signal $z(t)$ in the output of RLS spatial filter based on L-group operation according to algorithm (6.6.10), (6.6.9), (6.6.6)...(6.6.8) take their values in the interval $r[z(t), s(t)] = 0.79...0.94$. Using auxiliary filtering of the signal $z(t)$, quality of final processing can be improved, however this aspect is considered in Subsection 6.6.3. According to Theorem 6.2.1, the signal (6.6.10) completely repeats real part of the signal $y(t)$ in the output of adaptive classic RLS spatial filter (6.6.2), (6.6.9): $z(t) \equiv \mathrm{Re}[y(t)]$. $\triangledown$

6.6.3 Adaptive Spatial Filtering Algorithm Based on Method of Recursive Forming Measure of Statistical Interrelation Matrix Estimator

In some multichannel systems (for instance, in radar and sonar systems), exploiting a desired signal assumed to be present in LMS and RLS algorithms is impossible. Owing to confined computational resources of a system, it is desirable to avoid difficulties related to direct inversion of correlation matrix. For successful overcoming computational problems, one can use another class of algorithms based on recursive method of processing [173, 174]. Adaptive algorithm based on recursive inversion of correlation matrix [158, (7.29)] is defined by the relationships:

$$\mathbf{w}(t_m) = [w_1(t_m), \ldots, w_n(t_m)]^T =$$

$$\frac{1}{1 - \beta} \left[\mathbf{w}(t_{m-1}) - \mathbf{k}(t_m) \left(\overline{\mathbf{a}^T(t_m)} \mathbf{w}(t_{m-1}) \right) \right] ; \tag{6.6.11}$$

$$\mathbf{k}(t_m) = \frac{\beta \boldsymbol{\Phi}_a(t_{m-1}) \mathbf{a}(t_m)}{(1 - \beta) + \beta \overline{\mathbf{a}^T(t_m)} \boldsymbol{\Phi}_a(t_{m-1}) \mathbf{a}(t_m)}; \tag{6.6.11a}$$

$$\boldsymbol{\Phi}_a(t_m) = \frac{1}{1 - \beta} \left(\boldsymbol{\Phi}_a(t_{m-1}) - \mathbf{k}(t_m) \overline{\mathbf{a}^T(t_m)} \boldsymbol{\Phi}_a(t_{m-1}) \right) ; \tag{6.6.11b}$$

$$\boldsymbol{\Phi}_a(t_0) = \hat{\mathbf{R}}_a^{-1}(t_0) = c \cdot \mathbf{I}, \ \mathbf{I} = [\delta_{ik}]; \tag{6.6.11c}$$

$$\mathbf{w}(t_0) = [1, 1, \ldots, 1]^T; \tag{6.6.11d}$$

$$y(t) = \sum_{i=1}^{n} \overline{w_i(t)} a_i(t) = \overline{\mathbf{w}^T(t)} \mathbf{a}(t), \tag{6.6.12}$$

where $t = t_m = t_0 + m \cdot \Delta t$; $m = 0, 1, \ldots, N - 1$; Δt is a sampling interval; N is a number of samples used in processing; $\mathbf{a}(t_m) = [a_1(t_m), \ldots, a_n(t_m)]^T$ is vector (6.6.1) of the observed process $\mathbf{a}(t)$; the relationship (6.6.11a) determines a gain vector with dimensionality $n \times 1$; the relationship (6.6.11b) determines iterative modification of matrix $\boldsymbol{\Phi}_a(t) = \hat{\mathbf{R}}_a^{-1}(t)$; β is relative weight of current data that is chosen depending on the following parameters of correlation matrix $\hat{\mathbf{R}}_a$: $0 < \beta < 1/(10 \operatorname{tr}(\mathbf{R}_a))$ or $0 < \beta < 1/10\lambda_{\max}$, here $\operatorname{tr}(\mathbf{R}_a)$ is a trace of matrix $\hat{\mathbf{R}}_a$; $\lambda_{\max}$ is maximum eigenvalue of matrix $\hat{\mathbf{R}}_a$; the relationships (6.6.11c), (6.6.11d) determine initial conditions for autocorrelation matrix and vector of weight coefficients, respectively; $\mathbf{I}$ is identity matrix; δ_{ik} is Kronecker symbol: $\delta_{ik} = 1$, if $i = k$; $\delta_{ik} = 0$, if $i \neq k$; c=const.

Adaptive algorithm based on recursive method of matrix inversion applied to the matrix $\hat{\mathbf{N}}_{a,s}$ (6.3.6) of measure of statistical interrelation (MSI) (1.4.11) is synthesized on the basis of the following considerations. Let $\hat{\mathbf{A}}_{k+1}$ be recursive estimator of a matrix $\mathbf{A}$ with the dimensionality $n \times n$ at $(k+1)$-th step of iteration in the form:

$$\hat{\mathbf{A}}_{k+1} = \hat{\mathbf{A}}_k + \beta \mathbf{T}_k, \tag{6.6.13}$$

where $\mathbf{T}_k$ is a matrix of current data containing information on the estimated matrix $\mathbf{A}$; β=const, $0 < \beta << 1$.

Initial expression (6.6.13) can be rewritten in the following form:

$$\hat{\mathbf{A}}_{k+1} = \hat{\mathbf{A}}_k(\mathbf{I} + \beta\hat{\mathbf{A}}_k^{-1}\mathbf{T}_k), \tag{6.6.14}$$

where $\mathbf{I}$ is identity matrix with dimensionality $n \times n$.
Then recursive matrix estimator $\hat{\mathbf{A}}_{k+1}^{-1}$ that is inverse with respect to initial one $\hat{\mathbf{A}}_{k+1}$ takes the form:

$$\hat{\mathbf{A}}_{k+1}^{-1} = (\mathbf{I} + \beta\hat{\mathbf{A}}_k^{-1}\mathbf{T}_k)^{-1}\hat{\mathbf{A}}_k^{-1}. \tag{6.6.15}$$

For infinitely small linear mapping $\mathbf{I} + \beta\hat{\mathbf{A}}_k^{-1}\mathbf{T}_k$ figuring in (6.6.15), the following relationship holds [36, (14.4.20)]:

$$(\mathbf{I} + \beta\hat{\mathbf{A}}_k^{-1}\mathbf{T}_k)^{-1} = \mathbf{I} - \beta\hat{\mathbf{A}}_k^{-1}\mathbf{T}, \ |\beta|^2 << 1. \tag{6.6.16}$$

Substituting (6.6.16) into (6.6.15), obtain the resulting expression for recursive matrix estimator $\hat{\mathbf{A}}_{k+1}^{-1}$ that is inverse with respect to initial one $\hat{\mathbf{A}}_{k+1}$:

$$\hat{\mathbf{A}}_{k+1}^{-1} = (\mathbf{I} - \beta\hat{\mathbf{A}}_k^{-1}\mathbf{T}_k)\hat{\mathbf{A}}_k^{-1}. \tag{6.6.17}$$

As follows from the relationship (6.6.17), adaptive algorithm based on recursive method of matrix inversion applied to MSI matrix estimator $\hat{\mathbf{N}}_{a,s}$ (6.3.6) is defined by the following relationships:

$$\mathbf{w}(t_m) = [w_1(t_m), \ldots, w_n(t_m)]^T = [\mathbf{I} - \beta\boldsymbol{\Phi}_a(t_{m-1})\mathbf{T}(t_{m-1})] \cdot \mathbf{w}(t_{m-1}); \tag{6.6.18}$$

$$\boldsymbol{\Phi}_a(t_m) = [\mathbf{I} - \beta\boldsymbol{\Phi}_a(t_{m-1})\mathbf{T}(t_{m-1})] \cdot \boldsymbol{\Phi}_a(t_{m-1}); \tag{6.6.18a}$$

$$\mathbf{T}(t) = T[\text{Re}(a_i(t)), \text{Re}(a_k(t))] + T[\text{Im}(a_i(t)), \text{Im}(a_k(t))] -$$
$$-j \cdot (T[\text{Re}(a_i(t)), \text{Im}(a_k(t))] - T[\text{Im}(a_i(t)), \text{Re}(a_k(t))]); \tag{6.6.18b}$$

$$T[u, v] = |u| + |v| - |u - v| \cdot [\text{sgn}(u \vee v) - \text{sgn}(u \wedge v)]; \tag{6.6.18c}$$

$$\boldsymbol{\Phi}_a(t_0) = \hat{\mathbf{N}}_{a,s}^{-1}(t_0) = \mathbf{I}, \ \mathbf{I} = [\delta_{ik}]; \tag{6.6.18d}$$

$$\mathbf{w}(t_0) = [1, 1, \ldots, 1]^T; \tag{6.6.18e}$$

$$z(t) = \sum_{i=1}^{n} \overline{w_i(t)} a_i(t) = \overline{\mathbf{w}^T(t)}\mathbf{a}(t), \tag{6.6.19}$$

where $t = t_m = t_0 + m \cdot \Delta t$; $m = 0, 1, \ldots, N - 1$; Δt is a sampling interval; N is a number of samples used in processing; $\mathbf{a}(t_m) = [a_1(t_m), \ldots, a_n(t_m)]^T$ is vector (6.6.1) of the observed process $\mathbf{a}(t)$; the relationship (6.6.18a) determines iterative modification of matrix $\boldsymbol{\Phi}_a(t) = \hat{\mathbf{N}}_{a,s}^{-1}(t)$; $\hat{\mathbf{N}}_{a,s}(t)$ is MSI matrix estimator of vector $\mathbf{a}(t_m) = [a_1(t_m), \ldots, a_n(t_m)]^T$; $\mathbf{T}(t)$ is a matrix of current data containing information on the estimated MSI matrix $\hat{\mathbf{N}}_{a,s}^{-1}$; $T[u, v]$ is a measure of closeness of real variables u, v; β is a relative weight of current data; the relationships (6.6.18d), (6.6.18e) determine initial conditions for MSI matrix and vector of weight coefficients, respectively; $\mathbf{I}$ is identity matrix; δ_{ik} is Kronecker symbol: $\delta_{ik} = 1$, if $i = k$; $\delta_{ik} = 0$, if $i \neq k$; c=const; $j = \sqrt{-1}$.
Operation of adaptive L-group algorithm (6.6.18), (6.6.19) based on recursive method of matrix inversion applied to MSI matrix estimator $\hat{\mathbf{N}}_{a,s}$ is illustrated by the following example.

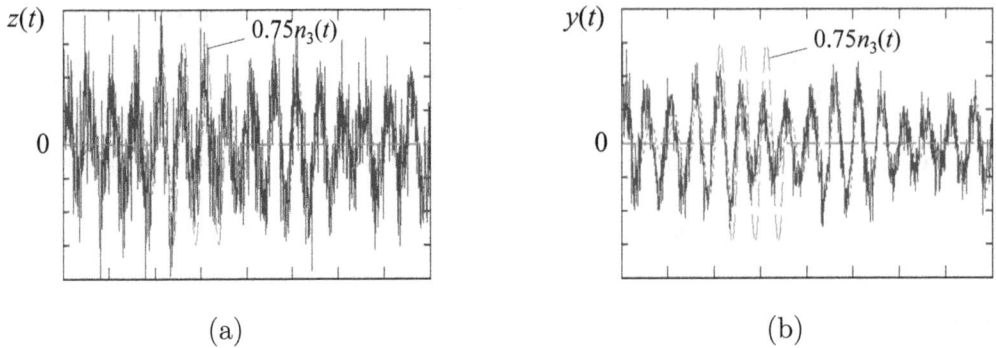

FIGURE 6.6.7   Output signals: (a) $z(t)$ defined by the equation (6.6.19), useful signal $s(t)$, and also pulse interference signal $u_3(t)$; (b) $y(t)$ defined by the equation (6.6.12), useful signal $s(t)$, and also pulse interference signal $u_3(t)$

Example 6.6.3. Consider again 13-elements linear antenna array (see Fig. 6.1.7a) that receives additive mixture of signals (6.6.1). The receiving conditions are similar as described in Example 6.6.1.

Fig. 6.6.7a depicts the resulting signal $z(t)$ in the output of array (solid line) defined by the relationship (6.6.19), useful signal $s(t)$ (dotted line), and also pulse interference signal $u_3(t)$ (dashed line) scaled down with coefficient 0.75. Correlation coefficients between useful signal $s(t)$ and the signal $z(t)$ in the output of spatial filter based on L-group operations according to algorithm (6.6.18), (6.6.19) take low values $r[z(t), s(t)]=0.51\ldots0.57$ owing to considerable level of noise related to remains of interference compensation, and also noise related to nonlinearity of processing when forming matrix of current data (6.6.18b,c). Fig. 6.6.7b depicts the resulting signal $y(t)$ (solid line) determined by the relationships (6.6.11), (6.6.12), useful signal $s(t)$ (dotted line), and also pulse interference signal $u_3(t)$ (dashed line) scaled down with coefficient 0.75. Correlation coefficients between useful signal $s(t)$ and the signal $y(t)$ in the output of spatial filter according to linear algorithm (6.6.11), (6.6.12) accepted the values in the interval $r[z(t), s(t)]=0.67\ldots0.73$. In both figures, there is some phase mismatching between, on the one hand, the signals $z(t)$ (6.6.19) and $y(t)$ (6.6.12), and on the other hand, and useful signal $s(t)$; this mismatching is compensated during the final stage of processing.

Fig. 6.6.8a,b illustrates the results of discrete Fourier transform of the signals $z(t)$ (6.6.19) and $y(t)$ (6.6.12), respectively. Spectrum of realization of the signal $s(t)$ is shown by solid line, and spectra $Z(f)$, $Y(f)$ of realizations of the signals $z(t)$ (6.6.19) and $y(t)$ (6.6.12), respectively, are shown by dotted line. A variance of noise component contained in the signal $z(t)$ is about an order greater than a variance of noise component of the signal $y(t)$, that can be seen in Fig. 6.6.7a,b. In the same time one can notice that spectrum of useful signal component is slightly changed. The last circumstance create a possibility for improving quality of the processed signal by known methods.

Fig. 6.6.9a depicts the signal $z'(t)$ (solid line) obtained by median filtering of the signal $z(t)$ (6.6.19) with simultaneous correction of its phase. Fig. 6.6.9a also

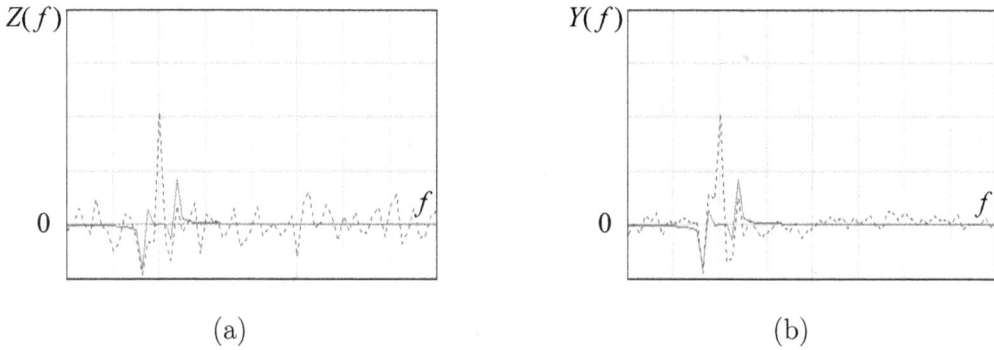

(a)   (b)

FIGURE 6.6.8   Results of discrete Fourier transforms of the signals: (a) $z(t)$ (6.6.19) (b) $y(t)$ (6.6.12)

demonstrates useful signal $s(t)$ (dotted line) and pulse interference $n_3(t)$ (dashed line) scaled down with coefficient 0.75. Correlation coefficients between useful signal $s(t)$ and the resulting signal $z'(t)$ after median filtering and phase correction take the values in the interval $r[z'(t), s(t)]=0.92\ldots0.95$. Fig. 6.6.9b illustrates the signal $y'(t)$ (solid line) obtained by bandpass filtering of the signal $y(t)$ (6.6.12) with simultaneous correction of its phase. Fig. 6.6.9b also demonstrates useful signal $s(t)$ (dotted line) and pulse interference $n_3(t)$ (dashed line) scaled down with coefficient 0.75. Correlation coefficients between useful signal $s(t)$ and the resulting signal $y'(t)$ after bandpass filtering and phase correction take their values in the interval $r[y'(t), s(t)]=0.92\ldots0.96$. A ratio of sample variances of filtering error for the signals $z'(t)$ and $y'(t)$ does not exceed 1 dB:

$$10\lg\left(\frac{D[z'(t) - s(t)]}{D[y'(t) - s(t)]}\right) \leqslant 1,$$

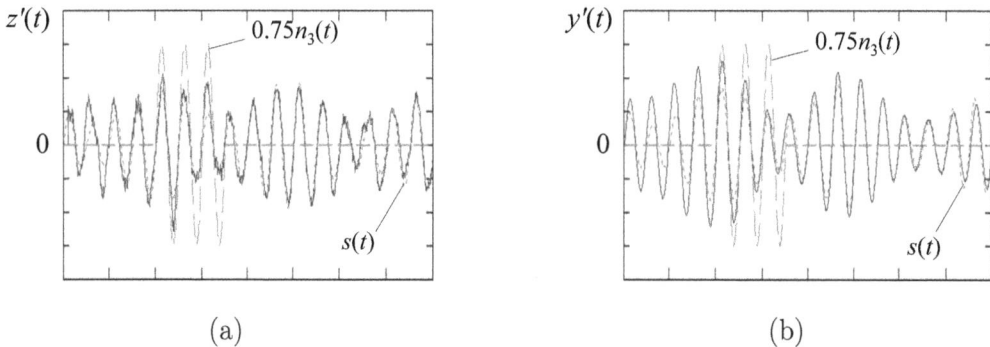

(a)   (b)

FIGURE 6.6.9   (a) Signal $z'(t)$ obtained by median filtering of the signal $z(t)$ (6.6.19) with simultaneous correction of its phase; (b) signal $y'(t)$ obtained by bandpass filtering of the signal $y(t)$ (6.6.12) with simultaneous correction of its phase

thus, exploiting lesser computational resources can make adaptive algorithm (6.6.18), (6.6.19) based on *L*-group operations more preferable as against classic linear adaptive algorithm based on recursive inversion of correlation matrix (6.6.11), (6.6.12).   ▽

It should be noted that adaptive *L*-group algorithm (6.6.18), (6.6.19) can be simplified by excluding the equation of recursive modification of a matrix (6.6.18a), so that the equation of recursive modification of vector of weight coefficients (6.6.18) takes the form:

$$\mathbf{w}(t_m) = [w_1(t_m), \ldots, w_n(t_m)]^T = [\mathbf{I} - \beta \mathbf{T}(t_{m-1})] \cdot \mathbf{w}(t_{m-1}), \qquad (6.6.20)$$

that insufficiently affect the resulting quality of signal filtering when preserving operations of median filtering and phase correction.

7

Signal Processing L-group Algorithms for Communication Systems and Networks

---

Multichannel and multi-station systems of information transmitting was known for a long time, despite the fact that some of them found their practical realization relatively recently [175–183]. Most of them use an idea on signal representation in the form of series of orthogonal functions (see Chapter 1) in linear spaces with scalar product that is practically realized on the basis of technologies with frequency division multiplexing/multiple access (FDM/FDMA); time division multiplexing/multiple access (TDM/TDMA); code division multiplexing/multiple access (CDM/CDMA), their variants and combinations.

In this chapter, along with known signal processing algorithms that exploit the corresponding orthogonal signal systems, we consider signal processing algorithms based on L-group operations and operating within the framework of statistical demultiplexing approach stated in Chapter 1.

7.1 Multichannel Communication Systems and Multi-station Networks

A goal of multichannel information transmitting systems is simultaneous message transmitting from a set of sources to a corresponding set of users (addressees). Generalized block diagram of multichannel information transmitting system is presented in Fig. 7.1.1. Messages $u_1(t), \ldots, u_n(t)$ coming from message sources $\text{Src}_1, \ldots, \text{Src}_n$ are fed in channel signal coders $\text{CC}_1, \ldots, \text{CC}_n$ that perform coding both a source and a channel. Channel coding is based on the signals that are produced by a coding signal generator (CSG). In the outputs of channel signal coders, modulated (coded) channel signals $s_1(t), \ldots, s_n(t)$ are united by additive mixer into a group signal $s_g(t)$. Group signal $s_g(t)$ arrives into common modulator (M) that modulates a carrier generated by a transmitter (Tx).

A received signal is processed by filtering, amplifying, etc. in a receiver (Rx) and is demodulated in common demodulator (D). An extracted estimate of a group signal $\hat{s}_g(t)$ comes in the inputs of channel signal selectors (CS), each of them extracts the corresponding estimate $\hat{s}_k(t)$ of a channel signal from the estimate $\hat{s}_g(t)$ of group signal. In some cases, operation of channel signal selectors (CS) is controlled by selector signals that are produced by a selector signal generator (SSG). Such a scheme is realized, for instance, in multichannel information transmitting

---

DOI: 10.1201/9781003275855-7

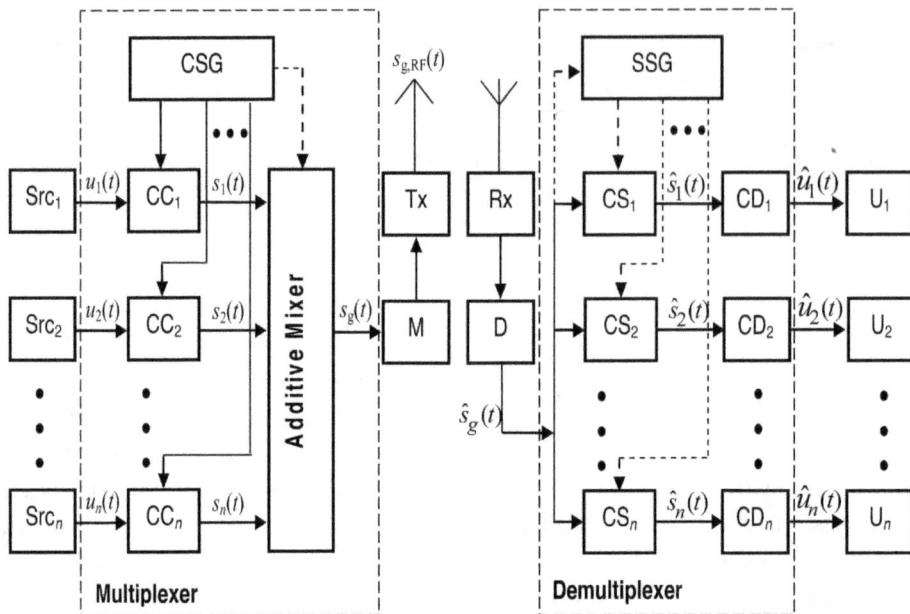

FIGURE 7.1.1 Generalized block diagram of multichannel information transmitting system

systems with TDM, CDM, multi-carrier modulation (MCM), and orthogonal frequency division multiplexing (OFDM). Estimates of channel signals $\hat{s}_k(t)$ are decoded in channel signal decoders (CD), as a result, estimates of messages $\hat{u}_k(t)$ are extracted and achieve users of messages U_k. Channel signal coders (CC), channel signal generator (CSG), and additive mixer form a multiplexer in the transmitting side of information transmitting system. Channel signal selectors (CS), selector signal generator (SSG), and channel decoders (CD) form a demultiplexer in the receiving side.

In multichannel information transmitting systems, one common path (the input of common modulator (M)—the output of common demodulator (D)) is used for message transmission from a set of sources to a set of users. As a rule, this common path including common modulator (M), transmitter (Tx), receiver (Rx), and common demodulator (D) is the most expensive part of multichannel information transmitting system. So, simultaneous (quasi-simultaneous) transmission of many messages by a multichannel information transmitting system is economically more expedient than a transmission of messages by a set of one-channel information transmitting systems.

Existing and being developed information transmitting systems must provide simultaneous information transmission between a large number of stationary and mobile users that are arbitrarily situated in some spatial area. In such systems, it is necessary to provide multiple access within common frequency band in which users transmit and/or receive information regardless of each other in such time intervals

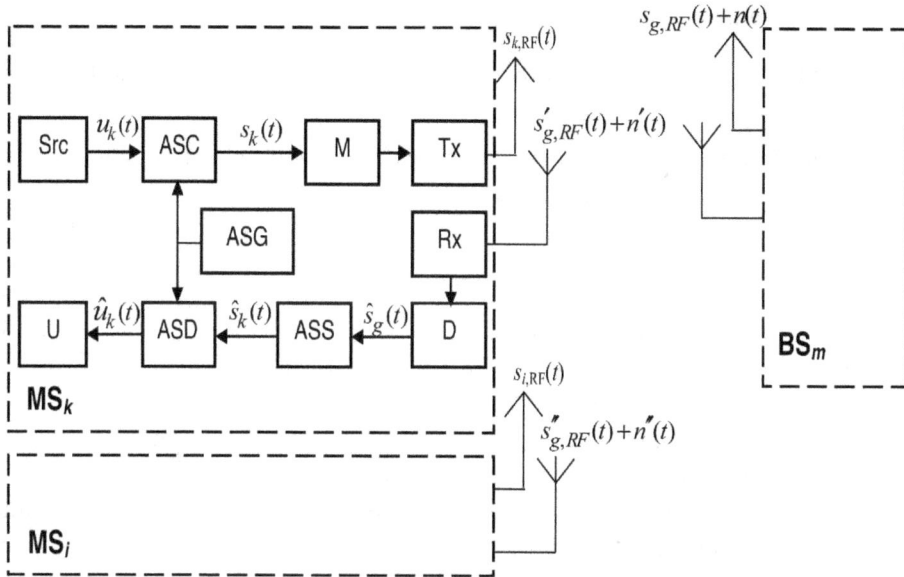

FIGURE 7.1.2 Generalized block diagram of multi-station information transmitting system (network)

when it is necessary. Multi-station networks play the main role in information transmission between mobile users. Multiple access within a common frequency band is the most expedient technology for constructing information transmitting system of various functionality, for instance, such as TDMA, CDMA, orthogonal frequency division multiple access (OFDMA) systems, and their combinations.

Within the further discussion, we take into account that main notions introduced for multichannel information transmitting systems can be used when considering multi-station information transmitting systems (networks). Usually, such a system includes some number of users, each of them is a source of discrete or continuous messages. A message of each user transforms into a signal $s_k(t)$, $k = 1, \ldots, n$, that is said to be address signal. However, multi-station (multiple access) systems have some distinctions with respect to multichannel systems. Thus, a group signal $s_g(t)$ is formed as a result of addition of RF signals directly in information transmitting channel (see Fig. 7.1.2); time synchronization of information sources is absent; levels of received signals can essentially differ, for instance, owing to various lengths of signal propagation paths.

In multiple access systems, all address signals $s_k(t)$, $k = 1, \ldots, n$ can be beforehand distributed and fixed between the corresponding users, or these signals can be distributed between the users just within a user session after which these signals are used by other users of the system. Method of address signal distribution between all users is determined by both interaction of stations in the multiple access system and users activity.

Multiple access systems can be constructed according to one of three variants of network creation:

1. information transmission between users is realized through the base (central) station;

2. information transmission between users is realized directly between mobile stations of users without exploiting base station;

3. information transmission between users can be realized both over base station and directly between mobile stations of users.

The examples corresponding to these three variants of multiple access information transmitting systems are well known mobile multi-station networks of GSM, WCDMA, LTE standards; some VHF and UHF communication systems of both commerce and military purposes operating under line-of-sight propagation conditions; and also trunking telecommunication systems.

Multi-station information transmitting system contains some set of mobile stations $\{MS_i\}$ and also some set of base stations $\{BS_l\}$ (see Fig. 7.1.2). In a mobile station $\{MS_k\}$, a message $u_k(t)$ arriving from a source of message (Src) enters an address signal coder (ASC) which performs both source coding and channel coding. Channel coding is based on the signals produced by address signal generator (ASG). Address signal $s_k(t)$ enters modulator (M) and modulates a carrier generated by a transmitter (Tx).

A received group signal $s_{g,RF}(t)$ together with interference $n(t)$ is processed (by filtering, amplification, etc.) in a receiver (Rx) and is demodulated in demodulator (D). An extracted estimate of the group signal $\hat{s}_g(t)$ enters an address signal selector (ASS) which extracts the corresponding estimate of address signal $\hat{s}_k(t)$ from the estimate $\hat{s}_g(t)$. Estimate of address signal $\hat{s}_k(t)$ is decoded in address signal decoder (ASD); on the basis of this estimate, a message estimate $\hat{u}_k(t)$ is extracted and enters to a user of a message (U).

In both multichannel and multi-station information transmitting systems, modulated channel signals possess such characteristics that provide these signals to be classified in a receiving part of the system. The most used methods providing this feature are linear methods, though nonlinear methods of channel multiplexing (multiple access) are also known [183].

When realizing linear channel multiplexing, a group signal is a sum of modulated channel signals:

$$s_g(t) = \sum_{k=1}^{n} s_k(t). \tag{7.1.1}$$

Operation of channel/address signal selectors (CS/ASS) (Figs. 7.1.1, 7.1.2) is determined by linear functions L_i:

$$L_i[s_g(t)] = L_i[\sum_{k=1}^{n} s_k(t)] = \sum_{k=1}^{n} L_i[s_k(t)] = \begin{cases} s_k(t), & k = i; \\ 0, & k \neq i. \end{cases} \tag{7.1.2}$$

The necessary and sufficient condition for fulfillment of the relationship (7.1.2) is a linear independence of channel signals $s_k(t)$. This means that the identity

$$\sum_{k=1}^{n} a_k \cdot s_k(t) \equiv 0, \ a_k = \text{const} \tag{7.1.3}$$

holds if and only if all a_k are equal to zero, i.e. if the signals are linearly independent, and none of them can be represented by linear combination of other signals.

If we take into account the case of classification of channel (address) signals when developing a multichannel (multi-station) information transmitting system and do not take into consideration other factors, then it is sufficient that modulated channel signals be linearly independent, i.e. the condition (7.1.2) holds. However, if one takes into account interference impact, then an advantage of orthogonal signals becomes obvious. Besides, processing units intended for orthogonal signals are easier to realize than those designed for nonorthogonal signals. Therefore, as a rule, channel (address) signals are chosen to be orthogonal. Some particular cases of orthogonal signals are those whose spectra are not overlapped, and also the signals that are not overlapped in time domain. The first of them are used as channel (address) signals in FDM(A) systems, and the second — in TDM(A) systems. Besides, there exist orthogonal signals overlapping in both frequency and time domains, but differing in their structure. Also, all possible combinations of aforementioned signals can be used as orthogonal signals.

7.2 Main Types of Orthogonal Signal Systems Used in Information Transmitting Systems

There is a great number of systems of orthogonal signals that are used in multi-channel (multi-station) information transmitting systems for the purposes of multi-plexing/demultiplexing channel signals. Conditionally, these systems of signals can be subdivided into the following:

- systems of orthogonal baseband pulses;

- systems of orthogonal multi-frequency (multi-carrier) signals;

- systems of orthogonal discrete phase-shift keying signals;

- systems of orthogonal discrete frequency-shift keying signals.

Basing on the combinations of these four main systems of signals, one can form other systems of orthogonal signals.

A system $\{p_k(t)\}$, $k = 1, \ldots, N$ of N orthogonal baseband pulses defined in the time interval $T_{\text{sys}} = [0, T]$ (see Fig. 7.2.1) is a set of the following functions:

$$p_k(t) = 1(t - (k \cdot T/N)) - 1(t - ((k + 1) \cdot T/N)), \tag{7.2.1}$$

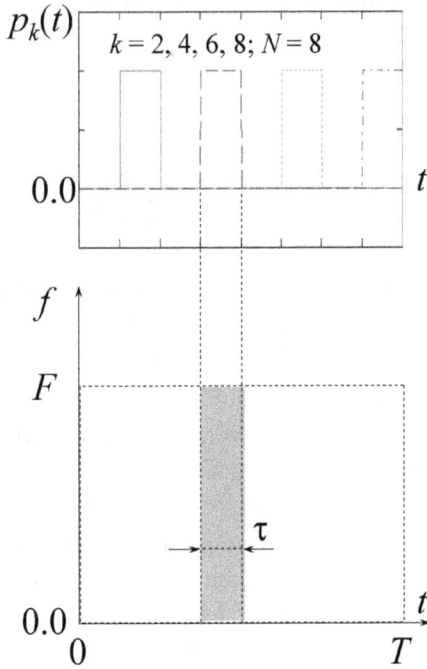

FIGURE 7.2.1   System $\{p_k(t)\}$ of $N = 8$ orthogonal baseband pulses (7.2.1)

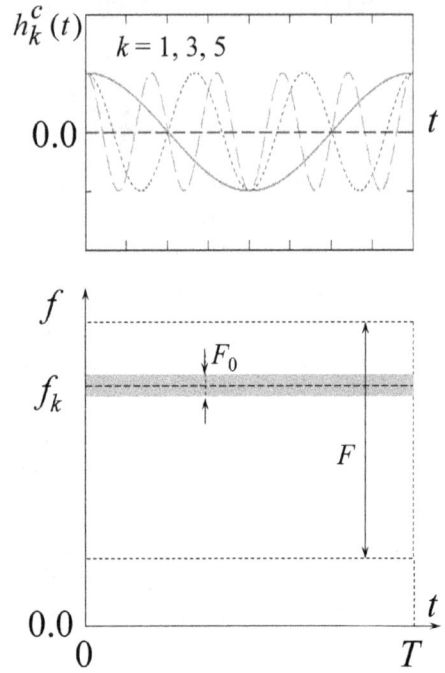

FIGURE 7.2.2   System of orthogonal multi-frequency signals (7.2.2a)

where $1(t)$ is Heaviside unit step function:

$$1(t) = \begin{cases} 1, & t \geqslant 0; \\ 0, & t < 0. \end{cases}$$

In time-frequency plane (Fig. 7.2.1), gray background depicts energy distribution of a single signal $p_k(t)$ from eight-element system $\{p_k(t)\}$, $N=8$. All the signals completely cover an allocated rectangle with the sides F and T, where signal spectrum bandwidth F of a single signal $p_k(t)$ is approximately equal to $F \approx N/T$.

A system of orthogonal multi-frequency (multi-carrier) signals (see Fig. 7.2.2) is a set of harmonic signals $\{h_k^c(t)\}$ ($\{h_k^s(t)\}$), whose amplitudes and phases are determined according to signal forming principles. The examples of such systems of orthogonal multi-frequency (multi-carrier) signals defined in the time interval $T_{\text{sys}} = [0, T]$ are the sets of signals $\{h_k^c(t)\}$ and $\{h_k^s(t)\}$:

$$h_k^c(t) = \cos(2\pi(a \cdot k + b)t/T + \alpha_c); \tag{7.2.2a}$$

$$h_k^s(t) = \sin(2\pi(a \cdot k + b)t/T + \alpha_s), \tag{7.2.2b}$$

where a, $b = \text{const}$, a, $b \in \mathbf{N}$, $\mathbf{N}$ is set of natural numbers; $\alpha_c, \alpha_s = \text{const}$, $\alpha_c, \alpha_s \in [0, 2\pi]$.

In time-frequency plane (Fig. 7.2.2), gray background depicts energy distribution of a harmonic signal from the system $\{h_k^c(t)\}$ ($\{h_k^s(t)\}$) in a frequency

$f_k = (a \cdot k + b)/T$. All the signals (harmonics) of the system $\{h_k^c(t)\}$ ($\{h_k^s(t)\}$) completely cover an allocated rectangle with the sides F and T, where a signal spectrum bandwidth F_0 of a single signal $h_k^c(t)$ ($h_k^s(t)$) is approximately equal to $F_0 \approx 1/T$, so that $F \cdot T = F/F_0 = N$.

A system of orthogonal phase-shift keying (PSK) signals defined in the time interval $T_{\text{sys}} = [0, T]$ (see Fig 7.2.3) is a set of phase-shift keying signals $\{\psi_k^{PSK}(t)\}$, $i = 1, \ldots, N$ whose phases are discretely changed according to a given dependence. In the case of using two-phase signals, a phase takes two values (for instance, 0 or π), so that a baseband signal consisting of positive and negative pulses corresponds to a RF PSK signal (see Fig. 7.2.3). If a number of elements of PSK signal is equal to N, then duration of a single element is equal to $\tau = T/N$, and its spectrum bandwidth is approximately equal to signal spectrum bandwidth: $F \approx N/T$. In time-frequency plane, in the figure, gray background depicts energy distribution of a single element of PSK signal. All the signals of the system $\{\psi_k^{PSK}(t)\}$ completely cover an allocated rectangle with the sides F and T, so that $F \cdot T = N$.

As an example, we bring the system $\{\psi_k^{BPSK}(t)\}$ of orthogonal binary PSK (BPSK) signals defined in the time interval $T_{\text{sys}} = [0, T]$ (see Fig. 7.2.3) based on Walsh functions:

$$\psi_k^{BPSK}(t) = wal_k^N(t) \cdot \cos(2\pi t/T_0); \tag{7.2.3}$$

$$wal_k^N(t) = H_{\left[\frac{t}{Tch}\right], k}^N, \tag{7.2.3a}$$

where $wal_k^N(t)$ is k-th Walsh function of order N, $[x]$ is integer part of a number x; T_{ch} is a duration of an elementary (chip) signal, $T_{ch} = T/N$; $T_{ch}/T_0 \in \mathbf{N}$; T_0 is oscillation period, $T_0 = 1/f_0$; $\mathbf{N}$ is set of natural numbers; H^N is Hadamard matrix of order N: $H^N = \|H_{r,c}^N\|$; $H_{r,c}^N$ is an element of Hadamard matrix of order N; r, c are indexes of row and column of a matrix element, respectively, so that Hadamard matrix of order $2n$ is formed iteratively according to the following relationship:

$$H^{2n} = \left\| \begin{array}{cc} H^n & H^n \\ H^n & -H^n \end{array} \right\|, \ H^1 = \|1\|. \tag{7.2.4}$$

As an alternative example, one can bring the system $\{\psi_k^{MPSK}(t)\}$ of quasi-orthogonal PSK signals defined in the time interval $T_{\text{sys}} = [0, T]$ and based on polyphase Frank codes (Frank R.L.) [71]:

$$\psi_k^{MPSK}(t) = \sum_{i=0}^{N-1} \cos(2\pi t/T_0 + \varphi_{i,k}) \times \tag{7.2.5}$$

$$\times \left[1(t - (i \cdot T_{ch})) - 1(t - ((i+1) \cdot T_{ch})) \right];$$

$$\varphi_{i,k} = \left(\frac{2\pi \cdot p}{N} \right) \cdot \text{mod}_N(p \cdot i \cdot k), \tag{7.2.5a}$$

T_{ch} is a duration of an elementary signal, $T_{ch} = T/N$; $T_{ch}/T_0 \in \mathbf{N}$; T_0 is an oscillation period, $T_0 = 1/f_0$; $\mathbf{N}$ is set of natural numbers; N and p are co-prime (mutually prime) integers; $i = 0, 1, \ldots, N - 1$; $k = 0, 1, \ldots, N - 1$; $\text{mod}_N(x)$ is a

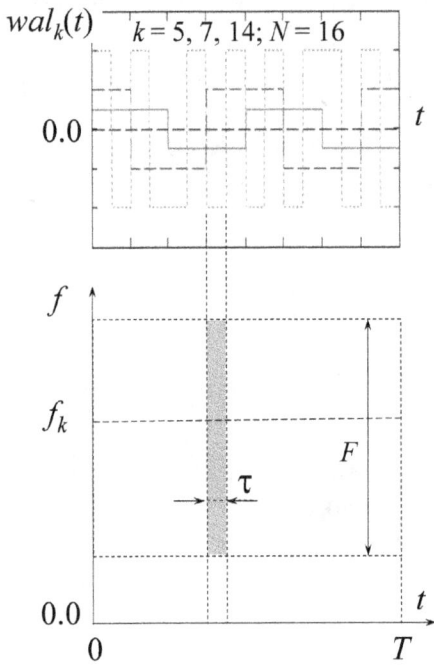

FIGURE 7.2.3 System $\{wal_k^N(t)\}$, $k = 1, \ldots, N$ ($N = 16$) of orthogonal Walsh functions (7.2.3a)

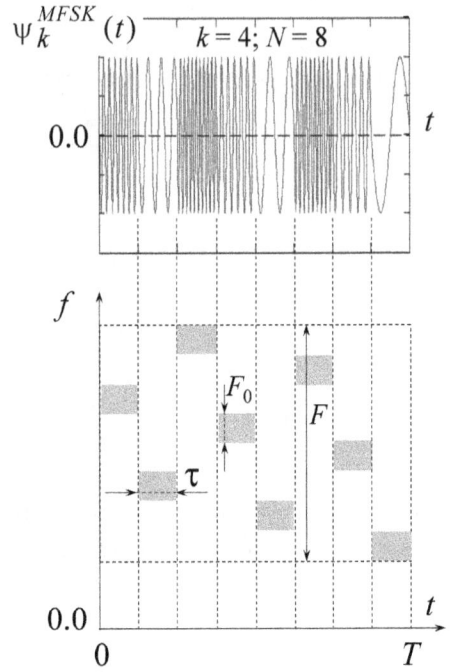

FIGURE 7.2.4 System of orthogonal M-ary frequency-shift keying signals $\{\psi_k^{MFSK}(t)\}$ (7.2.6)

function of x modulo N (modulo operation) that returns a remainder on dividing x by N; $1(t)$ is Heaviside unit step function.

A system of orthogonal discrete M-ary frequency-shift keying (MFSK) signals defined in the time interval $T_{sys} = [0, T]$ (see Fig. 7.2.4) is a set of M-ary FSK signals $\{\psi_k^{MFSK}(t)\}$, $k = 1, \ldots, N$ whose frequencies are changed according to a given dependence. If a number of elementary signals (chips) in discrete MFSK signal is equal to N, then their spectrum bandwidth F_0 is approximately equal to $F_0 \approx 1/\tau = N/T$. In time-frequency plane, in this figure, gray background depicts rectangles in which energies of discrete MFSK signal chips are distributed. As can be seen from Fig. 7.2.4, energy of discrete FSK signal is non-uniformly distributed in time-frequency plane. All the signals of the system $\{\psi_k^{MFSK}(t)\}$ completely cover an allocated rectangle with the sides F and T, so that $F \cdot T = NF_0 \cdot N\tau = N^2$.

As an example, one can bring the system $\{\psi_k^{MFSK}(t)\}$ of orthogonal discrete MFSK signals defined in the time interval $T_{sys} = [0, T]$ by the following relationships:

$$\psi_k^{MFSK}(t) = \cos\left(2\pi\left(M_0 - M_{[t/T_{ch}],k}\right)t/T_{ch} + \pi/2\right), \qquad (7.2.6)$$

where $[x]$ is an integer part of a number x; T_{ch} is a duration of elementary (chip) signal, $T_{ch} = T/N$; M_0 is an initial number of signal oscillation periods within the duration T_{ch}; $\mathbf{M}$ is a matrix of a modified number of oscillation periods: $\mathbf{M} =$

$\|M_{i,k}\|$; $M_{i,k}$ is an element of the matrix $\mathbf{M}$; i, k are indexes of row and column of matrix element, respectively; $M_0, M_{i,k} \in \mathbf{N}$, $\mathbf{N}$ is set of natural numbers, so that matrix of modified number of oscillation periods is formed according to the relationship:

$$\mathbf{M} = \text{submatrix}(\mathbf{M}'; 2, N+1, 2, N+1), \tag{7.2.7}$$

where operation submatrix($*$) assumes extracting a submatrix from initial matrix within pointed range of rows and columns, respectively; elements $M'_{i',k'}$ of auxiliary matrix $\mathbf{M}' = \left\| M'_{i',k'} \right\|$ are defined by the result of modulo b operation:

$$M'_{i',k'} = \text{mod}_b(i' \cdot k' + a), \tag{7.2.8}$$

$\text{mod}_b(x)$ is a function returning the remainder on dividing x by b; $i' = 1, \ldots, N+1$, $k' = 1, \ldots, N+1$; $a, b = \text{const}$, $a, b \in \mathbf{N}$; $\mathbf{N}$ is set of natural numbers; $b > N + 1$.

To illustrate the described variant of the system $\{\psi_k^{MFSK}(t)\}$ of orthogonal discrete MFSK signals that is defined by the relationship (7.2.6), we consider a case of its realization for $N=8$.

Example 7.2.1. Thus, setting the values of variables from the formulas (7.2.6), (7.2.7), (7.2.8):

$$M_0 = 11, \ N = 8, \ a = 0, \ b = 11,$$

obtain the auxiliary matrix $\mathbf{M}' = \left\| M'_{i',k'} \right\|$ (7.2.8) and, basing on it, obtain the matrix of modified number of oscillation periods $\mathbf{M}$ (7.2.7):

$$\mathbf{M}' = \begin{Vmatrix} 0 & 0 & 0 & 0 & 0 & 0 & 0 & 0 & 0 \\ 0 & 1 & 2 & 3 & 4 & 5 & 6 & 7 & 8 \\ 0 & 2 & 4 & 6 & 8 & 10 & 1 & 3 & 5 \\ 0 & 3 & 6 & 9 & 1 & 4 & 7 & 10 & 2 \\ 0 & 4 & 8 & 1 & 5 & 9 & 2 & 6 & 10 \\ 0 & 5 & 10 & 4 & 9 & 3 & 8 & 2 & 7 \\ 0 & 6 & 1 & 7 & 2 & 8 & 3 & 9 & 4 \\ 0 & 7 & 3 & 10 & 6 & 2 & 9 & 5 & 1 \\ 0 & 8 & 5 & 2 & 10 & 7 & 4 & 1 & 9 \end{Vmatrix}, \ \mathbf{M} = \begin{Vmatrix} 1 & 2 & 3 & 4 & 5 & 6 & 7 & 8 \\ 2 & 4 & 6 & 8 & 10 & 1 & 3 & 5 \\ 3 & 6 & 9 & 1 & 4 & 7 & 10 & 2 \\ 4 & 8 & 1 & 5 & 9 & 2 & 6 & 10 \\ 5 & 10 & 4 & 9 & 3 & 8 & 2 & 7 \\ 6 & 1 & 7 & 2 & 8 & 3 & 9 & 4 \\ 7 & 3 & 10 & 6 & 2 & 9 & 5 & 1 \\ 8 & 5 & 2 & 10 & 7 & 4 & 1 & 9 \end{Vmatrix}.$$

Then, for initial number of oscillation periods M_0 taken within the duration T_{ch} that is equal to $M_0=11$ (see the formula (7.2.6)), the fourth signal $\psi_4^{MFSK}(t)$ from the system $\{\psi_k^{MFSK}(t)\}$, whose number of periods is determined by the fourth column of the matrix $\mathbf{M}$ (taking into account the value of $M_0=11$), has the form shown in Fig. 7.2.4. System of discrete MFSK signals obtained in such a way is orthogonal since scalar product between the signals is equal to zero $(\psi_k^{MFSK}(t), \psi_i^{MFSK}(t)) = 0$, $k \neq i$, but it is equally important that a phase of the adjacent chips of MFSK signal determined by the relationship (7.2.6) is continuous. $\qquad \triangledown$

The systems of orthogonal signals considered in this section are further used when analyzing operation of the corresponding multichannel (multi-station) information transmitting systems with quadrature PSK (QPSK) signals. The choice of QPSK signals is based on convenience of visual perception of the demodulation results representation using I-Q diagrams.

7.3   Demultiplexing/Demodulation L-group Algorithms for TDM(A)-QPSK Systems

In information transmitting systems with TDM(A), information contained in a group signal $s_g(t)$ (7.1.1) is transmitted within the single frames T^{fr}. Each of them is a partition of N slots T_k^{sl} with the same duration. Each slot contains N_s symbols of a channel signal $s_k(t)$:

$$T^{fr} = \bigcup_{k=0}^{N-1} T_k^{sl}; \quad \bigcap_{k=0}^{N-1} T_k^{sl} = \emptyset, \tag{7.3.1}$$

where k is a slot number.

To avoid discrepancies in notations, hereinafter we denote some interval by the corresponding abbreviation in a superscript index, and its duration we denote by the corresponding abbreviation in a subscript index, for instance, T^{fr} is a frame interval, and T_{fr} is a frame duration, etc.

Since all processing from one frame to another is repeated, there is no need of considering a set of frames $\{T_i^{fr}\}$, so we confine our discussion on signal processing within a single frame.

Channel signal $s_k(t)^*$ in multichannel (multi-station) information transmitting systems with TDM(A)-QPSK is formed according to the relationships [38, (9.4.4)], [73, (12.13)], that can be rewritten in more compact form:

$$s_k(t) = \frac{1}{\sqrt{2}} \left[c^I_{\left[\frac{\mathrm{mod}\, T_{sl}(t)}{T_s}\right],k} \cdot M_k^I(t) + c^Q_{\left[\frac{\mathrm{mod}\, T_{sl}(t)}{T_s}\right],k} \cdot M_k^Q(t) \right], \ t \in T^{fr}; \tag{7.3.2}$$

$$M_k^I(t) = p_k(t) \cdot \cos\left(2\pi t/T_0 + \pi/4\right); \tag{7.3.2a}$$

$$M_k^Q(t) = p_k(t) \cdot \sin\left(2\pi t/T_0 + \pi/4\right), \tag{7.3.2b}$$

where $T^{fr} = [0, T_{fr}]$ is a frame of group signal $s_g(t)$; T_{fr} is a duration of the frame of group signal $s_g(t)$; $M_k^I(t)$, $M_k^Q(t)$ are modulating functions of in-phase and quadrature streams of QPSK signal, respectively; T_{sl} is a duration of a slot T_k^{sl} of channel signal $s_k(t)$, $T_{sl} = T_{fr}/N$; $T_s = T_{sl}/N_s$ is a symbol duration of QPSK signal; N_s is a number of symbols transmitted in a slot of channel signal $s_k(t)$;

---

*Hereinafter, we consider t as a discrete time parameter

$\{p_k(t)\}$ is the system of N orthogonal baseband pulses ($k = 0, \ldots, N-1$) defined by the function (7.2.1):

$$p_k(t) = 1(t - (k \cdot T_{fr}/N)) - 1(t - ((k+1) \cdot T_{fr}/N)); \qquad (7.3.3)$$

$1(t)$ is Heaviside unit step function; $[x]$ is an integer part of a quantity x; T_0 is a period of carrier; $\mathrm{mod}_y(x)$ is a function returning the remainder on dividing x by y; c^I, c^Q are matrices of transmitted symbols of in-phase and quadrature streams of QPSK signal, respectively: $c^I = \left\| c^I_{i,k} \right\|$, $c^Q = \left\| c^Q_{i,k} \right\|$; $c^I_{i,k}$, $c^Q_{i,k}$ are elements of matrices; i, k are the indexes of a row and a column of a matrix element, respectively, so that k-th column $\left(c^I\right)^{<k>} = [c^I_{0,k}, \ldots, c^I_{N_s-1,k}]^T$, $\left(c^Q\right)^{<k>} = [c^Q_{0,k}, \ldots, c^Q_{N_s-1,k}]^T$ of each matrix c^I, c^Q determines a message transmitted by k-th channel signal $s_k(t)$ within the corresponding slot whose position is given by the function (7.3.3); $N, N_s \in \mathbf{N}$, $\mathbf{N}$ is set of natural numbers.

In the formula (7.3.2), index i of a row of elements $c^I_{i,k}$, $c^Q_{i,k}$ of matrices c^I, c^Q that is equal to

$$i = \left[\frac{\mathrm{mod}_{T_{sl}}(t)}{T_s} \right],$$

points a number of a symbol in k-th slot that is a function of time.

Group TDM(A)-QPSK signal $s_{g/TDMA}(t)$ is a sum of modulated channel signals $s_k(t)$:

$$s_{g/TDMA}(t) = \sum_{k=0}^{N-1} s_k(t), \qquad (7.3.4)$$

each of them is determined by the formula (7.3.2).

Radiofrequency (RF) group TDM(A)-QPSK signal $s_{g/RF}(t)$ obtained by modulating a carrier with the signal (7.3.4)

$$s_{g,RF}(t) = s_{g/TDMA}(t)e^{j\omega_c t},$$

is radiated into propagation medium (space) at a carrier frequency f_c as it shown in Figs. 7.1.1, 7.1.2, and in receiving side, it additively interacts with complex quasi-white Gaussian noise $n(t)$ with power spectral density (PSD) $N(f)$:

$$x(t) = s_{g,RF}(t) + n(t), \qquad (7.3.5)$$

where $N(f) = N_0 \cdot [1(f) - 1(f - f_{\max})]$, $N_0 = \mathrm{const}$, $f_{\max} = \mathrm{const}$, $f_c << f_{\max}$.

Then, the estimators $\hat{c}^I_{i,k}$, $\hat{c}^Q_{i,k}$ of QPSK signal in-phase and quadrature stream elements (symbols) $c^I_{i,k}$, $c^Q_{i,k}$ from matrices c^I, c^Q, obtained by classic algorithm of signal demultiplexing/demodulation in linear space, are sign functions of scalar products between fragments of the observed stochastic process $x(t)$ (7.3.5) and modulating functions $M^I_k(t)$, $M^Q_k(t)$ (7.3.2a,b):

$$\hat{c}^I_{i,k} = \mathrm{sgn}(A_{i,k}), \ \hat{c}^Q_{i,k} = \mathrm{sgn}(B_{i,k}); \qquad (7.3.6)$$

$$A_{i,k} = 2\sqrt{2} \left(\mathrm{Re}[x(t)e^{-j\omega_c t}] \cdot \Delta_{i,k}(t), M^I_k(t) \cdot \Delta_{i,k}(t) \right)/T_s; \qquad (7.3.6a)$$

$$B_{i,k} = 2\sqrt{2} \left(\mathrm{Re}[x(t)e^{-j\omega_c t}] \cdot \Delta_{i,k}(t), M^Q_k(t) \cdot \Delta_{i,k}(t) \right)/T_s; \qquad (7.3.6b)$$

$$\Delta_{i,k}(t) = 1(t - (kT_{sl} + iT_s)) - 1(t - (kT_{sl} + (i+1)T_s)), \qquad (7.3.6c)$$

where $(a(t), b(t))$ is scalar product between functions $a(t)$ and $b(t)$; the function $\Delta_{i,k}(t)$ determines a time position of i-th symbol in k-th slot of group signal $s_{g/TDMA}(t)$ (7.3.4).

The relationships (7.3.6) define TDM(A)-QPSK signal demultiplexing/demodulation classic algorithm.

At the same time, estimators $\hat{c}^a_{i,k}$, $\hat{c}^b_{i,k}$ of QPSK signal in-phase and quadrature stream elements (symbols) $c^I_{i,k}$, $c^Q_{i,k}$ from the matrices c^I, c^Q, obtained by L-group algorithm of signal demultiplexing/demodulation in signal space with lattice properties based on measure of statistical interrelation (MSI) N_s (1.4.11) concerned with semimetric (1.3.26), are sign functions of MSIs between fragments of the observed stochastic process $x(t)$ (7.3.5) and modulating functions $M^I_k(t)$, $M^Q_k(t)$ (7.3.2a):

$$\hat{c}^a_{i,k} = \text{sgn}(a_{i,k}),\ \hat{c}^b_{i,k} = \text{sgn}(b_{i,k}); \tag{7.3.7}$$

$$a_{i,k} = \alpha \cdot \text{N}_s \left(\text{Re}[x(t)e^{-j\omega_c t}] \cdot \Delta_{i,k}(t), M^I_k(t) \cdot \Delta_{i,k}(t) \right); \tag{7.3.7a}$$

$$b_{i,k} = \alpha \cdot \text{N}_s \left(\text{Re}[x(t)e^{-j\omega_c t}] \cdot \Delta_{i,k}(t), M^Q_k(t) \cdot \Delta_{i,k}(t) \right); \tag{7.3.7b}$$

$$\Delta_{i,k}(t) = 1(t - (kT_{sl} + iT_s)) - 1(t - (kT_{sl} + (i+1)T_s)), \tag{7.3.7c}$$

where α is some constant coefficient, α=const.

The relationships (7.3.7) define TDM(A)-QPSK signal demultiplexing/demodulation L-group algorithm based on MSI N_s (1.4.11).

Estimators $\hat{c}^{a'}_{i,k}$, $\hat{c}^{b'}_{i,k}$ of QPSK signal in-phase and quadrature stream elements (symbols) $c^I_{i,k}$, $c^Q_{i,k}$ from the matrices c^I, c^Q, obtained by L-group algorithm of signal demultiplexing/demodulation in signal space with lattice properties and based on MSI N_p (1.4.10) concerned with pseudometric (1.3.25), are sign functions of MSIs between fragments of the observed stochastic process $x(t)$ (7.3.5) and modulating functions $M^I_k(t)$, $M^Q_k(t)$ (7.3.2a):

$$\hat{c}^{a'}_{i,k} = \text{sgn}(a'_{i,k}),\ \hat{c}^{b'}_{i,k} = \text{sgn}(b'_{i,k}); \tag{7.3.8}$$

$$a'_{i,k} = \beta \cdot \text{N}_p \left(\text{Re}[x(t)e^{-j\omega_c t}] \cdot \Delta_{i,k}(t), M^I_k(t) \cdot \Delta_{i,k}(t) \right); \tag{7.3.8a}$$

$$b'_{i,k} = \beta \cdot \text{N}_p \left(\text{Re}[x(t)e^{-j\omega_c t}] \cdot \Delta_{i,k}(t), M^Q_k(t) \cdot \Delta_{i,k}(t) \right); \tag{7.3.8b}$$

$$\Delta_{i,k}(t) = 1(t - (kT_{sl} + iT_s)) - 1(t - (kT_{sl} + (i+1)T_s)), \tag{7.3.8c}$$

where β is some constant coefficient, β=const.

The relationships (7.3.8) define TDM(A)-QPSK signal demultiplexing/demodulation L-group algorithm based on MSI N_p (1.4.10).

Similarly, estimators $\hat{c}^{a''}_{i,k}$, $\hat{c}^{b''}_{i,k}$ of QPSK signal in-phase and quadrature stream elements (symbols) $c^I_{i,k}$, $c^Q_{i,k}$ from the matrices c^I, c^Q, obtained by L-group algorithm of signal demultiplexing/demodulation in signal space with lattice properties and based on MSI N_{l_1} (1.4.12) concerned with l_1-metric (1.3.27), are sign functions of MSIs between fragments of the observed stochastic process $x(t)$ (7.3.5) and modulating functions $M_k^I(t)$, $M_k^Q(t)$ (7.3.2a):

$$\hat{c}_{i,k}^{a''} = \text{sgn}(a_{i,k}''), \ \hat{c}_{i,k}^{b''} = \text{sgn}(b_{i,k}''); \tag{7.3.9}$$

$$a_{i,k}'' = \gamma \cdot \text{N}_{l_1}\left(\text{Re}[x(t)e^{-j\omega_c t}] \cdot \Delta_{i,k}(t), M_k^I(t) \cdot \Delta_{i,k}(t)\right); \tag{7.3.9a}$$

$$b_{i,k}'' = \gamma \cdot \text{N}_{l_1}\left(\text{Re}[x(t)e^{-j\omega_c t}] \cdot \Delta_{i,k}(t), M_k^Q(t) \cdot \Delta_{i,k}(t)\right); \tag{7.3.9b}$$

$$\Delta_{i,k}(t) = 1(t - (kT_{sl} + iT_s)) - 1(t - (kT_{sl} + (i+1)T_s)), \tag{7.3.9c}$$

where γ is some constant coefficient, γ=const.

The relationships (7.3.9) define TDM(A)-QPSK signal demultiplexing/demodulation L-group algorithm based on MSI N_{l_1} (1.4.12).

Notice that coefficients α, β, γ figuring in the relationships (7.3.7), (7.3.8), (7.3.9), respectively, are used to provide a convenience of visual perception of I-Q diagrams shown below, but in real processing they could be omitted.

Example 7.3.1. Consider the results of using L-group algorithms (7.3.6), (7.3.7), (7.3.8) in the case of additive interaction of signal $s_{g/RF}(t)$ and quasi-white Gaussian noise $n(t)$ (7.3.5) obtained by simulation (statistical modeling). Simulation is realized under the following conditions.

1. Frame duration T_{fr} of a group signal $s_g(t)$ is equal to T_{fr}=4096; number of slots N in a frame is equal to N=8; slot duration T_{sl} of channel signal $s_k(t)$ is equal to T_{sl}=512; symbol duration T_s of QPSK signal is equal to $T_s = T_{sl}/N_s$=64; number of symbols N_s transmitted in a slot of channel signal is equal to N_s=8; oscillation period T_0 of QPSK signal is equal to T_0=32.

2. Interference $n(t)$ is quasi-white Gaussian noise (7.3.5) with PSD $N(f) = N_0 \cdot [1(f) - 1(f - f_{\max})]$, N_0=const; signal-to-noise ratio E_b/N_0 is equal to E_b/N_0=16, where E_b is energy per bit.

Fig. 7.3.1a,b illustrates channel QPSK signals $\{s_k(t)\}$ transmitted within the corresponding slots. In Fig. 7.3.1a, channel signals $s_0(t)$, $s_2(t)$, $s_4(t)$, $s_6(t)$ are not shown for convenience of visual perception.

Fig. 7.3.2a depicts realization $s_g^*(t)$ of group signal $s_g(t)$ determined by a sum (7.3.4), and Fig. 7.3.2b depicts Fourier transforms $N(f)$, $S_g(f)$ of realizations of interference $n(t)$ and group signal $s_g(t)$, respectively.

Fig. 7.3.3a illustrates I-Q diagram of demodulation results obtained by classic algorithm (7.3.6) of signal demultiplexing/demodulation in linear space. Fig. 7.3.3b depicts I-Q diagram of demodulation results obtained by L-group algorithm (7.3.7) of signal demultiplexing/demodulation based on MSI N_s (1.4.11) in semimetric space. Fig. 7.3.3c demonstrates I-Q diagram of demodulation results obtained by L-group algorithm (7.3.8) of signal demultiplexing/demodulation based on MSI N_p (1.4.10) in pseudometric space.

The results of comparative analysis of efficiency of algorithms (7.3.6), (7.3.7), (7.3.8) allow concluding that in the presence of interference $n(t)$ in the form of quasi-white Gaussian noise, L-group algorithm (7.3.7) is inferior to classic algorithm

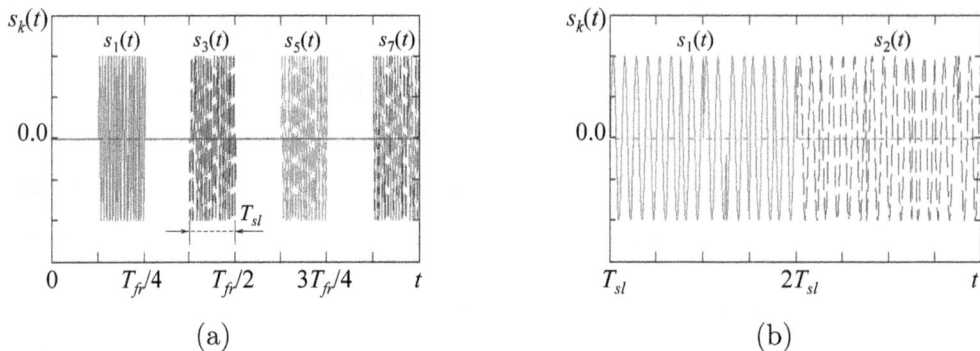

(a) (b)

FIGURE 7.3.1   Channel QPSK signals (a) $\{s_k(t)\}$, $k = 1, 3, 5, 7$; (b) $\{s_k(t)\}$, $k = 1, 2$

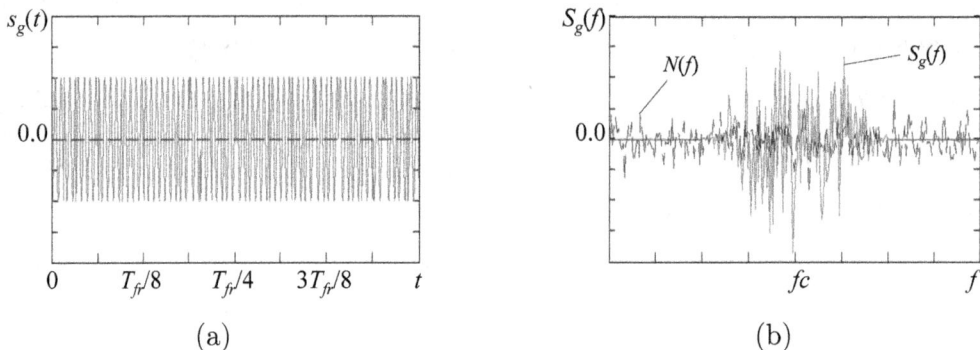

(a) (b)

FIGURE 7.3.2   (a) Realization $s_g^*(t)$ of group signal $s_g(t)$; (b) Fourier transforms $N(f)$ and $S_g(f)$ of realizations of interference $n(t)$ and group signal $s_g(t)$, respectively

(7.3.6) in 0.7...2.0 dB, L-group algorithm (7.3.8) is inferior to classic algorithm (7.3.6) 2.5...5.1 dB. These losses, in essential extent, are determined by a number of independent samples n of the observed stochastic process $x(t)$ (7.3.5), situated

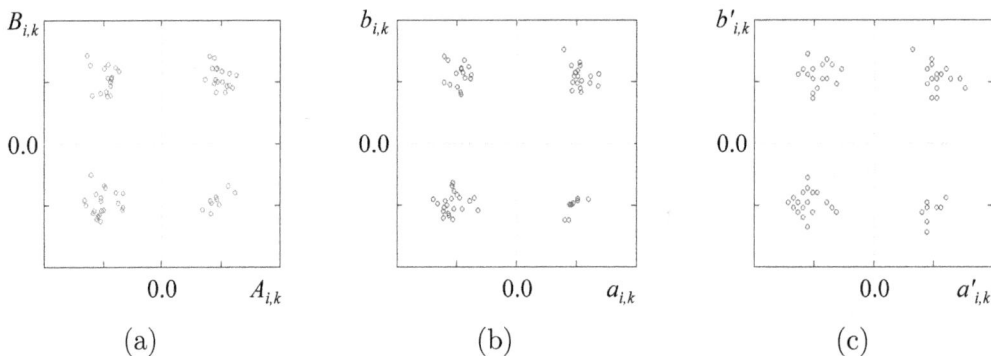

(a) (b) (c)

FIGURE 7.3.3   I-Q diagrams of TDM(A)-QPSK signal demodulation results obtained by (a) classic algorithm (7.3.6) (linear space); (b) L-group algorithm (7.3.7) (semimetric space); (c) L-group algorithm (7.3.8) (pseudometric space)

within a symbol duration T_s of QPSK signal, that are used when calculating MSIs N_s (1.4.11) and N_p (1.4.10). Dependences of losses $L(n)$ in SNR E_b/N_0 for L-group algorithms (7.3.7) and (7.3.8) based on MSIs N_s (1.4.11), N_p (1.4.10), respectively, are shown in Fig. 7.3.4. As follows from Fig. 7.3.4, asymptotically, level of losses $L(n)$ corresponds to the results obtained in Section 4.6 that are determined by the relationships (4.6.40).

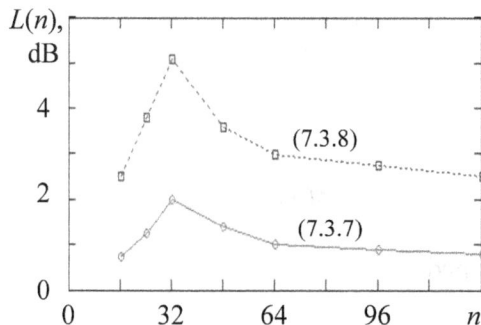

FIGURE 7.3.4   Dependences of losses $L(n)$ in SNR E_b/N_0 for L-group algorithms (7.3.7) and (7.3.8)

▽

L-group algorithms (7.3.7), (7.3.8), (7.3.9) based on MSIs N_s (1.4.11), N_p (1.4.10), N_{l_1} (1.4.12), respectively, do not require operation of multiplications for their computation, unlike classic linear algorithm (7.3.6). So, despite the aforementioned losses taking place when performing algorithms (7.3.7), (7.3.8), (7.3.9), they can be used in those applications that require a considerable signal processing rate, or when computational resources of processing system are rather confined.

7.4   Demultiplexing/Demodulation *L*-group Algorithms for DS-CDM(A)-QPSK Systems

Channel signal $s_k(t)$ in multichannel (multi-station) information transmitting systems with DS-CDM(A)-QPSK[†] is formed according to the relationships [38, (9.4.4)], [73, (12.22)], that can be rewritten in more compact form:

$$s_k(t) = \frac{1}{\sqrt{2}} \left[c_k^I \cdot M_k^I(t) + c_k^Q \cdot M_k^Q(t) \right], \ t \in T^s; \tag{7.4.1}$$

$$M_k^I(t) = wal_k^N(t) \cdot \cos\left(2\pi t/T_0 + \pi/4\right); \tag{7.4.1a}$$

$$M_k^Q(t) = wal_k^N(t) \cdot \sin\left(2\pi t/T_0 + \pi/4\right), \tag{7.4.1b}$$

[†]DS means direct sequence

where $T^s = [0, T_s]$ is a domain of QPSK signal symbol; T_s is a duration of QPSK signal symbol; $M_k^I(t)$, $M_k^Q(t)$ are modulating functions of in-phase and quadrature streams of QPSK signal, respectively; $T_{ch} = T_s/N_{ch}$ is an elementary signal (chip) duration; N_{ch} is a number of chips transmitted during a duration of QPSK signal symbol; $\{wal_k^N(t)\}$ is the system of N orthogonal Walsh functions $k = 0, \ldots, N-1$ determined by the relationship (7.2.3a), $N = N_{ch}$; T_0 is oscillation period of QPSK signal; N, $N_{ch} \in \mathbf{N}$, $\mathbf{N}$ is set of natural numbers; c^I, c^Q are symbols of in-phase $c^I = [c_0^I, \ldots, c_{N_{ch}-1}^I]^T$ and quadrature $c^Q = [c_0^Q, \ldots, c_{N_{ch}-1}^Q]^T$ streams of QPSK signal that are transmitted by k-th channel signal $s_k(t)$.

Group DS-CDM(A)-QPSK signal $s_{g/CDMA}(t)$ is a sum of modulated channel signals $s_k(t)$:

$$s_{g/CDMA}(t) = \sum_{k=0}^{N-1} s_k(t), \qquad (7.4.2)$$

each of them is determined by the formula (7.4.1).

RF group signal $s_{g,RF}(t)$ obtained by modulating a carrier with the signal (7.4.2)

$$s_{g,RF}(t) = s_{g/CDMA}(t)e^{j\omega_c t},$$

is radiated into propagation medium (space) at a carrier frequency f_c, as shown in Fig. 7.1.1, 7.1.2, and in receiving side of the system it additively interacts with complex quasi-white Gaussian noise $n(t)$ with PSD $N(f) = N_0 \cdot [1(f) - 1(f - f_{\max})]$, N_0=const, $f_{\max}$=const:

$$x(t) = s_{g,RF}(t) + n(t). \qquad (7.4.3)$$

Then, estimators $\hat{c}^I$, $\hat{c}^Q$ of symbols c^I, c^Q of QPSK signal in-phase $c^I = [c_0^I, \ldots, c_{N_{ch}-1}^I]^T$ and quadrature $c^Q = [c_0^Q, \ldots, c_{N_{ch}-1}^Q]^T$ streams, obtained by classic algorithm of signal demultiplexing/demodulation in linear space, are sign functions of scalar products between fragments of the observed stochastic process $x(t)$ (7.4.3) and modulating functions $M_k^I(t)$, $M_k^Q(t)$ (7.4.1a,b):

$$\hat{c}_k^I = \text{sgn}(A_k), \ \hat{c}_{i,k}^Q = \text{sgn}(\text{Re}\,B_{i,k}); \qquad (7.4.4)$$

$$A_k = 2\sqrt{2}\left(\text{Re}[x(t)e^{-j\omega_c t}], M_k^I(t)\right)/T_s; \qquad (7.4.4a)$$

$$B_k = 2\sqrt{2}\left(\text{Re}[x(t)e^{-j\omega_c t}], M_k^Q(t)\right)/T_s, \qquad (7.4.4b)$$

where $(a(t), b(t))$ is scalar product between functions $a(t)$ and $b(t)$.

The relationships (7.4.4) define DS-CDM(A)-QPSK signal demultiplexing/demodulation classic algorithm.

At the same time, estimators $\hat{c}^a$, $\hat{c}^b$ of symbols c^I, c^Q of QPSK signal in-phase $c^I = [c_0^I, \ldots, c_{N_{ch}-1}^I]^T$ and quadrature $c^Q = [c_0^Q, \ldots, c_{N_{ch}-1}^Q]^T$ streams, obtained by L-group algorithm of signal demultiplexing/demodulation in space with lattice properties and based on MSI N_s (1.4.11) concerned with semimetric (1.3.26), are sign functions of MSIs between fragments of the observed stochastic process $x(t)$

(7.4.3) and modulating functions $M_k^I(t)$, $M_k^Q(t)$ (7.4.1a,b):

$$\hat{c}_k^a = \text{sgn}(a_k), \; \hat{c}_k^b = \text{sgn}(b_k); \tag{7.4.5}$$

$$a_k = \alpha \cdot N_s \left(\text{Re}[x(t)e^{-j\omega_c t}], M_k^I(t) \right); \tag{7.4.5a}$$

$$b_k = \alpha \cdot N_s \left(\text{Re}[x(t)e^{-j\omega_c t}], M_k^Q(t) \right). \tag{7.4.5b}$$

where α is some constant coefficient, α=const.

The relationships (7.4.5) define DS-CDM(A)-QPSK signal demultiplexing/demodulation L-group algorithm based on MSI N_s (1.4.11) concerned with semimetric (1.3.26).

Similarly, estimators $\hat{c}^{a'}$, $\hat{c}^{b'}$ of symbols c^I, c^Q of QPSK signal in-phase $c^I = [c_0^I, \ldots, c_{N_{ch}-1}^I]^T$ and quadrature $c^Q = [c_0^Q, \ldots, c_{N_{ch}-1}^Q]^T$ streams, obtained within L-group algorithm of signal demultiplexing/demodulation and based on MSI N_{l_1} (1.4.12) concerned with l_1-metric (1.3.27), are sign functions of MSIs between fragments of the observed stochastic process $x(t)$ (7.4.3) and modulating functions $M_k^I(t)$, $M_k^Q(t)$ (7.4.1a,b):

$$\hat{c}_k^{a'} = \text{sgn}(a_k'), \; \hat{c}_k^{b'} = \text{sgn}(b_k'); \tag{7.4.6}$$

$$a_k' = \beta \cdot N_{l_1} \left(\text{Re}[x(t)e^{-j\omega_c t}], M_k^I(t) \right); \tag{7.4.6a}$$

$$b_k' = \beta \cdot N_{l_1} \left(\text{Re}[x(t)e^{-j\omega_c t}], M_k^Q(t) \right). \tag{7.4.6b}$$

where β is some constant coefficient, β=const.

Notice that coefficients α, β figuring in the relationships (7.4.5), (7.4.6), respectively, are used to provide a convenience of visual perception of I-Q diagrams shown below, but in real processing they could be omitted.

The relationships (7.4.6) define DS-CDM(A)-QPSK signal demultiplexing/demodulation L-group algorithm based on MSI N_{l_1} (1.4.12) concerned with l_1-metric.

L-group algorithm of signal demultiplexing/demodulation in space with lattice properties based on MSI N_p (1.4.10) seems too hard to be used in CDM(A) systems owing to difficulties taking place when providing orthogonality (in the sense of Definition 1.4.2) of channel signals $\{s_k(t)\}$, $k = 0, 1, \ldots, N-1$, if $N > 8$, so we do not discuss it.

Example 7.4.1. Consider the results of exploiting algorithms (7.4.4), (7.4.5) in the case of additive interaction between signal $s_{g,RF}(t)$ and quasi-white Gaussian noise $n(t)$ (7.4.3), obtained by simulation (statistical modeling). Simulation is realized under the following conditions.

    1. Symbol duration T_s of QPSK signal is equal to T_s=2048; a number of chips N_{ch} transmitted during QPSK signal symbol duration is equal to N_{ch}=64; chip duration $T_{ch} = T_s/N_{ch}$ is equal to $T_{ch} = 32$; number N of orthogonal Walsh functions from the system $\{wal_k^N(t)\}$, $k = 0, 1, \ldots, N-1$ determined by the relationship (7.2.3a) is equal to $N = N_{ch}$=64; oscillation period T_0 of QPSK signal is equal to T_0=16.

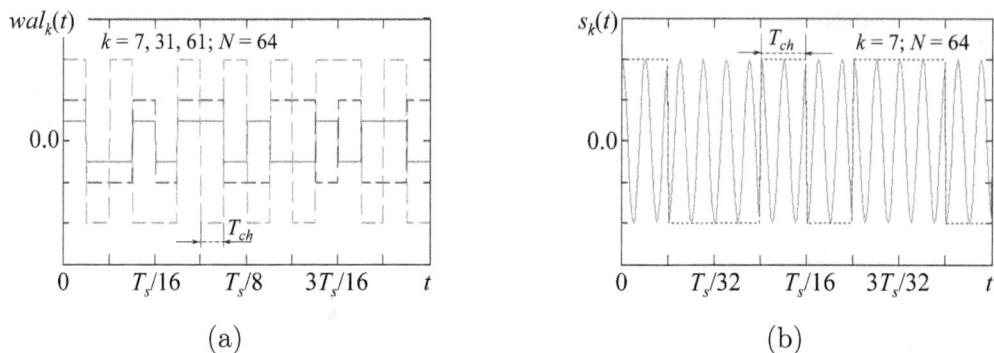

FIGURE 7.4.1   Fragments of (a) Walsh functions from the system $\{wal_k^N(t)\}$, $N = 64$, $k = 7, 31, 61$; (b) channel QPSK signal $s_k(t)$, $k = 7$

2. Interference $n(t)$ is quasi-white Gaussian noise (7.4.5) with PSD $N(f) = N_0 \cdot [1(f) - 1(f - f_{\max})]$, N_0=const; SNR E_b/N_0 is equal to E_b/N_0=3.5, where E_b is bit energy.

Fig. 7.4.1a illustrates fragments of Walsh functions $wal_7(t)$ (solid line), $wal_{31}(t)$ (dashed line), $wal_{61}(t)$ (dotted line) from the system $\{wal_k^N(t)\}$, $N = 64$. As follows from the figure, chip duration is equal to $T_{ch} = T_s/64$. Fig.7.4.1b depicts fragment of channel QPSK signal $s_k(t)$, $k = 7$ (solid line), and also the corresponding fragment of Walsh function $wal_k(t)$, $k = 7$ (dotted line).

Fig. 7.4.2a illustrates realization $s_g^*(t)$ of group signal $s_g(t)$ determined by the sum (7.4.2), and Fig. 7.4.2b depicts Fourier transforms $N(f)$, $S_g(f)$ of realizations of interference $n(t)$ and group signal $s_g(t)$, respectively.

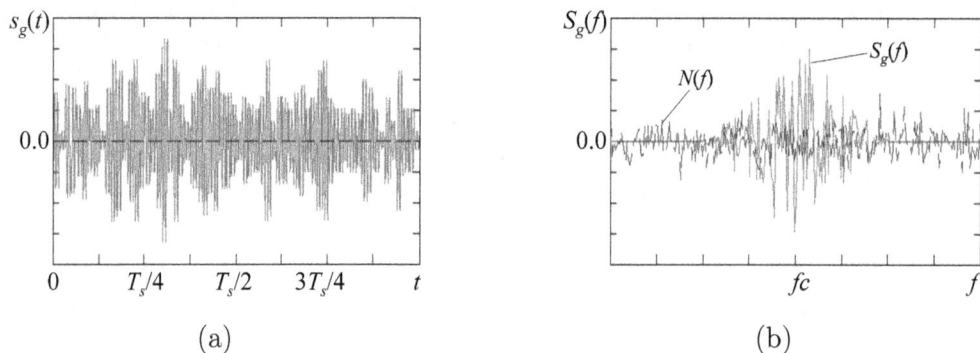

FIGURE 7.4.2   (a) Realization $s_g^*(t)$ of group signal $s_g(t)$; (b) Fourier transforms $N(f)$ and $S_g(f)$ of realizations of interference $n(t)$ and group signal $s_g(t)$, respectively

Fig. 7.4.3a illustrates I-Q diagram of demodulation results obtained by classic algorithm (7.4.4) of signal demultiplexing/demodulation in linear space. Fig. 7.4.3b depicts I-Q diagram of demodulation results obtained by L-group algorithm (7.4.5) of signal demultiplexing/demodulation based on MSI N_s (1.4.11). Fig. 7.4.3c shows

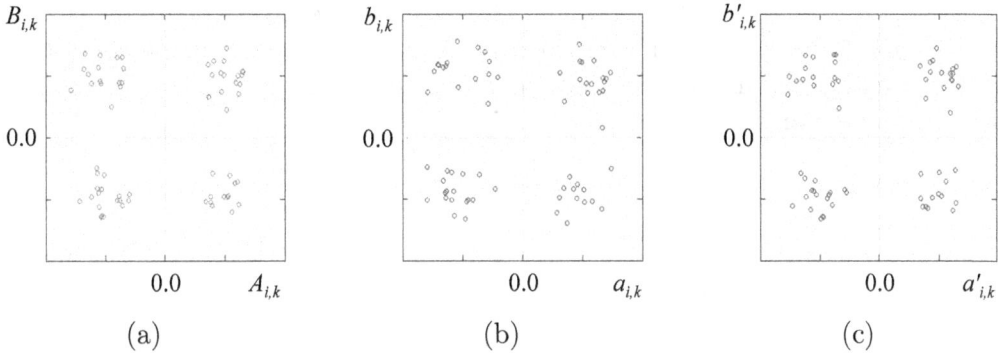

FIGURE 7.4.3 I-Q diagram of DS-CDM(A)-QPSK signal demodulation results obtained by (a) classic algorithm (7.4.4) (linear space); (b) *L*-group algorithm (7.4.5) (semimetric space); (c) *L*-group algorithm (7.4.6) (l_1-metric space)

I-Q diagram of demodulation results obtained by *L*-group algorithm (7.4.6) of signal demultiplexing/demodulation based on MSI N_{l_1} (1.4.12). ▽

The results of comparative analysis of efficiency of algorithms (7.4.4), (7.4.5), (7.4.6) allow concluding that in the presence of interference $n(t)$ in the form of quasi-white Gaussian noise, *L*-group algorithms (7.4.5) and (7.4.6) are inferior to classic one (7.4.4) in about 2.0 dB.

7.5 Demultiplexing/Demodulation *L*-group Algorithms for CP-MFSK-CDM(A)-QPSK Systems

Channel signal $s_k(t)$ in multichannel (multi-station) information transmitting systems with CP-MFSK-CDM(A)-QPSK[‡] is formed according to the relationships [38, (9.4.4)], [73, (12.64)], that can be rewritten in more compact form:

$$s_k(t) = \frac{1}{\sqrt{2}}\left[c_k^I \cdot M_k^I(t) + c_k^Q \cdot M_k^Q(t)\right], \ t \in T^s; \tag{7.5.1}$$

$$M_k^I(t) = \cos\left(2\pi\left(M_0 - M_{[t/T_{ch}],k}\right)t/T_{ch} + \pi/4\right); \tag{7.5.1a}$$

$$M_k^Q(t) = \sin\left(2\pi\left(M_0 - M_{[t/T_{ch}],k}\right)t/T_{ch} + \pi/4\right), \tag{7.5.1b}$$

where $T^s = [0, T_s]$ is a domain of QPSK signal symbol; T_s is a duration of QPSK signal symbol; $M_k^I(t)$, $M_k^Q(t)$ are modulating functions of QPSK signal in-phase and quadrature streams, respectively; c^I, c^Q are the symbols of QPSK signal in-phase $c^I = [c_0^I, \ldots, c_{N_{ch}-1}^I]^T$ and quadrature $c^Q = [c_0^Q, \ldots, c_{N_{ch}-1}^Q]^T$ streams that are transmitted by k-th channel signal $s_k(t)$; $[x]$ is integer part of a quantity x; $T_{ch} = T/N$ is a chip duration; M_0 is initial number of oscillation periods falling in a duration T_{ch}; **M** is matrix of modified number of oscillation periods with

---

[‡]CP — continuous phase

dimensionality $N \times N$: $\mathbf{M} = \|M_{i,k}\|$; $M_{i,k}$ is an element of matrix $\mathbf{M}$; i, k are indexes of a row and a column of a matrix element, respectively; $M_0, M_{i,k} \in \mathbf{N}$, $\mathbf{N}$ is set of natural numbers, so that the matrix of modified number of oscillation periods is based on the relationship (7.2.7):

$$\mathbf{M} = \mathrm{submatrix}(\mathbf{M}'; 2, N+1, 2, N+1), \tag{7.5.2}$$

where operation submatrix($*$) assumes extracting submatrix from initial matrix within pointed range of rows and columns, respectively; elements $M'_{i',k'}$ of auxiliary matrix $\mathbf{M}' = \left\| M'_{i',k'} \right\|$ are determined by the result of modulo b operation (7.2.8):

$$M'_{i',k'} = \mathrm{mod}_b(i' \cdot k' + a), \tag{7.5.3}$$

$\mathrm{mod}_b(x)$ is a function returning the remainder on division x by b; $i' = 1, \ldots, N+1$, $k' = 1, \ldots, N+1$; $a, b = \mathrm{const}$, $a, b \in \mathbf{N}$; $\mathbf{N}$ is set of natural numbers; $b > N+1$.

Group CP-MFSK-CDM(A)-QPSK signal $s_g(t)$ is the sum of modulated channel signals $s_k(t)$:

$$s_g(t) = \sum_{k=0}^{N-1} s_k(t), \tag{7.5.4}$$

each of them is determined by the formula (7.5.1).

RF group signal $s_{g,RF}(t)$ obtained by modulating a carrier with the signal (7.5.4):

$$s_{g,RF}(t) = s_g(t)e^{j\omega_c t}$$

is radiated into propagation medium (space), as shown in Fig. 7.1.1, 7.1.2, and in received side it additively interacts with complex quasi-white Gaussian noise $n(t)$ with PSD $N(f) = N_0 \cdot [1(f) - 1(f - f_{\max})]$, $N_0{=}\mathrm{const}$, $f_{\max}{=}\mathrm{const}$:

$$x(t) = s_{g,RF}(t) + n(t). \tag{7.5.5}$$

Then, estimators $\hat{c}_k^I$, $\hat{c}_k^Q$ of symbols c_k^I, c_k^Q of QPSK signal in-phase $c^I = [c_0^I, \ldots, c_{N_{ch}-1}^I]^T$ and quadrature $c^Q = [c_0^Q, \ldots, c_{N_{ch}-1}^Q]^T$ streams, obtained by classic algorithm of signal demultiplexing/demodulation in linear space, are sign functions of scalar products between fragments of the observed stochastic process $x(t)$ (7.5.5) and modulating functions $M_k^I(t)$, $M_k^Q(t)$ (7.5.1a,b):

$$\hat{c}_k^I = \mathrm{sgn}(A_k), \ \hat{c}_k^Q = \mathrm{sgn}(B_k); \tag{7.5.6}$$

$$A_k = 2\sqrt{2} \left(\mathrm{Re}[x(t)e^{-j\omega_c t}], M_k^I(t) \right)/T_s; \tag{7.5.6a}$$

$$B_k = 2\sqrt{2} \left(\mathrm{Re}[x(t)e^{-j\omega_c t}], M_k^Q(t) \right)/T_s, \tag{7.5.6b}$$

where $(a(t), b(t))$ is scalar product between functions $a(t)$ and $b(t)$.

The relationships (7.5.6) define CP-MFSK-CDM(A)-QPSK signal demultiplexing/demodulation classic algorithm.

At the same time, estimators $\hat{c}_k^a$, $\hat{c}_k^b$ of symbols c_k^I, c_k^Q of QPSK signal in-phase

$c^I = [c_0^I, \ldots, c_{N_{ch}-1}^I]^T$ and quadrature $c^Q = [c_0^Q, \ldots, c_{N_{ch}-1}^Q]^T$ streams, obtained by L-group algorithm of signal demultiplexing/demodulation in space with lattice properties and based on MSI N_s (1.4.11) concerned with semimetric (1.3.26), are sign function of MSIs between fragments of the observed stochastic process $x(t)$ (7.5.5) and modulating functions $M_k^I(t)$, $M_k^Q(t)$ (7.5.1a,b):

$$\hat{c}_k^a = \mathrm{sgn}(a_k), \ \hat{c}_k^b = \mathrm{sgn}(b_k); \tag{7.5.7}$$

$$a_k = \alpha \cdot N_s \left(\mathrm{Re}[x(t)e^{-j\omega_c t}], M_k^I(t) \right); \tag{7.5.7a}$$

$$b_k = \alpha \cdot N_s \left(\mathrm{Re}[x(t)e^{-j\omega_c t}], M_k^Q(t) \right). \tag{7.5.7b}$$

where α is some constant coefficient, α=const.

The relationships (7.5.7) define CP-MFSK-CDM(A)-QPSK signal demultiplexing/demodulation L-group algorithm based on MSI N_s (1.4.11).

Similarly, estimators $\hat{c}_k^{a'}$, $\hat{c}_k^{b'}$ of symbols c_k^I, c_k^Q of QPSK signal in-phase $c^I = [c_0^I, \ldots, c_{N_{ch}-1}^I]^T$ and quadrature $c^Q = [c_0^Q, \ldots, c_{N_{ch}-1}^Q]^T$ streams, obtained by L-group algorithm of signal demultiplexing/demodulation in space with lattice properties and based on MSI N_{l_1} (1.4.12) concerned with l_1-metric (1.3.27), are sign functions of MSIs between fragments of the observed stochastic process $x(t)$ (7.5.5) and modulating functions $M_k^I(t)$, $M_k^Q(t)$ (7.5.1a,b):

$$\hat{c}_k^{a'} = \mathrm{sgn}(a_k'), \ \hat{c}_k^{b'} = \mathrm{sgn}(b_k'); \tag{7.5.8}$$

$$a_k' = \beta \cdot N_{l_1} \left(\mathrm{Re}[x(t)e^{-j\omega_c t}], M_k^I(t) \right); \tag{7.5.8a}$$

$$b_k' = \beta \cdot N_{l_1} \left(\mathrm{Re}[x(t)e^{-j\omega_c t}], M_k^Q(t) \right). \tag{7.5.8b}$$

where β is some constant coefficient, β=const.

The relationships (7.5.8) define CP-MFSK-CDM(A)-QPSK signal demultiplexing/demodulation L-group algorithm based on MSI N_{l_1} (1.4.12).

Notice that coefficients α, β figuring in the relationships (7.5.7), (7.5.8), respectively, are used to provide a convenience of visual perception of I-Q diagrams shown below, but in real processing they could be omitted.

L-group algorithm of signal demultiplexing/demodulation in space with lattice properties based on MSI N_p (1.4.10) can be used in multichannel (multistation) information transmitting systems with CP-MFSK-CDM(A)-QPSK, unlike DS-CDM(A)-QPSK system, but owing to considerable losses in SNR E_b/N_0 with respect to classic algorithm (up to 6 dB), we do not consider it further.

Example 7.5.1. Consider the results of using algorithms (7.5.6), (7.5.7) in the case of additive interaction between the signal $s_{g,RF}(t)$ and quasi-white Gaussian noise $n(t)$ (7.5.5) obtained by simulation (statistical modeling). Simulation is realized under the following conditions.

1. QPSK signal symbol duration T_s is equal to T_s=4096; a number of chips N_{ch} transmitted during QPSK signal symbol duration is equal to N_{ch}=32; a chip duration $T_{ch} = T_s/N_{ch}$ is equal to $T_{ch} = 128$; a number N of orthogonal functions from the system $\{\psi_k^{MFSK}(t)\}$ ($k = 0, 1, \ldots, N-1$) determined by the relationship (7.2.6) is equal to $N = N_{ch}$=32.

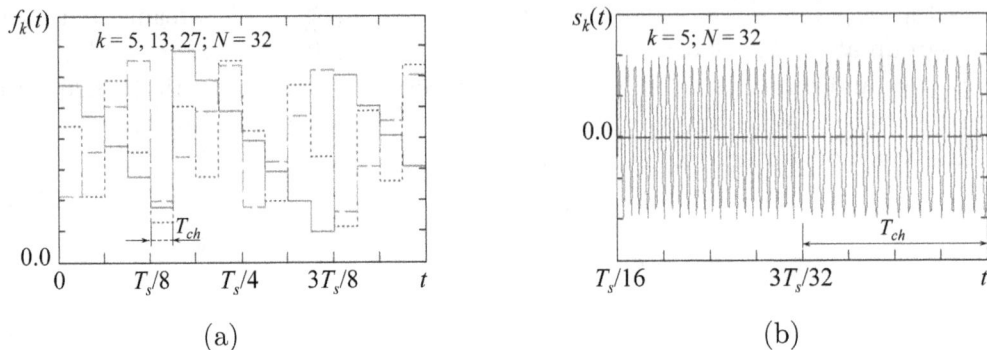

FIGURE 7.5.1   Fragments of (a) frequency dependences $f_k(t)$, $k = 5, 13, 27$, N=32; (b) channel QPSK signal $s_k(t)$, $k = 5$, N=32

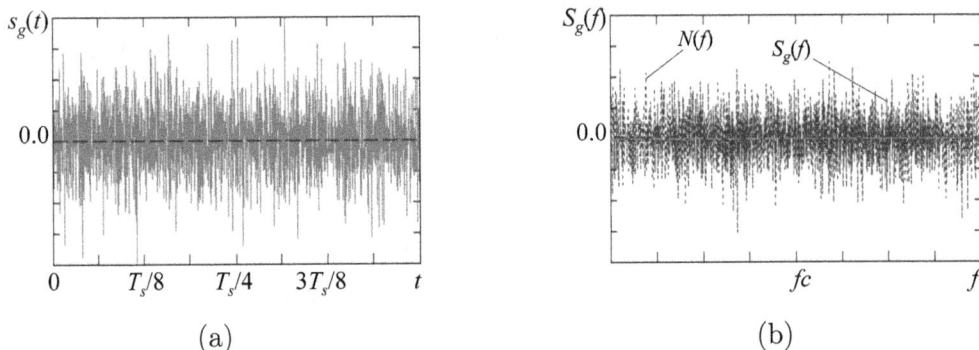

FIGURE 7.5.2   (a) Realization $s_g^*(t)$ of group signal $s_g(t)$; (b) Fourier transforms $N(f)$ and $S_g(f)$ of realizations of interference $n(t)$ and group signal $s_g(t)$, respectively

2. Values of constants figuring in the formulas (7.5.1a,b), (7.5.2a) are chosen to be equal to: M_0=42; a=1, b=37.

3. Interference $n(t)$ is quasi-white Gaussian noise (7.5.5) with PSD $N(f) = N_0 \cdot [1(f) - 1(f - f_{max})]$, N_0=const; SNR E_b/N_0 is equal to E_b/N_0=5, where E_b is energy per bit.

Fig. 7.5.1a illustrates fragments of frequency dependences $f_k(t)$ of orthogonal functions of the system $\{\psi_k^{MFSK}(t)\}$, $k = 0, 1, \ldots, N-1$, $N = N_{ch}$=32: $f_5(t)$ (solid line), $f_{13}(t)$ (dotted line), $f_{27}(t)$ (dashed line). As can be seen from the figure, chip duration is equal to $T_{ch} = T_s/32$. Fig. 7.5.1b depicts fragment of channel QPSK signal $s_k(t)$, $k = 5$ (solid line), and also the corresponding fragment of frequency dependence $f_k(t)$ of orthogonal function $\psi_k^{MFSK}(t)$, $k = 5$ (dotted line).

Fig. 7.5.2a illustrates realization $s_g^*(t)$ of group signal $s_g(t)$ determined by the sum (7.5.4), and Fig. 7.5.2b depicts Fourier transforms $N(f)$, $S_g(f)$ of realizations of interference $n(t)$ and group signal $s_g(t)$, respectively.

Fig. 7.5.3a illustrates I-Q diagram of CP-MFSK-CDM(A)-QPSK signal demodulation results obtained by classic algorithm (7.5.6) of signal demultiplexing/demodulation in linear space. Fig. 7.4.3b depicts I-Q diagram of CP-MFSK-

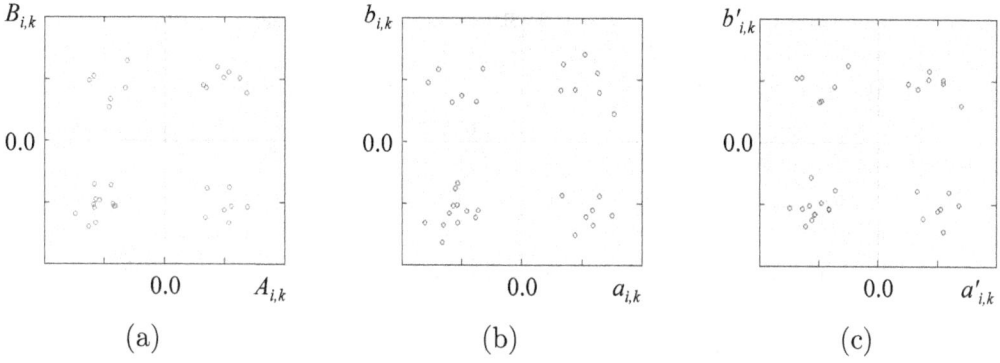

FIGURE 7.5.3   I-Q diagram of CP-MFSK-CDM(A)-QPSK signal demodulation results obtained by (a) classic algorithm (7.5.6) (linear space); (b) L-group algorithm (7.5.7) (semimetric space); (c) L-group algorithm (7.5.8) (l_1-metric space)

CDM(A)-QPSK signal demodulation results obtained by L-group algorithm (7.5.7) of signal demultiplexing/demodulation in semimetric space. Fig. 7.4.3c shows I-Q diagram of CP-MFSK-CDM(A)-QPSK signal demodulation results obtained by L-group algorithm (7.5.8) of signal demultiplexing/demodulation in l_1-metric space.

$\triangledown$

Results of comparative analysis of efficiency of algorithms (7.5.6), (7.5.7), (7.5.8) allow concluding that in the presence of interference $n(t)$ in the form of quasi-white Gaussian noise, L-group algorithms (7.5.7) and (7.5.8) are inferior to the classic one (7.5.6) in about 2.0 dB.

7.6   Demultiplexing/Demodulation L-group Algorithms for OFDM(A)-QPSK Systems

Channel signal $s_k(t)$ in multichannel (multi-station) information transmitting systems with OFDM(A)-QPSK is formed according to the relationships [38, (9.4.4)], [73, (12.27)] [184, (3.232)] that can be rewritten in more compact form:

$$s_k(t) = \frac{1}{\sqrt{2}} \left[c_k^I \cdot F_k^c(t) + c_k^Q \cdot F_k^s(t) \right], \ t \in T^s; \tag{7.6.1}$$

$$F_k^c(t) = \cos\left(2\pi \left(\frac{a \cdot p_k + b}{N \cdot T} \right) t + \pi/4 \right); \tag{7.6.1a}$$

$$F_k^s(t) = \sin\left(2\pi \left(\frac{a \cdot p_k + b}{N \cdot T} \right) t + \pi/4 \right), \tag{7.6.1b}$$

where $T^s = [0, T_s]$ is a domain of a symbol of channel OFDM(A)-QPSK signal $s_k(t)$; T_s is a duration of symbol of channel OFDM(A)-QPSK signal $s_k(t)$, $T_s = T \cdot N$; c^I, c^Q are the symbols of QPSK signal in-phase $c^I = [c_0^I, \ldots, c_{N_{ch}-1}^I]^T$ and quadrature $c^Q = [c_0^Q, \ldots, c_{N_{ch}-1}^Q]^T$ streams that are transmitted by k-th channel OFDM signal

$s_k(t)$; T is a duration of a signal transmitting the symbols c^I and c^Q; $\{F_k^c(t)\}$, $\{F_k^s(t)\}$ are the systems of N orthogonal functions $k = 0, 1, \ldots, N-1$; $N \in \mathbf{N}$, $\mathbf{N}$ is set of natural numbers; $\mathbf{p} = [2, 3, 5, \ldots]^T$ is a vector composed of prime numbers $\{p_k\}$, $p_k \in \mathbf{p}$; a, b=const, a, $b \in \mathbf{N}$, constants a, b are chosen basing on providing L-group orthogonality of the functions $F_k^c(t)$, $F_k^s(t)$.

Here we notice, that in classic variant of OFDM(A) information transmitting system, the systems $\{F_k^c(t)\}$, $\{F_k^s(t)\}$ of N orthogonal functions ($k = 0, 1, \ldots, N-1$) are defined by the following relationships (see, for instance [185, (19.3)]):

$$F_k^c(t) = \cos\left(2\pi\left(\frac{k}{N \cdot T}\right)t + \pi/4\right);$$

$$F_k^s(t) = \sin\left(2\pi\left(\frac{k}{N \cdot T}\right)t + \pi/4\right).$$

Functions of such a kind are not orthogonal in terms of MSIs N_s (1.4.11), N_{l_1} (1.4.12), so we use the functions (7.6.1a,b) where L-group orthogonality is provided due to the absence of multiple frequencies.

Group OFDM(A)-QPSK signal $s_g(t)$ is the sum of modulated channel signals $s_k(t)$:

$$s_{g/OFDM}(t) = \sum_{k=0}^{N-1} s_k(t), \tag{7.6.2}$$

each of them is determined by the formula (7.6.1).

RF group signal $s_{g,RF}(t)$ obtained by modulating a carrier with the signal (7.6.2)

$$s_{g,RF}(t) = s_{g/OFDM}(t)e^{j\omega_c t}$$

is radiated into propagation medium (space) as shown in Fig. 7.1.1, 7.1.2, and in receiving side of the system, it additively interacts with complex quasi-white Gaussian noise $n(t)$ with PSD $N(f) = N_0 \cdot [1(f) - 1(f - f_{\max})]$, N_0=const, $f_{\max}$=const:

$$x(t) = s_{g,RF}(t) + n(t). \tag{7.6.3}$$

Then, estimators $\hat{c}_k^I$, $\hat{c}_k^Q$ of symbols c_k^I, c_k^Q of QPSK signal in-phase $c^I = [c_0^I, \ldots, c_{N_{ch}-1}^I]^T$ and quadrature $c^Q = [c_0^Q, \ldots, c_{N_{ch}-1}^Q]^T$ streams, obtained by classic algorithm of signal demultiplexing/demodulation in linear space, are sign functions of scalar products between fragments of the observed stochastic process $x(t)$ (7.6.3) and modulating orthogonal functions $F_k^c(t)$, $F_k^s(t)$ (7.6.1a,b):

$$\hat{c}_k^I = \text{sgn}(A_k), \quad \hat{c}_{i,k}^Q = \text{sgn}(B_{i,k}); \tag{7.6.4}$$

$$A_k = 2\sqrt{2}\left(\text{Re}[x(t)e^{-j\omega_c t}], F_k^c(t)\right)/T_s; \tag{7.6.4a}$$

$$B_k = 2\sqrt{2}\left(\text{Re}[x(t)e^{-j\omega_c t}], F_k^s(t)\right)/T_s, \tag{7.6.4b}$$

where $(a(t), b(t))$ is scalar product between functions $a(t)$ and $b(t)$.

The relationships (7.6.4) define OFDM(A)-QPSK signal demultiplexing/demodulation classic algorithm.

At the same time, estimators $\hat{c}_k^a$, $\hat{c}_k^b$ of symbols c_k^I, c_k^Q of QPSK signal in-phase $c^I = [c_0^I, \ldots, c_{N_{ch}-1}^I]^T$ and quadrature $c^Q = [c_0^Q, \ldots, c_{N_{ch}-1}^Q]^T$ streams, obtained by L-group algorithm of signal demultiplexing/demodulation in space with lattice properties and based on MSI N_s (1.4.11) concerned with semimetric (1.3.26), are sign functions of MSIs between the observed stochastic process $x(t)$ (7.6.3) and modulating orthogonal functions $F_k^c(t)$, $F_k^s(t)$ (7.6.1a,b):

$$\hat{c}_k^a = \text{sgn}(a_k), \ \hat{c}_k^b = \text{sgn}(b_k); \tag{7.6.5}$$

$$a_k = \alpha \cdot N_s \left(\text{Re}[x(t)e^{-j\omega_c t}], F_k^c(t) \right); \tag{7.6.5a}$$

$$b_k = \alpha \cdot N_s \left(\text{Re}[x(t)e^{-j\omega_c t}], F_k^s(t) \right). \tag{7.6.5b}$$

where α is some constant coefficient, α=const.

The relationships (7.6.5) define OFDM(A)-QPSK signal demultiplexing/demodulation L-group algorithm based on MSI N_s (1.4.11).

Similarly, estimators $\hat{c}_k^{a'}$, $\hat{c}_k^{b'}$ of symbols c_k^I, c_k^Q of QPSK signal in-phase $c^I = [c_0^I, \ldots, c_{N_{ch}-1}^I]^T$ and quadrature $c^Q = [c_0^Q, \ldots, c_{N_{ch}-1}^Q]^T$ streams, obtained by L-group algorithm of signal demultiplexing/demodulation in space with lattice properties and based on MSI N_{l_1} (1.4.12) concerned with l_1-metric (1.3.27), are sign functions of MSIs between the observed stochastic process $x(t)$ (7.6.3) and modulating orthogonal functions $F_k^c(t)$, $F_k^s(t)$ (7.6.1a,b):

$$\hat{c}_k^{a'} = \text{sgn}(a_k'), \ \hat{c}_k^{b'} = \text{sgn}(b_k'); \tag{7.6.6}$$

$$a_k' = \beta \cdot N_{l_1} \left(\text{Re}[x(t)e^{-j\omega_c t}], F_k^c(t) \right); \tag{7.6.6a}$$

$$b_k' = \beta \cdot N_{l_1} \left(\text{Re}[x(t)e^{-j\omega_c t}], F_k^s(t) \right). \tag{7.6.6b}$$

where β is some constant coefficient, β=const.

The relationships (7.6.6) define OFDM(A)-QPSK signal demultiplexing/demodulation L-group algorithm based on MSI N_{l_1} (1.4.12).

Notice that coefficients α, β figuring in the relationships (7.6.5), (7.6.6), respectively, are used to provide a convenience of visual perception of I-Q diagrams shown below, but in real processing they could be omitted.

Like in CDM(A) information transmitting systems, L-group algorithm of signal demultiplexing/demodulation in space with lattice properties based on MSI N_p (1.4.10) seems to be hardly used in OFDM(A) information transmitting systems owing to problematic providing L-group orthogonality (in the sense of Definition 1.4.2) of channel signals $\{s_k(t)\}$, so we do not consider this algorithm further.

Example 7.6.1. Consider the results of using algorithms (7.6.4), (7.6.5) in the case of additive interaction between the signal $s_{g,RF}(t)$ and quasi-white Gaussian noise $n(t)$ (7.6.3) obtained by simulation (statistical modeling). Simulation is realized under the following conditions.

    1. Symbol duration T_s of OFDM(A)-QPSK signal $s_k(t)$ is equal to T_s=6144; a number of elementary signals N transmitted during OFDM(A)-QPSK signal symbol duration is equal to N=128; duration

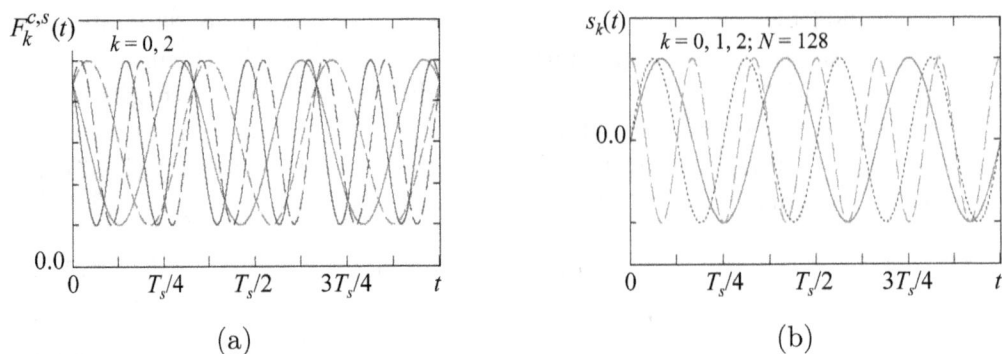

(a)   (b)

FIGURE 7.6.1   (a) Orthogonal functions $F_0^c(t)$, $F_2^c(t)$ and $F_0^s(t)$, $F_2^s(t)$; (b) channel OFDM(A)-QPSK signals $s_0(t)$, $s_1(t)$, $s_2(t)$, $N=128$

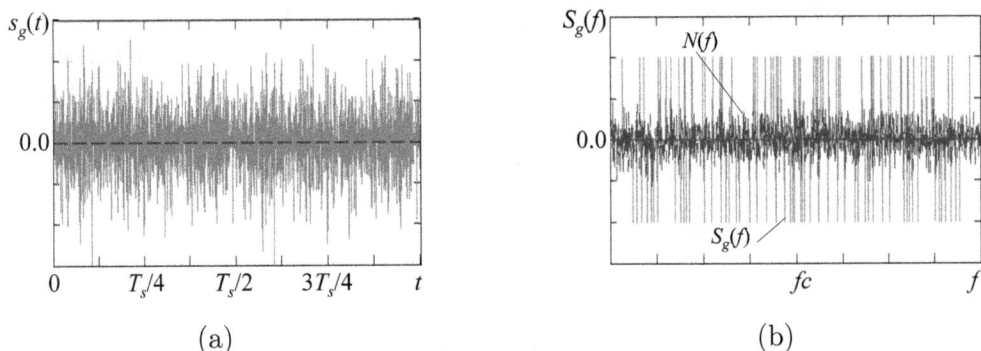

(a)   (b)

FIGURE 7.6.2   (a) Realization $s_g^*(t)$ of group signal $s_g(t)$; (b) Fourier transforms $N(f)$ and $S_g(f)$ of realizations of interference $n(t)$ and group signal $s_g(t)$, respectively

of elementary signal $T = T_s/N$ is equal to $T=48$; constants a, b figuring in (7.6.1a,b) are chosen to be equal to $a = 1$, $b = 1$.

2. Interference $n(t)$ is quasi-white Gaussian noise (7.6.3) with PSD $N(f) = N_0 \cdot [1(f) - 1(f - f_{max})]$, N_0=const; SNR E_b/N_0 is equal to $E_b/N_0=5$, where E_b is energy per bit.

Fig. 7.6.1a illustrates orthogonal functions $F_0^c(t)$, $F_2^c(t)$ (solid line), $F_0^s(t)$, $F_2^s(t)$ (dashed line) from the systems of orthogonal functions $\{F_k^c(t)\}$, $\{F_k^s(t)\}$, $k = 0, 1, \dots, N - 1$ (7.6.1a,b), respectively. Fig. 7.6.1b depicts channel OFDM(A)-QPSK signals $s_0(t)$, $s_1(t)$, $s_2(t)$ (solid, dotted, and dashed lines, respectively), so that, as follows from the formulas (7.6.1a,b) and shown in the figure, in the symbol interval $T^s = [0, T_s]$ of OFDM(A) signal $s_k(t)$, these signals contain 2+1=3; 3+1=4; 5+1=6 oscillation periods.

Fig. 7.6.2a illustrates realization $s_g^*(t)$ of group signal $s_g(t)$ determined by the sum (7.6.2), and Fig. 7.6.2b depicts Fourier transforms $N(f)$, $S_g(f)$ of realizations of interference $n(t)$ and group signal $s_g(t)$, respectively. Peak factor of group OFDM(A)-QPSK signal is equal to 10 dB.

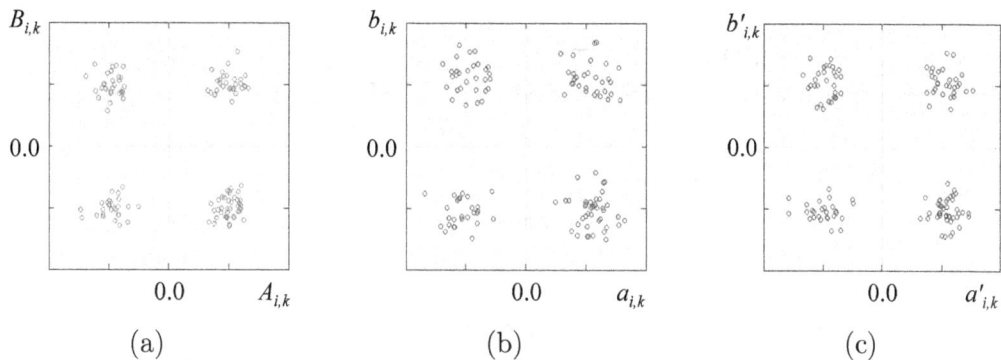

FIGURE 7.6.3 I-Q diagrams of OFDM(A)-QPSK signal demodulation results obtained by (a) classic algorithm (7.6.4) (linear space); (b) *L*-group algorithm (7.6.5) (semimetric space); (c) *L*-group algorithm (7.6.6) (l_1-metric space)

Fig. 7.6.3a illustrates I-Q diagram of OFDM(A)-QPSK signal demodulation results obtained by classic algorithm (7.6.4) of signal demultiplexing/demodulation in linear space. Fig. 7.6.3b depicts I-Q diagram of OFDM(A)-QPSK signal demodulation results obtained by *L*-group algorithm (7.6.5) of signal demultiplexing/demodulation in semimetric space. Fig. 7.6.3c shows I-Q diagram of OFDM(A)-QPSK signal demodulation results obtained by *L*-group algorithm (7.6.6) of signal demultiplexing/demodulation in l_1-metric space. ▽

The results of comparative analysis of efficiency of algorithms (7.6.4), (7.6.5), (7.6.6) allow concluding that in the presence of interference $n(t)$ in the form of quasi-white Gaussian noise, *L*-group algorithms (7.6.5) and (7.6.6) are inferior to the classic one (7.6.4) in 1...2 dB.

7.7 Demultiplexing/Demodulation *L*-group Algorithms for OFDM-CDM(A)-QPSK Systems

Channel signal $s_{i,k}(t)$ in multichannel (multi-station) information transmitting systems with OFDM-CDM(A)-QPSK is formed according to the relationships [38, (9.4.4)], [73, (12.28)] that can be rewritten in more compact form:

$$s_{i,k}(t) = \frac{1}{\sqrt{2}} \left[c_{i,k}^I \cdot M_{i,k}^I(t) + c_{i,k}^Q \cdot M_{i,k}^Q(t) \right], \ t \in T^s; \tag{7.7.1}$$

$$M_{i,k}^I(t) = wal_k^N(t) \cdot F_i^c(t); \tag{7.7.1a}$$

$$M_{i,k}^Q(t) = wal_k^N(t) \cdot F_i^s(t); \tag{7.7.1b}$$

$$F_k^c(t) = \cos\left(2\pi \left(\frac{a \cdot p_k + b}{N \cdot T}\right) t + \pi/4\right); \tag{7.7.1c}$$

$$F_k^s(t) = \sin\left(2\pi \left(\frac{a \cdot p_k + b}{N \cdot T}\right) t + \pi/4\right), \tag{7.7.1d}$$

where $T^s = [0, T_s]$ is a domain of a symbol of OFDM-CDM(A)-QPSK channel signal $s_{i,k}(t)$; T_s is an OFDM-CDM(A)-QPSK signal symbol duration; $M^I_{i,k}(t)$, $M^Q_{i,k}(t)$ are modulating functions of OFDM-CDM(A)-QPSK signal in-phase and quadrature streams, respectively; c^I, c^Q are the matrices of OFDM-CDM(A)-QPSK signal in-phase and quadrature streams, respectively: $c^I = \left\| c^I_{i,k} \right\|$, $c^Q = \left\| c^Q_{i,k} \right\|$; $c^I_{i,k}$, $c^Q_{i,k}$ are elements of the matrices c^I, c^Q; i, k are indexes of a row and a column, respectively; $\{wal^N_k(t)\}$ is the system of N orthogonal Walsh functions ($k = 0, 1, \ldots, N-1$) determined by the relationship (7.2.3a); T_{ch} is a chip duration, $T_{ch} = T_s/N$, $N = N_{ch}$; N_{ch} is a number of chips transmitted during the symbol duration T_s; $N, N_{ch} \in \mathbf{N}$, $\mathbf{N}$ is set of natural numbers; $\{F^c_i(t)\}$, $\{F^s_i(t)\}$ are the systems of M orthogonal functions ($i = 0, 1, \ldots, M-1$) determined by the relationships (7.7.1c,d); $M \in \mathbf{N}$; $\mathbf{p} = [2, 3, 5, \ldots]^T$ is a vector composed of elements $\{p_k\}$ of prime numbers series, $p_k \in \mathbf{p}$; a, b=const, $a, b \in \mathbf{N}$, constants a, b are chosen basing on providing L-group orthogonality of the functions $F^c_i(t)$, $F^s_i(t)$.

Group OFDM-CDM(A)-QPSK signal $s_g(t)$ is the sum of modulated channel signals $s_{i,k}(t)$:

$$s_g(t) = \sum_{i=0}^{M-1} \sum_{k=0}^{N-1} s_{i,k}(t), \qquad (7.7.2)$$

each of them is determined by the formula (7.7.1).

RF group signal $s_{g,RF}(t)$ obtained by modulating a carrier with the signal (7.7.2)

$$s_{g,RF}(t) = s_g(t)e^{j\omega_c t}$$

is radiated into propagation medium (space) as shown in Figs. 7.1.1, 7.1.2, and in receiving side it additively interacts with complex quasi-white Gaussian noise $n(t)$ with PSD $N(f) = N_0 \cdot [1(f) - 1(f - f_{max})]$, N_0=const, f_{max}=const:

$$x(t) = s_{g,RF}(t) + n(t). \qquad (7.7.3)$$

Then, estimators $\hat{c}^I_{i,k}$, $\hat{c}^Q_{i,k}$ of elements $c^I_{i,k}$, $c^Q_{i,k}$ of the matrices c^I, c^Q of OFDM-CDM(A)-QPSK signal in-phase and quadrature streams, obtained by classic algorithm of signal demultiplexing/demodulation in linear space, are sign functions of scalar products between fragments of the observed stochastic process $x(t)$ (7.7.3) and modulating functions $M^I_{i,k}(t)$, $M^Q_{i,k}(t)$ (7.7.1a,b):

$$\hat{c}^I_{i,k} = \operatorname{sgn}(A_{i,k}), \ \hat{c}^Q_{i,k} = \operatorname{sgn} \operatorname{Re}(B_{i,k}); \qquad (7.7.4)$$

$$A_{i,k} = 2\sqrt{2} \left(\operatorname{Re}[x(t)e^{-j\omega_c t}], M^I_{i,k}(t) \right) /T_s; \qquad (7.7.4a)$$

$$B_{i,k} = 2\sqrt{2} \left(\operatorname{Re}[x(t)e^{-j\omega_c t}], M^Q_{i,k}(t) \right) /T_s, \qquad (7.7.4b)$$

where $(a(t), b(t))$ is scalar product between functions $a(t)$ and $b(t)$.

The relationships (7.7.4) define OFDM-CDM(A)-QPSK signal demultiplexing/demodulation classic algorithm.

At the same time, estimator $\hat{c}^a_{i,k}$, $\hat{c}^b_{i,k}$ of elements $c^I_{i,k}$, $c^Q_{i,k}$ of the matrices c^I, c^Q of transmitted OFDM-CDM(A)-QPSK signal in-phase and quadrature streams, obtained by L-group algorithm of signal demultiplexing/demodulation in space with lattice properties and based on MSIs N_s (1.4.11) concerned with semimetric (1.3.26), are sign functions of MSIs between fragments of the observed stochastic process $x(t)$ (7.7.3) and modulating functions $M_{i,k}^I(t)$, $M_{i,k}^Q(t)$ (7.7.1a,b):

$$\hat{c}_{i,k}^a = \text{sgn}(a_{i,k}), \ \hat{c}_{i,k}^b = \text{sgn}(b_{i,k}); \tag{7.7.5}$$

$$a_{i,k} = \alpha \cdot N_s \left(\text{Re}[x(t)e^{-j\omega_c t}], M_{i,k}^I(t)\right); \tag{7.7.5a}$$

$$b_{i,k} = \alpha \cdot N_s \left(\text{Re}[x(t)e^{-j\omega_c t}], M_{i,k}^Q(t)\right). \tag{7.7.5b}$$

where α is some constant coefficient, α=const.

The relationships (7.7.5) define OFDM-CDM(A)-QPSK signal demultiplexing/demodulation L-group algorithm based on MSI N_s (1.4.11).

Similarly, estimators $\hat{c}_{i,k}^{a'}$, $\hat{c}_{i,k}^{b'}$ of elements $c_{i,k}^I$, $c_{i,k}^Q$ from the matrices c^I, c^Q of OFDM-CDM(A)-QPSK signal in-phase and quadrature streams, obtained by L-group algorithm of signal demultiplexing/demodulation in space with lattice properties and based on MSI N_{l_1} (1.4.12) concerned with l_1-metric (1.3.27), are sign functions of MSIs between fragments of the observed stochastic process $x(t)$ (7.7.3) and modulating functions $M_{i,k}^I(t)$, $M_{i,k}^Q(t)$ (7.7.1a,b):

$$\hat{c}_{i,k}^{a'} = \text{sgn}(a_{i,k}'), \ \hat{c}_{i,k}^{b'} = \text{sgn}(b_{i,k}'); \tag{7.7.6}$$

$$a_{i,k}' = \beta \cdot N_{l_1} \left(\text{Re}[x(t)e^{-j\omega_c t}], M_{i,k}^I(t)\right); \tag{7.7.6a}$$

$$b_{i,k}' = \beta \cdot N_{l_1} \left(\text{Re}[x(t)e^{-j\omega_c t}], M_{i,k}^Q(t)\right). \tag{7.7.6b}$$

where β is some constant coefficient, β=const.

The relationships (7.7.6) define OFDM-CDM(A)-QPSK signal demultiplexing/demodulation L-group algorithm based on MSI N_{l_1} (1.4.12).

Notice that coefficients α, β figuring in the relationships (7.7.5), (7.7.6), respectively, are used to provide a convenience of visual perception of I-Q diagrams shown below, but in real processing they could be omitted.

Like in CDM(A) information transmitting systems, L-group algorithm of signal demultiplexing/demodulation in space with lattice properties based on MSI N_p (1.4.10) seems to be hardly used in OFDM-CDM(A) systems due to problematic providing L-group orthogonality (in the sense of Definition 1.4.2) of channel signals $\{s_{i,k}(t)\}$, so we do not discuss this algorithm further.

Example 7.7.1. Consider the results of using algorithms (7.7.4), (7.7.5) obtained by simulation (statistical modeling) in the case of additive interaction between the signal $s_{g,RF}(t)$ and quasi-white Gaussian noise $n(t)$ (7.7.3). Simulation is realized under the following conditions.

    1. Symbol duration T_s of channel OFDM-CDM(A)-QPSK signal $s_{i,k}(t)$ is equal to T_s=4096; number of chips N_{ch} transmitted during symbol duration T_s of channel OFDM-CDM(A)-QPSK signal is equal to N_{ch}=16; chip duration $T_{ch} = T_s/N_{ch}$ is equal to $T_{ch} = 256$; number N of orthogonal functions from the system $\{wal_k^N(t)\}$ ($k = 0, 1, \ldots, N-1$) determined

(a)

(b)

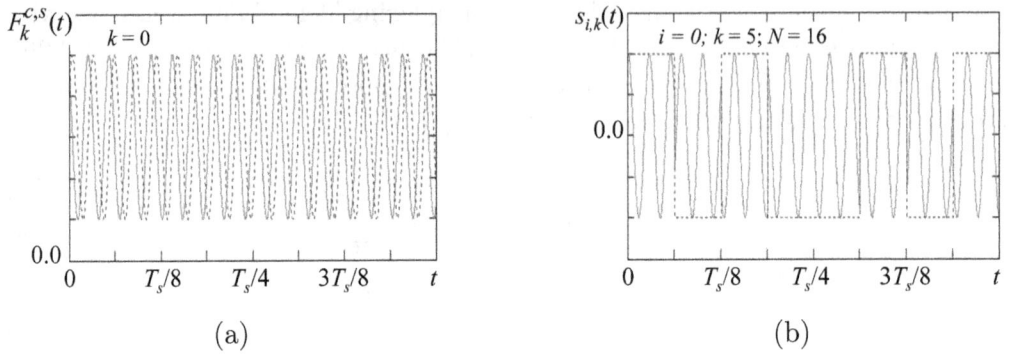

FIGURE 7.7.1 Fragments of (a) orthogonal functions $F_0^c(t)$ (solid line), $F_0^s(t)$ (dotted line); (b) channel OFDM-CDM(A)-QPSK signal $s_{0,5}(t)$, $i = 0$, $k = 5$ (solid line) and also the corresponding Walsh function $wal_5^{N=16}(t)$ (dotted line)

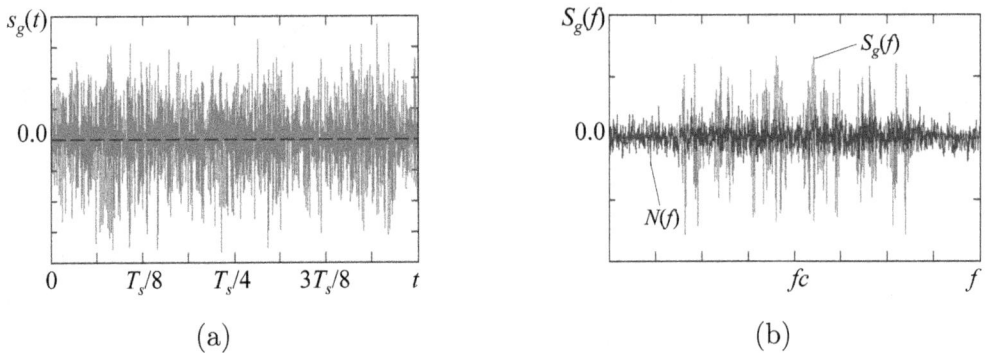

(a)

(b)

FIGURE 7.7.2 (a) Realization $s_g^*(t)$ of group signal $s_g(t)$; (b) Fourier transforms $N(f)$ and $S_g(f)$ of realizations of interference $n(t)$ and group signal $s_g(t)$, respectively

by the relationship (7.2.3a) is equal to $N = N_{ch}=16$; number M of orthogonal functions from the system $\{F_i^c(t)\}$, $\{F_i^s(t)\}$ ($i = 0, 1, \ldots, M-1$) determined by the relationships (7.7.1c,d) is equal to $M = 8$; constants $a\,b$ figuring in (7.7.1c,d) are equal to $a = N$, $b = 3$.

2. Interference $n(t)$ is quasi-white Gaussian noise (7.7.3) with PSD $N(f) = N_0 \cdot [1(f) - 1(f - f_{\max})]$, N_0=const; SNR E_b/N_0 is equal to E_b/N_0=5, where E_b is energy per bit.

Fig. 7.7.1a illustrates fragments of orthogonal functions $F_0^c(t)$ (solid line), $F_0^s(t)$ (dotted line) from the system $\{F_i^c(t)\}$, $\{F_i^s(t)\}$, $i = 0, 1, \ldots, M-1$, $M = 8$ determined by the relationships (7.7.1c,d). Fig 7.7.1b depicts fragment of channel OFDM-CDM(A)-QPSK signal $s_{0,5}(t)$, $i = 0$, $k = 5$ (solid line) and also the corresponding Walsh function $wal_5^{N=16}(t)$ (dotted line).

Fig. 7.7.2a illustrates realization $s_g^*(t)$ of group signal $s_g(t)$ determined by the sum (7.7.2), and Fig. 7.7.2b depicts Fourier transforms $N(f)$, $S_g(f)$ of realizations of interference $n(t)$ and group signal $s_g(t)$, respectively.

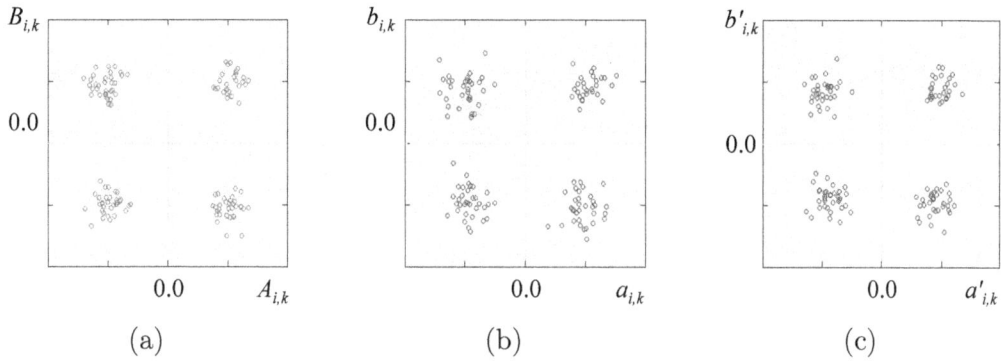

FIGURE 7.7.3   I-Q diagrams of OFDM-CDM(A)-QPSK signal demodulation results obtained by: (a) classic algorithm (7.7.4) (linear space); (b) L-group algorithm (7.7.5) (semi-metric space); (c) L-group algorithm (7.7.6) (l_1-metric space)

Fig. 7.7.3a illustrates I-Q diagram of OFDM-CDM(A)-QPSK signal demodulation results obtained by classic algorithm (7.7.4) of signal demultiplexing/demodulation in linear space. Fig. 7.7.3b depicts I-Q diagram of OFDM-CDM(A)-QPSK signal demodulation results obtained by L-group algorithm (7.7.5) of signal demultiplexing/demodulation based on MSI N_s (1.4.11). Fig. 7.7.3c depicts I-Q diagram of OFDM-CDM(A)-QPSK signal demodulation results obtained by L-group algorithm (7.7.6) of signal demultiplexing/demodulation based on MSI N_{l_1} (1.4.12).   ▽

The results of comparative analysis of efficiency of algorithms (7.7.4), (7.7.5), (7.7.6) allow concluding that in the presence of interference $n(t)$ in the form of quasi-white Gaussian noise, L-group algorithms (7.7.5) and (7.7.6) are inferior to the classic one (7.7.4) in $1 \ldots 2$ dB.

Despite the fact that in all considered above cases, algorithms of signal demultiplexing/demodulation based on L-group operations in multichannel (multi-station) information transmitting systems are slightly inferior to their classic analogues in SNR, their application seems to be expedient due to multiplication-free approach lying in their basis. These property of L-group algorithms makes them preferable for exploiting in applications that require a considerable computational rate, or in applications in which using considerable computational resources is impossible or difficult.

8

Wavelets on L-groups

8.1 Wavelet Signal Analysis on L-groups: Introduction

The emergence of Wavelet Theory is considered to be one of the most important event in mathematics of last decades, since, being relatively a new mathematical concept, it becomes a new tool for applied researches in many branches of science. Recently, wavelets are widely used in signal analysis and synthesis, in image processing (including image compression), in pattern recognition, etc. In this section we consider some problems of signal analysis that are solved by wavelet signal processing methods based on measures of statistical interrelations (MSIs) $N_p(x,y), N_s(x,y), N_{l_1}(x,y)$ (1.4.10...1.4.12) between realizations $x = (x_1, x_2, \ldots, x_n)$, $y = (y_1, y_2, \ldots, y_n)$ of samples X, Y (discrete observations) in sample space $\mathcal{L}(\mathcal{X}, \mathcal{B}_{\mathcal{X}}; +, \vee, \wedge)$ with L-group properties.

Any signal carrying information is, as a rule, a non-stationary stochastic process. Classic methods of statistical data processing (for instance, correlation and spectral methods, cumulant analysis method) are the tools that developed a good reputation in researches related to stationary stochastic processes; their application for analyzing non-stationary stochastic processes (time series) can cause different problems in interpretation of obtained results.

Often, when exploring experimental data, researchers use an idea of analyzing systems with slowly changed parameters: it is assumed that during small time intervals a stochastic process properties are slightly changed; such a process is considered to be stationary that allows using classic apparatus of statistical data processing. Such an approach can be efficient just within baseband signal spectrum domain. If signal properties can essentially changed even within relatively small time intervals, then there are two variants of further actions: either to get rid of classic methods of time series analysis and use special methods or to realize careful preliminary data processing, choosing just those intervals of signals in which these signals can be considered (with a certain degree of approximation) as stationary. In any case, signal analysis should be performed on the basis of the most universal methods that can be used regardless of prior information on signal stationarity. Wavelet analysis is just such a multi-purpose mean.

8.1.1 Continuous Wavelet Transform and Its Properties

Unlike harmonic functions used in classic spectral analysis, wavelet transform exploits localized functions, i.e. wavelets. These functions can be dually transformed,

DOI: 10.1201/9781003275855-8

namely, in scale (by either compression or dilation) and in shift along time axis. This allows, first, studying local properties of signals within different observation scales, second, indicating various changes of a signal in time.

Unlike classic Fourier analysis, where signal spectrum is a function of one variable (frequency), wavelet transform operates with two variables: (1) scale (or frequency); (2) shift along a time axis. A feature of wavelet analysis is that both frequency and time are considered as independent variables, this circumstance allows speaking on «simultaneous» (or time-frequency) spectral analysis. Having placed a wavelet in a chosen time coordinate of a signal and having performed a wavelet scaling, one can get information on frequency components of researched process in a neighborhood of a fixed signal sample. Toward this goal, scalar product between wavelet and a signal is calculated, so that the greater correlation between them is, the greater scalar product is obtained, and vice versa, the less correlation between wavelet and a signal is, the less scalar product is. Further, analogous mathematical operations are repeated for other scales and other time instants. As a result, some bivariate function of univariate signal is obtained. A choice of scale and shift can be arbitrary (when performing continuous wavelet transform), or these quantities can take their values from the corresponding finite sets (in the case of discrete wavelet transform).

One of the most important aspects when analyzing signals with help of wavelets is a choice of concrete wavelet $\psi(t)$, that is called *basis* or *mother wavelet function*. Choice of a wavelet is not a mathematical problem, since intuition and a practice experience of a researcher play here a considerable role, as well as the existing traditions of using these or those functions for solving concrete problems.

Basis wavelet function $\psi(t)$ must meet some requirements: (1) localization; (2) zero mean; (3) finite energy; (4) self-similarity.

Localization condition assumes localization of a basis function $\psi(t)$ in time and in frequency. This means that the function $\psi(t)$ decreases down to zero out of some interval, so that for practical applications the more quickly it decreases the better. To provide it, it is sufficient the following inequalities hold:

$$|\psi(t)| \leq C(1 + |t|)^{-1-\varepsilon};$$

$$\mathcal{F}[\psi(t)] \leq C(1 + |\omega|)^{-1-\varepsilon}; \ \varepsilon > 0,$$

where $\mathcal{F}[\psi(t)]$ is Fourier transform of a function $\psi(t)$.

Zero mean condition assumes the following equality holds:

$$\int\limits_{-\infty}^{\infty} \psi(t)dt = 0,$$

i.e. the function must be alternating with respect to zero.

To ignore slow non-stationarity, it is necessary the following equation holds:

$$\int\limits_{-\infty}^{\infty} t^m \psi(t)dt = 0,$$

that requires the first m moments of a function $\psi(t)$ to be equal to zero.

The condition of *finite energy* requires fulfillment of the inequality:

$$\int\limits_{-\infty}^{\infty} |\psi(t)|^2 dt < \infty.$$

The condition of *self-similarity* means that a function $\psi(t)$ possesses a self-similarity property with respect to scale and shift transformations.

In practice, basis wavelet functions that are Gauss function derivatives are rather frequently used, in particular, wave-wavelet (derivative of the 1-st order):

$$\psi(t) = t \exp(-t^2/2),$$

mhat-wavelet (derivative of the 2-nd order):

$$\psi(t) = (1 - t^2) \exp(-t^2/2),$$

n-th order derivative:

$$\psi(t) = (-1)^n \frac{d^n}{dt^n} \left[\exp(-t^2/2)\right].$$

Among complex functions, Morlet wavelet (Jean Morlet) is rather frequently used:

$$\psi(t) = \exp(j2\pi f_0 t) \exp(-t^2/2), \tag{8.1.1}$$

where f_0 is a central frequency of signal spectrum, $f_0 >> 0$.

After choosing a basis function $\psi(t)$, a set of such functions $\{\psi_{a,b}(t)\}$ is constructed that are based on the former:

$$\psi_{a,b}(t) = \frac{1}{\sqrt{b}} \psi((t - a)/b), \tag{8.1.2}$$

where a is a time shift parameter; b is a scale parameter (compression or dilation); $a, b \in \mathbb{R}$; $\mathbb{R}$ is set of real numbers. Multiplier $1/\sqrt{b}$ provides independence of this function norm on a scale parameter b.

Here it should be noted, that in the framework of classic spectral analysis, a power of a signal $s(t)$ is equal to a square under a curve of power spectral density $S(f)$, so that a quantity of spectral peak allows determining an amplitude of oscillation with corresponding frequency. In the case of wavelet analysis, a situation seems to be a little bit more complicated. When providing a correct energy estimation by the obtained wavelet transform, it is impossible to estimate amplitude of harmonics signals accurately, and vice versa. For a correct estimation of relationships between the amplitudes of harmonic signals it is necessary to perform a special normalization. Toward this goal one can replace a multiplier $1/\sqrt{b}$ in the formula (8.1.2) by a multiplier $1/b$, that allows correctly determining the amplitudes of harmonic signals with different frequencies but leads to a violation of energy relations:

$$\psi_{a,b}(t) = \frac{1}{b} \psi((t - a)/b). \tag{8.1.3}$$

Continuous wavelet transform $X(a,b)$ of a signal $x(t)$ takes the form of scalar product between a signal and wavelet function:

$$X(a,b) = CWT_x(a,b) = \int\limits_{-\infty}^{\infty} x(t)\bar{\psi}_{a,b}(t)dt, \qquad (8.1.4)$$

where symbol «$\bar{\psi}$» denotes complex conjugate of initial function.

One of the most important aspects when comparing wavelet transform and Fourier transform is different time-frequency localization. Fourier transform operates with harmonic functions that are infinite in time, so in order to obtain spectral information at a chosen frequency, it is necessary to calculate an integral in infinite range $]-\infty, \infty[$. Fourier transform does not take into account possible change of a frequency of an observed signal. In particular, Fourier transform does not allow distinguishing a signal that is a sum of harmonic functions with different frequencies from a frequency shift keying (FSK) signal. Using a set of basis functions $\{\psi_{a,b}(t)\}$ (8.1.2), (8.1.3) allows solving this problem.

Wavelets can be considered as "window" functions possessing their widths in both time (σ_t) and frequency (σ_f) domains, respectively. This width is defined as a standard deviation (square root of a variance) of the functions $|\psi_{a,b}(t)|$ and $|\Psi_{a,b}(f)|$, where $\Psi_{a,b}(f)$ is a Fourier-image of a wavelet $\psi_{a,b}(t)$. Time-frequency resolution of wavelet transform can be described by a rectangle with sides σ_t, σ_f in a plane (t, f). This rectangle is sometimes called a *Heisenberg rectangle* reflecting the known uncertainty principle that characterizes a minimum area occupied by this rectangle (see, for instance, [186, Fig 4.1], [187, Fig 1.2]. In the case of wavelet transform, this area cannot be less then $1/(4\pi)$: $\sigma_t \cdot \sigma_f \geq 1/(4\pi)$.

The existence of uncertainty principle means that when analyzing signals, one can not simultaneously improve both frequency and time resolution, inasmuch as decreasing wavelet width in time domain is accompanied by increasing wavelet width of a spectral peak in frequency domain. Conversely, aspiration for achieving narrow spectrum lines leads to the necessity of processing more prolonged fragments of the observed signal. This makes a time-frequency window of wavelet transform more agile: Heisenberg rectangle is constricted in frequency and expanded in time within a low frequency range, and vice versa, is expanded in frequency and constricted in time within high frequency range.

This property of wavelet transform is very important. Since a signal frequency is inversely proportional to oscillation period, to obtain high frequency information with a required accuracy it is necessary to extract this information within relatively small time intervals, whereas for extracting low frequency spectral information it is necessary to consider more lengthy intervals.

When performing windowed Fourier transform, using a window function in time domain cause confining spectral resolution in frequency domain and appearing of Heisenberg rectangle in the plane (t, f), so that a size of Heisenberg rectangle does not change when changing a frequency, so both time and frequency resolutions are the same for all frequencies.

8.1.2 Discrete Wavelet Transform

Discrete wavelet transform $DWT_x(u,s)$ of a signal $x(t_j)$ that is represented by its own samples $x = (x_0, x_1, \ldots, x_{J-1})$, $x(t_j) = x(\Delta t \cdot j) = x_j$ has a form which is analogous to continuous wavelet transform (8.1.4):

$$DWT_x(u,s) = \sum_{j=0}^{J-1} [x_j \cdot \bar{\psi}(j; u, s)]$$

$$= \sum_{j=0}^{J-1} [x_j \sqrt{s \cdot \Delta f} \cdot \bar{\psi}(s \cdot \Delta f \cdot (j-u) \cdot \Delta t)]; \quad (8.1.5)$$

$$\psi(j; u, s) = \sqrt{s \cdot \Delta f} \cdot \psi(s \cdot \Delta f \cdot (j-u) \cdot \Delta t), \quad (8.1.5a)$$

where $x(t_j) = x(\Delta t \cdot j) = x_j$; $j = 0, 1, \ldots, J-1$; u is discrete time shift parameter, $u = 0, 1, \ldots, J-1$; s is discrete scale (frequency change) parameter, $s = 0, 1, \ldots, S-1$; $J, S \in \mathbb{N}$; $\mathbb{N}$ is set of natural numbers; a symbol «$\bar{\psi}$» denotes complex conjugate of ψ; Δt is a sampling interval in time domain; Δf is a sampling interval in frequency domain; $t_j = \Delta t \cdot j \in T_{sp}$; T_{sp} is signal processing interval; $\Delta t \cdot (J-1) = \Delta T_{sp}$ is a duration of signal processing interval.

Sometimes, for practical applications, it is convenient using discrete wavelet transform $DWT_x(u,k)$ of the signal $x(t)$, where a scale parameter s figuring in the formula (8.1.5) is some exponential function, for instance, $s = 2^k$:

$$DWT_x(u,k) = \sum_{j=0}^{J-1} [x_j \cdot \bar{\psi}(j; u, s)] = \sum_{j=0}^{J-1} [x_j \sqrt{2^k \Delta f} \bar{\psi}(2^k \Delta f (j-u) \cdot \Delta t)], \quad (8.1.6)$$

where $k = 0, 1, \ldots, K-1$, $K \in \mathbb{N}$.

Hereinafter we use real wavelet functions for real signals, so that:

$$\bar{\psi}(j; u, s) = \psi(j; u, s).$$

8.1.3 Discrete Wavelet Transform in Sample Spaces with Lattice Properties

In addition to known formula defining wavelet transform for linear space with scalar product (8.1.5), we define three variants of wavelet transforms based on the corresponding three *measures of statistical interrelation* (MSIs) determined by the relationships (1.4.10), (1.4.11), (1.4.12) in sample space $\mathcal{L}(\mathcal{X}, \mathcal{B}_{\mathcal{X}}; +, \vee, \wedge)$ with generalized metrics (1.3.25), (1.3.26), (1.3.27), respectively:

$$N_p(x,y) = J - \sum_{j=0}^{J-1} [\text{sgn}(x_j \vee y_j) - \text{sgn}(x_j \wedge y_j)]; \quad (8.1.7)$$

$$N_s(x,y) = \|x\| + \|y\| - \sum_{j=0}^{J-1} |x_j - y_j| \cdot [\text{sgn}(x_j \vee y_j) - \text{sgn}(x_j \wedge y_j)]; \quad (8.1.8)$$

$$N_{l_1}(x, y) = \sum_{j=0}^{J-1} [|x_j + y_j| - |x_j - y_j|], \qquad (8.1.9)$$

where $\|x\|$, $\|y\|$ are l_1-norms of realizations (observations) $x = (x_0, x_1, \ldots, x_{J-1})$, $y = (y_0, y_1, \ldots, y_{J-1})$ of the samples X, Y, that are determined by the relationships under the formula (1.4.7).

Definition 8.1.1. Wavelet transform $X_p(u, s)$ of a discrete signal $x(t_j)$, $t_j \in T_{sp}$ that is represented by its own samples $x = (x_0, x_1, \ldots, x_{J-1})$, $x(t_j) = x(\Delta t \cdot j) = x_j$ in sample space $\mathcal{L}(\mathcal{X}, \mathcal{B}_{\mathcal{X}}; +, \vee, \wedge)$ with pseudometric (1.3.25) is MSI (8.1.7) between this signal $x(t_j)$ and wavelet function $\psi(j; u, s)$ (8.1.5a):

$$x(t_j) \xrightarrow{DWT_p} X_p(u, s) :$$

$$X_p(u, s) = J - \sum_{j=0}^{J-1} [\mathrm{sgn}(x_j \vee \psi(j; u, s)) - \mathrm{sgn}(x_j \wedge \psi(j; u, s))]. \qquad (8.1.10)$$

Definition 8.1.2. Wavelet transform $X_s(u, s)$ of a discrete signal $x(t_j)$, $t_j \in T_{sp}$ that is represented by its own samples $x = (x_0, x_1, \ldots, x_{J-1})$, $x(t_j) = x(\Delta t \cdot j) = x_j$ in sample space $\mathcal{L}(\mathcal{X}, \mathcal{B}_{\mathcal{X}}; +, \vee, \wedge)$ with semimetric (1.3.26) is MSI (8.1.8) between this signal $x(t_j)$ and wavelet function $\psi(j; u, s)$ (8.1.5a):

$$x(t_j) \xrightarrow{DWT_s} X_s(u, s) :$$

$$X_s(u, s) = \|x\| + \|\psi(u, s)\|$$

$$- \sum_{j=0}^{J-1} |x_j - \psi(j; u, s)| \cdot [\mathrm{sgn}(x_j \vee \psi(j; u, s)) - \mathrm{sgn}(x_j \wedge \psi(j; u, s))], \qquad (8.1.11)$$

where $\|x\| = \sum_{j=0}^{J-1} |x_j|$; $\|\psi(u, s)\| = \sum_{j=0}^{J-1} |\psi(j; u, s)|$.

Definition 8.1.3. Wavelet transform $X_{l_1}(u, s)$ of a discrete signal $x(t_j)$, $t_j \in T_{sp}$ that is represented by its own samples $x = (x_0, x_1, \ldots, x_{J-1})$, $x(t_j) = x(\Delta t \cdot j) = x_j$ in sample space $\mathcal{L}(\mathcal{X}, \mathcal{B}_{\mathcal{X}}; +, \vee, \wedge)$ with l_1-metric (1.3.27) is MSI (8.1.9) between this signal $x(t_j)$ and wavelet function $\psi(j; u, s)$ (8.1.5a):

$$x(t_j) \xrightarrow{DWT_{l_1}} X_{l_1}(u, s) :$$

$$X_{l_1}(u, s) = \sum_{j=0}^{J-1} [|\varepsilon_1 \cdot x_j + \psi(j; u, s)| - |\varepsilon_1 \cdot x_j - \psi(j; u, s)|], \qquad (8.1.12)$$

where $\varepsilon_1 = 2^m$, $m \in \mathbb{Z}$ is a positive constant that provides equalizing l_1-norms of a signal $x(t_j)$ and wavelet function $\psi(j; u, s)$: $\varepsilon_1 = \arg\min_{\varepsilon} (|\varepsilon \cdot \|x\|_1 - \|\psi(j; u, s)\|_1|)$.

Here, as before, indexes p, s, and l_1 in $X_p(u, s)$, $X_s(u, s)$, and $X_{l_1}(u, s)$ denote the relation to pseudometric (1.3.25), semimetric (1.3.26), and l_1-metric (1.3.27) spaces $\mathcal{L}(\mathcal{X}, \mathcal{B}_{\mathcal{X}}; +, \vee, \wedge)$, respectively, where the transforms (8.1.10), (8.1.11), and (8.1.12) are performed. Hereinafter, we relate wavelet transforms $X_p(u, s)$, $X_s(u, s)$, and $X_{l_1}(u, s)$ defined by relationships (8.1.10), (8.1.11), and (8.1.12), respectively, to a class of L-group discrete wavelet transforms (DWT).

In the relationships (8.1.10), (8.1.11), and (8.1.12), we assume that $t_j = \Delta t \cdot j \in T_{sp}$; T_{sp} is signal processing interval; $\Delta t \cdot (J - 1) = \Delta T_{sp}$ is a duration of signal processing interval; $\Delta T_{sp} \geqslant T$, T is a duration of the observed signal. In the relationships (8.1.11) and (8.1.12) we bear in mind that function of modulus $|x|$ is calculated basing on the relationship (1.1.7): $|x| = x \vee -x$, and function of modulus of a difference is calculated as $|x - y| = (x \vee y) - (x \wedge y)$.

Taking into account the fact that a difference between sign functions figuring in the relationships (8.1.8) accepts its values in the set $\{0, 1, 2\}$, calculating MSIs (8.1.10), (8.1.11), and (8.1.12) can be performed without operation of multiplication, that is a doubtless advantage when organizing calculations, first, in applications that do not assume exploiting considerable computational resources, second, in applications requiring high computational rate.

It is not a complete list of advantages of wavelet transforms that are performed in L-group sample space $\mathcal{L}(\mathcal{X}, \mathcal{B}_{\mathcal{X}}; +, \vee, \wedge)$. Further we consider main features of wavelet transforms defined by the formulas (8.1.10), (8.1.11), and (8.1.12), that, as it is shown below, can essentially differ from classic discrete wavelet transform defined by the relationship (8.1.5).

8.1.4 Discrete Wavelet Transform in Sample Spaces with Pseudometric and Semimetric: Harmonic Signals

Let $x(t_j)$ be additive interaction between two harmonic signals $a(t_j)$, $b(t_j)$ (see Fig. 8.1.1):

$$x(t_j) = a(t_j) + b(t_j); \tag{8.1.13}$$

$$a(t_j) = A_a \cdot \cos(2\pi f_a \cdot \Delta t \cdot j + \varphi_a); \tag{8.1.13a}$$

$$b(t_j) = A_b \cdot \cos(2\pi f_b \cdot \Delta t \cdot j + \varphi_b), \tag{8.1.13b}$$

where $t_j = \Delta t \cdot j$ is time parameter, $j = 0, 1, \ldots, J - 1$; Δt is a sampling interval; $A_{a,b}$ are amplitudes, $A_{a,b}$=const; $f_{a,b}$ are frequencies, $f_{a,b}$=const; $\varphi_{a,b}$ are initial phases ($\varphi_{a,b}$=const) of discrete harmonic signals $a(t_j)$, $b(t_j)$, respectively, so that the frequencies f_a, f_b of these signals $a(t_j)$, $b(t_j)$ are related to each other at one to two ratio $f_a/f_b = 2/1$.

Fig. 8.1.2a,b illustrate the results of wavelet transforms $DWT_x(u, s)$, $X_s(u, s)$ of the sum $x(t_j)$ (8.1.13) of two harmonic signals $a(t_j)$, $b(t_j)$ based on the relationships (8.1.5) and (8.1.11), respectively, using Morlet wavelet function (8.1.1):

$$\psi(j; u, s) = 2\pi\Delta f \cdot s \cdot \cos[C \cdot (2\pi\Delta f \cdot s)\Delta t(j - u)] \times \tag{8.1.14}$$

$$\times \exp[-(2\pi\Delta f \cdot s \cdot \Delta t(j - u))^2/2],$$

where C=const.

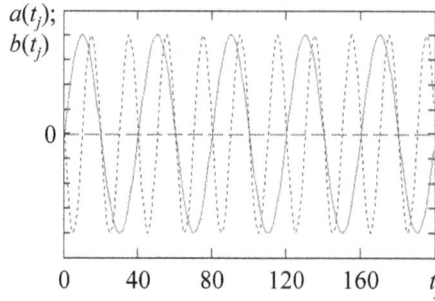

FIGURE 8.1.1   Harmonic signals $a(t_j)$, $b(t_j)$

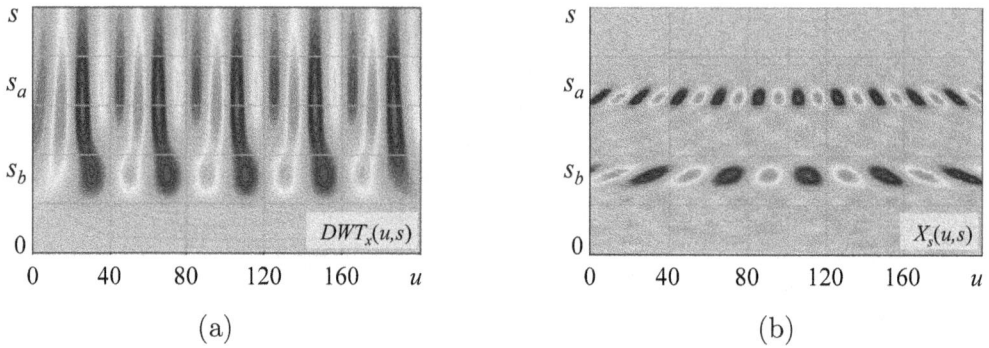

(a)   (b)

FIGURE 8.1.2   Wavelet transforms of the sum (8.1.13) of two harmonic signals: (a) $DWT_x(u, s)$ (8.1.5); (b) $X_s(u, s)$ (8.1.11)

These results are represented in the form of contour plots of bivariate functions. Scale parameters s_a, s_b shown in the figures correspond to the frequencies f_a, f_b of the signals $a(t_j)$, $b(t_j)$. Remind that the relationships (8.1.5), (8.1.11) assume using linear scale parameter $s \cdot \Delta f$, $s = 0, 1, \ldots, S - 1$, unlike the formula (8.1.6) exploiting exponential scale parameter $2^k \cdot \Delta f$, $k = 0, 1, \ldots, K - 1$. Red-yellow tint in both figures corresponds to positive semi-periods of harmonic signals, while dark blue-light blue hue corresponds to negative semi-periods of oscillations. Despite evident nonlinearity of L-group transform (8.1.11), the first (and the main) feature characterizing wavelet transform $x(t_j) \xrightarrow{DWT_s} X_s(u, s)$ (8.1.11) lies in the fact that here we deal with so-called *quasi-linearity* (see Chapter 1) assuming fulfillment of superposition principle which takes place with a sufficient degree of accuracy (for probable applications):

$$x(t_j) = a(t_j) + b(t_j) \Rightarrow X_s(u, s) \approx A_s(u, s) + B_s(u, s), \qquad (8.1.15)$$

where $a(t_j), b(t_j), x(t_j)$ are initial signals, $A_s(u, s), B_s(u, s), X_s(u, s)$ are the results of their transforms (8.1.11):

$$a(t_j) \xrightarrow{DWT_s} A_s(u, s); \quad b(t_j) \xrightarrow{DWT_s} B_s(u, s); \quad x(t_j) \xrightarrow{DWT_s} X_s(u, s).$$

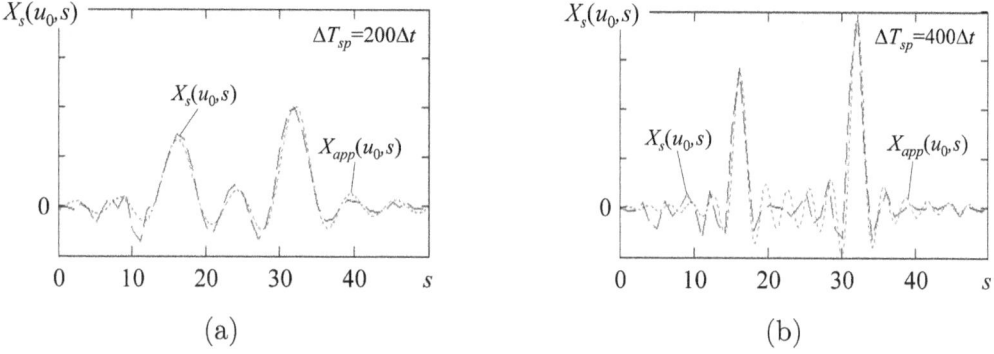

FIGURE 8.1.3 Sections $X_s(u,s)\,|_{u=u_0}$ of wavelet transform of the sum of two harmonic signals (8.1.13) for two cases: (a) $u_0 = 95$, $\Delta T_{sp} = 200\Delta t$; (b) $u_0 = 215$, $\Delta T_{sp} = 400\Delta t$

Obtaining an analytical relationship for wavelet transform (8.1.11) is rather complicated problem even for the sum of two harmonic signals (8.1.13). At the same time, one can easily obtain an approximating expression $X_{app}(u_0, s)$ for an arbitrary section $X_s(u, s)\,|_{u=u_0}$, u_0=const of wavelet transform (8.1.11) of the sum (8.1.13) of two harmonic signals (see Fig. 8.1.2b) in the following form:

$$X_{app}(u_0, s) = k_a(u_0) \cdot S_a(s, f_a, \Delta T_{sp}) + k_b(u_0) \cdot S_b(s, f_b, \Delta T_{sp}); \qquad (8.1.16)$$

$$S_{a,b}(s, f_{a,b}, \Delta T_{sp}) = k \cdot A_{a,b} \cdot \Delta T_{sp} \cdot \frac{\sin[2\pi k_T \Delta T_{sp}(s \cdot \Delta f - f_{a,b})]}{2\pi k_T \Delta T_{sp}(s \cdot \Delta f - f_{a,b})}; \qquad (8.1.16a)$$

$$k_{a,b}(u_0) = \cos(2\pi f_{a,b} \cdot u_0 \cdot \Delta t + \varphi_{a,b}), \qquad (8.1.16b)$$

where u_0 is a concrete value of a shift parameter u determining a position of the section $X_s(u, s)\,|_{u=u_0}$; u is a shift parameter, $u = 0, 1, \ldots, J-1$; s is a scale parameter, $s = 0, 1, \ldots, S-1$; $J, S \in \mathbb{N}$; $\mathbb{N}$ is set of natural numbers; Δt is a sampling interval in time domain; Δf is a sampling interval in frequency domain; T_{sp} is a signal processing interval; $\Delta t \cdot (J-1) = \Delta T_{sp}$ is a duration of signal processing interval, so that $\Delta T_{sp} \geqslant T$, T is a duration of the observed signal; k=const, k_T=const; $A_{a,b}$ are amplitudes; $f_{a,b}$ are frequencies; $\varphi_{a,b}$ are initial phases of discrete signals $a(t_j)$, $b(t_j)$, respectively.

Fig. 8.1.3a,b depicts the sections $X_s(u, s)\,|_{u=u_0}$ of wavelet transform of the sum of two harmonic signals (8.1.13) shown in Fig. 8.1.2b, for the cases $u_0 = 95$, $\Delta T_{sp} = 200\Delta t$ and $u_0 = 215$, $\Delta T_{sp} = 400\Delta t$ (dashed line), and also their approximation $X_{app}(u_0, s)$ (dotted line), respectively. As follows from the figures, the function $X_{app}(u_0, s)$ (8.1.16) provides quite satisfactory approximation of an arbitrary section $X_s(u, s)\,|_{u=u_0}$ of wavelet transform that allow its using for estimating a resolution of wavelet transform (8.1.11).

The relationship (8.1.16a) implies the second feature of L-group DWT (8.1.11) lying in a fact that a frequency resolution *does not depend on a frequency of the observed signal*, unlike classic wavelet transform $DWT_x(u, s)$ (8.1.5) in linear space with scalar product. Also, the relationship (8.1.16a) implies that when performing L-group DWT (8.1.11), a frequency resolution Δ_f is determined by a quantity that

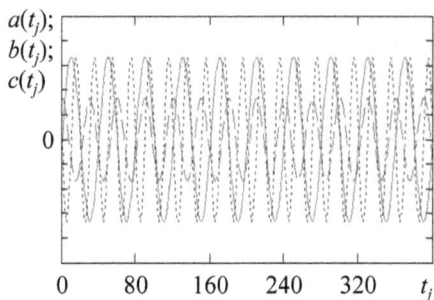

FIGURE 8.1.4   Harmonic signals $a(t_j)$, $b(t_j)$, $c(t_j)$

is inversely proportional to a duration ΔT_{sp} of signal processing interval T_{sp}:

$$\Delta_f = 1/(2k_T \Delta T_{sp}),\qquad\qquad(8.1.17)$$

where k_T is some constant.

Let $x(t_j)$ be additive interaction of three harmonic signals $a(t_j)$, $b(t_j)$, $c(t_j)$ (see Fig. 7.1.4):

$$x(t_j) = a(t_j) + b(t_j) + c(t_j);\qquad\qquad(8.1.18)$$
$$a(t_j) = A_a \cdot \cos(2\pi f_a \cdot \Delta t \cdot j + \varphi_a);\qquad\qquad(8.1.18a)$$
$$b(t_j) = A_b \cdot \cos(2\pi f_b \cdot \Delta t \cdot j + \varphi_b);\qquad\qquad(8.1.18b)$$
$$c(t_j) = A_c \cdot \cos(2\pi f_c \cdot \Delta t \cdot j + \varphi_c),\qquad\qquad(8.1.18c)$$

where $t_j = \Delta t \cdot j$ is time parameter, $j = 0, 1, \ldots, J - 1$; Δt is a sampling interval in time domain; $A_{a,b,c}$ are amplitudes, $A_{a,b,c}$=const; $f_{a,b,c}$ are frequencies, $f_{a,b,c}$=const; $\varphi_{a,b,c}$ are initial phases ($\varphi_{a,b,c}$=const) of discrete signals $a(t_j)$, $b(t_j)$, $c(t_j)$, respectively, so that the frequencies f_a, f_b, f_c of these signals are related to each other in such a way that the following inequality $f_a > f_c > f_b$ and the ratio $f_a/f_b = 2/1$ hold.

Fig. 8.1.5a,b illustrates the results of wavelet transforms $DWT_x(u, s)$, $X_p(u, s)$ of the sum $x(t_j)$ (8.1.18) of three harmonic signals $a(t_j)$, $b(t_j)$, $c(t_j)$ based on the relationships (8.1.5) and (8.1.10), respectively, using Morlet wavelet function (8.1.1), (8.1.14). These results are represented in the form of contour plots of bivariate functions. Scale parameters s_a, s_b, s_c shown in the figures correspond to the frequencies f_a, f_b, f_c of the signals $a(t_j)$, $b(t_j)$, $c(t_j)$. Also, as before, we use linear scale parameter $s \cdot \Delta f$, $s = 0, 1, \ldots, S - 1$. Red-yellow tint in both figures corresponds to positive semi-periods of harmonic signals, and dark blue-light blue hue corresponds to negative semi-periods of oscillation. As was mentioned with respect to L-group DWT $x(t_j) \xrightarrow{DWT_s} X_s(u, s)$ (8.1.11), in the case of transform $x(t_j) \xrightarrow{DWT_p} X_p(u, s)$ (8.1.10) we also deal with quasi-linearity assuming fulfillment of superposition principle which takes place with a sufficient degree of accuracy (for probable applications):

$$x(t_j) = a(t_j) + b(t_j) + c(t_j) \Rightarrow X_p(u, s) \approx A_p(u, s) + B_p(u, s) + C_p(u, s), \quad(8.1.19)$$

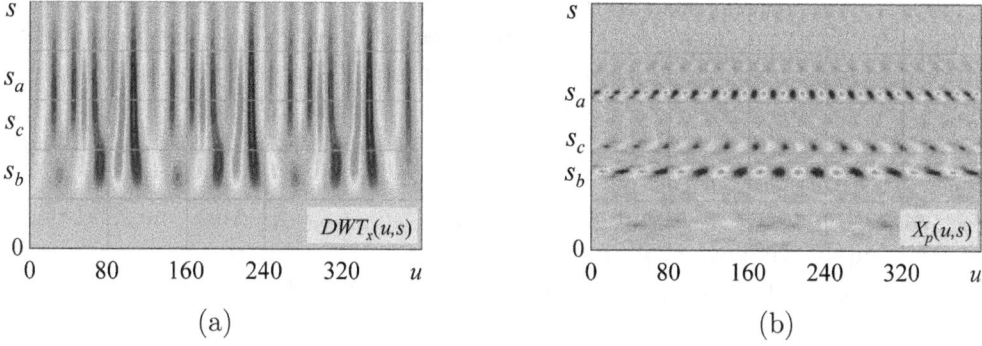

(a)

(b)

FIGURE 8.1.5 Wavelet transforms of the sum (8.1.18) of three harmonic signals (a) $DWT_x(u, s)$ (8.1.5); (b) $X_p(u, s)$ (8.1.10)

where $a(t_j)$, $b(t_j)$, $c(t_j)$, $x(t_j)$ are initial signals, $A_p(u, s)$, $B_p(u, s)$, $C_p(u, s)$, $X_p(u, s)$ are their L-group DWTs (8.1.10), respectively:

$$a(t_j) \xrightarrow{DWT_p} A_p(u, s); \quad b(t_j) \xrightarrow{DWT_p} B_p(u, s);$$

$$c(t_j) \xrightarrow{DWT_p} C_p(u, s); \quad x(t_j) \xrightarrow{DWT_p} X_p(u, s).$$

Obtaining an analytical relationship for L-group DWT (8.1.10) is rather complicated problem even for the sum of two harmonic signals (8.1.13). At the same time, by the analogy with the relationship (8.1.16), one can easily obtain an approximating expression $X_{app}(u_0, s)$ for an arbitrary section $X_p(u, s)|_{u=u_0}$, u_0=const of wavelet transform (8.1.10) of the sum (8.1.18) of three harmonic signals (see Fig. 8.1.5b):

$$X_{app}(u_0, s) = k_a(u_0) \cdot S_a(s, f_a, \Delta T_{sp})$$
$$+ k_b(u_0) \cdot S_b(s, f_b, \Delta T_{sp}) + k_c(u_0) \cdot S_c(s, f_c, \Delta T_{sp}); \qquad (8.1.20)$$

$$S_{a,b,c}(s, f_{a,b,c}, \Delta T_{sp}) = k' \cdot g_{a,b,,c}(A_a, A_b, A_c)$$
$$\times \Delta T_{sp} \cdot \frac{\sin[2\pi k_T \Delta T_{sp}(s \cdot \Delta f - f_{a,b,,c})]}{2\pi k_T \Delta T_{sp}(s \cdot \Delta f - f_{a,b,c})}; \qquad (8.1.20a)$$

$$k_{a,b,c}(u_0) = \cos(2\pi f_{a,b,c} \cdot u_0 \cdot \Delta t + \varphi_{a,b,c}), \qquad (8.1.20b)$$

where u is a shift parameter, $u = 0, 1, \ldots, J - 1$; u_0 is a concrete value of a shift parameter u that determines a position of the section $X_p(u, s)|_{u=u_0}$; s is a scale parameter, $s = 0, 1, \ldots, S - 1$; $J, S \in \mathbb{N}$; $\mathbb{N}$ is set of natural numbers; Δt is a sampling interval in time domain; Δf is a sampling interval in frequency domain; T_{sp} is a signal processing interval; $\Delta t \cdot (J - 1) = \Delta T_{sp}$ is a duration of signal processing interval, so that $\Delta T_{sp} \geqslant T$, T is a duration of the observed signal; k'=const, k_T=const; $g_{a,b,c}(A_a, A_b, A_c)$ is some function of amplitudes $A_{a,b,c}$ of the signals that takes into account nonlinearity of the transform (8.1.10), so that $g_{a,b,c}(A_a, A_b, A_c) = 1$ if $A_a = A_b = A_c$; $f_{a,b,c}$ are frequencies; $\varphi_{a,b,c}$ are initial phases of discrete signals $a(t_j)$, $b(t_j)$, $c(t_j)$, respectively.

The relationship (8.1.20a) implies the second feature of L-group DWT (8.1.10) lying in the fact that frequency resolution does not depend on a frequency of the observed signal, unlike classic wavelet transform $DWT_x(u,s)$ (8.1.5) in linear space with scalar product, and is just determined by a duration ΔT_{sp} of signal processing interval T_{sp}. As follows from the Fig. 8.1.5a, in the considered case, classic wavelet transform (8.1.5) based on scalar product does not provide a resolution of the signal $c(t_j)$ with a frequency satisfying the inequality $f_a > f_c > f_b$. Also, the relationship (8.1.20a) implies that when performing L-group DWT (8.1.10), a frequency resolution Δ_f is equal to a resolution of L-group DWT (8.1.11) and also is determined by the formula (8.1.17).

One more feature of L-group DWT (8.1.10) is a weak dependence of amplitude distribution of the partial bivariate signals $A_p(u,s)$, $B_p(u,s)$, $C_p(u,s)$ within common bivariate output signal $X_p(u,s)$ on amplitudes of input signals $a(t_j)$, $b(t_j)$, $c(t_j)$, under condition, that their amplitudes are slightly differ from each other. This property is explained by the fact that L-group DWT (8.1.10) $A_p(u,s)$ of a sole signal $a(t_j)$ is identically equal to wavelet transform of its amplified copy $\alpha \cdot a(t_j)$:

$$a(t_j) \xrightarrow{\ DWT_p\ } A_p(u,s); \quad \alpha \cdot a(t_j) \xrightarrow{\ DWT_p\ } A_p(u,s). \qquad (8.1.21)$$

In the case, when amplitudes A_a, A_b, A_c of harmonic signals $a(t_j)$, $b(t_j)$, $c(t_j)$ (8.1.18a,b,c) are noticeably differ from each other, a weak signal is suppressed by a stronger one, then the relationship (8.1.19) does not hold. In this sense the transform (8.1.10) is essentially more nonlinear than the transform (8.1.11).

Possessing better frequency resolution as against classic wavelet transform (8.1.5) based on scalar product, L-group DWTs (8.1.10) and (8.1.11) have a common considerable disadvantage: they do not satisfy causality principle in the case when a duration of the observed signal T is less than a duration of signal processing interval ΔT_{sp}: $T < \Delta T_{sp}$ and when using known wavelet functions $\{\psi_{a,b}(t)\}$ (8.1.2) determined in infinite interval $t \in]-\infty, \infty[$. Further we discuss possible variants of overcoming this disadvantage.

8.1.5 Discrete Wavelet Transform in Sample Space with l_1-metric: Harmonic Signals. Improving a Resolution by a Combination with Pseudometric ans Semimetric Spaces

Let $x(t_j)$ be additive interaction of two harmonic signals $a(t_j)$, $b(t_j)$ with Gaussian envelopes (see Fig. 8.1.6):

$$x(t_j) = a(t_j) + b(t_j); \qquad (8.1.22)$$

$$a(t_j) = A_a(t_j) \cdot \cos(2\pi f_a \cdot \Delta t \cdot j + \varphi_a);$$
$$b(t_j) = A_b(t_j) \cdot \cos(2\pi f_b \cdot \Delta t \cdot j + \varphi_b); \qquad (8.1.22a)$$

$$A_{a,b}(t_j) = \exp\left[-\frac{(\Delta t \cdot j - u_{a,b})^2}{\tau_{a,b}^2}\right], \qquad (8.1.22b)$$

where $t_j = \Delta t \cdot j$ is time parameter, $j = 0, 1, \ldots, J-1$; Δt is a sampling interval in time domain; $u_{a,b}$ is a signal shift in time, $u_{a,b}$=const; $\tau_{a,b}$ is time parameter

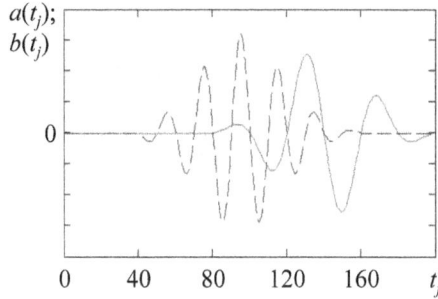

FIGURE 8.1.6   Harmonic signals $a(t_j)$, $b(t_j)$

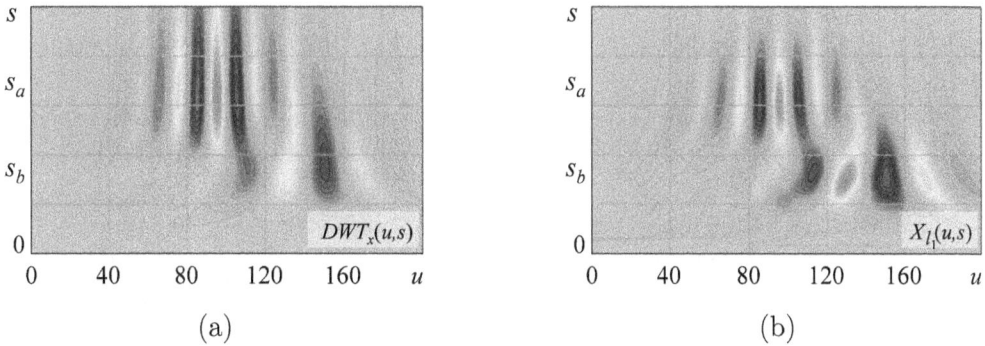

(a)   (b)

FIGURE 8.1.7   Wavelet transforms of the sum (8.1.22) of two signals (a) $DWT_x(u,s)$ (8.1.5); (b) $X_{l_1}(u,s)$ (8.1.12)

determining a signal duration, $\tau_{a,b}$=const; $f_{a,b}$ are frequencies, $f_{a,b}$=const; $\varphi_{a,b}$ are initial phases ($\varphi_{a,b}$=const) of discrete signals $a(t_j)$, $b(t_j)$, respectively, so that the frequencies f_a, f_b of these signals $a(t_j)$, $b(t_j)$ are related at one to two ratio $f_a/f_b = 2/1$.

Fig. 8.1.7a,b illustrates the results of wavelet transforms $DWT_x(u,s)$, $X_{l_1}(u,s)$ of the sum $x(t_j)$ (8.1.22) of two harmonic signals $a(t_j)$, $b(t_j)$ with Gaussian envelopes based on the relationships (8.1.5) and (8.1.12), respectively, using Morlet wavelet functions (8.1.1), (8.1.14). These results are represented in the form of contour plots of bivariate functions. Scale parameters s_a, s_b shown in the figures correspond to the frequencies f_a, f_b of the signals $a(t_j)$, $b(t_j)$. Remind, that the relationships (8.1.5), (8.1.12) assume using a linear scale parameter $s \cdot \Delta f$, $s = 0, 1, \ldots, S-1$, unlike the formula (8.1.6) using an exponential scale parameter $2^k \cdot \Delta f$, $k = 0, 1, \ldots, K-1$. Red-yellow tint in both figures corresponds to positive semi-periods of harmonic signals, and dark blue-light blue hue corresponds to the negative semi-periods of oscillations.

Quite similarly, as in previous cases of L-group DWTs (8.1.10) and (8.1.11), despite obvious nonlinearity of the mapping (8.1.12), main feature of wavelet transform $x(t_j) \xrightarrow{DWT_{l_1}} X_{l_1}(u,s)$ (8.1.12) is characterized by *quasi-linearity*, that assumes fulfillment of superposition principle taking place with a sufficient degree of accuracy (for probable applications):

$$x(t_j) = a(t_j) + b(t_j) \Rightarrow X_{l_1}(u, s) \approx A_{l_1}(u, s) + B_{l_1}(u, s), \tag{8.1.23}$$

where $a(t_j)$, $b(t_j)$, $x(t_j)$ are initial signals, and $A_{l_1}(u, s)$, $B_{l_1}(u, s)$, $X_{l_1}(u, s)$ are their wavelet transforms (8.1.12):

$$a(t_j) \xrightarrow{DWT_{l_1}} A_{l_1}(u, s); \quad b(t_j) \xrightarrow{DWT_{l_1}} B_{l_1}(u, s); \quad x(t_j) \xrightarrow{DWT_{l_1}} X_{l_1}(u, s).$$

As follows from Fig. 8.1.7b, unlike the mappings (8.1.10) and (8.1.11), when performing wavelet transform (8.1.12) as well as wavelet transform $DWT_x(u, s)$ (8.1.5) in linear space with scalar product, frequency resolution depends on a frequency of the observed signal. At the same time, comparing Fig. 8.1.7a and 8.1.7b, one can claim that L-group DWT $x(t_j) \xrightarrow{DWT_{l_1}} X_{l_1}(u, s)$ (8.1.12) provides better frequency resolution than classic wavelet transform $DWT_x(u, s)$ (8.1.5). In the case of processing harmonic signals, a frequency resolution provided by wavelet transform (8.1.12) can be improved on the basis of information contained in either (8.1.10) or (8.1.11) transforms.

Consider a method of combining information obtained in different spaces, i.e., on the one hand, in sample space with l_1-metric (l_1-metric space), on the other hand, either in sample space with pseudometric (or pseudometric space) or in sample space with semimetric (or semimetric space). Method of combining information from the corresponding pairs of spaces is defined by the following relationships:

$$X_{l_1 + l_p}(u, s) = (X_{l_1}(u, s) \vee 0) \wedge (\varepsilon_p \cdot X_{l_p}(u, s) \vee 0)$$
$$+ (X_{l_1}(u, s) \wedge 0) \vee (\varepsilon_p \cdot X_{l_p}(u, s) \wedge 0); \tag{8.1.24a}$$

$$X_{l_1 + l_s}(u, s) = (X_{l_1}(u, s) \vee 0) \wedge (\varepsilon_s \cdot X_{l_s}(u, s) \vee 0)$$
$$+ (X_{l_1}(u, s) \wedge 0) \vee (\varepsilon_s \cdot X_{l_s}(u, s) \wedge 0), \tag{8.1.24b}$$

where $X_{l_p}(u, s)$, $X_{l_s}(u, s)$, $X_{l_1}(u, s)$ are discrete wavelet transforms (8.1.10), (8.1.11), (8.1.12), respectively; $\varepsilon_p = 2^{m_1}$, $m_1 \in \mathbb{Z}$; $\varepsilon_s = 2^{m_2}$, $m_2 \in \mathbb{Z}$ are positive constants that provide a pairwise equalization of l_1-norms of bivariate functions $X_{l_p}(u, s), X_{l_1}(u, s); X_{l_s}(u, s), X_{l_1}(u, s)$, respectively:

$$\varepsilon_p = \arg\min_\varepsilon \left(\left| \|X_{l_1}(u, s)\|_1 - \varepsilon \cdot \|X_{l_p}(u, s)\|_1 \right| \right);$$

$$\varepsilon_s = \arg\min_\varepsilon \left(\left| \|X_{l_1}(u, s)\|_1 - \varepsilon \cdot \|X_{l_s}(u, s)\|_1 \right| \right).$$

Fig. 8.1.8a,b illustrates the results $X_{l_1 + l_p}(u, s)$, $X_{l_1 + l_s}(u, s)$ of combining information obtained in l_1-metric and pseudometric spaces (8.1.24a), and also in l_1-metric and semimetric spaces (8.1.24b), respectively. These results are represented in the form of contour plots of bivariate functions. Scale parameters s_a, s_b shown in the figures correspond to the frequencies f_a, f_b of the signals $a(t_j)$, $b(t_j)$. Red-yellow tint in both figures corresponds to positive semi-periods of harmonic signals, and dark blue-light blue hue corresponds to the negative semi-periods of oscillations.

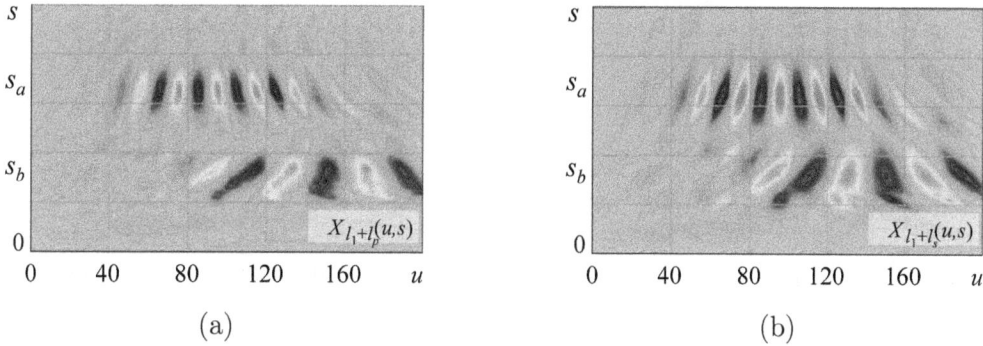

(a)   (b)

FIGURE 8.1.8   Results of combining information obtained in (a) l_1-metric and pseudometric spaces (8.1.24a); (b) l_1-metric and semimetric spaces (8.1.24b)

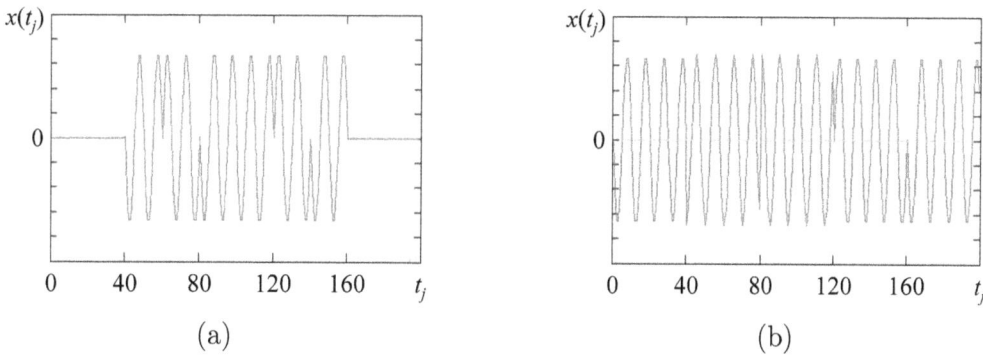

(a)   (b)

FIGURE 8.1.9   (a) BPSK signal; (b) QPSK signal

Comparing the results of the compiled wavelet transforms shown in Fig. 8.1.8a,b and the results of wavelet transforms depicted in Fig. 8.1.7a,b, one can conclude that using method of combining information obtained in different sample spaces allows improving frequency resolution, so that the higher a frequency of the observed signal is, the better such an improvement is.

8.1.6   Discrete Wavelet Transform in Sample Space with l_1-metric: BPSK, QPSK, V-LFM, and FSK Signals

In this section, we compare classic wavelet transform realized in linear space with scalar product that is defined by the relationship (8.1.5), and, on the other hand, L-group DWT realized in l_1-metric space that is defined by the relationship (8.1.12). For performing a comparative analysis we use BPSK, QPSK, V-LFM, and FSK signals.

Fig. 8.1.9a,b illustrates initial BPSK and QPSK signals, respectively, that should to be converted by wavelet transform in both linear space with scalar product and l_1-metric space.

Fig. 8.1.10a,b illustrates the results of wavelet transforms $DWT_x(u, s)$, $X_{l_1}(u, s)$ of BPSK signal $x(t_j)$ (see Fig. 8.1.9a), and Fig. 8.1.11a,b depict the results of

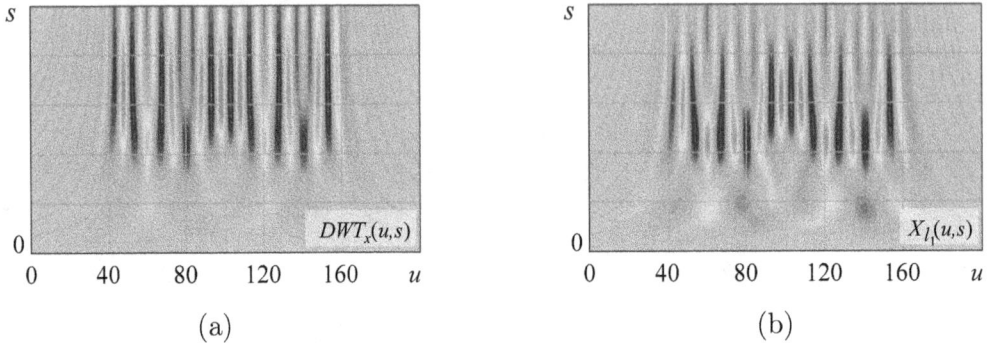

(a)   (b)

FIGURE 8.1.10   Results of wavelet transforms of BPSK signal: (a) $DWT_x(u,s)$ (8.1.5);
(b) $X_{l_1}(u,s)$ (8.1.12)

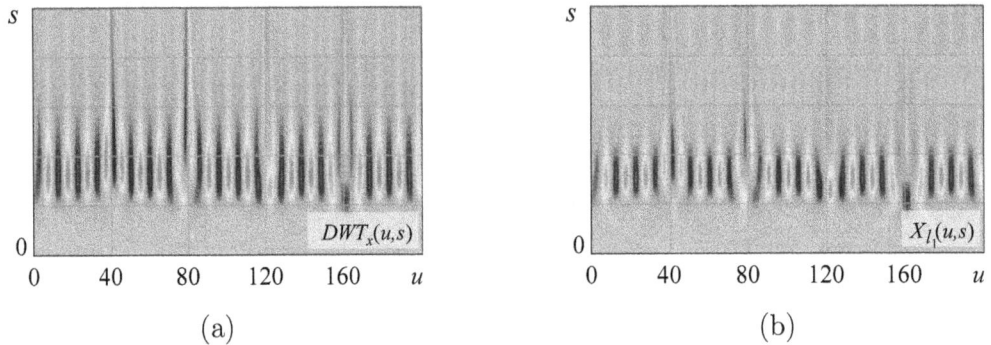

(a)   (b)

FIGURE 8.1.11   Results of wavelet transforms of QPSK signal: (a) $DWT_x(u,s)$ (8.1.5);
(b) $X_{l_1}(u,s)$ (8.1.12)

wavelet transforms $DWT_x(u,s)$, $X_{l_1}(u,s)$ of QPSK signal $x(t_j)$ (see Fig. 8.1.9b) based on the relationships (8.1.5) and (8.1.12), respectively, using Morlet wavelet function (8.1.1), (8.1.14).

In all the cases, phase shifts in initial signals are well reproduced in the results of wavelet transforms (8.1.5) and (8.1.12). Like in the case of harmonic signals, wavelet transform of the considered signals in l_1-metric space provides better frequency resolution than wavelet transform performed in linear space with scalar product.

Fig. 8.1.12a,b illustrates initial V-LFM and FSK signals, respectively, that should to be converted by wavelet transform in both linear space with scalar product and l_1-metric space.

Fig. 8.1.13a,b illustrates the results of wavelet transforms $DWT_x(u,s)$, $X_{l_1}(u,s)$ of V-LFM signal $x(t_j)$ (see Fig. 8.1.12a), and Fig. 8.1.14a,b depicts the results of wavelet transforms $DWT_x(u,s)$, $X_{l_1}(u,s)$ of FSK signal $x(t_j)$ (see Fig. 8.1.12b) based on the relationships (8.1.5) and (8.1.12), respectively, using Morlet wavelet function (8.1.14).

As follows from Fig. 8.1.13a,b and 8.1.14a,b, all changes in initial signal frequencies are well reproduced in the result of wavelet transforms (8.1.5) and (8.1.12).

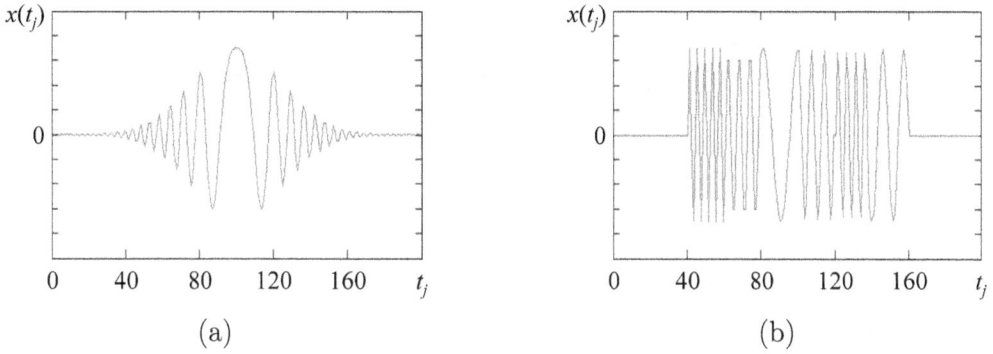

(a)   (b)

FIGURE 8.1.12   (a) V-LFM signal; (b) FSK signal

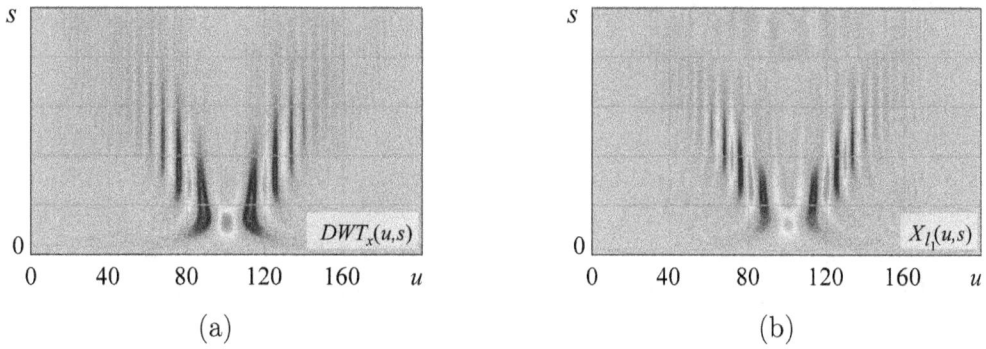

(a)   (b)

FIGURE 8.1.13   Results of wavelet transforms of V-LFM signal: (a) $DWT_x(u, s)$ (8.1.5);
(b) $X_{l_1}(u, s)$ (8.1.12)

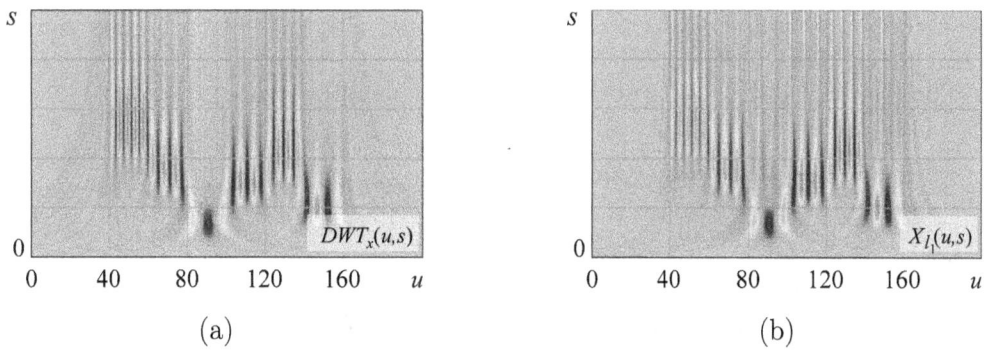

(a)   (b)

FIGURE 8.1.14   Results of wavelet transforms of FSK signal: (a) $DWT_x(u, s)$ (8.1.5); (b)
$X_{l_1}(u, s)$ (8.1.12)

Comparative analysis of wavelet transforms in both linear space with scalar product and l_1-metric space allows concluding on a high degree of their similarity when performing transformations of various types of signals.

Wavelet transform (8.1.5) in linear space with scalar product requires performing $N_\times = S \cdot J^2$ operations of multiplications and $N_+ = S \cdot J^2$ operations of additions, where J, S are numbers of values of shift parameter $u = 0, 1, \ldots, J - 1$ and scale parameter $s = 0, 1, \ldots, S - 1$, respectively. Wavelet transform (8.1.12) in l_1-space requires performing $N_+ = S \cdot J(2J+1)$ operations of addition, $N_{\vee,\wedge} = 2J^2 S$ operations of join/ meet and $N_{s\_inv} = J^2 S$ operations of sign inverting, or equivalently $N_+ = S \cdot J(2J + 1)$ operations of addition and $N_{|*|} = 2J^2 S$ operations of modulus. As follows from the aforementioned numerical relationships determining the required number of algebraic operations, basing on providing a higher data processing rate on a given computational performance of processing system, it is preferable to perform wavelet transform in l_1-metric space, that can provide better signal processing quality indices as compared with linear space with scalar product.

8.1.7 Discrete Wavelet Transform in Sample Spaces with Pseudometric and Semimetric: Signals with Finite Durations

Within a discussion in Subsection 8.1.4, we noted that wavelet transforms (8.1.10) and (8.1.11) possess a common disadvantage: they do not satisfy *causality principle* in the case when a duration of the observed signal T is less than a duration of signal processing interval ΔT_{sp}: $T < \Delta T_{sp}$, and when using known wavelet functions $\{\psi_{a,b}(t)\}$ (8.1.2), (8.1.3) determined in infinite interval $t \in]-\infty, \infty[$. In this subsection, we discuss possible variants of overcoming this disadvantage.

Consider a discrete signal $x(t_j)$ that should to be converted by wavelet transforms (8.1.10) or (8.1.11) in the form of the sum of elementary signals $x_m(t_j, f_m)$:

$$x(t_j) = \sum_{m=0}^{M-1} x_m(t_j, f_m), \ t_j \in T_m, \tag{8.1.25}$$

where T_m is such an interval that: $x_m(t_j, f_m) = 0$ if $t_j \notin T_m$; ΔT_m is a duration of elementary signal $x_m(t_j, f_m)$; f_m is a frequency that is considered to be constant in the interval T_m.

Then, to provide fulfillment of causality principle for wavelet transforms (8.1.10), (8.1.11), i.e.:

$$X_p(u \cdot \Delta t < t_0, s) = 0;$$

$$X_s(u \cdot \Delta t < t_0, s) = 0;$$

$$\text{if } x(t_j = \Delta t \cdot j) = 0, t_j < t_0,$$

it is necessary and sufficient that a duration T_ψ of wavelet function $\psi(j; u, s)$ should be less or equal to a duration ΔT_m of elementary signal $x_m(t_j, f_m)$ contained in the signal $x(t_j)$ (8.1.25):

$$T_\psi \leqslant \min_m \{\Delta T_m\}. \tag{8.1.26}$$

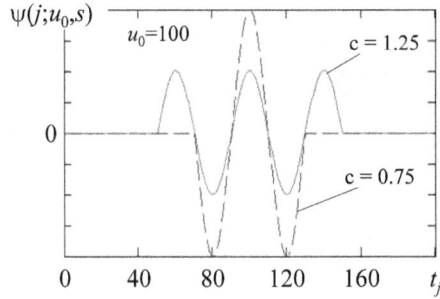

FIGURE 8.1.15   Appearance of wavelet functions $\psi(j; u, s)$ (8.1.27) for $c = 0.75, 1.25$

In this section, as a wavelet function for the transforms (8.1.10) and (8.1.11), we use a part of harmonic function of the following form:

$$\psi(j; u, s) = 2\pi\Delta f \cdot s \cdot \cos[C \cdot (2\pi\Delta f \cdot s)\Delta t(j - u)] \times$$
$$\times [1(\Delta t \cdot j - (\Delta t \cdot u - cT(s))) - 1(\Delta t \cdot j - (\Delta t \cdot u + cT(s)))], \quad (8.1.27)$$

where C=const; c=const; $T(s) = 1/[C\Delta f(s \vee 1)]$, $t_j = \Delta t \cdot j$.

Appearance of wavelet functions $\psi(j; u, s)$ determined by (8.1.27) for different values of a constant $c = 0.75, 1.25$ is shown in Fig. 8.1.15. For a convenience of visual perception, two functions are shown with different amplitudes.

Hereinafter we use wavelet functions $\psi(j; u, s)$ with an integer number of oscillation semi-periods. To improve visual perception, we use modified wavelet transforms $X'_p(u, s)$, $X'_s(u, s)$ based on the mappings (8.1.10) and (8.1.11), respectively:

$$X'_p(u, s) = DWT'_p[x(t_j), \psi(j; u, s)] =$$
$$= s \cdot (DWT_p[x(t_j), \psi(j; u, s)] - DWT_p[x(t_j), -\psi(j; u, s)]); \quad (8.1.28)$$

$$X'_s(u, s) = DWT'_s[x(t_j), \psi(j; u, s)] = s \cdot DWT_s[(t_j), \psi(j; u, s)]. \quad (8.1.29)$$

Notice that in order to compensate an influence of difference between positive and negative semi-periods of wavelet function $\psi(j; u, s)$ (8.1.27) (see Fig. 8.1.15), in the relationship (8.1.28) we use a difference between wavelet transforms (8.1.10) of the opposite functions $\psi(j; u, s)$ and $-\psi(j; u, s)$.

Consider again additive interaction $x(t_j)$ between two harmonic signals $a(t_j)$, $b(t_j)$ with Gaussian envelopes determined by the relationship (8.1.22) (see Fig. 8.1.6).

Fig. 8.1.16a,b illustrates the results of wavelet transforms $X'_p(u, s)$, $X'_s(u, s)$ of the sum $x(t_j)$ (8.1.22) of two signals $a(t_j)$, $b(t_j)$ based on the relationships (8.1.28) and (8.1.29), respectively, and wavelet function (8.1.27). These results are represented in the form of contour plots of bivariate functions. Scale parameters s_a, s_b shown in the figures correspond to the frequencies f_a, f_b of the signals $a(t_j)$, $b(t_j)$.

As follows from Fig. 8.1.16a,b, images of the signals $a(t_j)$, $b(t_j)$ (8.1.22a) obtained by the transforms (8.1.28) and (8.1.29), respectively, correspond to the images of such signals obtained by the mapping (8.1.5) (see Fig. 8.1.7a).

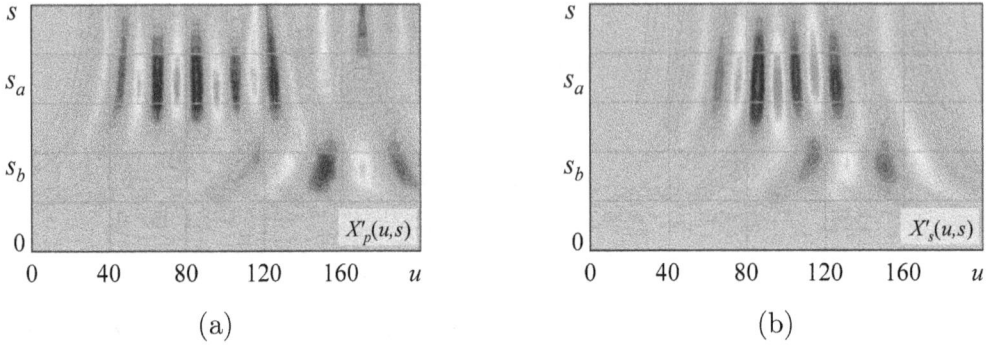

FIGURE 8.1.16   Results of wavelet transforms of the sum $x(t_j)$ (8.1.22) of two signals: (a) $X'_p(u,s)$ (8.1.28); (b) $X'_s(u,s)$ (8.1.29)

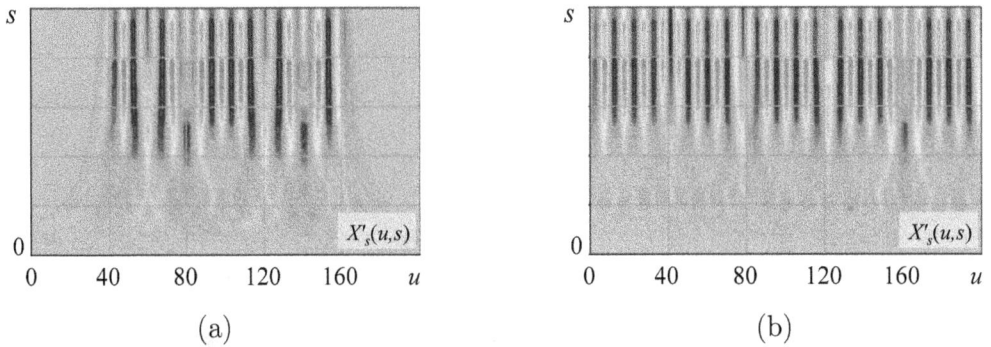

FIGURE 8.1.17   Results of wavelet transforms of (a) BPSK signal (see Fig. 8.1.9a); (b) QPSK signal (see Fig. 8.1.9b)

In this section we also compare wavelet transforms obtained in semimetric space by the relationship (8.1.29) and also in linear space with scalar product based on the relationship (8.1.5). For comparative analysis, we use BPSK, QPSK, V-LFM, and FSK signals.

Fig. 8.1.9a,b illustrates initial BPSK and QPSK signals, and Fig. 8.1.12a,b depict initial V-LFM and FSK signals, respectively, that should to be converted by wavelet transforms in both linear space with scalar product (8.1.5) and semimetric space (8.1.29).

Fig. 8.1.17a,b illustrates the results of wavelet transforms $X'_s(u,s)$ of BPSK and QPSK signals $x(t_j)$ (see Fig. 8.1.9a,b, respectively), and Fig. 8.1.18a,b depicts the results of wavelet transforms $X'_s(u,s)$ of V-LFM and FSK signals $x(t_j)$ (see Fig. 8.1.12a,b, respectively) based on the relationship (8.1.29) and wavelet function (8.1.27).

As follows from Fig. 8.1.17a,b and 8.1.18a,b, and also from Fig. 8.1.10a, 8.1.11a, 8.1.13a, 8.1.14a, in all cases, frequency and phase changes in the initial signals are well reproduced in the result of wavelet transforms (8.1.29) and (8.1.5). Comparative analysis of wavelet transforms in both linear space with scalar product and

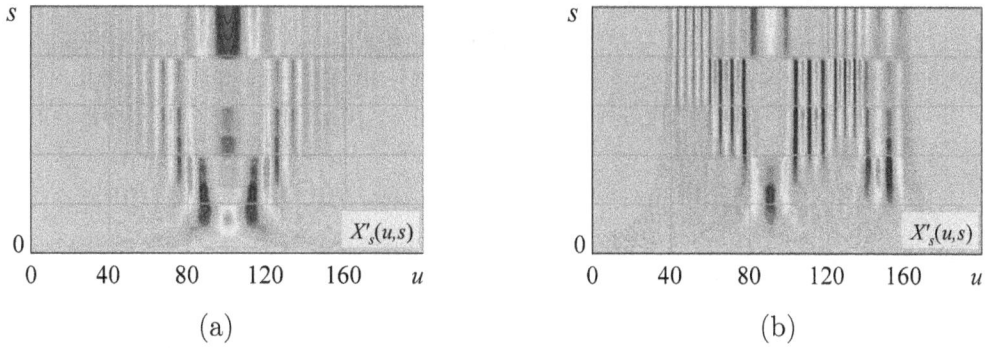

FIGURE 8.1.18   Results of wavelet transforms of (a) V-LFM signal (see Fig. 8.1.12a); (b) FSK signal (see Fig. 8.1.12b)

semimetric space allows concluding on a high degree of their similarity when performing transformations of various types of signals. Here we notice that the images of BPSK, QPSK, V-LFM, and FSK signals (see Fig. 8.1.9a,b; 8.1.12a,b, respectively) obtained by wavelet transform (8.1.28) correspond to the images of these signals obtained by the relationship (8.1.29).

Remind that wavelet transform (8.1.5) in linear space with scalar product requires performing $N_\times = S \cdot J^2$ operations of multiplication and $N_+ = S \cdot J^2$ operations of addition, where J, S are the numbers of values of shift parameter $u = 0, 1, \ldots, J - 1$ and scale parameter $s = 0, 1, \ldots, S - 1$, respectively.

Wavelet transform (8.1.10) in pseudometric space require performing $N_+ = 2J^2S$ operations of addition, $N_{\vee,\wedge} = 6J^2S$ operations of join/meet, or equivalently, $N_+ = 2J^2S$ operations of addition, $N_{\vee,\wedge} = 2J^2S$ operations of join/meet, and $N_{sgn} = 2J^2S$ operations of sign function.

Wavelet transform (8.1.11) in semimetric space requires performing $N_+ = 4J^2S$ operations of addition, $N_{\vee,\wedge} = 9J^2S$ operations of join/meet, $N_{s\_inv} = 3J^2S$ operations of sign inverting, and up to $N_{shift} = J^2S$ operations of binary digit shift, or equivalently, $N_+ = 4J^2S$ operations of addition, $N_{|*|} = 3J^2S$ operations of modulus, $N_{\vee,\wedge} = 2J^2S$ operations of join/meet, $N_{sgn} = 2J^2S$ operations of sign function, and up to $N_{shift} = J^2S$ operations of binary digit shift.

Wavelet transform (8.1.12) in l_1-metric space requires performing $N_+ = 3J^2S$ operations of addition, $N_{\vee,\wedge} = 3J^2S$ operations of join/meet, $N_{s\_inv} = J^2S$ operations of sign inverting, or equivalently, $N_+ = 3J^2S$ operations of addition and $N_{|*|} = 2J^2S$ operations of modulus.

Generalized data concerning a required number of algebraic operations for performing wavelet transform with dimensionality $J \times S$ (time-frequency) in sample spaces of different types: linear space with scalar product (8.1.5), pseudometric space (8.1.10), semimetric space (8.1.11), and l_1-metric space (8.1.12) are contained in the Table 8.1.1.

Notice, that a computational rate of performing wavelet transforms in sample spaces with L-group properties (8.1.10), (8.1.11), (8.1.12), unlike their analogue in linear space (8.1.5), does not depend on a processed data width. Numerical relationships from the Table 8.1.1 determining the required number of algebraic operations allow concluding that it is preferable (from a standpoint of providing a higher data processing rate on a given computational performance of a system) to perform wavelet transform in spaces with L-group properties, so that an achieved quality of signal processing in these spaces can be better than in linear spaces with scalar product.

TABLE 8.1.1   Required number of algebraic operations for performing wavelet transform in sample spaces of different types

Operation	Linear space	Pseudometric space	Semimetric space	l_1-metric space
Multiplication	$J^2 S$			
Addition	$J^2 S$	$2J^2 S$	$4J^2 S$	$3J^2 S$
Join/meet		$6J^2 S$	$9J^2 S$	$3J^2 S$
Sign inverting			$3J^2 S$	$J^2 S$
Binary digit shift			up to $J^2 S$	

8.2   Multiscale Image Decomposition on L-groups

In the end of 80s of the last century, it was found out that wavelets can form a basis of outstanding method of image processing called *multiresolution (multiscale) signal analysis* [188]. This approach deals with analysis and representation of signals (uni- and bivariate) using different scales and resolutions.

First (in the first half of 80s), as a rather simple structure for multiscale image representation there appeared an *image pyramid* [189]. Image pyramid developed for exploiting in machine vision and image compression applications is a set of images arranged in a decreasing scale and organized in the form of a pyramid. The pyramid basis is composed by an image of high resolution with a dimensionality $M \times N = k_M 2^J \times k_N 2^J$ $(k_M, k_N, J \in \mathbb{N})$ that assumed to be processed. Intermediate level of processing j in image pyramid has a dimensionality $k_M 2^j \times k_N 2^j$, $0 \leqslant j \leqslant J$.

In this section, we consider image mappings based on wavelets from the standpoint of multiscale analysis. We start with a brief discussion on known methods of multiscale image decomposition, and on this basis we consider their analogues based on L-group operations.

8.2.1 Multiscale Image Decompositions Based on Hadamard Matrix

The goal of image compression algorithms is to perform an image transformation providing the smallest converted image file on a given quality. Real images, such as, for instance, photos, have a feature lying in a fact that intensities of adjacent pixels are usually differ from one another in rather small quantity, i.e., intensity values of these pixels are correlated. As a rule, contrast steps take place within a small part of an image.

Let $c = \{c_0, c_1, \ldots, c_{2n-1}\}$ be a numerical sequence. Split all elements of the sequence c into pairs and find half-sums a_i and half-differences b_i of even c_{2i} and odd c_{2i+1} elements:

$$a_i = (c_{2i} + c_{2i+1})/2; \; b_i = (c_{2i} - c_{2i+1})/2; \tag{8.2.1}$$

$$i = 0, 1, \ldots, n - 1.$$

Knowing a half-sum a_i and a half-difference b_i, one can find values of the initial sequence c:

$$c_{2i} = a_i + b_i; \; c_{2i+1} = a_i - b_i.$$

Data stream

$$a = \{a_0, a_1, \ldots, a_{n-1}\} \tag{8.2.2}$$

is called a *main data stream* one, and data stream

$$b = \{b_0, b_1, \ldots, b_{n-1}\} \tag{8.2.3}$$

is called an *auxiliary data stream*.

The obtained main data stream (8.2.2) can be considered as a result of compression of the initial sequence c, and the auxiliary data stream (8.2.3) can be considered as a correction to the main data stream that allows restoring the initial stream.

If the stream (8.2.2) is too large to be transmitted then it can be split by the similar procedure into two streams preserving a possibility for its further splitting.

Consider pairs of adjacent (even and odd) pixels of a central row of the initial 8-bit Mona Lisa image with size 512×512 pixels shown in Fig. 8.2.1a, so that each pair is represented by a point of a graph (see Fig. 8.2.1b).

In all actual images, points corresponding to adjacent pixels draw up along a straight line (regression line). Correlation coefficient for even and odd pixels of a central row of image shown in Fig. 8.2.1a, is equal to 0.999. Upper left and lower right corners of the figure are always empty.

Now consider a graph shown in Fig. 8.2.1c whose points are defined by half-sums and half-differences (8.2.1). Half-differences not correlated with half-sums (since a regression line is inclined to abscissa axis by $0°$ angle) are concentrated in more narrow range than half-sums. Half-differences coding requires relatively small bit number. Sets of points shown in both figures contain same information. The difference lies in regression line rotation by $45°$. Thus, in order to encode compactly required information, it is necessary to perform affine mapping (rotation) of initial image.

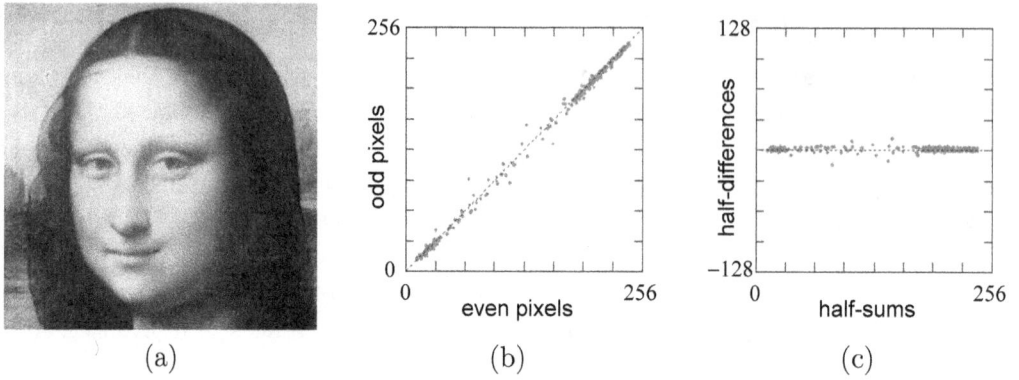

FIGURE 8.2.1 (a) Initial image; (b) dependence between intensity values of even and odd pixels; (c) dependence between intensity values of half-sum a_i and half-difference b_i pixels (8.2.1)

Let $(x\ y)^T$ be a pair of pixels (vector). It is necessary to obtain a pair

$$(x + y\ x - y)^T/2$$

of a half-sum $(x + y)/2$ and half-difference $(x - y)/2$ of initial values x, y. Such a transformation is described by Hadamard matrix (7.2.4):

$$H = \left\| \begin{array}{cc} 1 & 1 \\ 1 & -1 \end{array} \right\|;$$

$$\frac{1}{2} H \left(\begin{array}{c} x \\ y \end{array} \right) = \frac{1}{2} \left(\begin{array}{c} x + y \\ x - y \end{array} \right). \tag{8.2.4}$$

The direct mapping (8.2.4) implies the inverse mapping:

$$2H^{-1} \frac{1}{2} \left(\begin{array}{c} x + y \\ x - y \end{array} \right) = H \frac{1}{2} \left(\begin{array}{c} x + y \\ x - y \end{array} \right) = \left(\begin{array}{c} x \\ y \end{array} \right).$$

In the following subsubsection, we consider an approach concerning multiscale image decomposition that is close to stated in the work [189] but simpler and based on Hadamard matrix using.

8.2.1.1 Linear Multiscale Image Decomposition Based on Hadamard Matrix

Linear multiscale image decomposition based on Hadamard matrix relies on the stated above idea on a decorrelation of adjacent pixel values that is realized by bivariate sum-difference processing information contained in four neighbor pixels.

Here we use the same notations applied in the approach based on image pyramid $\{W_j(m, n)\}$ [189], an image $W_J(m, n)$ of high resolution has a dimensionality $M \times N = k_M 2^J \times k_N 2^J$ $(k_M, k_N, J \in \mathbb{N})$. An image $W_j(m, n)$ of an intermediate processing level j from image pyramid $\{W_j(m, n)\}$ has a dimensionality $M_j \times N_j = k_M 2^j \times k_N 2^j$, $0 \leqslant j \leqslant J$, where m, n are coordinates of an image

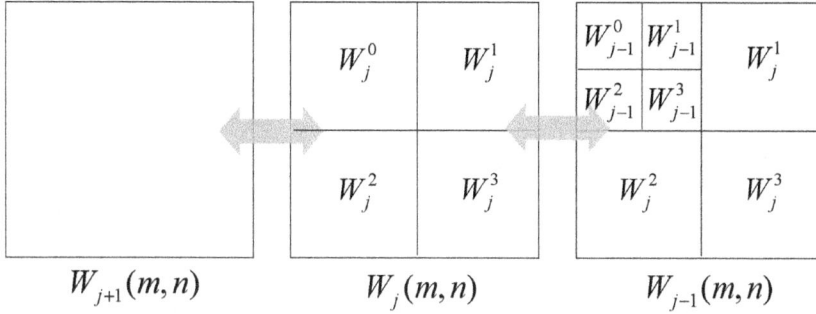

FIGURE 8.2.2 Scheme of mappings (both direct and inverse) between adjacent processing levels $j + 1$; j; $j - 1$

element (pixel) $m = 0, 1, \ldots, M - 1$, $n = 0, 1, \ldots, N - 1$. Within the framework of multiscale image decomposition, scheme of mappings (both direct and inverse) between adjacent processing levels $j + 1; j; j - 1$ is shown in the Fig. 8.2.2, so that an image $W_j(m, n)$ of intermediate processing level j in image pyramid $\{W_j(m, n)\}$ is represented by a partition of the corresponding subsets:

$$W_j(m, n) = W_j^0 \cup W_j^1 \cup W_j^2 \cup W_j^3;$$

$$W_j^0 \cap W_j^1 \cap W_j^2 \cap W_j^3 = \emptyset;$$

$$W_{j-1}(m, n) = W_{j-1}^0 \cup W_{j-1}^1 \cup W_{j-1}^2 \cup W_{j-1}^3 \cup W_j^1 \cup W_j^2 \cup W_j^3;$$

$$W_{j-1}^0 \cap W_{j-1}^1 \cap W_{j-1}^2 \cap W_{j-1}^3 \cap W_j^1 \cap W_j^2 \cap W_j^3 = \emptyset;$$

$$W_{j-2}(m, n) = W_{j-2}^0 \cup W_{j-2}^1 \cup W_{j-2}^2 \cup W_{j-2}^3 \cup W_{j-1}^1 \cup W_{j-1}^2 \cup W_{j-1}^3 \cup W_j^1 \cup W_j^2 \cup W_j^3;$$

$$W_{j-2}^0 \cap W_{j-2}^1 \cap W_{j-2}^2 \cap W_{j-2}^3 \cap W_{j-1}^1 \cap W_{j-1}^2 \cap W_{j-1}^3 \cap W_j^1 \cap W_j^2 \cap W_j^3 = \emptyset;$$

$$\ldots\ldots\ldots\ldots\ldots\ldots\ldots\ldots\ldots$$

Subsets (submatrices) W_j^0, W_{j-1}^0 contain reduced copies of images $W_{j+1}(m, n)$, $W_j(m, n)$, respectively. Subsets (submatrices) W_j^1, W_{j-1}^1 contain averaged differences of pixel value pairs along a horizontal. Subsets (submatrices) W_j^2, W_{j-1}^2 contain averaged differences of pixel value pairs along a vertical. Subsets (submatrices) W_j^3, W_{j-1}^3 contain averaged differences of pixel value pairs along a diagonal.

Relationships between the subsets of image pyramid $\{W_j(m, n)\}$ characterizing a direct transform DT are defined by the following expressions:

$$W_{j-1}^0 = DT[W_j^0, H^{\langle 0 \rangle}]; \tag{8.2.5a}$$

$$W_{j-1}^1 = DT[W_j^0, H^{\langle 1 \rangle}]; \tag{8.2.5b}$$

$$W_{j-1}^2 = DT[W_j^0, H^{\langle 2 \rangle}]; \tag{8.2.5c}$$

$$W_{j-1}^3 = DT[W_j^0, H^{\langle 3 \rangle}]; \tag{8.2.5d}$$

$$DT[a, H^{\langle i \rangle}] = W^i_{j-1;m,n} = \frac{1}{4}(a, H^{\langle i \rangle}); \tag{8.2.6}$$

$$a = [a_{2m,2n}\ a_{2m,2n+1}\ a_{2m+1,2n}\ a_{2m+1,2n+1}]^T, \tag{8.2.7}$$

where $m = 0, 1, \ldots, M_j/2 - 1$, $n = 0, 1, \ldots, N_j/2 - 1$ are coordinates of an image element (pixel) $W^i_{j-1;m,n}$; $i = 0, 1, 2, 3$; (a, h) is a scalar product of two vectors: $(a, h) = a^T h$; a is a column vector containing information on four adjacent pixels in the image W^0_j: $W^0_{j;i,k} = a_{i,k}$; $i = 2m, 2m+1$, $k = 2n, 2n+1$; $H^{\langle i \rangle}$ is i-th column of Hadamard matrix of order 4:

$$H = \left\|\begin{array}{cccc} 1 & 1 & 1 & 1 \\ 1 & -1 & 1 & -1 \\ 1 & 1 & -1 & -1 \\ 1 & -1 & -1 & 1 \end{array}\right\|. \tag{8.2.8}$$

Relationships between the subsets of image pyramid $\{W_j(m,n)\}$ characterizing inverse transform IT are defined by the following equations:

$$W^0_j = IT[W^0_{j-1}, W^1_{j-1}, W^2_{j-1}, W^3_{j-1}]; \tag{8.2.9}$$

$$IT[W^0_{j-1}, W^1_{j-1}, W^2_{j-1}, W^3_{j-1}] = \begin{cases} a_{2m,2n} & = (c, H^{\langle 0 \rangle}); \\ a_{2m,2n+1} & = (c, H^{\langle 1 \rangle}); \\ a_{2m+1,2n} & = (c, H^{\langle 2 \rangle}); \\ a_{2m+1,2n+1} & = (c, H^{\langle 3 \rangle}). \end{cases} \tag{8.2.10}$$

where $m = 0, 1, \ldots, M_j/2 - 1$, $n = 0, 1, \ldots, N_j/2 - 1$; $H^{\langle i \rangle}$ is i-th column of Hadamard matrix of order 4 (8.2.8), $i = 0, 1, 2, 3$; (c, h) is scalar product of two vectors: $(c, h) = c^T h$; c is a column vector containing information on four adjacent pixels of the images W^0_{j-1}, W^1_{j-1}, W^2_{j-1}, W^3_{j-1}:

$$c = [W^0_{j-1;m,n}, W^1_{j-1;m,n}, W^2_{j-1;m,n}, W^3_{j-1;m,n}].$$

Let there be an initial 8-bit 512×512 image shown in Fig. 8.2.1a. Applying the transform (8.2.5) 4 times successively to initial image, we obtain four 256×256 matrices (see Fig. 8.2.3a); three 256×256 matrices, and four 128×128 matrices (see Fig. 8.2.3b); three 256×256 matrices, three 128×128 matrices, and four 64×64 matrices (see Fig. 8.2.3c); three 256×256 matrices, three 128×128 matrices, three 64×64 matrices, and four 32×32 matrices (are not shown).

Applying the inverse transform (8.2.9) successively a required number of times (in this case 4 times) to the results of decomposition $W_{j-3}(m,n)$, $W_{j-2}(m,n)$, $W_{j-1}(m,n)$, $W_j(m,n)$ of initial image, we obtain the estimate $\hat{W}_{j+1}(m,n)$ of initial image $W_{j+1}(m,n)$ shown in Fig. 8.2.1a that is identical to the latter:

$$\hat{W}^0_{j-2} = IT[W^0_{j-3}, W^1_{j-3}, W^2_{j-3}, W^3_{j-3}]; \tag{8.2.11a}$$

$$\hat{W}^0_{j-1} = IT[W^0_{j-2}, W^1_{j-2}, W^2_{j-2}, W^3_{j-2}]; \tag{8.2.11b}$$

$$\hat{W}^0_j = IT[W^0_{j-1}, W^1_{j-1}, W^2_{j-1}, W^3_{j-1}]; \tag{8.2.11c}$$

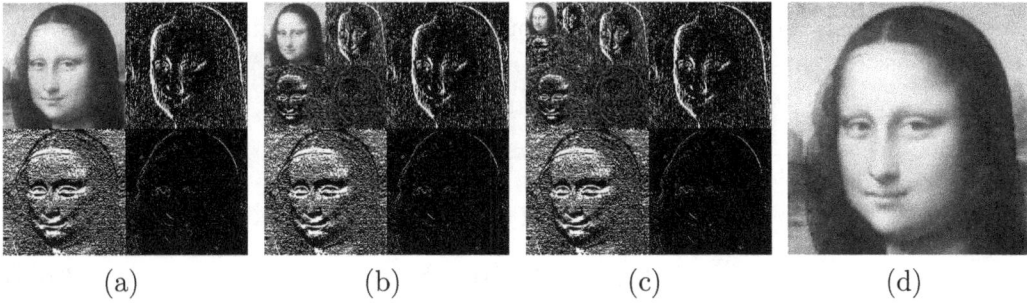

(a) (b) (c) (d)

FIGURE 8.2.3 Linear multiscale image decomposition based on Hadamard matrix: (a) $W_j(m,n)$; (b) $W_{j-1}(m,n)$; (c) $W_{j-2}(m,n)$; (d) image $\hat{W}_{j+1}(m,n)$ restored after compression

$$\hat{W}_{j+1}^0 = IT[W_j^0, W_j^1, W_j^2, W_j^3]; \tag{8.2.11d}$$

$$W_{j+1}(m,n) = \hat{W}_{j+1}(m,n). \tag{8.2.12}$$

The last relationship allows relating the transform (8.2.5) to a class of one-to-one mappings.

If in the relationships (8.2.11c) and (8.2.11d), we replace matrices W_{j-1}^1, W_{j-1}^2, W_{j-1}^3 and W_j^1, W_j^2, W_j^3 by the corresponding matrices $\mathbf{0}_{j-1}^1$, $\mathbf{0}_{j-1}^2$, $\mathbf{0}_{j-1}^3$ and $\mathbf{0}_j^1$, $\mathbf{0}_j^2$, $\mathbf{0}_j^3$ containing zero elements, we obtain approximate estimate $\hat{W}_{j+1}(m,n)$ of initial image $W_{j+1}(m,n)$ shown in Fig. 8.2.1a and based on information contained in four images of the last level W_{j-3}^0, W_{j-3}^1, W_{j-3}^2, W_{j-3}^3, and also three images W_{j-2}^1, W_{j-2}^2, W_{j-2}^3 of the penultimate level:

$$\hat{W}_j^0 = IT[\hat{W}_{j-1}^0, \mathbf{0}_{j-1}^1, \mathbf{0}_{j-1}^2, \mathbf{0}_{j-1}^3]; \tag{8.2.13a}$$

$$\hat{W}_{j+1} = IT[\hat{W}_j^0, \mathbf{0}_j^1, \mathbf{0}_j^2, \mathbf{0}_j^3]; \tag{8.2.13b}$$

$$W_{j+1}(m,n) \approx \hat{W}_{j+1}(m,n). \tag{8.2.14}$$

For the stated above example, the result of joint using the formulas (8.2.11a), (8.2.11b), (8.2.13a), and (8.2.13b) is shown in Fig. 8.2.3d. Using a modified mapping based on the formulas (8.2.11a), (8.2.11b), (8.2.13a), and (8.2.13b) allow compressing information contained in initial image (see Fig. 8.2.1a) by 16 times, so that correlation coefficient $r[W_{j+1}(m,n), \hat{W}_{j+1}(m,n)]$ between the images $W_{j+1}(m,n)$, $\hat{W}_{j+1}(m,n)$ is equal to $r[W_{j+1}(m,n), \hat{W}_{j+1}(m,n)]=0.997$.

8.2.1.2 Multiscale Image Decomposition Based on Hadamard Matrix and L-group Operations

Multiscale image decomposition based on Hadamard matrix and L-group operations uses the stated above idea on a decorrelation of adjacent pixel values that is realized by bivariate sum-difference processing information contained in four neighbor pixels. We use notations applied in the previous subsubsection.

Relationships between the subsets of image pyramid $\{W_j(m,n)\}$ characterizing the direct transform DT_L based on L-group operations are defined by the following expressions:

$$W^0_{j-1} = DT_L[W^0_j, H^{\langle 0 \rangle}]; \tag{8.2.15a}$$

$$W^1_{j-1} = DT_L[W^0_j, H^{\langle 1 \rangle}]; \tag{8.2.15b}$$

$$W^2_{j-1} = DT_L[W^0_j, H^{\langle 2 \rangle}]; \tag{8.2.15c}$$

$$W^3_{j-1} = DT_L[W^0_j, H^{\langle 3 \rangle}]; \tag{8.2.15d}$$

$$DT_L[a, H^{\langle i \rangle}] = W^i_{j-1;m,n} = k_1 N_{l_1}(a, K \cdot H^{\langle i \rangle}); \tag{8.2.16}$$

$$a = [a_{2m,2n} \; a_{2m,2n+1} \; a_{2m+1,2n} \; a_{2m+1,2n+1}]^T; \tag{8.2.17}$$

$$N_{l_1}(a, b) = \sum_i [|a_i + b_i| - |a_i - b_i|], \tag{8.2.18}$$

where $m = 0, 1, \ldots, M_j/2 - 1$, $n = 0, 1, \ldots, N_j/2 - 1$ are coordinates of an image element (pixel) $W^i_{j-1;m,n}$; $i = 0, 1, 2, 3$; $N_{l_1}(a,b)$ is a sample measure of statistical interrelation (MSI) between two vectors (1.4.12); a is a column vector containing information on four adjacent pixels in image W^0_j: $W^0_{j;i,k} = a_{i,k}$; $i = 2m, 2m + 1$, $k = 2n, 2n+1$; $H^{\langle i \rangle}$ is i-th column of Hadamard matrix of order 4 (8.2.8); $k_1 = 1/8$; $K = 256$ (for 8-bit image).

Relationships between the subsets of image pyramid $\{W_j(m,n)\}$ characterizing the inverse transform IT_L are defined by the following expressions:

$$W^0_j = IT_L[W^0_{j-1}, W^1_{j-1}, W^2_{j-1}, W^3_{j-1}]; \tag{8.2.19}$$

$$IT_L[W^0_{j-1}, W^1_{j-1}, W^2_{j-1}, W^3_{j-1}] = \begin{cases} a_{2m,2n} & = k_2 N_{l_1}(c, K \cdot H^{\langle 0 \rangle}); \\ a_{2m,2n+1} & = k_2 N_{l_1}(c, K \cdot H^{\langle 1 \rangle}); \\ a_{2m+1,2n} & = k_2 N_{l_1}(c, K \cdot H^{\langle 2 \rangle}); \\ a_{2m+1,2n+1} & = k_2 N_{l_1}(c, K \cdot H^{\langle 3 \rangle}), \end{cases} \tag{8.2.20}$$

where $m = 0, 1, \ldots, M_j/2 - 1$, $n = 0, 1, \ldots, N_j/2 - 1$; $H^{\langle i \rangle}$ is i-th column of Hadamard matrix of order 4 (8.2.8), $i = 0, 1, 2, 3$; $N_{l_1}(a, b)$ is sample MSI between two vectors (1.4.12); $k_2 = 1/2$; $K = 256$ (for 8-bit image); c is a column vector containing information concerning four pixels from the images W^0_{j-1}, W^1_{j-1}, W^2_{j-1}, W^3_{j-1}:

$$c = [W^0_{j-1;m,n}, \; W^1_{j-1;m,n}, \; W^2_{j-1;m,n}, \; W^3_{j-1;m,n}].$$

Applying the inverse transform (8.2.20) successively a required number of times (in this case 4 times) to the results of decomposition $W_{j-3}(m,n)$, $W_{j-2}(m,n)$, $W_{j-1}(m,n)$, $W_j(m,n)$ of initial image, we obtain the estimate $\hat{W}_{j+1}(m,n)$ of initial image $W_{j+1}(m,n)$ shown in Fig. 8.2.1a that is identical to the latter:

$$\hat{W}^0_{j-2} = IT_L[W^0_{j-3}, W^1_{j-3}, W^2_{j-3}, W^3_{j-3}]; \tag{8.2.21a}$$

$$\hat{W}^0_{j-1} = IT_L[W^0_{j-2}, W^1_{j-2}, W^2_{j-2}, W^3_{j-2}]; \tag{8.2.21b}$$

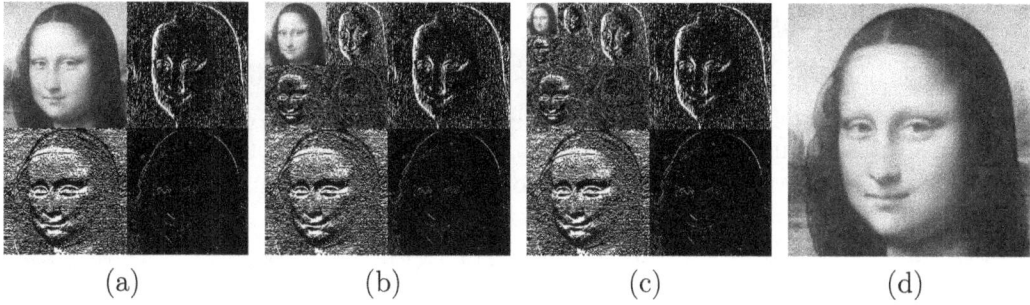

FIGURE 8.2.4   Multiscale image decomposition based on Hadamard matrix and L-group operations: (a) $W_j(m,n)$; (b) $W_{j-1}(m,n)$; (c) $W_{j-2}(m,n)$; (d) image $\hat{W}_{j+1}(m,n)$ restored after compression

$$\hat{W}_j^0 = IT_L[W_{j-1}^0, W_{j-1}^1, W_{j-1}^2, W_{j-1}^3]; \tag{8.2.21c}$$

$$\hat{W}_{j+1}^0 = IT_L[W_j^0, W_j^1, W_j^2, W_j^3]; \tag{8.2.21d}$$

$$W_{j+1}(m,n) = \hat{W}_{j+1}(m,n). \tag{8.2.22}$$

The last relationship allows relating the transform (8.2.15) to a class of one-to-one mappings. Here we draw attention of the reader to the fact that the transforms (8.2.15) and (8.2.19) are nonlinear mappings.

The results of successive applying the transform (8.2.15) with respect to initial 8-bit 512x512 image (see Fig. 8.2.1a) are shown in Fig. 8.2.4a,b,c.

If in the relationships (8.2.21c) and (8.2.21d) we replace matrices W_{j-1}^1, W_{j-1}^2, W_{j-1}^3 and W_j^1, W_j^2, W_j^3 by the corresponding matrices $\mathbf{0}_{j-1}^1$, $\mathbf{0}_{j-1}^2$, $\mathbf{0}_{j-1}^3$ and $\mathbf{0}_j^1$, $\mathbf{0}_j^2$, $\mathbf{0}_j^3$ containing zero elements, we obtain an approximate estimate $\hat{W}_{j+1}(m,n)$ of initial image $W_{j+1}(m,n)$ shown in Fig. 8.2.1a and based on information contained in four images of the last level W_{j-3}^0, W_{j-3}^1, W_{j-3}^2, W_{j-3}^3, and also three images W_{j-2}^1, W_{j-2}^2, W_{j-2}^3 of the penultimate level:

$$\hat{W}_j^0 = IT_L[\hat{W}_{j-1}^0, \mathbf{0}_{j-1}^1, \mathbf{0}_{j-1}^2, \mathbf{0}_{j-1}^3]; \tag{8.2.23a}$$

$$\hat{W}_{j+1} = IT_L[\hat{W}_j^0, \mathbf{0}_j^1, \mathbf{0}_j^2, \mathbf{0}_j^3]; \tag{8.2.23b}$$

$$W_{j+1}(m,n) \approx \hat{W}_{j+1}(m,n). \tag{8.2.24}$$

For the stated above example, the result of joint using the formulas (8.2.21a), (8.2.21b), (8.2.23a), and (8.2.23b) is shown in Fig. 8.2.3d. Using a modified mapping based on the formulas (8.2.21a), (8.2.21b), (8.2.23a), and (8.2.23b) allow compressing information contained in initial image (see Fig. 8.2.1a) by 16 times, so that correlation coefficient $r[W_{j+1}(m,n), \hat{W}_{j+1}(m,n)]$ between the images $W_{j+1}(m,n)$, $\hat{W}_{j+1}(m,n)$ is equal to $r[W_{j+1}(m,n), \hat{W}_{j+1}(m,n)]$=0.998.

8.2.2   Fast 2D Discrete Wavelet Transform Based on L-group Operations

One-dimensional (1D) wavelet transforms considered in the previous section can be generalized with respect to two-dimensional (2D) signals (images). In 2D case, four bivariate wavelet functions $g^{(i)}(x, y)$, $i = 0, 1, 2, 3$ must be formed. Each function is a product of univariate scaling function $\varphi(x)$ and univariate directing function $\psi(x)$, so that:

$$g^{(0)}(x, y) = \varphi(x)\varphi(y); \qquad\qquad (8.2.25a)$$

$$g^{(1)}(x, y) = \psi(x)\varphi(y); \qquad\qquad (8.2.25b)$$

$$g^{(2)}(x, y) = \varphi(x)\psi(y); \qquad\qquad (8.2.25c)$$

$$g^{(3)}(x, y) = \psi(x)\psi(y). \qquad\qquad (8.2.25d)$$

Bivariate wavelet directing functions $g^{(i)}(x, y)$, $i = 1, 2, 3$ determine changes of pixel intensities along different directions: $g^{(1)}(x, y)$ determines intensity variations along a horizontal, $g^{(2)}(x, y)$ determines intensity variations along a vertical, and $g^{(3)}(x, y)$ determines intensity variations along a diagonal.

If scaling (8.2.25a) and directing (8.2.25b,c,d) functions are given, then one can define fast discrete wavelet transform (DWT), having determined a family of basis functions based on operation of a shift:

$$g_{m,n}^{(i)}(x, y) = g^{(i)}(x - 2m, y - 2n). \qquad\qquad (8.2.26)$$

8.2.2.1   Fast 2D Discrete Wavelet Transform Based on Linear Algorithms

Basing on the formula (8.2.26), we define fast DWT of a bivariate function (image) $W_j^0(x, y)$ with a dimensionality $M \times N$:

$$W_{j-1}^{(i)}(m, n) = \frac{1}{\sqrt{MN}} \sum_{x=0}^{M-1} \sum_{y=0}^{N-1} W_j^0(x, y) g_{m,n}^{(i)}(x, y), \qquad\qquad (8.2.27)$$

where j is an arbitrary level of image representation; $i = 0, 1, 2, 3$; coefficients $W_{j-1}^{(0)}(m, n)$ determine an approximation of bivariate function $W_j^0(m, n)$ in a reduced scale ($m = 0, 1, \ldots, M/2 - 1$; $n = 0, 1, \ldots, N/2 - 1$); coefficients $W_{j-1}^{(i)}(m, n)$, $i = 1, 2, 3$ determine horizontal, vertical, and diagonal details for the next level of decomposition $j - 1$.

Further, providing a conciseness and simplicity of notation, we denote direct DWT (8.2.27) in the abbreviated form:

$$W_{j-1}^{(i)} = DWT[W_j^0, g^{(i)}]. \qquad\qquad (8.2.28)$$

Initial function $W_j^0(m, n)$ can be restored over the given coefficients (8.2.27) by inverse fast DWT:

$$W_j^0(x, y) = \frac{k_j}{\sqrt{MN}} \sum_{i=0}^{3} \sum_{m=0}^{M/2-1} \sum_{n=0}^{N/2-1} W_{j-1}^{(i)}(m, n) g_{m,n}^{(i)}(x, y), \qquad\qquad (8.2.29)$$

where k_j is some normalizing factor.

Similarly, for brevity, we denote inverse DWT (8.2.29) by the following abbreviated form:

$$W_j^0 = IDWT[W_{j-1}^0, W_{j-1}^1, W_{j-1}^2, W_{j-1}^3]. \tag{8.2.30}$$

For the following example, as a univariate scaling function $\varphi(x)$ and univariate directing function $\psi(x)$ we use modified Haar functions (see, for instance, [190, (7.4.11), (7.4.12)], that we rewrite in the form of Kronecker function $\delta(x,a)$:

$$\varphi(x) = \delta(x,0) + \delta(x,1); \tag{8.2.31a}$$

$$\psi(x) = \delta(x,0) - \delta(x,1), \tag{8.2.31b}$$

where $\delta(x,a)=1$, if $x = a$, $\delta(x,a)=0$, if $x \neq a$, $x,a \in \mathbb{N}$.

Let there be an initial 8-bit 512×512 image shown in Fig. 8.2.5a. Applying the transform (8.2.27) to initial image, we obtain four 256×256 matrices (see Fig. 8.2.5b). Repeating this transform, we obtain three 256×256 matrices and four 128×128 matrices (see Fig. 8.2.5c). Repeating this transform the 3d time, we obtain as a result: four 64×64 matrices, three 128×128 matrices, and three 256×256 matrices (see Fig. 8.2.5d).

Applying the inverse transform (8.2.30) successively a required number of times (in this case 4 times) to the results of decomposition $W_{j-3}(m,n)$, $W_{j-2}(m,n)$, $W_{j-1}(m,n)$, $W_j(m,n)$ of initial image, we obtain the estimate $\hat{W}_{j+1}(m,n)$ of initial image $W_{j+1}(m,n)$ shown in Fig. 8.2.5a that is identical to the latter:

$$\hat{W}_{j-2}^0 = IDWT[W_{j-3}^0, W_{j-3}^1, W_{j-3}^2, W_{j-3}^3]; \tag{8.2.32a}$$

$$\hat{W}_{j-1}^0 = IDWT[\hat{W}_{j-2}^0, W_{j-2}^1, W_{j-2}^2, W_{j-2}^3]; \tag{8.2.32b}$$

$$\hat{W}_j^0 = IDWT[\hat{W}_{j-1}^0, W_{j-1}^1, W_{j-1}^2, W_{j-1}^3]; \tag{8.2.32c}$$

$$\hat{W}_{j+1} = IDWT[\hat{W}_j^0, W_j^1, W_j^2, W_j^3]; \tag{8.2.32d}$$

$$W_{j+1}(m,n) = \hat{W}_{j+1}(m,n). \tag{8.2.33}$$

The last relationship allows relating the transform (8.2.27) to a class of one-to-one mappings.

8.2.2.2 Fast 2D Discrete Wavelet Transform Based on *L*-group Operations

Using the formula (8.2.18), we define fast DWT of a bivariate function (image) $W_j^0(x,y)$ with a dimensionality $M \times N$ based on measure of statistical interrelation (MSI) (1.4.12):

$$W_{j-1}^{(i)}(m,n) = \frac{1}{\sqrt{MN}} \sum_{x=0}^{M-1} \sum_{y=0}^{N-1} [|W_j^0(x,y) + K \cdot g_{m,n}^{(i)}(x,y)| -$$

$$- |W_j^0(x,y) - K \cdot g_{m,n}^{(i)}(x,y)|], \quad (8.2.34)$$

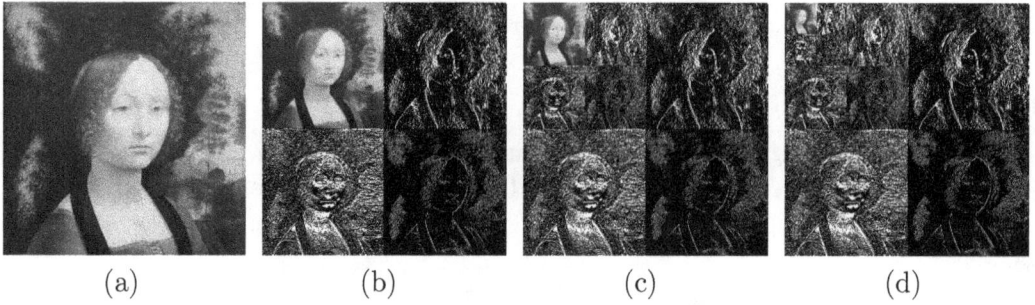

(a)   (b)   (c)   (d)

FIGURE 8.2.5   Fast 2D DWT based on linear algorithms: (a) initial image $W_{j+1}(m,n)$; (b) $W_j(m,n)$; (c) $W_{j-1}(m,n)$; (d) $\hat{W}_{j-2}(m,n)$

where $i = 0, 1, 2, 3$; $K = 256$ (for 8-bit image); j is an arbitrary level of image representation; coefficients $W_{j-1}^{(0)}(m,n)$ determine an approximation of bivariate function $W_j^0(x,y)$ in a reduced scale ($m = 0, 1, \ldots, M/2 - 1$; $n = 0, 1, \ldots, N/2 - 1$); coefficients $W_{j-1}^{(i)}(m,n)$, $i = 1, 2, 3$ determine horizontal, vertical, and diagonal details for the next level of decomposition $j - 1$.

The relationship (8.2.34) defines direct fast 2D discrete wavelet transform based on L-group operations. Further, providing a conciseness and simplicity of notation, we denote DWT (8.2.34) based on L-group operations in the abbreviated form:

$$W_{j-1}^{(i)} = DWT_L[W_j^0, g^{(i)}]. \tag{8.2.35}$$

Initial function $W_j^0(x,y)$ can be restored over the given coefficients (8.2.30) by inverse fast DWT based on MSI:

$$W_j^0(x,y) = \frac{k}{\sqrt{MN}} \sum_{i=0}^{3} \sum_{m=0}^{M/2-1} \sum_{n=0}^{N/2-1} [|W_{j-1}^{(i)}(m,n) + K \cdot g_{m,n}^{(i)}(x,y)|$$
$$- |W_{j-1}^{(i)}(m,n) - K \cdot g_{m,n}^{(i)}(x,y)|], \tag{8.2.36}$$

where k is some normalizing factor.

Similarly, for brevity, we denote inverse DWT (8.2.36) based on L-group operations by the following abbreviated form:

$$W_j^0 = IDWT_L[W_{j-1}^0, W_{j-1}^1, W_{j-1}^2, W_{j-1}^3]. \tag{8.2.37}$$

For the following example as a univariate scaling function $\varphi(x)$ and univariate directing function $\psi(x)$ we use modified discrete Haar functions that we rewrite in the form of Kronecker function $\delta(x,a)$ (8.2.31a,b).

Let there be an initial 8-bit 512×512 image shown in Fig. 8.2.5a. Applying the transform (8.1.34) to initial image, we obtain four 256×256 matrices (see Fig. 8.2.6a). Repeating this transform, we obtain three 256×256 matrices and four 128×128 matrices (see Fig. 8.2.6b). Repeating this transform the 3d time, we obtain as a result: four 64×64 matrices, three 128×128 matrices, and three 256×256 matrices (see Fig. 8.2.6c).

Applying the inverse transform (8.2.37) successively a required number of times (in this case 4 times) to the results of decomposition $W_{j-3}(m,n)$, $W_{j-2}(m,n)$, $W_{j-1}(m,n)$, $W_j(m,n)$ of initial image, we obtain the estimate $\hat{W}_{j+1}(m,n)$ of initial image $W_{j+1}(m,n)$ shown in Fig. 8.2.5a that is identical to the latter:

$$\hat{W}_{j-2}^0 = IDWT_L[W_{j-3}^0, W_{j-3}^1, W_{j-3}^2, W_{j-3}^3]; \qquad (8.2.38a)$$

$$\hat{W}_{j-1}^0 = IDWT_L[\hat{W}_{j-2}^0, W_{j-2}^1, W_{j-2}^2, W_{j-2}^3]; \qquad (8.2.38b)$$

$$\hat{W}_j^0 = IDWT_L[\hat{W}_{j-1}^0, W_{j-1}^1, W_{j-1}^2, W_{j-1}^3]; \qquad (8.2.38c)$$

$$\hat{W}_{j+1} = IDWT_L[\hat{W}_j^0, W_j^1, W_j^2, W_j^3]; \qquad (8.2.38d)$$

$$W_{j+1}(m,n) = \hat{W}_{j+1}(m,n). \qquad (8.2.39)$$

The last relationship allows relating the transform (8.2.34) to a class of one-to-one mappings. Meanwhile, the questions related to one-to-one property of the mapping (8.2.34) when using wavelet functions differing from the functions used in (8.2.31) require special researches.

If in the relationships (8.2.38c) and (8.2.38d) we replace matrices W_{j-1}^1, W_{j-1}^2, W_{j-1}^3 and W_j^1, W_j^2, W_j^3 by the corresponding matrices $\mathbf{0}_{j-1}^1$, $\mathbf{0}_{j-1}^2$, $\mathbf{0}_{j-1}^3$ and $\mathbf{0}_j^1$, $\mathbf{0}_j^2$, $\mathbf{0}_j^3$ containing zero elements, we obtain an approximate estimate $\hat{W}_{j+1}(m,n)$ of initial image $W_{j+1}(m,n)$ shown in Fig. 8.2.1a and based on information contained in four images of the last level W_{j-3}^0, W_{j-3}^1, W_{j-3}^2, W_{j-3}^3, and also three images W_{j-2}^1, W_{j-2}^2, W_{j-2}^3 of the penultimate level:

$$\hat{W}_j^0 = IT_L[\hat{W}_{j-1}^0, \mathbf{0}_{j-1}^1, \mathbf{0}_{j-1}^2, \mathbf{0}_{j-1}^3]; \qquad (8.2.40a)$$

$$\hat{W}_{j+1} = IT_L[\hat{W}_j^0, \mathbf{0}_j^1, \mathbf{0}_j^2, \mathbf{0}_j^3]; \qquad (8.2.40b)$$

$$W_{j+1}(m,n) \approx \hat{W}_{j+1}(m,n). \qquad (8.2.41)$$

For the stated example, the result of joint using the formulas (8.2.38a), (8.2.38b), (8.2.40a), and (8.2.40b) is shown in Fig. 8.2.6d. Using a modified mapping based on the formulas (8.2.38a), (8.2.38b), (8.2.40a), and (8.2.40b) allow compressing information contained in initial image (see Fig. 8.2.5a) by 16 times, so that correlation coefficient $r[W_{j+1}(m,n), \hat{W}_{j+1}(m,n)]$ between the images $W_{j+1}(m,n)$, $\hat{W}_{j+1}(m,n)$ is equal to $r[W_{j+1}(m,n), \hat{W}_{j+1}(m,n)]=0.996$.

Taking into account a relation between difference of moduli, figuring in (8.2.34) and (8.2.36) and determining MSI (1.4.12), and Huber function $HF(x,a)$ (1.2.8):

$$\begin{cases} |x + k \cdot a| - |x - k \cdot a| = HF(2x, |2k \cdot a|) \cdot \operatorname{sgn}(a); & \text{(a)} \\ HF(x, b) = -b \vee (x \wedge b); & \text{(b)} \\ k > 0. & \text{(c)} \end{cases} \qquad (8.2.42)$$

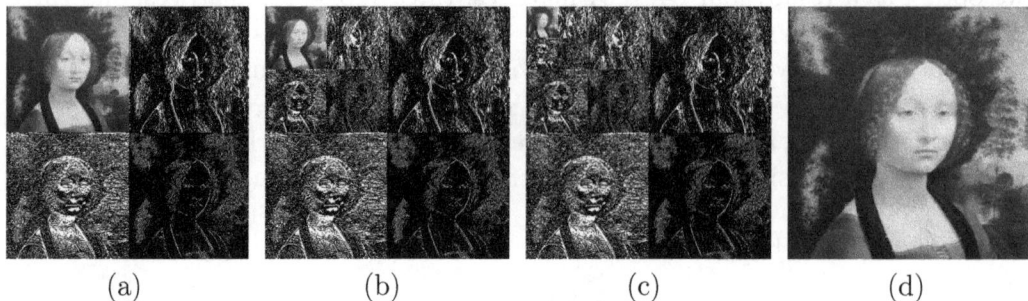

FIGURE 8.2.6 Fast 2D DWT based on MSI: (a) $W_j(m,n)$; (b) $W_{j-1}(m,n)$; (c) $W_{j-2}(m,n)$; (d) image $\hat{W}_{j+1}(m,n)$ restored after compression

the relationships defining direct (8.2.34) and inverse (8.2.36) DWT take the following forms:

$$W_{j-1}^{(i)}(m,n) = \sum_{x=0}^{M-1} \sum_{y=0}^{N-1} HF[2W_j^0(x,y), |2K \cdot g_{m,n}^{(i)}(x,y)|]$$

$$\times \operatorname{sgn}(g_{m,n}^{(i)}(x,y))/\sqrt{MN}; \quad (8.2.43)$$

$$W_j^0(x,y) = \sum_{i=0}^{3} \sum_{m=0}^{M/2-1} \sum_{n=0}^{N/2-1} HF[2W_{j-1}^{(i)}(m,n), |2K \cdot g_{m,n}^{(i)}(x,y)|]$$

$$\times \operatorname{sgn}(g_{m,n}^{(i)}(x,y)) \cdot k/\sqrt{MN}, \quad (8.2.44)$$

where k is some normalizing factor; $HF(x,b) = -b \vee (x \wedge b)$; $K = 256$ (for 8-bit image).

The formulas defining direct (8.2.43) and inverse (8.2.44) DWT by Huber function allow realizing computations based on join and meet operations of a lattice (8.2.42b) without using operation of multiplication. Hear we bear in mind that operation of multiplication by an integer power of 2 is equivalent to operation of shift over the corresponding number of binary digits, and operation of multiplication by a sign function is equivalent to repeating a quantity, changing its sign, or its zeroing.

Conclusion

Having collected in one place information concerning signal processing in generalized metric spaces with lattice properties and having stated the principles of the most widely used algorithms whose analogues were formulated in terms of L-group operations, the author expects he managed to acquaint the reader with relatively new approach to signal processing. As it is shown in Fig. C.1, this approach substantially relies on such well investigated algebraic systems as groups, lattices, and lattice-ordered groups (L-groups). It is extremely difficult, almost impossible to synthesize and analyze signal processing algorithms without introducing some measure of distance in a generalized metric space. In this work we explored L-group signal processing algorithms, based on measures of statistical interrelation that are defined by l_1-metric, pseudometric, and semimetric, in comparison with their known classic analogues. These measures of statistical interrelation being the functions of order statistics are robust or nonparametric statistics.

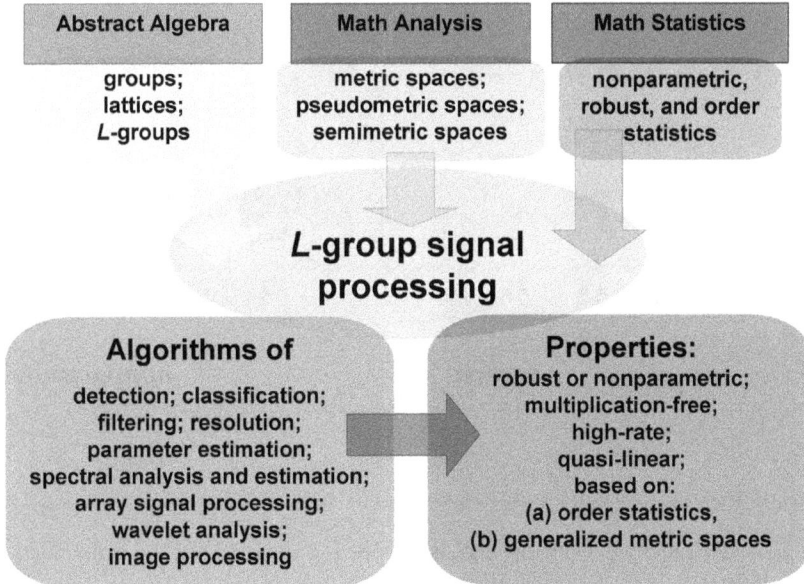

Abstract Algebra	Math Analysis	Math Statistics
groups; lattices; *L*-groups	metric spaces; pseudometric spaces; semimetric spaces	nonparametric, robust, and order statistics

L-group signal processing

Algorithms of	Properties:
detection; classification; filtering; resolution; parameter estimation; spectral analysis and estimation; array signal processing; wavelet analysis; image processing	robust or nonparametric; multiplication-free; high-rate; quasi-linear; based on: (a) order statistics, (b) generalized metric spaces

FIGURE C.1   *L*-group signal processing algorithms: sources and properties

In the book we discussed L-group signal processing algorithms that allow solving the problems of signal detection, classification, resolution, filtering, parameter estimation (including spectral and DOA estimation), antenna array signal processing, demultiplexing and demodulation, and also algorithms of wavelet analysis (including discrete wavelet transform L-group algorithms of 1D/2D signals). The main properties of considered algorithms are: robustness or nonparametricity; multiplication-free; high signal processing rate; quasi-linearity; direct relation to generalized metrics and the functions of order statistics.

In the book we stated a general approach based on generalized metric spaces that allow overcoming prior uncertainty when solving main problems of signal processing, so that interferences (noises) can be described on a rather wide class of distributions.

Depending on qualitative kind of dependence of efficiency (relative efficiency), L-group signal processing algorithms discussed in the book relate to four groups (types): *robust* (type I) and *nonparametric* (type II) (see Fig. C.2a,b, respectively); *quasi-linear* (type III) and *linear* (type IV) (see Fig. C.3a,b, respectively). Here under efficiency we mean the ratio between maximum signal-to-noise ratio (SNR) and SNR (in dB) that provide the required value(s) of one or several quality indices (QI) of signal processing characterizing a given algorithm:

$$\text{efficiency} = 10 \cdot \lg(\text{SNR}_{\max}/\text{SNR}) \text{ (dB)},$$

where $\text{SNR}_{\max}$ corresponds to the worst case of interference (noise) distribution, $\lg(*) = \log_{10}(*)$.

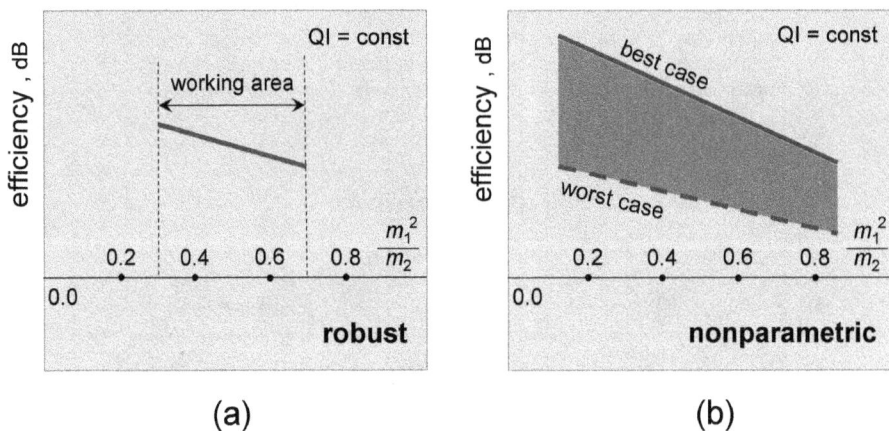

FIGURE C.2 Efficiency of L-group signal processing algorithms: (a) robust (type I algorithms); (b) nonparametric (type II algorithms)

Under relative efficiency we mean the ratio between SNRs (in dB) that provide the required value(s) of one or several quality indices (QI) of signal processing and relate to a developed algorithm (SNR) and to a known linear algorithm (SNR_{opt}), assuming that the latter is optimal in the presence of interference (noise) with

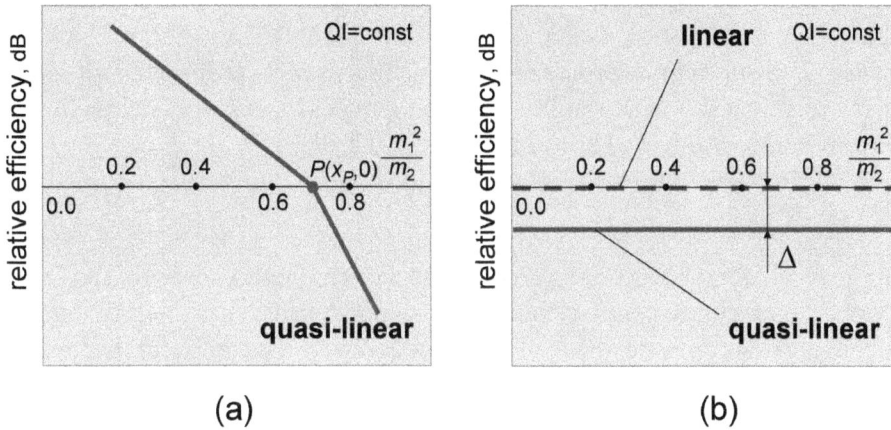

FIGURE C.3 Relative efficiency of L-group signal processing algorithms with respect to optimal linear algorithms: (a) quasi-linear (type III algorithms) (b) quasi-linear and linear (type III and IV algorithms, respectively)

normal distribution:

$$\text{relative efficiency} = 10 \cdot \lg(\text{SNR}_{\text{opt}}/\text{SNR}) \text{ (dB)}.$$

In all cases we consider dependences of the ratio m_1^2/m_2 of squared expectation m_1 to the second moment m_2 of interference (noise) envelope within some class of distributions. Depending on statistical properties of interference (noise) distribution within a given generalized distribution, the efficiencies of L-group signal processing algorithm and optimal linear algorithm could change to one or another side. Thus, the efficiencies of robust and nonparametric L-group signal processing algorithms are, as a rule, nonincreasing functions of m_1^2/m_2 within a working area (see Fig. C.2a,b), though there exist some exceptions.

Robust and nonparametric L-group signal processing algorithms possess the property of quasi-linearity. When operating in the presence of impulse type noise (with heavy tails of distribution), i.e. when the ratio m_1^2/m_2 is defined by the inequality $m_1^2/m_2 < x_P$, the relative efficiency of these algorithms is greater than zero (see Fig. C.3a). On the contrary, when operating in the presence of harmonic type noise (with relatively light tails of distribution), i.e. when the ratio m_1^2/m_2 is determined by the inequality $m_1^2/m_2 > x_P$, the relative efficiency of these algorithms is less than zero. An abscissa x_P of a point of intersection $P(x_P, 0)$ can belong to the interval $x_P \in [0.65, 0.75]$ depending on algorithm and distribution family.

Theorem 3.2.1 and its analogues stated in other chapters establishes the existence of L-group signal processing algorithms, that are equivalent to existing known linear algorithms that are optimal from the standpoint of one or another criterion (see Fig. C.3b, dash line), so that their relative efficiency is equal to zero.

Linear L-group signal processing algorithms represent degenerated case, when superposition principle holds (1.7.2a), unlike approximate relationship (1.7.7) defining quasi-linearity.

At last, L-group signal processing algorithms based on measure of statistical interrelation associated with semimetric, being quasi-linear, are slightly inferior to optimal linear algorithms in $\Delta = |10\lg(8/\pi^2)| < 1$ dB.

Some suggestions concerning the directions of further research based on this book are given below.

- Apparently, all signal processing algorithms, existing and known by that time and developed for linear systems within linear signal space, have their own analogues in each generalized metric space. Thus, a separate direction of further research is obtaining L-group algorithms that are pithy analogues of known linear algorithms but not included in the book.

- Since the dependences shown in Fig. C.2, C.3 are of a qualitative nature, establishing analogous theoretical dependences for concrete distribution families and signal processing algorithms is definitely of interest.

- Generalized metric spaces investigated in the book, obviously, are not the only ones, so searching other spaces, where signal processing could be more efficient, makes eminent sense.

Finally, it is quite appropriate paraphrasing the quotation of British mathematician Arthur Cayley on projective geometry*: **signal processing based on lattice-ordered groups is all signal processing.**

The author hopes that by this moment the reader managed to perceive a great diversity of possible approaches to the truths that formerly seemed to be firm and steadfast.

---

*"projective geometry is all geometry"

Bibliography

[1] Amari, S. and Nagaoka, H. *Methods of information geometry*. Oxford University Press, Oxford, 2000.

[2] Püschel, M. and Moura, J.M.F. Algebraic signal processing theory: foundation and 1-D time. *IEEE Trans. on Signal Processing*, 56(8):3572–3585, 2008.

[3] Püschel, M. and Moura, J.M.F. Algebraic signal processing theory: 1-D space. *IEEE Trans. on Signal Processing*, 56(8):3586–3599, 2008.

[4] Besl, P.J. Geometric signal processing. In Jain, R.C., Jain A.K., editor, *Analysis and Interpretation of Range Images. Springer Series in Perception Engineering*. Springer, New York, 1990.

[5] De Lathauwer, L. Signal Processing based on Multilinear Algebra. PhD thesis, Catholic University, Leuven, Belgium, 1997.

[6] Virendra, P. Sinha. *Symmetries and Groups in Signal Processing: An Introduction*. Springer, 2010.

[7] Hegde, C. Nonlinear Signal Models: Geometry, Algorithms, and Analysis. PhD thesis, Rice University, Houston, Texas, 2012.

[8] Byrne, C.L. *Signal Processing: A Mathematical Approach*. CRC Press, 2nd edition, 2014.

[9] Robinson, M. *Topological Signal Processing*. Springer, 2014.

[10] Tao, R., Li, B., and Sun, H. Research progress of the algebraic and geometric signal processing. *Defence Technology*, 9:40–47, 2013.

[11] Huber, P. J. *Robust Statistics*. John Wiley & Sons, New York, 1981.

[12] Gordienko, V.I. Detection algorithm based on the notion of "scalar intersection". *Radioelektronika (Izv. VUZov)*, 7:8–18, 1992 (in Russian).

[13] Merhav, N. Multiplication-free approximate algorithms for compressed domain linear operations on images. *IEEE Trans. on Image Processing*, 8(2):247–253, 1999.

[14] May, F., Klappenecker, A., Baumgarte, V., Nückel, A., and Beth, T. A high through-output multiplication-free approximation to arithmetic coding. In *Proc. of Int. Symp. on Information Theory and its Applications*, pages 845–847. IEEE, Victoria, Canada, 1996.

[15] Scaglione, A., Barbarossa, S., Porchia, A., and Scarano, G. Nonlinear time-frequency distributions with multiplication-free kernels. In *IEEE Signal Processing Workshop on Statistical Signal and Array Processing*, pages 456–459. 1996.

[16] Gordienko, V.I., Dubrovsky, S.E., Ryumshin, R.I., and Fenev, D.V. Universal multifunction element of information processing system. *Radioelektronika (Izv. VUZov)*, 3:12–20, 1998 (in Russian).

[17] Badawy, W.M. and Byoumi, M.A. A multiplication-free algorithms and a parallel architecture for affine transformation. *Journal of VLSI Signal Processing Systems for Signal, Image and video Technology*, 31:173–184, 2002.

[18] Suhre, A., Keskin, F., Ersahin, T., Cetin-Atalay, R., Ansari, R., and Cetin, A. A multiplication-free framework for signal processing and applications in biomedical image analysis. In *ICASSP*, pages 1123–1127. IEEE, 2013.

[19] Volkov, A.V. Models and Algorithms of Multiplication-Free Signal Processing in Radiofrequency Monitoring System. PhD thesis, Voronezh State Technical University, Voronezh, Russia, 2013 (in Russian).

[20] Mahdavi, M. and Shabany, M. A 13 Gbps, 0.13 μm CMOS, multiplication-free MIMO detector. *Journal of Signal processing Systems*, 88:273–285, 2017.

[21] Ekblom, H. and Henriksson, S. L_p-criteria for the estimation of location parameters. *SIAM J. on Appl. Math.*, 17(6):1130–1141, 1969.

[22] Andrews D.F., Bickel P.J., Hampel F.R. Huber P.J. Tukey J.W. *Robust Estimates of Location: Survey and Advances.* Princeton University Press, 1972.

[23] Dodge, Y. (editor). Statistical data analysis based on the L_1-norm and related methods. In *Proceedings of I...IV International Conference.* Springer, 1987, 1992, 1997, 2002.

[24] Popoff, A.A. *Fundamentals of Signal Processing in Metric Spaces with Lattice Properties: Algebraic Approach.* CRC Press, 2018.

[25] Tukey J.W. *Exploratory Data Analysis.* Addison-Wesley, Reading, MA, 1977.

[26] Püschel, M. A discrete signal processing framework for meet/join lattices with applications to hypergraphs and trees. In *Proc. of Int. Conf. on Acoustics, Speech and Signal Processing*, pages 5371–5375. IEEE, 2019.

[27] Seifert, B., Wendler, C., and Püschel, M. Wiener filter on meet/join lattices. In *Proc. of Int. Conf. on Acoustics, Speech and Signal Processing*, pages 5355–5359. IEEE, 2021.

[28] Ritter G.X., Urcid G. *Introduction to Lattice Algebra With Applications in AI, Pattern Recognition, Image Analysis, and Biomimetic Neural Networks.* Chapman and Hall/CRC Press, 2021.

[29] Birkhoff, G. *Lattice Theory.* American Mathematical Society, Providence, 1967.

[30] Artamonov, V.A., Saliy, V.N., Skornyakov, L.A., Shevrin, L.N., and Shulgeyfer, E.G. *General Algebra*, volume 2. Nauka, Moscow, 1991 (in Russian).

[31] Artamonov, V.A., Saliy, V.N., Skornyakov, L.A., Shevrin, L.N., and Shulgeyfer, E.G. *General Algebra*, volume 1. Nauka, Moscow, 1990 (in Russian).

[32] Anderson, M. and Feil, T. *Lattice-Ordered Groups: An Introduction.* Springer, 1988.

[33] Glass, A.M.W. and Holland, W.C. (eds.). *Lattice-Ordered Groups: Advances and Techniques.* Kluwer Academic, 1989.

[34] Darnel, M.R. *Theory of Lattice-Ordered Groups.* CRC Press, 1994.

[35] Kopytov, V.M. and Medvedev, N.Ya. *Theory of Lattice-Ordered Groups.* Springer, 1994.

[36] Korn, G.A and Korn, T.M. *Mathematical Handbook for Scientists and Engineers.* McGraw-Hill, 1968.

[37] Hampel, F.R., Ronchetti, E.M., Rousseeuw, P.J., and Stahel, W.A. *Robust Statistics: The Approach Based on Influence Functions.* Wiley, New York, 1986.

[38] Sklar, B. *Digital Communications: Fundamentals and Applications.* Prentice-Hall, Upper Saddle River, New Jersey, 2nd edition, 2001.

[39] Menger, K. Untersuchungen über allgemeine metrik. *Math. Ann.*, 100(1):75–163, 1928.

[40] Wilson, W.A. On semimetric spaces. *American J. Math.*, 53(2):361–373, 1931.

[41] Cedar, J.G. Some generalizations of metric spaces. *Pacific J. Math.*, 11:105–124, 1963.

[42] Copson, E.T. *Metric Spaces*. Cambridge University Press, Cambridge, 1968.

[43] Giles, J.R. *Introduction to the Analysis of Metric Spaces*. Cambridge University Press, Cambridge, 1987.

[44] Deza, M.M. and Deza, E. *Dictionary of Distances*. Elsevier Science, 2006.

[45] Deza, M.M. and Deza, E. *Encyclopedia of Distances*. Springer, Berlin, 2009.

[46] Greenhoe, D.J. Properties of distance spaces with power triangle inequalities. *Carpathian Math. Publ.*, 8(1):51–82, 2016.

[47] Deza, M.M. and Laurent, M. *Geometry of cuts and metrics*. Springer, Berlin, 1997.

[48] Blumenthal, L.M. *Theory and applications of distance geometry*. Oxford University Press, Oxford, 1953.

[49] Grätzer, G.A. *General Lattice Theory*. Birkhäuser Verlag, Basel, 2nd edition, 2003.

[50] Davey, B.A. and Priestley, H.A. *Introduction to Lattice and Order*. Cambridge University Press, Cambridge, 2nd edition, 2002.

[51] Borovkov, A.A. *Mathematical Statistics*. Nauka, Novosibirsk, 1997 (in Russian).

[52] Tihonov, V.I. *Statistical Radio Engineering*. Radio i Svyaz, Moscow, 1982 (in Russian).

[53] Levin, B.R. *Theoretical Foundations of Statistical Radio Engineering*, volume 1. Soviet Radio, Moscow, 1969 (in Russian).

[54] Kendall, M.G. and Stuart, A. *The Advanced Theory of Statistics. Inference and Relationship*. Charles Griffin, London, 1961.

[55] Cramer, H. *Mathematical Methods of Statistics*. Princeton University Press, 1946.

[56] Popoff, A.A. Invariant of group of random samples mappings in sample space with lattice properties. *Radioelectronics and Communications Systems*, 58:417–425, 2015.

[57] Franks, L.E. *Signal Theory*. Prentice-Hall, Englewood Cliffs, New Jersey, 1969.

[58] Apostol, T.M. *Mathematical Analysis*. Addisson-Wesley, Reading, MA, 2nd edition, 1975.

[59] Kudryavtsev, L.D. *Course of Mathematical Analysis*, volume 3. Vysshaya shkola, Moscow, 2nd edition, 1989 (in Russian).

[60] Zorich, V.A. *Mathematical analysis II*. Springer, 2016.

[61] Ahlfors, L. *Complex Analysis*. McGraw-Hill, 3d edition, 1979.

[62] Henrici, P. *Applied and Computational Complex Analysis*, volume 1, 2, 3. Wiley, 1974, 1977, 1986.

[63] Rudin, W. *Real and Complex Analysis*. McGraw-Hill, 3d edition, 1986.

[64] Marshall, D.E. *Complex Analysis*. Cambridge University Press, Cambridge, 2019.

[65] Aliprantis, C.D. and Burkinshaw, O. *Principles of Real Analysis*. Academic Press, London, 3d edition, 1998.

[66] Carothers, N. *Real Analysis*. Cambridge University Press, Cambridge, 2000.

[67] Brychkov, Yu.A. *Integral Transforms of Generalized Functions.* CRC Press, 1989.

[68] Ahmed, N., Natarajan, T., and Rao, K.R. Discrete cosine transform. *IEEE Transactions on Computers*, C-32:90–93, 1974.

[69] Rao, K.R. and Yip, P. *Discrete Cosine Transform: Algorithms, Advantages, Applications.* Academic Press, Boston, MA, 1990.

[70] Bracewell, R.N. *The Hartley Transform.* Oxford Univ. Press, New York, 1986.

[71] Cook, C.E. and Bernfeld, M. *Radar signals.* Academic Press, New York, 1967.

[72] Minkoff, J. *Signal processing fundamentals and applications for communications and sensing systems.* Artech House, Norwood, MA, 2002.

[73] Zhang, K.Q.T. *Wireless Communications: Principles, Theory, and Methodology.* Wiley and Sons, 2016.

[74] Zacks, S. *Theory of Statistical Inference.* Wiley, New York, 1971.

[75] Lehmann, E.L. *Theory of Point Estimation.* Wiley, New York, 1983.

[76] Fuchs, L. *Partially Ordered Algebraic Systems.* Pergamon, 1963.

[77] Conrad, P.F. *Lattice Ordered Groups.* Tulane University, 1970.

[78] Martinez, J. (ed.). Ordered algebraic structures. In *Proceedings of the Caribbean Mathematics Foundation Conference on Ordered Algebraic Structures.* Kluwer Academic, Curacao, 1989.

[79] Mudrov, V.I. and Kushko, V.L. *Least Moduli Method.* Znanie, Moscow, 1971 (in Russian).

[80] Tukey, J.W. A survey of sampling from contaminated distributions. In *Contributions to Probability and Statistics*, pages 448–485. Stanford University Press, Redwood City, CA, 1960.

[81] Huber, P.J. Robust estimation of a location parameter. *Ann. Math. Statist.*, 35(1):73–101, 1964.

[82] David, H.A. *Order Statistics.* Wiley, New York, 1970.

[83] Efimov, A.N. *Order Statistics: Their Properties and Applications.* Znanie, Moscow, 1980 (in Russian).

[84] Berztiss, A.T. *Structures of Data.* Statistica, Moscow, 1974.

[85] Kung, H.T. and Leiserson, C.E. Systolic arrays (for VLSI). In *Sparse Matrix Proc.*, pages 256–282. SIAM, 1979.

[86] Swartzlander, E. *Systolic Signal Processing Systems.* CRC Press, 1987.

[87] McCanney, H.J. *Systolic Array Processors.* Prentice Hall, 1989.

[88] Jaeckel, L.A. Some flexible estimates of location. *Ann. Math. Statist.*, 42(5):1540–1552, 1971.

[89] Hodges, J.L. and Lehmann, E.L. Estimates of location based on rank tests. *Ann. Math. Statist.*, 34:598–611, 1963.

[90] Jaeckel, L.A. Robust estimates of location. PhD thesis, University of California, Berkely, 1969.

[91] Hodges, J.L. Efficiency in normal samples and tolerance of extreme values for some estimates of location. In *Proc. of the 5th Berkely Symp. on Mathematical Statistics and Probability.*, volume 1, pages 163–186. Univercity of California Press, Berkely, CA, 1967.

[92] Smith, S.W. *Digital Signal Processing: A Practical Guide for Engineers and Scientists.* Newnes, 2003.

[93] Akimov, P.S., Bakut, P.A., Bogdanovich, V.A., et al. *Signal Detection Theory.* Radio i Svyaz, Moscow, 1984 (in Russian).

[94] Antoniou, A. *Digital Signal Processing: Signal, Systems, and Filters.* McGraw-Hill, 2006.

[95] Vaseghi, S.V. *Advanced Digital Signal Processing and Noise Reduction.* Wiley and Sons, 2nd edition, 2000.

[96] Kay, S.M. *Modern Spectral Estimation: Theory and Application.* Prentice Hall, Englewoods Cliffs, New Jersey, 1988.

[97] Kay, S.M. *Fundamentals of Statistical Signal Processing. Vol. I: Estimation Theory.* Prentice Hall, Englewood Cliffs, N.J., 1993.

[98] Wiener, N. *Extrapolation, Interpolation and Smoothing of Stationary Time Series.* MIT Press, Cambridge, MA, 1949.

[99] Kalman, R.E. A new approach to linear filtering and prediction problem. *J. Basic Eng. (Trans. ASME)*, 82D:35–45, 1960.

[100] Ifeacher, E.C. and Jervis, B.W. *Digital Signal Processing: A Practical Approach.* Prentice Hall, Englewoods Cliffs, New Jersey, 2nd edition, 2002.

[101] Widrow, B. and Stearns, S.D. *Adaptive Signal Processing.* Prentice Hall, Englewoods Cliffs, New Jersey, 1985.

[102] Buckley, K.M., Douglass, S.C., Sayed, A.H., and Van Veen, B. et al. *Digital Signal Processing Handbook.* CRC Press, 1999.

[103] Kay, S.M. *Fundamentals of Statistical Signal Processing, Vol. II: Detection Theory.* Pearson, 1998.

[104] Middleton, D. *An introduction to statistical communication theory.* IEEE, New York, 1996.

[105] Helstrom, C.W. *Elements of Signal Detection and Estimation.* Prentice-Hall, Englewood Cliffs, N.J., 1994.

[106] Van-Trees, H.L. *Detection, Estimation, and Modulation Theory, Part I: Detection, Estimation, and Filtering Theory.* Wiley and Sons, New York, 2nd edition, 1968.

[107] Poor, H.V. *An Introduction to Signal Detection and Estimation.* Springer-Verlag, New York, 2nd edition, 1994.

[108] Scharf, L.L. *Statistical Signal Processing: Detection, Estimation, and Time Series Analysis.* Addison-Wesley, Reading, MA, 1991.

[109] Tuzlukov, V.P. *Signal Detection Theory.* Springer, 2001.

[110] Sosulin, Yu.G. *Theory of Detection and Estimation of Stochastic Signals.* Soviet Radio, Moscow, 1978 (in Russian).

[111] Hayek, J. and Sidak, Z. *Theory of Rank Tests.* Academic Press, New York, 1967.

[112] Kassam, S.A. *Signal Detection in non-Gaussian Noise.* Springer, 1988.

[113] Kassam, S.A. A bibliography on nonparametric detection. *IEEE Trans.*, IT-26(5):595–692, 1980.

[114] Levin, B.R. *Theoretical Foundations of Statistical Radio Engineering.* Radio i Svyaz, Moscow, 1989 (in Russian).

[115] Forbes, C., Evans, M., Hastings, N., and Peacock, B. *Statistical Distributions*. Wiley, 4th edition, 2011.

[116] Everitt, B. and Hand, D. *Finite Mixture Distributions*. Chapman and Hall, New York, 1981.

[117] Titterington, D.M., Smith, A., and Makov, U. *Statistical Analysis of Finite Mixture Distributions*. Wiley, 1985.

[118] Lindsay, B.G. *Mixture Models: Theory, Geometry and Application*. Inst. Math. Stat., Hayward, CA, 1995.

[119] McLachlan, G.J. and Peel, D. *Finite Mixture Models*. Wiley, 2000.

[120] Fruhwirth-Schnatter, S. *Finite Mixture and Markov Switching Models*. Springer, 2006.

[121] McNicholas, P.D. *Mixture Model-Based Classification*. CRC Press, 2017.

[122] Viterby, A.J. *Principles of Coherent Communications*. McGraw-Hill, New York, 1966.

[123] Lindsey, W.C. and Simon, M.K. *Telecommunication Systems Engineering*. Prentice-Hall, Englewood Cliffs, N.J., 1973.

[124] Couch, L.W. *Digital and Analog Communication Systems*. Macmillan Publishing Company, New York, 1983.

[125] Ziemer, R.E. and Peterson, R.L. *Digital Communications and Spread Spectrum Systems*. Macmillan Publishing Company, New York, 1985.

[126] Bogdanovich, V.A. and Vostretsov, A.G. *Theory of Robust Detection, Classification and Estimation of Signals*. Fiz. Math. Lit, Moscow, 2004 (in Russian).

[127] Trifonov, A.P. and Shinakov, Yu.S. *Joint Classification of Signals and Estimation of Their Parameters in Noise Background*. Radio i Svyaz, Moscow, 1986 (in Russian).

[128] Miao, G.J. and Clements, M.A. *Digital Signal Processing and Statistical Classification*. Artech House, 2002.

[129] Tihonov, V.I. *Optimal Signal Reception*. Radio i Svyaz, Moscow, 1983 (in Russian).

[130] Woodward, P.M. *Probability and Information Theory, with Application to Radar*. Pergamon Press, Oxford, 1953.

[131] Barton, D.K. *Radar System Analysis and Modeling*. Artech House, 2005.

[132] Shirman, Y.D. and Manzhos, V.N. *Theory and Technique of Radar Information Processing in Interference Background*. Radio i Svyaz, Moscow, 1981 (in Russian).

[133] Grishin, Y.P., Ipatov, V.P., Kazarinov, Y.M., and Ulyanitsky, Y.D. *Radioengineering Systems*. Vysshaya Shkola, Moscow, 1990 (in Russian).

[134] Blackman, R.B. and Tukey, J.W. The measurement of power spectra from the point of view of communications engineering. *Bell Syst. Tech. J.*, 33:185–282, 485–569, 1958.

[135] Jenkins, G.M. and Watts, D. G. *Spectral Analysis and Its Applications*. Holden-Day, San Francisco, 1965.

[136] Koopmans, L.H. *Spectral Analysis of Time Series*. Academic Press, New York, 1974.

[137] Papoulis, A. *Signal Analysis*. McGraw-Hill, New York, 1977.

[138] Childers, D.G. *Modern Spectrum Analysis*. IEEE Press, New York, 1978.

[139] Priestley, M.B. *Spectral Analysis and Time Series*. Academic Press, London, 1981.

[140] Marple, S.L. *Digital Spectral Analysis with Applications*. Prentice Hall, Englewoods Cliffs, New Jersey, 1987.

[141] Haykin, S. *Advances in Spectrum Analysis and Array Processing. Vol. 1,2,3.* Prentice Hall, Englewoods Cliffs, New Jersey, 1991,1995.

[142] Naidu, P.S. *Modern Spectrum Analysis of Time Series.* CRC Press, 1996.

[143] Stoica, P. and Moses, R. *Spectral Analysis of Signals.* Prentice Hall, Upper Saddle River, New Jersey, 2005.

[144] Bartlett, M.S. Smoothing periodograms from time series with continuous spectra. *Nature*, 161:686–687, 1948.

[145] Welch, P.D. Use of fft for estimation of power spectra: a method based on time averaging over short, modified periodograms. *IEEE Trans. on Audio and Electroacoustics*, AU-15(2):70–73, 1967.

[146] Tukey, J.W. The sampling theory of power spectrum estimates. In *Proceedings Symposium on Applied Autocorrelation Analysis of Physical Problems*, pages 47–67. U.S. Office of Naval Research, 1949.

[147] Schuster, A. On the investigation of hidden periodicities with application to a supposed twenty-six-day period of meteorological phenomena. *Terr. Mag.*, 3(1):13–41, 1898.

[148] Capon, J. High-resolution frequency-wavenumber spectrum analysis. *Proc. IEEE*, 57:1408–1418, 1969.

[149] Johnson, D.H. Application of spectral estimation methods in bearing estimation problems. *Proc. IEEE*, 70:1018–1028, 1982.

[150] Schmidt, R.O. A Signal Subspace Approach to Multiple Emitter Location and Spectral Estimation. PhD thesis, Department of Electrical Engineering, Stanford University, Stanford, 1981.

[151] Kumaresan, R. and Tufts, D.W. Estimating the angles of arrival of multiple plane waves. *IEEE Trans. on Aerospace and Electronic systems*, AES-19:134–139, 1983.

[152] Paulraj, A., Roy, R., and Kailath, T. A subspace rotation approach to signal parameter estimation. *Proceedings of IEEE*, 74(7):1044–1046, 1986.

[153] Roy, R. and Kailath, T. ESPRIT—estimation of signal parameters via rotational invariance techniques. *IEEE Trans. on Acoustics, Speech, and Signal Processing*, ASSP-37(7):984–995, 1989.

[154] Kocherzhevsky, G.N., Yerohin, G.A., and Kozyrev, N.D. *Antennas and Feeders.* Radio i Svyaz, Moscow, 1989 (in Russian).

[155] Burin, L.I., Vasilyev, V.P., and Kaganov, V.I. *Handbook of Radioelectronic Sets*, volume 1. Energia, Moscow, 1978 (in Russian).

[156] Bryn, F. Optimum signal processing of three-dimensional arrays operating on gaussian signals and noise. *J. Acoust. Soc. Am.*, 34(3):289–297, 1962.

[157] Widrow, B., Mantey, P.E., Griffits, L.J., and Goode, B.B. Adaptive antenna systems. *Proc. IEEE*, 55:2143–2159, 1967.

[158] Monzingo, R.A and Miller, T.W. *Introduction to Adaptive Arrays.* John Wiley and Sons, 1980.

[159] Zhuravlev, A.K., Lukoshkin, A.P., and Poddubny, S.S. *Signal Processing in Adaptive Antenna Arrays.* Leningrad State University, Leningrad, 1983.

[160] Furse, C.M., Grant, C., Huang, J., Imbriale, W.A., and Nakano, H. et al. *Modern Antenna Handbook.*

[161] Balanis, C.A. *Antenna Theory: Analysis and Design.* John Wiley and Sons, 1997.

[162] Johnson, D.H. and Dudgeon, D.E. *Array Signal Processing—Concepts and Methods*. Prentice Hall, Englewood Cliffs, N.J., 1992.

[163] Samarsky, A.A and Gulin, A.V. *Numerical Methods*. Nauka, Moscow, 1989 (in Russian).

[164] Bangs, W.J. Array Processing with Generalized Beamformers. PhD thesis, Yale University, New Haven, CT, 1971.

[165] Schmidt, R.O. Multiple emitter location and signal parameter estimation. In *Proc. RADC Spectral Estimation Workshop*, pages 243–258. Saxpy Computer Corporation, New York, 1979.

[166] Barabel, A.J. Improving the resolution performance of eigenstructure-based direction finding algorithms. In *Proceedings of the International Conference on Acoustics, Speech, and Signal Processing*, pages 336–339. IEEE, Boston, MA, 1983.

[167] Bienvenu, G. Influence of the spatial coherence of the background noise on high resolution passive methods. In *Proceedings of the International Conference on Acoustics, Speech, and Signal Processing*, pages 306–309. IEEE, Washington, DC, 1979.

[168] Van Veen, B.D. and Buckley, K.M. Beamforming: A versatile approach to special filtering. *IEEE ASSP Magazine*, 5(2):4–24, 1988.

[169] Pillai, S.U. *Array Signal Processing*. Springer, 1989.

[170] Viberg, M. and Ottersten, B. Sensor array processing based on subspace fitting. *IEEE Trans. on Signal Processing*, 39(5):1110–1121, 1991.

[171] Doron, M., Doron, E., and Weiss, A. Coherent wide-band processing for arbitrary array geometry. *IEEE Trans. on Signal Processing*, 41(1):414–417, 1993.

[172] Viberg, M. Subspace-based methods for the identification of linear time-invariant systems. *Automatica*, 31(12):1835–1851, 1995.

[173] Mantey, P.E. and Griffits, L.J. Iterative least-squares algorithm for signal extraction. In *Second Hawaii International Conference on System Science*, pages 767–770. 1969.

[174] Baird, C.A. Recursive processing for adaptive arrays. In *Proceedings of the Adaptive Antenna Systems Workshop*, volume 1, pages 163–182. Naval Research Laboratory, Washington, DC, 1974.

[175] Ageev, D.V. Foundations of linear selection theory. *Scientific Journal of Leningrad Institute of Communication Engineers*, 10:8–28, 1935 (in Russian).

[176] Chang, R.W. Synthesis of band-limited orthogonal signals for multichannel data transmission. *Bell Sys. Tech. J.*, 45:1775–1796, 1966.

[177] Weinstein, S. and Ebert, P. Data transmission by frequency-division multiplexing using the discrete fourier transform. *IEEE Trans. Comm.*, 19:628–634, 1971.

[178] Kleinrock, L. and Tobagi, F. Packet switching in radio channels: Part I – carrier sense multiple access modes and their throughout-delay characteristics. *IEEE Trans. Comm.*, 23:1400–1416, 1975.

[179] Scholtz, R.A. The origins of spread spectrum communications. *IEEE Trans. Comm.*, 30:822–854, 1982.

[180] Cimini, L.J. Analysis and simulation of a digital mobile channel using orthogonal frequency division multiplexing. *IEEE Trans. Comm.*, 33:665–675, 1985.

[181] Maric, S.V. and Titlebaum, E.L. A class of frequency hop codes with nearly ideal characteristics for use in multiple-access spread-spectrum communications and radar and sonar systems. *IEEE Trans. Comm.*, 40:1442–1447, 1992.

[182] Varakin, L.E. *Communication Systems with Noise-Like Signals.* Radio i Svyaz, Moscow, 1985 (in Russian).

[183] Gitlits, M.V. and Lev, A.Y. *Theoretical Foundations of Multichannel Communication.* Radio i Svyaz, Moscow, 1985 (in Russian).

[184] Vitetta, G.M., Taylor, D.P., Colavolpe, G., Pancaldi, F., and Martin, P.A. *Wireless Communications: Algorithmic Techniques.* Wiley and Sons, 2013.

[185] Molish, A.F. *Wireless Communications.* Wiley and Sons, 2nd edition, 2011.

[186] Hlawatsch, F. and Auger, F. *Time-Frequency Analysis: Concepts and Methods.* John Wiley and Sons, 2008.

[187] Mallat, S. *A Wavelet Tour of Signal Processing: The Sparse Way.* Academic Press, 3rd edition, 2009.

[188] Mallat, S. A theory for multiresolution signal decomposition: the wavelet representation. *IEEE Trans. Patt. Anal. and Mach. Intell.*, 11(7):674–693, 1989.

[189] Burt, P.J. and Adelson, E.H. The laplasian pyramid as a compact image code. *IEEE Trans. on Comm.*, Com-31(4):532–540, 1983.

[190] Gonzales, R.C. and Woods, R.E. *Digital Image Processing.* Prentice Hall, Upper Saddle River, N.J., 2nd edition, 2002.

[191] Gilbo, E.P. and Chelpanov, I.B. *Signal Processing on the Basis of Ordered Selection: Majority Transformation and Others That are Close to it.* Soviet Radio, Moscow, 1977 (in Russian).

[192] Tihonov, V.I. and Harisov, V.N. *Statistical Analysis and Synthesis of Radioengineering Sets and Systems.* Radio and Svyaz, Moscow, 1991 (in Russian).

[193] Prudnikov, A.P., Brychkov, Yu.A., and Marichev, O.I. *Integrals and Series: Elementary Functions.* Gordon & Breach, New York, 1986.

[194] Hua, Y., Gershman, A., and Cheng, Q. *High-Resolution and Robust Signal Processing.* CRC Press, 2017.

[195] Tuzlukov, V.P. *Signal Processing in Radar Systems.* CRC Press, 2012.

[196] Frazier, M.W., Heil, C., Feichtinger, H.G., Coifman, R.R., and Jawerth, B. et al. *Wavelets: Mathematics and Applications.* CRC Press, 1994.

[197] Chui, C.K. *An Introduction to Wavelets.* Academic Press, New York, 1992.

[198] Meyer, Y. and Roques, S., (editors). Progress in wavelet analysis and applications. In *Proc. of Int. Conf. "Wavelets and Applications".* Frontieres, Toulouse, France, 1993.

[199] Holschneider, M. *Wavelets: An Analysis Tool. Oxford Mathematical Monographs.* Clarendon Press, Oxford, 1995.

[200] Mehra, M. *Wavelets Theory and its Applications.* Springer, 2018.

[201] Ryan, O. *Linear Algebra, Signal Processing, and Wavelets: A Unified Approach.* Springer, 2019.

[202] Cohen, A. and Ryan, R.D. *Wavelets and Multiscale Signal Processing.* Springer, 2019.

[203] Walnut, D.F. *An Introduction to Wavelet Analysis.* Birkhauser, 2013.

[204] Debnath, L. *Wavelet Transforms and Their Applications.* Birkhauser, 2011.

[205] Heil, C., Walnut, D.F., and Daubechies, I. *Fundamental Papers in Wavelet Theory.* Princeton University Press, 2009.

[206] Resnikoff, H.L. and Wells, R.O. *Wavelet Analysis.* Springer, 2012.

[207] Sundararajan, D. *Discrete Wavelet Transform. A Signal Processing Approach*. Wiley, 2015.

[208] Lim, J.S. *Two-Dimensional Signal and Image Processing*. Prentice Hall, Englewood Cliffs, N.J., 1989.

[209] Pratt, W.K. *Digital Image Processing*. John Wiley & Sons, 3d edition, 2001.

[210] Diop, H., Radjesvarane, A., and Moisan, L. Intrinsic nonlinear multiscale image decomposition: A 2D empirical mode decomposition-like tool. *Computer Vision and Image Understanding*, 116:102–119, 2012.

[211] Mitra, S.K. and Sicuranza, G.L. (eds.). *Nonlinear Image Processing*. Academic Press, New York, 2000.

[212] Astola, J. and Kuosmanen, P. *Fundamentals of Nonlinear Digital Filtering*. CRC Press, 1997.

[213] Ritter, G.X. and Wilson, J.N. *Handbook of Computer Vision Algorithms in Image Algebra*. CRC Press, 2001.

[214] Russ, J.C. *The Image Processing Handbook*. CRC Press, 3rd edition, 1999.

[215] Starck, J.L., Mallat, S., Daubechies, I., Donodo, D., and Adelson, E.H. et al. Nonlinear multiscale transforms. In *Multiscale and Multiresolution Methods. Lecture notes in Computational Science and Engineering, vol. 20*. Springer, Berlin, 2002.

[216] Donodo, D.I. Nonlinear pyramid transforms based on median interpolation. *SIAM J. Math. Anal.*, 60(4):1137–1156, 2000.

[217] Guotsias, J. and Heijmans, H.J. Nonlinear multiresolution signal decomposition schemes. Part 1: Morphological pyramids. *IEEE Trans. on Image Processing*, 9(11):1862–1876, 2000.

Index

For Product Safety Concerns and Information please contact our EU
representative GPSR@taylorandfrancis.com
Taylor & Francis Verlag GmbH, Kaufingerstraße 24, 80331 München, Germany

www.ingramcontent.com/pod-product-compliance
Lightning Source LLC
Chambersburg PA
CBHW080134220326
41598CB00032B/5067